SRA
Connecting Math Concepts

Level F Presentation Book 2

COMPREHENSIVE EDITION

A DIRECT INSTRUCTION PROGRAM

Mc Graw Hill Education

Bothell, WA • Chicago, IL • Columbus, OH • New York, NY

MHEonline.com

 **Education**

Send all inquiries to:
McGraw-Hill Education
8787 Orion Place
Columbus, OH 43240

ISBN: 978-0-02-103642-4
MHID: 0-02-103642-X

Printed in the United States of America.

2 3 4 5 6 7 8 9 DOH 16

Table of Contents

Table of Contents (continued)

Lessons 61–65 Planning Page

	Lesson 61	Lesson 62	Lesson 63	Lesson 64	Lesson 65
Student Learning Objectives	**Exercises** 1. Solve division problems with remainders using mental math **2. Solve division problems that have decimals** 3. Solve for equivalent fractions **4. Multiply fractions using cross cancellation** 5. Write repeated multiplication problems using exponents 6. Solve ratio word problems with classification (two questions) Complete work independently	**Exercises** 1. Solve division problems with remainders using mental math 2. Multiply fractions using cross cancellation 3. Write repeated multiplication problems using exponents 4. Solve for equivalent fractions 5. Solve division problems that have decimals **6. Simplify fractions using the greatest common factor** Complete work independently	**Exercises** 1. Solve division problems with remainders using mental math 2. Solve for equivalent fractions 3. Solve division problems that have decimals 4. Simplify fractions using the greatest common factor **5. Subtract by renaming thousands numbers with zeros** **6. Change decimals to whole numbers** **7. Solve exponents using repeated multiplication** 8. Multiply fractions Complete work independently	**Exercises** 1. Solve division problems with remainders using mental math **2. Identify equivalent fractions on a coordinate system** **3. Rewrite fractions that have a decimal number in the denominator** 4. Round numbers 5. Simplify fractions using the greatest common factor 6. Subtract by renaming thousands numbers with zeros 7. Solve exponents using repeated multiplication Complete work independently	**Exercises** 1. Solve division problems with remainders using mental math 2. Identify equivalent fractions on a coordinate system 3. Rewrite fractions that have a decimal number in the denominator 4. Round numbers 5. Simplify fractions using the greatest common factor 6. Subtract by renaming thousands numbers with zeros 7. Solve exponents using repeated multiplication Complete work independently

Common Core State Standards for Mathematics

	Lesson 61	Lesson 62	Lesson 63	Lesson 64	Lesson 65
5.OA 1	✔	✔	✔	✔	✔
5.NBT 1			✔	✔	✔
5.NBT 2			✔	✔	✔
5.NBT 3	✔	✔	✔	✔	✔
5.NBT 6	✔	✔		✔	✔
5.NBT 7	✔	✔	✔	✔	✔
5.NF 4	✔	✔	✔	✔	✔
5.NF 5	✔	✔	✔	✔	✔
5.NF 6	✔	✔		✔	
5.G 1		✔			
5.G 2		✔		✔	✔

Teacher Materials	Presentation Book 1, Board Displays CD or chalk board
Student Materials	Textbook, Workbook, pencil, lined paper, ruler
Additional Practice	• Student Practice Software: Block 3: Activity 1 (5.NBT 7), Activity 2, Activity 3 (5.NF 4), Activity 4, Activity 5 (5.NBT 6), Activity 6 (5.NF 4, 5.NF 6, and 5.OA 1) • Provide needed fact practice with Level D or E Math Fact Worksheets.
Mastery Test	

Lesson 61

EXERCISE 1: MENTAL MATH
REMAINDERS

a. You're going to do mental math with division problems that have remainders.
(Display:) [61:1A]

$$3\overline{)29} \quad 3\overline{)25} \quad 3\overline{)13} \quad 3\overline{)17}$$

- (Point to **29.**) 29 divided by 3. (Pause.) What's the answer? (Signal.) *9 with a remainder of 2.*
- (Point to **25.**) 25 divided by 3. (Pause.) What's the answer? (Signal.) *8 with a remainder of 1.*
- (Point to **13.**) 13 divided by 3. (Pause.) What's the answer? (Signal.) *4 with a remainder of 1.*
- (Point to **17.**) 17 divided by 3. (Pause.) What's the answer? (Signal.) *5 with a remainder of 2.*
(Repeat until firm.)

b. (Display:) [61:1B]

$$3\overline{)11} \quad 3\overline{)17} \quad 2\overline{)7} \quad 2\overline{)15}$$

- (Point to **11.**) 11 divided by 3. (Pause.) What's the answer? (Signal.) *3 with a remainder of 2.*
- (Point to **17.**) 17 divided by 3. (Pause.) What's the answer? (Signal.) *5 with a remainder of 2.*
- (Point to **7.**) 7 divided by 2. (Pause.) What's the answer? (Signal.) *3 with a remainder of 1.*
- (Point to **15.**) 15 divided by 2. (Pause.) What's the answer? (Signal.) *7 with a remainder of 1.*
(Repeat until firm.)

c. (Display:) [61:1C]

$$4\overline{)15} \quad 4\overline{)13} \quad 4\overline{)18} \quad 4\overline{)37}$$

- (Point to **15.**) 15 divided by 4. (Pause.) What's the answer? (Signal.) *3 with a remainder of 3.*
- (Point to **13.**) 13 divided by 4. (Pause.) What's the answer? (Signal.) *3 with a remainder of 1.*
- (Point to **18.**) 18 divided by 4. (Pause.) What's the answer? (Signal.) *4 with a remainder of 2.*
- (Point to **37.**) 37 divided by 4. (Pause.) What's the answer? (Signal.) *9 with a remainder of 1.*
(Repeat until firm.)

d. (Display:) [61:1D]

$$4\overline{)33} \quad 4\overline{)19} \quad 3\overline{)19} \quad 2\overline{)19}$$

- (Point to **33.**) 33 divided by 4. (Pause.) What's the answer? (Signal.) *8 with a remainder of 1.*
- (Point to $4\overline{)19}$.) 19 divided by 4. (Pause.) What's the answer? (Signal.) *4 with a remainder of 3.*
- (Point to $3\overline{)19}$.) 19 divided by 3. (Pause.) What's the answer? (Signal.) *6 with a remainder of 1.*
- (Point to $2\overline{)19}$.) 19 divided by 2. (Pause.) What's the answer? (Signal.) *9 with a remainder of 1.*
(Repeat until firm.)

EXERCISE 2: DECIMAL OPERATIONS
DIVISION

a. (Display:) [61:2A]

$$2\overline{)14.2}$$

- Read this problem. (Signal.) *14 and 2 tenths divided by 2.*
This problem has a decimal value. So we will show a decimal point in the answer. We write the decimal point right above the decimal point in 14 and 2 tenths.
(Add to show:) [61:2B]

$$2\overline{)1\overset{\cdot}{4}.2}$$

b. Now we work the problem. What's 14 ÷ 2? (Signal.) *7.*
We write 7 above the last digit of 14.
(Add to show:) [61:2C]

$$2\overline{)1\overset{7.}{4}.2}$$

- The next problem is 2 divided by 2. What's 2 ÷ 2? (Signal.) *1.*
(Add to show:) [61:2D]

$$2\overline{)1\overset{7.1}{4}.2}$$

- What's 14.2 ÷ 2? (Signal.) *7 and 1 tenth.*

c. (Display:) [61:2E]

$$3\overline{)1.26}$$

- Read this problem. (Signal.) *1 and 26 hundredths divided by 3.*
  We write the decimal point in the answer right above the other decimal point.
  (Add to show:) [61:2F]

$$3\overline{)1.26}$$

d. Now we work the problem the way we normally would for the digits.
- What's 12 ÷ 3? (Signal.) *4.*
  I write 4 above the last digit of 12.
  (Add to show:) [61:2G]

$$3\overline{)1.\overset{.4}{26}}$$

- Say the problem for 6. (Signal.) *6 ÷ 3.*
- What's the answer? (Signal.) *2.*
  (Add to show:) [61:2H]

$$3\overline{)1.\overset{.42}{26}}$$

- Read the problem and the answer. (Signal.)
  1 and 26 hundredths ÷ 3 = 42 hundredths.
e. (Display:) [61:2I]

$$2\overline{)1.06}$$

- Read this problem. (Signal.) *1 and 6 hundredths divided by 2.*
  We write the decimal point above the other decimal point.
  (Add to show:) [61:2J]

$$2\overline{)1.06}$$

f. Now we work the problem the way we normally would for the digits.
- Say the problem for 10. (Signal.) *10 ÷ 2.*
- What's the answer? (Signal.) *5.*
  (Add to show:) [61:2K]

$$2\overline{)1.\overset{.5}{06}}$$

- Say the problem for 6. (Signal.) *6 ÷ 2.*

- What's the answer? (Signal.) *3.*
  (Add to show:) [61:2L]

$$2\overline{)1.\overset{.53}{06}}$$

- Read the problem and the answer. (Signal.)
  1 and 6 hundredths ÷ 2 = 53 hundredths.

━━━━━━━ **WORKBOOK PRACTICE** ━━━━━━━

a. Open your workbook to Lesson 61 and find part 1. ✔
  (Teacher reference:)

  a. $3\overline{)69.03}$ b. $5\overline{)2.50}$ c. $4\overline{)1.68}$

b. Read problem A. (Signal.) *69 and 3 hundredths divided by 3.*
- Write the decimal point in the answer. Then work the problem the way you normally would. (Observe students and give feedback.)
- Everybody, read the problem and the answer. (Signal.) *69.03 ÷ 3 = 23.01.*
  (Display:) [61:2M]

$$\textbf{a. } 3\overline{)69.\overset{23.01}{03}}$$

Here's what you should have.
c. Read problem B. (Signal.) *2.50 ÷ 5.*
- Write the decimal point in the answer. Then work the problem the way you normally would. ✔
- Everybody, read the problem and the answer. (Signal.) *2.50 ÷ 5 = 50 hundredths.*
  (Display:) [61:2N]

$$\textbf{b. } 5\overline{)2.\overset{.50}{50}}$$

Here's what you should have.
d. Read problem C. (Signal.) *1.68 ÷ 4.*
- Write the decimal point in the answer and work the problem. ✔
- Everybody, read the problem and the answer. (Signal.) *1.68 ÷ 4 = 42 hundredths.*
  (Display:) [61:2O]

$$\textbf{c. } 4\overline{)1.\overset{.42}{68}}$$

Here's what you should have.

EXERCISE 3: EQUIVALENT FRACTIONS [REMEDY]

a. Find part 2 in your workbook. ✔
- Pencils down. ✔
  (Teacher reference:) R Part F

$$\frac{9}{10} = \frac{\square}{50} = \frac{36}{\square} = \frac{\square}{70} = \frac{\square}{30} = \frac{72}{\square}$$
a.　　b.　　c.　　d.　　e.

You're going to figure out fractions that are equal. All the fractions will equal 9/10.

b. Fraction A has a missing numerator. So you figure out the fraction that equals 1.
- Say the problem you'll work to figure out the fraction of 1. (Signal.) *10 times what number equals 50.*
- What's the fraction of 1? (Signal.) *5/5.*
- Raise your hand when you know the missing numerator. ✔
- What's the missing numerator? (Signal.) *45.*
  (Repeat until firm.)

c. Touch fraction B. ✔
- Say the problem you'll work to figure out the fraction of 1. (Signal.) *9 times what number equals 36.*
- What's the fraction of 1? (Signal.) *4/4.*
- Raise your hand when you know the missing denominator. ✔
- What's the missing denominator? (Signal.) *40.*
  (Repeat until firm.)

d. Touch fraction C. ✔
- Say the problem you'll work to figure out the fraction of 1. (Signal.) *10 times what number equals 70.*
- What's the fraction of 1? (Signal.) *7/7.*
- Raise your hand when you know the missing numerator. ✔
- What's the missing numerator? (Signal.) *63.*
  (Repeat until firm.)

e. Complete all the fractions in part 2.
  (Observe students and give feedback.)

f. Check your work.
- Everybody, what's fraction A? (Signal.) *45/50.*
- What's fraction B? (Signal.) *36/40.*
- What's fraction C? (Signal.) *63/70.*
- What's fraction D? (Signal.) *27/30.*
- What's fraction E? (Signal.) *72/80.*

g. What fraction do all of those fractions equal? (Signal.) *9/10.*

EXERCISE 4: FRACTION OPERATIONS
MULTIPLICATION—CROSS CANCELLATION [REMEDY]

a. (Display:)　　　　　　　　　　　[61:4A]

$$\frac{5}{3} \times \frac{9}{20} =$$

We're going to work this problem the fast way. We're going to simplify fractions that have multiples.
- Read the problem. (Signal.) *5/3 × 9/20.*

b. The first numerator is 5.
- Is there a multiple of 5 in one of the denominators? (Signal.) *Yes.*
- What multiple of 5? (Signal.) *20.*
- 20 is how many times 5? (Signal.) *4.*
  So I cross out 20 and write 4 in the denominator, and I write 1 in the numerator.
  (Add to show:)　　　　　　　[61:4B]

$$\frac{\overset{1}{\cancel{5}}}{3} \times \frac{9}{\underset{4}{\cancel{20}}} =$$

c. The other fraction is 9/3. Can that fraction be simplified? (Signal.) *Yes.*
- What does that fraction equal? (Signal.) *3.*
  So I write 3 in the numerator.
  (Add to show:)　　　　　　　[61:4C]

$$\frac{\overset{1}{\cancel{5}}}{\cancel{3}} \times \frac{\overset{3}{\cancel{9}}}{\underset{4}{\cancel{20}}} =$$

The answer is 3/4.
(Add to show:)　　　　　　　[61:4D]

$$\frac{\overset{1}{\cancel{5}}}{\cancel{3}} \times \frac{\overset{3}{\cancel{9}}}{\underset{4}{\cancel{20}}} = \boxed{\frac{3}{4}}$$

a. Open your textbook to Lesson 61 and find part 1. ✔
(Teacher reference:)

a. $\frac{7}{5} \times \frac{40}{35}$ b. $\frac{3}{5} \times \frac{40}{27}$ c. $\frac{15}{8} \times \frac{64}{3}$ d. $\frac{3}{28} \times \frac{4}{21}$

b. Problem A: 7/5 × 40/35.
• Copy the problem. Cross out the fractions that have multiples and write the simplified values. Then write the answer.
(Observe students and give feedback.)
c. Check your work.
You crossed out 7/35.
• What did you write in the numerator? (Signal.) *1.*
• What did you write in the denominator? (Signal.) *5.*
You crossed out 40/5.
• What did you write in the numerator? (Signal.) *8.*
• What's 1 × 8/5? (Signal.) *8/5.*
(Display:)   [61:4E]

$$a. \quad \frac{\cancel{7}^{1}}{\cancel{5}} \times \frac{\cancel{40}^{8}}{\cancel{35}_{5}} = \boxed{\frac{8}{5}}$$

Here's what you should have.
d. Problem B: 3/5 × 40/27.
• Copy and work the problem.
(Observe students and give feedback.)
e. Check your work.
You crossed out 3/27.
• What did you write in the numerator? (Signal.) *1.*
• What did you write in the denominator? (Signal.) *9.*
You crossed out 40/5.
• What did you write in the numerator? (Signal.) *8.*
• What does 1 × 8/9 equal? (Signal.) *8/9.*
(Display:)   [61:4F]

$$b. \quad \frac{\cancel{3}^{1}}{\cancel{5}} \times \frac{\cancel{40}^{8}}{\cancel{27}_{9}} = \boxed{\frac{8}{9}}$$

Here's what you should have.

f. Problem C: 15/8 × 64/3.
• Copy and work the problem.
(Observe students and give feedback.)
g. Check your work.
You crossed out 15/3.
• What does it equal? (Signal.) *5.*
• You crossed out 64/8.
• What does it equal? (Signal.) *8.*
• What does 5 × 8 equal? (Signal.) *40.*
(Display:)   [61:4G]

$$c. \quad \frac{\cancel{15}^{5}}{\cancel{8}} \times \frac{\cancel{64}}{\cancel{3}} = \boxed{40}$$

Here's what you should have.
h. Problem D: 3/28 × 4/21.
• Copy and work the problem.
(Observe students and give feedback.)
i. Check your work.
You crossed out 3/21.
• What does it equal? (Signal?) *1/7.*
You crossed out 4/28.
• What does it equal? (Signal.) *1/7.*
• What does 1/7 × 1/7 equal? (Signal.) *1/49.*
(Display:)   [61:4H]

$$d. \quad \frac{\cancel{3}^{1}}{\cancel{28}_{7}} \times \frac{\cancel{4}^{1}}{\cancel{21}_{7}} = \boxed{\frac{1}{49}}$$

Here's what you should have.

EXERCISE 5: EXPONENTS
FOR REPEATED MULTIPLICATION   `REMEDY`

a. (Display:)   [61:5A]

$$4 \times 4 \times 4$$
$$1 \times 1 \times 1 \times 1 \times 1 \times 1$$
$$3 + 3 + 3$$

You learned how to write repeated multiplication as a base number and an exponent. Remember, the base number is the number that is shown in the repeated multiplication. The exponent tells how many times the number is shown.

b. (Point to **4**.) Here's 4 times 4 times 4.
- What's the base number? (Signal.) *4.*
- How many times is 4 shown? (Signal.) *3.*
- So what's the exponent? (Signal.) *3.*
(Add to show:) [61:5B]

$$4 \times 4 \times 4 = 4^3$$
$$1 \times 1 \times 1 \times 1 \times 1 \times 1$$
$$3 + 3 + 3$$

(Point to **4³**.) This says 4 to the third power.
- What does it say? (Signal.) *4 to the third power.*
c. (Point to **1**.) Here's a repeated multiplication for 1.
- What's the base number? (Signal.) *1.*
- Raise your hand when you know the exponent. ✔
- What's the exponent? (Signal.) *6.*
(Add to show:) [61:5C]

$$4 \times 4 \times 4 = 4^3$$
$$1 \times 1 \times 1 \times 1 \times 1 \times 1 = 1^6$$
$$3 + 3 + 3$$

(Point to **1⁶**.) This says one to the sixth power.
- What does it say? (Signal.) *1 to the sixth power.*
d. (Point to **4³**.) What does this say? (Signal.) *4 to the third power.*
e. (Point to **3**.) Here's 3 plus 3 plus 3.
- Does this show repeated multiplication? (Signal.) *No.*
- So can we write it as a base number and an exponent? (Signal.) *No.*
Right. We can only show a base and exponent for repeated multiplication.

a. Find part 2 in your textbook. ✔
(Teacher reference:)

a. 12 × 12 × 12 × 12   b. 2 × 2 × 2
c. 8 × 8   d. 5 × 5 × 5 × 5 × 5

b. Problem A. Read it. (Signal.) *12 × 12 × 12 × 12.*
- Does it show repeated multiplication? (Signal.) *Yes.*
- Write the base and exponent. ✔
- Everybody, what's the base? (Signal.) *12.*
- What's the exponent? (Signal.) *4.*
(Display:) [61:5D]

a. 12^4

Here's what you should have.
- What does this say? (Signal.) *12 to the fourth power.*
c. Write the base and exponent for the rest of the items.
(Observe students and give feedback.)
d. Check your work.
- Item B: 2 × 2 × 2. What's the base? (Signal.) *2.*
- What's the exponent? (Signal.) *3.*
- Read it. (Signal.) *2 to the third power.*
e. Item C: 8 × 8. What's the base? (Signal.) *8.*
- What's the exponent? (Signal.) *2.*
- Read it. (Signal.) *8 to the second power.*
f. Item D: 5 × 5 × 5 × 5 × 5. What's the base? (Signal.) *5.*
- What's the exponent? (Signal.) *5.*
- Read it. (Signal.) *5 to the fifth power.*

EXERCISE 6: RATIO AND PROPORTION
CLASSIFICATION WITH 2 QUESTIONS

a. Find part 3 in your textbook. ✔
   (Teacher reference:)

> **a.** The ratio of hens to roosters was 9 to 2. There were 110 birds.
> How many roosters were there? How many hens were there?
>
> **b.** On the beach walk, there were 3 broken shells for every 5 shells
> we found. We found 16 unbroken shells. How many broken
> shells did we find? How many shells did we find in all?

To work these problems you need three ratio numbers.

b. Problem A: The ratio of hens to roosters was 9 to 2. There were 110 birds. How many roosters were there? How many hens were there?

• Write the ratio numbers and work the problems for both questions.
  (Observe students and give feedback.)

c. Check your work.
   (Display:)   [61:6A]

```
a.  | h   9  |
    | r   2  |
    | b  11  |
```

Here are the ratio numbers.
• How many roosters were there? (Signal.)
  20 roosters.
• How many hens were there? (Signal.)
  90 hens.

d. Problem B: On the beach walk, there were 3 broken shells for every 5 shells we found. We found 16 unbroken shells. How many broken shells did we find? How many shells did we find in all?

• Write the ratio numbers and work the problems for both questions.
  (Observe students and give feedback.)

e. Check your work.
   (Display:)   [61:6B]

```
b.  | b   3  |
    | s   5  |
    | u   2  |
```

Here are the ratio numbers.
• How many broken shells were there? (Signal.)
  24 broken shells.
• How many shells were there in all? (Signal.)
  40 shells.

> Assign Independent Work, Textbook parts 4–12 and Workbook part 3.

Optional extra math-fact practice worksheets are available on ConnectED.

Lesson 62

EXERCISE 1: MENTAL MATH
REMAINDERS

a. You're going to do mental math with division problems that have remainders.
   (Display:)   [62:1A]

$$4\overline{)15} \quad 4\overline{)18} \quad 4\overline{)11} \quad 4\overline{)7}$$

- (Point to **15**.) 15 divided by 4. (Pause.) What's the answer? (Signal.) *3 with a remainder of 3.*
- (Point to **18**.) 18 divided by 4. (Pause.) What's the answer? (Signal.) *4 with a remainder of 2.*
- (Point to **11**.) 11 divided by 4. (Pause.) What's the answer? (Signal.) *2 with a remainder of 3.*
- (Point to **7**.) 7 divided by 4. (Pause.) What's the answer? (Signal.) *1 with a remainder of 3.*
  (Repeat until firm.)

b. (Display:)   [62:1B]

$$4\overline{)13} \quad 2\overline{)13} \quad 5\overline{)13} \quad 4\overline{)21}$$

- (Point to $4\overline{)13}$.) 13 divided by 4. (Pause.) What's the answer? (Signal.) *3 with a remainder of 1.*
- (Point to $2\overline{)13}$.) 13 divided by 2. (Pause.) What's the answer? (Signal.) *6 with a remainder of 1.*
- (Point to $5\overline{)13}$.) 13 divided by 5. (Pause.) What's the answer? (Signal.) *2 with a remainder of 3.*
- (Point to **21**.) 21 divided by 4. (Pause.) What's the answer? (Signal.) *5 with a remainder of 1.*
  (Repeat until firm.)

c. (Display:)   [62:1C]

$$4\overline{)25} \quad 3\overline{)25} \quad 3\overline{)19} \quad 4\overline{)19}$$

- (Point to $4\overline{)25}$.) 25 divided by 4. (Pause.) What's the answer? (Signal.) *6 with a remainder of 1.*
- (Point to $3\overline{)25}$.) 25 divided by 3. (Pause.) What's the answer? (Signal.) *8 with a remainder of 1.*
- (Point to $3\overline{)19}$.) 19 divided by 3. (Pause.) What's the answer? (Signal.) *6 with a remainder of 1.*
- (Point to $4\overline{)19}$.) 19 divided by 4. (Pause.) What's the answer? (Signal.) *4 with a remainder of 3.*
  (Repeat until firm.)

d. (Display:)   [62:1D]

$$5\overline{)19} \quad 2\overline{)19} \quad 3\overline{)23} \quad 3\overline{)25} \quad 3\overline{)29}$$

- (Point to $5\overline{)19}$.) 19 divided by 5. (Pause.) What's the answer? (Signal.) *3 with a remainder of 4.*
- (Point to $2\overline{)19}$.) 19 divided by 2. (Pause.) What's the answer? (Signal.) *9 with a remainder of 1.*
- (Point to **23**.) 23 divided by 3. (Pause.) What's the answer? (Signal.) *7 with a remainder of 2.*
- (Point to **25**.) 25 divided by 3. (Pause.) What's the answer? (Signal.) *8 with a remainder of 1.*
- (Point to **29**.) 29 divided by 3. (Pause.) What's the answer? (Signal.) *9 with a remainder of 2.*
  (Repeat until firm.)

EXERCISE 2: FRACTION OPERATIONS
MULTIPLICATION—CROSS CANCELLATION   **REMEDY**

a. (Display:)   [62:2A]

$$\frac{3}{5}\left(\frac{65}{6}\right) =$$

- Read this problem. (Signal.) *3/5 × 65/6.*
  We'll simplify fractions that have multiples.
b. Is 65 a multiple of 2? (Signal.) *No.*
- So is it a multiple of 6? (Signal.) *No.*
- Is 65 a multiple of 5? (Signal.) *Yes.*
  So we can simplify 65/5.
- Raise your hand when you know the simplified value. ✔
- What does 65/5 equal? (Signal.) *13.*
  (Add to show:)   [62:2B]

$$\frac{3}{\cancel{5}}\left(\frac{\cancel{65}^{13}}{6}\right) =$$

c. The other fraction is 3/6. Is 6 a multiple of 3? (Signal.) *Yes.*
  So we can simplify 3/6.
- Raise your hand when you know the simplified fraction. ✔

- What does 3/6 equal? (Signal.) *1/2.*
  (Add to show:)   [62:2C]

$$\frac{\cancel{3}^{1}}{5}\left(\frac{\cancel{65}^{13}}{\cancel{6}_{2}}\right)=\boxed{\frac{13}{2}}$$

So the answer is 13/2.

━━━━━━━━━━ **TEXTBOOK PRACTICE** ━━━━━━━━━━

a. Open your textbook to Lesson 62 and find
   part 1. ✔
   (Teacher reference:)

a. $\frac{21}{4}\left(\frac{8}{3}\right)=\blacksquare$ b. $\frac{3}{15}\left(\frac{2}{20}\right)=\blacksquare$ c. $\frac{7}{9}\left(\frac{27}{28}\right)=\blacksquare$

b. Read problem A. (Signal.) *21/4 × 8/3.*
- Copy the problem on your lined paper.
  Simplify and write the answer.
  (Observe students and give feedback.)
c. Check your work.
- You simplified 21/3. What does it equal?
  (Signal.) *7.*
- You simplified 8/4. What does it equal?
  (Signal.) *2.*
- What's 7 × 2? (Signal.) *14.*
  (Display:)   [62:2D]

$$a.\ \frac{\cancel{21}^{7}}{\cancel{4}_{}}\left(\frac{\cancel{8}^{2}}{\cancel{3}_{}}\right)=\boxed{14}$$

Here's what you should have.
d. Read problem B. (Signal.) *3/15 × 2/20.*
- Copy the problem. Simplify and write the answer.
  (Observe students and give feedback.)
e. Check your work.
- You simplified 3/15. What does it equal?
  (Signal.) *1/5.*
- You simplified 2/20. What does it equal?
  (Signal.) *1/10.*
- What's 1/5 × 1/10? (Signal.) *1/50.*
  (Display:)   [62:2E]

$$b.\ \frac{\cancel{3}^{1}}{\cancel{15}_{5}}\left(\frac{\cancel{2}^{1}}{\cancel{20}_{10}}\right)=\boxed{\frac{1}{50}}$$

Here's what you should have.

f. Read problem C. (Signal.) *7/9 × 27/28.*
- Copy the problem. Simplify and write the answer.
  (Observe students and give feedback.)
g. Check your work.
- You simplified 7/28. What does it equal?
  (Signal.) *1/4.*
- You simplified 27/9. What does it equal?
  (Signal.) *3.*
- What's the answer to the problem? (Signal.) *3/4.*
  (Display:)   [62:2F]

$$c.\ \frac{\cancel{7}^{1}}{\cancel{9}_{}}\left(\frac{\cancel{27}^{3}}{\cancel{28}_{4}}\right)=\boxed{\frac{3}{4}}$$

Here's what you should have.

EXERCISE 3: EXPONENTS
FOR REPEATED MULTIPLICATION   `REMEDY`

a. (Display:)   [62:3A]

$$4 \times 4$$

I'm going to write this as a base and an
exponent.
- What's the base? (Signal.) *4.*
- What's the exponent? (Signal.) *2.*
  (Add to show:)   [62:3B]

$$4 \times 4 = 4^{2}$$

We read the base and exponent as 4 to the
second power, or we just say 4 to the second.
- Read the base and exponent the short way.
  (Signal.) *4 to the second.*
  Yes, 4 to the second.
b. (Display:)   [62:3C]

$$4$$

- It's not 4 to the second now. It's 4 to the first.
  What is it? (Signal.) *4 to the first.*
  (Add to show:)   [62:3D]

$$4 = 4^{1}$$

- Read the base and exponent. (Signal.) *4 to
  the first.*

c. (Display:) [62:3E]

10

- New number. What's the base? (Signal.) *10.*
- What's the exponent? (Signal.) *1.*
(Add to show:) [62:3F]

$10 = 10^1$

- Read it. (Signal.) *10 to the first.*
d. (Display:) [62:3G]

406

- New number. What's the base? (Signal.) *406.*
- What's the exponent? (Signal.) *1.*
(Add to show:) [62:3H]

$406 = 406^1$

- Read it. (Signal.) *406 to the first.*

=== **TEXTBOOK PRACTICE** ===

a. Find part 2 in your textbook. ✔
(Teacher reference:)

a. 5 × 5   b. 9   c. 6 × 6 × 6
d. 1 × 1 × 1 × 1   e. k × k × k × k × k   f. b × b

- Write the base and exponent for each item.
(Observe students and give feedback.)
b. Check your work.
- Item A. Read it. (Signal.) *5 to the second.*
- Item B. Read it. (Signal.) *9 to the first.*
- C. Read it. (Signal.) *6 to the third.*
- D. Read it. (Signal.) *1 to the fourth.*
- E. Read it. (Signal.) *K to the fifth.*
- F. Read it. (Signal.) *B to the second.*

EXERCISE 4: EQUIVALENT FRACTIONS   REMEDY

a. Open your workbook to Lesson 62 and find part 1. ✔
(Teacher reference:)   R Part G

$$\frac{5}{2} = \frac{20}{} = \frac{15}{} = \frac{}{14} = \frac{}{10} = \frac{45}{}$$
   a.   b.   c.   d.   e.

b. Complete the fractions that equal 5/2.
(Observe students and give feedback.)

c. Check your work.
- What's fraction A? (Signal.) *20/8.*
- What's fraction B? (Signal.) *15/6.*
- What's fraction C? (Signal.) *35/14.*
- What's fraction D? (Signal.) *25/10.*
- What's fraction E? (Signal.) *45/18.*
d. What do all of these fractions equal?
(Signal.) *5/2.*

EXERCISE 5: DECIMAL OPERATIONS
DIVISION   REMEDY

a. (Display:) [62:5A]

5⟌.455

You worked division problems that have a decimal number. You write a decimal point in the answer.
- Where does it go? (Call on a student. Idea: *Above the decimal point in the problem.*)
Yes, right above the other decimal point.
(Add to show:) [62:5B]

5⟌̇.455

b. Now we work the problem.
- What's 45 ÷ 5? (Signal.) *9.*
So I write 9 above the last digit of 45.
(Add to show:) [62:5C]

. 9
5⟌.455

- I need a digit in front of 9. What digit do I write? (Signal.) *Zero.*
(Add to show:) [62:5D]

.09
5⟌.455

c. Say the problem for the last digit of 455.
(Signal.) *5 ÷ 5.*
- What's the answer? (Signal.) *1.*
(Add to show:) [62:5E]

.091
5⟌.455

- Read the problem and the answer. (Signal.)
455 thousandths ÷ 5 = 91 thousandths.

a. Find part 2 in your workbook. ✔
   (Teacher reference:)   R Part H

   a. $3\overline{).24}$ b. $6\overline{).546}$ c. $7\overline{)8.421}$ d. $2\overline{).058}$

b. Read problem A. (Signal.) *24 hundredths ÷ 3.*
   • Can you work the problem 2 ÷ 3? (Signal.) *No.*
   • Can you work the problem 24 ÷ 3? (Signal.) *Yes.*
   • Work that part of the problem. Write the
     decimal point and a zero in the answer. Then
     write what 24 divided by 3 equals. ✔
     (Display:)   [62:5F]

   $$\text{a. } 3\overline{).24}\;\;\overset{.08}{}$$

   Here's what you should have.
   • Everybody, read the problem and the answer.
     (Signal.) *.24 ÷ 3 = .08.*

c. Work the rest of the problems in part 2.
   Remember, write the decimal point in the
   answer right above the other decimal point.
   Make sure there is a digit in every decimal
   place of the answer.
   (Observe students and give feedback.)

d. Check your work.
   • Problem B: 546 thousandths ÷ 6.
   • What's the answer? (Signal.) *.091.*
     (Display:)   [62:5G]

   $$\text{b. } 6\overline{).546}\;\;\overset{.091}{}$$

   Here's what you should have.

e. Problem C: 8 and 421 thousandths ÷ 7.
   • What's the answer? (Signal.) *1.203.*
     (Display:)   [62:5H]

   $$\text{c. } 7\overline{)8.421}\;\;\overset{1.203}{}\quad 7\;{}^{1}0$$

   Here's what you should have.

f. Problem D: 58 thousandths ÷ 2.
   • What's the answer? (Signal.) *.029.*
     (Display:)   [62:5I]

   $$\text{d. } 2\overline{).05{}_{1}8}\;\;\overset{.029}{}\quad 4$$

   Here's what you should have.

EXERCISE 6: FRACTION SIMPLIFICATION
GREATEST COMMON FACTOR

a. We're going to simplify fractions the fast way.
   (Display:)   [62:6A]

   $$\frac{14}{12} =$$

   • Read this fraction. (Signal.) *14/12.*
     Both 14 and 12 have a common factor.
   • Raise your hand when you know that factor. ✔
   • What's the common factor? (Signal.) *2.*
   • 14 is 2 times what number? (Signal.) *7.*
     (Add to show:)   [62:6B]

   $$\frac{14}{12} = \frac{2 \times 7}{}$$

   • 12 is 2 times what number? (Signal.) *6.*
     (Add to show:)   [62:6C]

   $$\frac{14}{12} = \frac{2 \times 7}{2 \times 6}$$

   So 14/12 = 7/6.
   (Add to show:)   [62:6D]

   $$\frac{14}{12} = \frac{\cancel{2} \times 7}{\cancel{2} \times 6} = \boxed{\frac{7}{6}}$$

b. (Display:)   [62:6E]

   $$\frac{8}{12} =$$

   • Read this fraction. (Signal.) *8/12.*
     8 and 12 have more than one common factor.
   • Is 2 a factor of 8 and 12? (Signal.) *Yes.*
   • Raise your hand when you know the other
     common factor of 8 and 12. ✔
   • What's the other common factor of 8 and 12?
     (Signal.) *4.*
     The greatest common factor is 4. So that's the
     factor we use.
   • 8 is 4 times what number? (Signal.) *2.*
     (Add to show:)   [62:6F]

   $$\frac{8}{12} = \frac{4 \times 2}{}$$

   • 12 is 4 times what number? (Signal.) *3.*

(Add to show:)   [62:6G]

$$\frac{8}{12} = \frac{4 \times 2}{4 \times 3}$$

- What does 8/12 equal? (Signal.) *2/3.*
  Yes, 2/3.
  (Add to show:)   [62:6H]

$$\frac{8}{12} = \frac{\cancel{4} \times 2}{\cancel{4} \times 3} = \boxed{\frac{2}{3}}$$

c. (Display:)   [62:6I]

$$\frac{10}{25} =$$

- Read this fraction. (Signal.) *10/25.*
  Both 10 and 25 have a common factor.
- Raise your hand when you know that factor. ✔
- What's the common factor? (Signal.) *5.*
- 10 is 5 times what number? (Signal.) *2.*
- 25 is 5 times what number? (Signal.) *5.*
  (Add to show:)   [62:6J]

$$\frac{10}{25} = \frac{5 \times 2}{5 \times 5}$$

- What does 10/25 equal? (Signal.) *2/5.*
  (Add to show:)   [62:6K]

$$\frac{10}{25} = \frac{\cancel{5} \times 2}{\cancel{5} \times 5} = \boxed{\frac{2}{5}}$$

Yes, 2/5.

d. (Display:)   [62:6L]

$$\frac{3}{18} = \frac{}{} = \frac{}{}$$

- Copy the fraction in part A. Write the multiplication for the common factor and write the simplified fraction.
  (Observe students and give feedback.)
- Everybody, what's the common factor? (Signal.) *3.*
- 3 is 3 times what number? (Signal.) *1.*
- 18 is 3 times what number? (Signal.) *6.*
- What does 3/18 equal? (Signal.) *1/6.*
  (Add to show:)   [62:6M]

$$\frac{3}{18} = \frac{\cancel{3} \times 1}{\cancel{3} \times 6} = \boxed{\frac{1}{6}}$$

Here's what you should have.

e. (Display:)   [62:6N]

$$\frac{6}{8} = \frac{}{} = \frac{}{}$$

- Copy the fraction. Write the multiplication for the common factor and write the simplified fraction.
  (Observe students and give feedback.)
- Everybody, what's the common factor? (Signal.) *2.*
- 6 is 2 times what number? (Signal.) *3.*
- 8 is 2 times what number? (Signal.) *4.*
- What does 6/8 equal? (Signal.) *3/4.*
  (Add to show:)   [62:6O]

$$\frac{6}{8} = \frac{\cancel{2} \times 3}{\cancel{2} \times 4} = \boxed{\frac{3}{4}}$$

Here's what you should have.

f. (Display:)   [62:6P]

$$\frac{12}{18} = \frac{}{} = \frac{}{}$$

12 and 18 have more than one common factor.
- Raise your hand when you know both common factors of 12 and 18. ✔
- What are the two common factors of 12 and 18? (Signal.) *2 and 6.*
- Which is the greatest common factor—2 or 6? (Signal.) *6.*
  That's the factor you use to work the problem.
- Copy the fraction. Write the multiplication for the greatest common factor and write the simplified fraction. ✔
  The greatest common factor is 6.
- What's the simplified fraction? (Signal.) *2/3.*
  (Add to show:)   [62:6Q]

$$\frac{12}{18} = \frac{\cancel{6} \times 2}{\cancel{6} \times 3} = \boxed{\frac{2}{3}}$$

Here's what you should have.

Assign Independent Work, Workbook parts 3 and 4 and Textbook parts 3–8.

Optional extra math-fact practice worksheets are available on ConnectED.

Lesson 63

EXERCISE 1: MENTAL MATH
REMAINDERS

a. (Display:) [63:1A]

$$4\overline{\smash{)}17}^{R} \qquad 5\overline{\smash{)}17}^{R} \qquad 3\overline{\smash{)}14}^{R}$$

Here are division problems. I'll write the whole-number part of the answer and the remainder after the R.

b. (Point to $4\overline{\smash{)}17}^{R}$.) 17 divided by 4.
• What's the answer? (Signal.) *4 with a remainder of 1.*
  (Add to show:) [63:1B]

$$4\overline{\smash{)}17}^{4\,R\,1} \qquad 5\overline{\smash{)}17}^{R} \qquad 3\overline{\smash{)}14}^{R}$$

Here's 4 with a remainder of 1.

c. (Point to $5\overline{\smash{)}17}^{R}$.) 17 divided by 5.
• What's the answer? (Signal.) *3 with a remainder of 2.*
  (Add to show:) [63:1C]

$$4\overline{\smash{)}17}^{4\,R\,1} \qquad 5\overline{\smash{)}17}^{3\,R\,2} \qquad 3\overline{\smash{)}14}^{R}$$

d. (Point to $3\overline{\smash{)}14}^{R}$.) 14 divided by 3.
• What's the answer? (Signal.) *4 with a remainder of 2.*
  (Add to show:) [63:1D]

$$4\overline{\smash{)}17}^{4\,R\,1} \qquad 5\overline{\smash{)}17}^{3\,R\,2} \qquad 3\overline{\smash{)}14}^{4\,R\,2}$$

━━━━━━ **WORKBOOK PRACTICE** ━━━━━━

a. Open your workbook to Lesson 63 and find part 1. ✔
  (Teacher reference:)

$$\text{a. } 5\overline{\smash{)}14}^{R} \quad \text{d. } 4\overline{\smash{)}33}^{R} \quad \text{g. } 4\overline{\smash{)}23}^{R} \quad \text{j. } 2\overline{\smash{)}13}^{R}$$

$$\text{b. } 3\overline{\smash{)}29}^{R} \quad \text{e. } 3\overline{\smash{)}22}^{R} \quad \text{h. } 4\overline{\smash{)}26}^{R} \quad \text{k. } 3\overline{\smash{)}13}^{R}$$

$$\text{c. } 4\overline{\smash{)}29}^{R} \quad \text{f. } 5\overline{\smash{)}29}^{R} \quad \text{i. } 4\overline{\smash{)}13}^{R}$$

Your turn to write answers that show the whole number and the remainder.

You'll write the whole-number part of the answer. Then you'll write the remainder after the R.

b. Read problem A. (Signal.) *14 ÷ 5.*
• Write the answer. ✔
• Everybody, what's the whole-number part of the answer? (Signal.) *2.*
• What's the remainder? (Signal.) *4.*
  (Display:) [63:1E]

$$\text{a. } 5\overline{\smash{)}14}^{2\,R\,4}$$

Here's what you should have.

c. Read problem B. (Signal.) *29 ÷ 3.*
• Write the answer. ✔
• Everybody, what's the whole-number? (Signal.) *9.*
• What's the remainder? (Signal.) *2.*
d. Read problem C. (Signal.) *29 ÷ 4.*
• Write the answer. ✔
• What's the whole-number? (Signal.) *7.*
• What's the remainder? (Signal.) *1.*
e. Write answers for the rest of the problems in part 1.
  (Observe students and give feedback.)
f. Check your work.
• Problem D: 33 ÷ 4.
• What's the whole-number? (Signal.) *8.*
• What's the remainder? (Signal.) *1.*
g. E: 22 ÷ 3.
• What's the whole-number? (Signal.) *7.*
• What's the remainder? (Signal.) *1.*
h. F: 29 ÷ 5.
• Whole-number? (Signal.) *5.*
• Remainder? (Signal.) *4.*
i. G: 23 ÷ 4.
• Whole-number? (Signal.) *5.*
• Remainder? (Signal.) *3.*
j. H: 26 ÷ 4.
• Whole-number? (Signal.) *6.*
• Remainder? (Signal.) *2.*
k. I: 13 ÷ 4.
• Whole-number? (Signal.) *3.*
• Remainder? (Signal.) *1.*

l. J: 13 ÷ 2.
  * Whole-number? (Signal.) *6.*
  * Remainder? (Signal.) *1.*
m. K: 13 ÷ 3.
  * Whole-number? (Signal.) *4.*
  * Remainder? (Signal.) *1.*

EXERCISE 2: EQUIVALENT FRACTIONS

a. Find part 2 in your workbook. ✔
   (Teacher reference:)

$$\boxed{\dfrac{8}{3}} = \dfrac{}{12} = \dfrac{48}{} = \dfrac{}{15} = \dfrac{16}{} = \dfrac{}{21}$$
  a.   b.   c.   d.   e.

b. Complete the fractions that equal 8/3.
   (Observe students and give feedback.)
c. Check your work.
  * What's fraction A? (Signal.) *32/12.*
  * What's fraction B? (Signal.) *48/18.*
  * What's fraction C? (Signal.) *40/15.*
  * What's fraction D? (Signal.) *16/6.*
  * What's fraction E? (Signal.) *56/21.*
d. What do all these fractions equal? (Signal.) *8/3.*

EXERCISE 3: DECIMAL OPERATIONS
DIVISION

a. Find part 3 in your workbook. ✔
   (Teacher reference:)

  a. $6\overline{).042}$ b. $4\overline{)9.28}$ c. $9\overline{).189}$ d. $5\overline{)10.45}$

   The answers to all these problems have a
   decimal point. Remember, that decimal point
   goes right above the other decimal point. And
   you must have a digit in every decimal place
   of the answer.
b. Work problem A.
   (Observe students and give feedback.)
  * Check your work. What's 42 thousandths ÷ 6?
   (Signal.) *.007.*
c. Work the rest of the problems in part 3.
   (Observe students and give feedback.)
d. Check your work.
  * Problem B: 9 and 28 hundredths ÷ 4.
  * What's the answer? (Signal.) *2.32.*
e. Problem C: 189 thousandths ÷ 9.
  * What's the answer? (Signal.) *.021.*
f. Problem D: 10 and 45 hundredths ÷ 5.
  * What's the answer? (Signal.) *2.09.*

EXERCISE 4: FRACTION SIMPLIFICATION
GREATEST COMMON FACTOR

a. (Display:)   [63:4A]

$$\dfrac{12}{16} =$$

  * Read the fraction. (Signal.) *12/16.*
   We're going to simplify this fraction the fast way.
   12 and 16 have more than one common factor.
  * Is 2 a factor of 12 and 16? (Signal.) *Yes.*
  * Raise your hand when you know the other
   common factor of 12 and 16. ✔
  * What's the other common factor of 12 and 16?
   (Signal.) *4.*
   The greatest common factor is 4. So that's the
   factor we use.
  * 12 is 4 times what number? (Signal.) *3.*
   (Add to show:)   [63:4B]

$$\dfrac{\overset{3}{\cancel{12}}}{16} =$$

  * 16 is 4 times what number? (Signal.) *4.*
   (Add to show:)   [63:4C]

$$\dfrac{\overset{3}{\cancel{12}}}{\underset{4}{\cancel{16}}} =$$

  * What does 12/16 equal? (Signal.) *3/4.*
   (Add to show:)   [63:4D]

$$\dfrac{\overset{3}{\cancel{12}}}{\underset{4}{\cancel{16}}} = \boxed{\dfrac{3}{4}}$$

   Yes, 3/4.
b. (Display:)   [63:4E]

$$\dfrac{10}{14} =$$

  * Read the fraction. (Signal.) *10/14.*
  * Raise your hand when you know the common
   factor. ✔
  * What's the common factor? (Signal.) *2.*
  * 10 is 2 times what number? (Signal.) *5.*
  * 14 is 2 times what number? (Signal.) *7.*

(Add to show:)   [63:4F]

$$\frac{\frac{5}{\cancel{10}}}{\cancel{14}}=$$
$$\frac{}{7}$$

- What does 10/14 equal? (Signal.) *5/7.*
  (Add to show:)   [63:4G]

$$\frac{\frac{5}{\cancel{10}}}{\cancel{14}}=\boxed{\frac{5}{7}}$$
$$\frac{}{7}$$

Yes, 5/7.

═══════════ **TEXTBOOK PRACTICE** ═══════════

a. Open your textbook to Lesson 63 and find
   part 1. ✔
   (Teacher reference:)

   a. $\frac{16}{24}$ b. $\frac{20}{25}$ c. $\frac{45}{18}$ d. $\frac{28}{35}$

b. Copy fraction A. Then stop. ✔
   16 and 24 have more than one common factor.
- Raise your hand when you know the greatest
  common factor of 16 and 24. ✔
- What's the greatest common factor of 16 and
  24? (Signal.) *8.*
- Write the simplified fraction.
  (Observe students and give feedback.)
  The greatest common factor is 8.
- 16 is 8 times what number? (Signal.) *2.*
- 24 is 8 times what number? (Signal.) *3.*
- So what does 16/24 equal? (Signal.) *2/3.*
c. Copy fraction B. Find the greatest common
   factor and write the simplified fraction. ✔
- Everybody, what's the greatest common factor
  for 20 and 25? (Signal.) *5.*
- 20 is 5 times what number? (Signal.) *4.*
- 25 is 5 times what number? (Signal.) *5.*
- So what does 20/25 equal? (Signal.) *4/5.*
d. Simplify the rest of the fractions in part 1.
   (Observe students and give feedback.)
e. Check your work.
- Fraction C: 45/18. The greatest common factor
  is 9. What's the simplified fraction? (Signal.) *5/2.*
- Fraction D: 28/35. The greatest common factor
  is 7. What's the simplified fraction? (Signal.) *4/5.*

EXERCISE 5: SUBTRACTION
RENAMING WITH ZERO   `REMEDY`

a. (Display:)   [63:5A]

3002	6009	4007	5008

- (Point to **3002**.) Read this number. (Signal.) *3
  thousand 2.*
  3002 is 300 tens plus 2.
- How many tens are in 3002? (Signal.) *300.*
b. My turn to say the new tens plus ones for
   3002: 299 tens plus 12.
- Say the new tens plus ones. (Signal.) *299 tens
  plus 12.*
c. (Point to **6009**.) Read this number. (Signal.) *6009.*
- Say the new tens plus ones for 6009. (Signal.)
  599 tens plus 19.
d. (Point to **4007**.) Read this number. (Signal.) *4007.*
- Say the new tens plus ones for 4007. (Signal.)
  399 tens plus 17.
e. (Point to **5008**.) Read this number. (Signal.) *5008.*
- Say the new tens plus ones for 5008. (Signal.)
  499 tens plus 18.
  (Repeat until firm.)
f. (Display:)   [63:5B]

$$\begin{array}{r} 5\,0\,0\,3 \\ -\quad 2\,5 \\ \hline \end{array}$$

  Here's a new kind of problem.
- Read it. (Signal.) *5003 – 25.*
- Say the problem for the ones. (Signal.) *3 – 5.*
- Can you work that problem? (Signal.) *No.*
  So we rewrite the top number to show the new
  tens plus ones.
- Say the new tens plus ones. (Signal.) *499 tens
  plus 13.*
  (Add to show:)   [63:5C]

$$\begin{array}{r} \overset{4\,9\,9}{\cancel{5\,0\,0}3} \\ -\quad 2\,5 \\ \hline \end{array}$$

  Now we can work the problem.

a. Find part 2 in your textbook. ✔
(Teacher reference:)

a. 2004   b. 1007   c. 5003   d. 2005
  – 165   – 59   –1426   –1806

For all these problems, you have to rewrite the top number to show the new tens plus ones.

b. Copy problem A. Then stop. ✔
• Read problem A. (Signal.) *2004 – 165.*
• Say the problem for the ones. (Signal.) *4 – 5.*
• Can you work the problem? (Signal.) *No.*
• Rewrite the top number. Stop when you've done that much. ✔
(Display:) [63:5D]

$$\begin{array}{r} 1\ 9\ 9 \\ a.\quad \overset{}{2}\,\overset{}{0}\,0\,4 \\ -\ \ 1\ 6\ 5 \\ \hline \end{array}$$

Here's what you should have.
• Work the problem.
(Observe students and give feedback.)
• Everybody, read the problem you started with and the answer. (Signal.) *2004 – 165 = 1839.*
c. Problem B.
• Say the problem for the ones. (Signal.) *7 – 9.*
• Can you work that problem? (Signal.) *No.*
d. Copy problem B. Rewrite the top number and work the problem. Then work the rest of the problems in part 2.
(Observe students and give feedback.)
e. Check your work.
• Problem B. Read the problem you started with and the answer. (Signal.) *1007 – 59 = 948.*
• C. Read the problem you started with and the answer. (Signal.) *5003 – 1426 = 3577.*
• D. Read the problem you started with and the answer. (Signal.) *2005 – 1806 = 199.*

EXERCISE 6: DECIMAL OPERATIONS
MISSING FACTOR

a. (Display:) [63:6A]

$$.74 \times \underline{} = 74$$
$$.8 \times \underline{} = 8$$
$$.3 \times \underline{} = 3$$
$$.013 \times \underline{} = 13$$

We're going to multiply each decimal number by something to change it into a whole number.

b. (Point to **.74**) 74 hundredths times something equals 74.
Listen: 74 hundredths is hundredths. So if we multiply 74 hundredths by 100 we end up with 74.
• What do we multiply 74 hundredths by? (Signal.) *100.*
(Add to show:) [63:6B]

$$.74 \times \underline{100} = 74$$
$$.8 \times \underline{} = 8$$
$$.3 \times \underline{} = 3$$
$$.013 \times \underline{} = 13$$

c. (Point to **.8**) 8 tenths is tenths. So what do we multiply 8 tenths by to get 8? (Signal.) *10.*
(Add to show:) [63:6C]

$$.74 \times \underline{100} = 74$$
$$.8 \times \underline{10} = 8$$
$$.3 \times \underline{} = 3$$
$$.013 \times \underline{} = 13$$

d. (Point to **.3**) What do we multiply 3 tenths by to get 3? (Signal.) *10.*
(Add to show:) [63:6D]

$$.74 \times \underline{100} = 74$$
$$.8 \times \underline{10} = 8$$
$$.3 \times \underline{10} = 3$$
$$.013 \times \underline{} = 13$$

e. (Point to **.013.**) What do we multiply 13 thousandths by to get 13? (Signal.) *1000.*
(Add to show:) [63:6E]

$$.74 \times \underline{100} = 74$$
$$.8 \times \underline{10} = 8$$
$$.3 \times \underline{10} = 3$$
$$.013 \times \underline{1000} = 13$$

(Repeat until firm.)

f. Let's see if we're right. We'll multiply
    74 hundredths by 100.
    (Add to show:)   [63:6F]

$$.74 \times \underline{100} = 74$$
$$.8 \times \underline{10} = 8$$
$$.3 \times \underline{10} = 3$$
$$.013 \times \underline{1000} = 13$$

$$\begin{array}{r} .74 \\ \times\ 100 \\ \hline \end{array}$$

We write two zeros because we multiply by 100.
(Add to show:)   [63:6G]

$$.74 \times \underline{100} = 74$$
$$.8 \times \underline{10} = 8$$
$$.3 \times \underline{10} = 3$$
$$.013 \times \underline{1000} = 13$$

$$\begin{array}{r} .74 \\ \times\ 100 \\ \hline 0\,0 \end{array}$$

• What's 1 times 74? (Signal.) *74.*
  (Add to show:)   [63:6H]

$$.74 \times \underline{100} = 74$$
$$.8 \times \underline{10} = 8$$
$$.3 \times \underline{10} = 3$$
$$.013 \times \underline{1000} = 13$$

$$\begin{array}{r} .74 \\ \times\ 100 \\ \hline 7\,4\,0\,0 \end{array}$$

• How many decimal places are in the numbers
  we multiply? (Signal.) *2.*
• So how many are in the answer? (Signal.) *2.*
  (Add to show:)   [63:6I]

$$.74 \times \underline{100} = 74$$
$$.8 \times \underline{10} = 8$$
$$.3 \times \underline{10} = 3$$
$$.013 \times \underline{1000} = 13$$

$$\begin{array}{r} .74 \\ \times\ 100 \\ \hline 74.00 \end{array}$$

• Does 74 hundredths times 100 equal 74?
  (Signal.) *Yes.*
g. Remember, if the decimal value is tenths, you
    multiply by 10 to change it to a whole number.
• If the decimal value is hundredths, what
  do you multiply by to get a whole number?
  (Signal.) *100.*
• If the decimal value is thousandths, what
  do you multiply by to get a whole number?
  (Signal.) *1000.*

EXERCISE 7: EXPONENTS
EVALUATION   REMEDY

a. (Display:)   [63:7A]

$$3 \times 3 \times 3 \times 3$$

You're going to do repeated multiplication.
• To do repeated multiplication for 3, you first
  figure out what 3×3 equals. What's the
  answer? (Signal.) *9.*
• Then you figure out what 9×3 equals. What's
  the answer? (Signal.) *27.*
  Then you figure out what 27×3 equals.

b. (Display:)   [63:7B]

$$2 \times 2 \times 2 \times 2$$

• Figure out what $2 \times 2 \times 2 \times 2$ equals.
  Remember, first figure out what 2×2 equals
  then figure out what 4×2 equals. Then
  multiply again. Raise your hand when you
  have the answer. ✔
• Everybody, what does $2 \times 2 \times 2 \times 2$ equal?
  (Signal.) *16.*
  Yes, 16.
  (Add to show:)   [63:7C]

$$2 \times 2 \times 2 \times 2 = \boxed{16}$$

c. (Display:)   [63:7D]

$$5^3$$

• Here's a base with an exponent. Read it.
  (Signal.) *5 to the third.*
  I'm going to show the repeated multiplication.
• What number do I multiply by? (Signal.) *5.*
• How many times do I show 5? (Signal.)
  3 times.
  (Add to show:)   [63:7E]

$$5^3 = 5 \times 5 \times 5$$

The equation says that 5 to the third equals
5 times 5 times 5.

d. We have to figure out the number for $5 \times 5 \times 5$.
- What's 5×5? (Signal.) *25.*
- Raise your hand when you know what 25 times 5 is. ✔
- What's 25×5? (Signal.) *125.*
(Add to show:) [63:7F]

$$5^3 = 5 \times 5 \times 5 = \boxed{125}$$

So 5 to the third equals 125.
e. (Display:) [63:7G]

$$10^2$$

- Read it. (Signal.) *10 to the second.*
- What's the number we multiply by? (Signal.) *10.*
- How many times do we show 10? (Signal.) *2 times.*
(Add to show:) [63:7H]

$$10^2 = 10 \times 10$$

10 to the second equals 10 times 10.
- What does 10×10 equal? (Signal.) *100.*
(Add to show:) [63:7I]

$$10^2 = 10 \times 10 = \boxed{100}$$

- Read the equation. (Signal.) *10 to the second = 10 × 10 = 100.*

━━━━━━━ **TEXTBOOK PRACTICE** ━━━━━━━

a. Find part 3 in your textbook. ✔
(Teacher reference:)

a. 12^2 b. 6^3 c. 2^5 d. 10^3 e. 3^1

b. Read item A. (Signal.) *12 to the second.*
- Copy the base and exponent. Show the multiplication and write what 12 to the second equals.
(Observe students and give feedback.)

(Display:) [63:7J]

$$a. \quad 12^2 = 12 \times 12 = \boxed{144}$$

Here's what you should have.
12 to the second equals $12 \times 12 = 144$.
c. Read item B. (Signal.) *6 to the third.*
- Copy 6 to the third. Show the multiplication and write what it equals.
(Observe students and give feedback.)
- Everybody, say the repeated multiplication for 6 to the third. (Signal.) *6 × 6 × 6.*
- What does 6 to the third equal? (Signal.) *216.*
(Display:) [63:7K]

$$b. \quad 6^3 = 6 \times 6 \times 6 = \boxed{216}$$

Here's what you should have.
d. Write the equation for the rest of the items. Remember to show the repeated multiplication. Then show what it equals.
(Observe students and give feedback.)
e. Check your work.
- Item C. Say the repeated multiplication for 2 to the fifth. (Signal.) *2 × 2 × 2 × 2 × 2.*
- What does 2 to the fifth equal? (Signal.) *32.*
- D. Say the repeated multiplication for 10 to the third. (Signal.) *10 × 10 × 10.*
- What does 10 to the third equal? (Signal.) *1000.*
- E. What does 3 to the first equal? (Signal.) *3.*
(Display:) [63:7L]

$$c. \quad 2^5 = 2 \times 2 \times 2 \times 2 \times 2 = \boxed{32}$$
$$d. \quad 10^3 = 10 \times 10 \times 10 = \boxed{1000}$$
$$e. \quad 3^1 = \boxed{3}$$

Here's what you should have.
You didn't need to show 3 multiplied because it is shown only one time.

EXERCISE 8: INDEPENDENT WORK
FRACTION MULTIPLICATION

a. Find part 4 in your textbook. ✔
(Teacher reference:)

a. $\frac{3}{5} \times \frac{10}{27}$ b. $\frac{7}{4}$ (32)   c. $\frac{3}{15}\left(\frac{2}{4}\right)$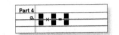

To work these problems, you find multiples.

b. Read problem A. (Signal.) *3/5 × 10/27.*
The numerator of one fraction is 10.
- Is 10 a multiple of 27? (Signal.) *No.*
- Is 10 a multiple of 5? (Signal.) *Yes.*
So you can simplify 10/5.

c. The denominator of the other fraction is 27.
- Is 27 a multiple of 10? (Signal.) *No.*
- Is 27 a multiple of 3? (Signal.) *Yes.*
So you can simplify 3/27.

d. Copy problem A. Simplify the fractions and multiply.
(Observe students and give feedback.)

e. Check your work.
- What's the answer? (Signal.) *2/9.*
(Display:)   [63:8A]

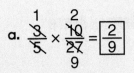

Here's what you should have.

f. You'll work the other problems in part 4 as part of your independent work. Remember, first simplify. Then multiply.

Assign Independent Work, the rest of Textbook part 4 and parts 5–10 and Workbook part 4.

Optional extra math-fact practice worksheets are available on ConnectED.

Lesson 64

EXERCISE 1: MENTAL MATH
REMAINDERS

a. Open your workbook to Lesson 64 and find part 1. ✔

• Pencils down. ✔
(Teacher reference:)

$$\begin{array}{llll} \text{a. } 2\overline{)9}^{\,R} & \text{d. } 3\overline{)14}^{\,R} & \text{g. } 3\overline{)22}^{\,R} & \text{j. } 3\overline{)28}^{\,R} \\[2mm] \text{b. } 4\overline{)9}^{\,R} & \text{e. } 4\overline{)17}^{\,R} & \text{h. } 5\overline{)23}^{\,R} & \text{k. } 4\overline{)30}^{\,R} \\[2mm] \text{c. } 5\overline{)9}^{\,R} & \text{f. } 4\overline{)21}^{\,R} & \text{i. } 3\overline{)25}^{\,R} \end{array}$$

You're going to write answers to division problems. You'll show the whole number and the remainder.

b. Problem A: 9 ÷ 2. (Pause.) What's the answer? (Signal.) *4 with a remainder of 1.*

c. Problem B: 9 ÷ 4. (Pause.) What's the answer? (Signal.) *2 with a remainder of 1.*

d. Problem C: 9 ÷ 5. (Pause.) What's the answer? (Signal.) *1 with a remainder of 4.*

e. Write answers to all the problems in part 1. (Observe students and give feedback.)

f. Check your work.

• Problem A. What's 9 ÷ 2? (Signal.) *4 with a remainder of 1.*

• Problem B. What's 9 ÷ 4? (Signal.) *2 with a remainder of 1.*

• C. What's 9 ÷ 5? (Signal.) *1 with a remainder of 4.*

• D. What's 14 ÷ 3? (Signal.) *4 with a remainder of 2.*

• E. What's 17 ÷ 4? (Signal.) *4 with a remainder of 1.*

• F. What's 21 ÷ 4? (Signal.) *5 with a remainder of 1.*

• G. What's 22 ÷ 3? (Signal.) *7 with a remainder of 1.*

• H. What's 23 ÷ 5? (Signal.) *4 with a remainder of 3.*

• I. What's 25 ÷ 3? (Signal.) *8 with a remainder of 1.*

• J. What's 28 ÷ 3? (Signal.) *9 with a remainder of 1.*

• K. What's 30 ÷ 4? (Signal.) *7 with a remainder of 2.*

EXERCISE 2: COORDINATE SYSTEM
RATIO AND PROPORTION

a. (Display:)   [64:2A]

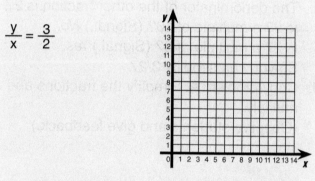

We're going to show equivalent fractions on the coordinate system.

• Listen: Each fraction will show Y over X. What will it show? (Signal.) *Y over X.*

b. We start with the fraction 3/2. The equation shows that Y over X equals 3/2.

• What's the Y value? (Signal.) *3.*

• What's the X value? (Signal.) *2.*
(Add to show:)   [64:2B]

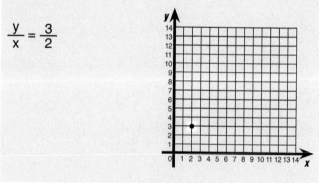

Here's the point for Y over X = 3/2.
The point shows the Y value of 3 and the X value of 2. Now we draw a line that starts at zero and goes through the point.

(Add to show:) [64:2C]

$$\frac{y}{x} = \frac{3}{2}$$

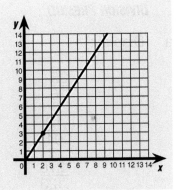

c. Y is 3. So we can make points on the line to show multiples of 3.
The first point is at 3. The next point is at 6.
(Add to show:) [64:2D]

$$\frac{y}{x} = \frac{3}{2}$$

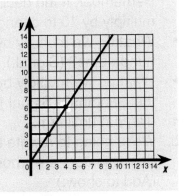

d. Where's the next point? **(Signal.)** *9.*
• Where's the next point? **(Signal.)** *12.*
(Add to show:) [64:2E]

$$\frac{y}{x} = \frac{3}{2}$$

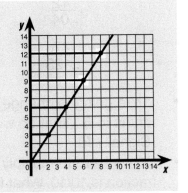

e. The points show multiples of Y. They also show multiples of X. The first point shows an X value of 2.
• What's the X value for the first point? **(Signal.)** *2.*

(Add to show:) [64:2F]

$$\frac{y}{x} = \frac{3}{2}$$

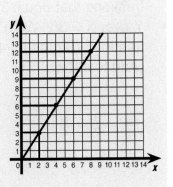

• What's the X value for the next point? **(Signal.)** *4.*
(Add to show:) [64:2G]

$$\frac{y}{x} = \frac{3}{2}$$

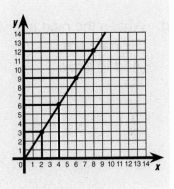

• What's the X value for the next point? **(Signal.)** *6.*
• What's the X value for the next point? **(Signal.)** *8.*
(Add to show:) [64:2H]

$$\frac{y}{x} = \frac{3}{2}$$

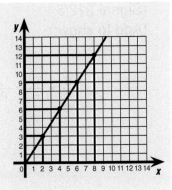

f. We can look at these points and write the fractions that equal 3 halves.

• Y over X = 3/2. What's the next equivalent fraction? (Signal.) *6/4.*
   (Add to show:) [64:2I]

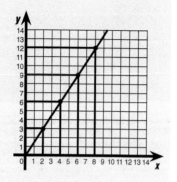

$$\frac{y}{x} = \frac{3}{2} = \frac{6}{4}$$

g. What's the next equivalent fraction? (Signal.) *9/6.*
   (Add to show:) [64:2J]

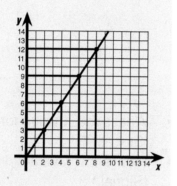

$$\frac{y}{x} = \frac{3}{2} = \frac{6}{4} = \frac{9}{6}$$

h. What's the next equivalent fraction? (Signal.) *12/8.*
   (Add to show:) [64:2K]

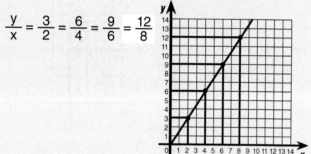

$$\frac{y}{x} = \frac{3}{2} = \frac{6}{4} = \frac{9}{6} = \frac{12}{8}$$

• Everybody, read the equation. (Signal.)
   Y over X = 3/2 = 6/4 = 9/6 = 12/8.

i. Remember, a straight line that starts at zero on the coordinate system shows equivalent fractions. All the fractions show the Y value over the X value.

Exercise 3: Fraction Operations
Division Preskill

a. (Display:) [64:3A]

$$\frac{35}{.6}$$

$$\frac{12}{.04}$$

$$\frac{276}{.123}$$

These fractions have a decimal value in the denominator. We're going to rewrite the fractions so they don't have decimal values. Remember, if the decimal value is tenths, we multiply by 10 to change it to a whole number.

• What do we multiply by if the decimal value is hundredths? (Signal.) *100.*
• What do we multiply by if the decimal value is thousandths? (Signal.) *1000.*
   (Repeat until firm.)

b. We're going to rewrite these fractions so the denominators are whole numbers.
   (Add to show:) [64:3B]

$$\frac{35}{.6} = \frac{}{6}$$

$$\frac{12}{.04} = \frac{}{4}$$

$$\frac{276}{.123} = \frac{}{123}$$

• We're going to change the denominator from 6 tenths to 6. What do we multiply 6 tenths by? (Signal.) *10.*
• So we have to multiply by a fraction of 1. That fraction is 10/10. What fraction? (Signal.) *10/10.*
   (Add to show:) [64:3C]

$$\frac{35}{.6}\left(\frac{10}{10}\right) = \frac{}{6}$$

$$\frac{12}{.04} = \frac{}{4}$$

$$\frac{276}{.123} = \frac{}{123}$$

• Say the problem for the numerators. (Signal.) *35 × 10.*
• What's the answer? (Signal.) *350.*

(Add to show:) [64:3D]

$$\frac{35}{.6}\left(\frac{10}{10}\right) = \frac{350}{6}$$

$$\frac{12}{.04} = \frac{}{4}$$

$$\frac{276}{.123} = \frac{}{123}$$

We can read both fractions as division problems.
- 35 ÷ 6 tenths = 350 ÷ 6. What does 35 ÷ 6 tenths equal? (Signal.) *350 ÷ 6.*
c. (Point to $\frac{12}{.04}$.) We're going to rewrite this fraction so the denominator is 4.
- What do we multiply 4 hundredths by to get 4? (Signal.) *100.*
- So what fraction of 1 do we multiply by? (Signal.) *100/100.*
(Add to show:) [64:3E]

$$\frac{35}{.6}\left(\frac{10}{10}\right) = \frac{350}{6}$$

$$\frac{12}{.04}\left(\frac{100}{100}\right) = \frac{}{4}$$

$$\frac{276}{.123} = \frac{}{123}$$

- Raise your hand when you know what we get when we multiply 12 by 100. ✔
- What does 12 × 100 equal? (Signal.) *1200.*
(Add to show:) [64:3F]

$$\frac{35}{.6}\left(\frac{10}{10}\right) = \frac{350}{6}$$

$$\frac{12}{.04}\left(\frac{100}{100}\right) = \frac{1200}{4}$$

$$\frac{276}{.123} = \frac{}{123}$$

- Everybody, what does 12 ÷ 4 hundredths equal? (Signal.) *1200 ÷ 4.*

d. (Point to $\frac{276}{.123}$.) What do we multiply 123 thousandths by to get 123? (Signal.) *1000.*
- So what fraction of 1 do we multiply by? (Signal.) *1000/1000.*
(Add to show:) [64:3G]

$$\frac{35}{.6}\left(\frac{10}{10}\right) = \frac{350}{6}$$

$$\frac{12}{.04}\left(\frac{100}{100}\right) = \frac{1200}{4}$$

$$\frac{276}{.123}\left(\frac{1000}{1000}\right) = \frac{}{123}$$

- Raise your hand when you know the numerator we get when we multiply. ✔
- What's the numerator? (Signal.) *276,000.*
(Add to show:) [64:3H]

$$\frac{35}{.6}\left(\frac{10}{10}\right) = \frac{350}{6}$$

$$\frac{12}{.04}\left(\frac{100}{100}\right) = \frac{1200}{4}$$

$$\frac{276}{.123}\left(\frac{1000}{1000}\right) = \frac{276,000}{123}$$

- What does 276 ÷ 123 thousandths equal? (Signal.) *276,000 ÷ 123.*

═══ **WORKBOOK PRACTICE** ═══

a. Find part 2 in your workbook. ✔
(Teacher reference:)

a. $\frac{7}{.16}\left(\frac{}{}\right) = \frac{}{16}$ b. $\frac{15}{.5}\left(\frac{}{}\right) = \frac{}{5}$

b. Complete equation A. Show the fraction that equals 1 and the fraction with a denominator of 16.
(Observe students and give feedback.)
- Check your work. What does 7 ÷ 16 hundredths equal? (Signal.) *700 ÷ 16.*
c. Complete equation B. Show the fraction that equals 1 and the fraction with a denominator of 5.
(Observe students and give feedback.)
- Check your work. What does 15 ÷ 5 tenths equal? (Signal.) *150 ÷ 5.*

EXERCISE 4: ROUNDING

a. When you round 9 up, you change two digits.
   Listen: When you round 39 up, it's 40.
   - What do you get if you round 89 up?
     (Signal.) *90.*
   - What do you get if you round 19 up?
     (Signal.) *20.*
   - What do you get if you round 59 up?
     (Signal.) *60.*
     (Repeat until firm.)

b. (Display:)   [64:4A]

2 3 9 5

Here's 2395. If we round 9 up, we change the
part that shows 39.
   - What do we change it to? (Signal.) *40.*
   - Once more: If we round 9 up, what part do we
     change? (Signal.) *39.*
     (Add to show:)   [64:4B]

2 3̲9̲ 5

   - What does 39 round up to? (Signal.) *40.*
     Yes, 40.

c. (Display:)   [64:4C]

2 1,9 6 0

   - Read this number. (Signal.) *21,960.*
   - Read the part that changes when we round 9
     up. (Signal.) *19.*
     (Add to show:)   [64:4D]

2 1̲,9̲ 6 0

   - What does 19 round up to? (Signal.) *20.*

d. (Display:)   [64:4E]

6 9 8 2

   - Read this number. (Signal.) *6982.*
   - Read the part that changes when we round 9
     up. (Signal.) *69.*
     (Add to show:)   [62:4F]

6̲ 9̲ 8 2

   - What does 69 round up to? (Signal.) *70.*

========= **WORKBOOK PRACTICE** =========

a. Find part 3 in your workbook. ✔
   (Teacher reference:)

   a. 5 1 6 , 9 6 2 ↑   b. 8 9 5 , 4 2 6 ↑   c. 2 4 5 , 9 9 8 ↑
   _____   _____   _____

   d. 4 9 8 , 6 2 8 ↑   e. 1 2 9 , 5 4 2 ↑
   _____   _____

b. Read number A. (Signal.) *516,962.*
   - Underline the part that changes when you
     round 9 up. ✔
   - What part did you underline? (Signal.) *69.*
   - Write the rounded number below. ✔
   - Everybody, read the rounded number.
     (Signal.) *517,000.*

c. Read number B. (Signal.) *895,426.*
   - Underline the part that changes when you
     round up. ✔
   - What part did you underline? (Signal.) *89.*
   - Write the rounded number below. ✔
   - Everybody, read the rounded number.
     (Signal.) *900,000.*

d. Read number C. (Signal.) *245,998.*
   - Underline the part that changes when you
     round up. ✔
   - What part did you underline? (Signal.) *59.*
   - Write the rounded number below. ✔
   - Everybody, read the rounded number.
     (Signal.) *246,000.*

e. Read number D. (Signal.) *498,628.*
   - Underline the part that changes when you
     round up. ✔
   - What part did you underline? (Signal.) *49.*
   - Write the rounded number below. ✔
   - Everybody, read the rounded number.
     (Signal.) *500,000.*

f. Read number E. (Signal.) *129,542.*
   - Underline the part that changes when you
     round up. ✔
   - What part did you underline? (Signal.) *29.*
   - Write the rounded number below. ✔
   - Everybody, read the rounded number.
     (Signal.) *130,000.*

g. Remember, if 9 rounds up, you have to
   change the digit before the 9.

EXERCISE 5: FRACTION SIMPLIFICATION
GREATEST COMMON FACTOR

a. Open your textbook to Lesson 64 and find part 1. ✔
    (Teacher reference:)

a. $\frac{6}{9}$ b. $\frac{30}{24}$ c. $\frac{15}{25}$ d. $\frac{24}{16}$ e. $\frac{16}{20}$

You're going to write simplified fractions. Remember, find the greatest common factor and simplify.

b. Read fraction A. (Signal.) *6/9.*
•   Copy the fraction and simplify. ✔
•   Everybody, the common factor is 3. What's the simplified fraction? (Signal.) *2/3.*
c. Copy and simplify fraction B: 30/24. ✔
•   Everybody, the greatest common factor is 6. What's the simplified fraction? (Signal.) *5/4.*
    If you didn't identify the greatest common factor of 30 and 24, you could have this answer.
    (Display:)   [64:5A]

$$b. \quad \frac{\overset{10}{\cancel{30}}}{\underset{8}{\cancel{24}}} = \frac{10}{8}$$

•   How do you know that 10/8 is not the correct simplified fraction? (Call on a student. Idea: *10/8 can be simplified.*)
    Yes, 10/8 can be simplified to 5/4.
    (Add to show:)   [64:5B]

$$b. \quad \frac{\overset{10}{\cancel{30}}}{\underset{8}{\cancel{24}}} = \frac{10}{8} = \boxed{\frac{5}{4}}$$

    Remember, if your answer is a fraction that can be simplified, simplify again.
d. Copy and simplify fraction C: 15/25. ✔
    The greatest common factor is 5.
•   What's the simplified fraction? (Signal.) *3/5.*
e. Copy and simplify fraction D: 24/16. ✔
    The greatest common factor is 8.
•   What's the simplified fraction? (Signal.) *3/2.*
f. Copy and simplify fraction E: 16/20. ✔
    The greatest common factor is 4.
•   What's the simplified fraction? (Signal.) *4/5.*

EXERCISE 6: SUBTRACTION
RENAMING WITH ZERO   `REMEDY`

a. (Display:)   [64:6A]

$$\begin{array}{r} 8005 \\ -\ 124 \\ \hline \end{array}$$

•   Read the problem. (Signal.) *8005 – 124.*
•   Read the problem for the ones. (Signal.) *5 – 4.*
•   Can you work that problem? (Signal.) *Yes.*
•   What's the answer? (Signal.) *1.*
    (Add to show:)   [64:6B]

$$\begin{array}{r} 8005 \\ -\ 124 \\ \hline 1 \end{array}$$

b. Read the problem for the tens. (Signal.) *0 – 2.*
•   Can you work that problem? (Signal.) *No.*
    So you rewrite the first two digits of the top number.
    (Add to show:)   [64:6C]

$$\begin{array}{r} {\scriptstyle 7\ 9} \\ {\scriptstyle \cancel{8}\cancel{0}}10\ 5 \\ -\ 124 \\ \hline 1 \end{array}$$

•   Say the new problem for the tens. (Signal.) *10 – 2.*
•   What's the answer? (Signal.) *8.*
•   Say the new problem for the hundreds. (Signal.) *9 – 1.*
•   What's the answer? (Signal.) *8.*
    (Add to show:)   [64:6D]

$$\begin{array}{r} {\scriptstyle 7\ 9} \\ {\scriptstyle \cancel{8}\cancel{0}}10\ 5 \\ -\ 124 \\ \hline \boxed{7881} \end{array}$$

•   Read the answer to the whole problem. (Signal.) *7881.*
c. (Display:)   [64:6E]

$$\begin{array}{r} 8005 \\ -\ 126 \\ \hline \end{array}$$

•   Read this problem. (Signal.) *8005 – 126.*
•   Read the problem for the ones. (Signal.) *5 – 6.*
•   Can you work that problem? (Signal.) *No.*
    So you have to rewrite the first three digits of the top number.

(Add to show:) [64:6F]

$$\begin{array}{r} 7\ 9\ 9 \\ \cancel{8\ 0\ 0}{}^{1}5 \\ -\quad 1\ 2\ 6 \end{array}$$

- Say the new problem for the ones. (Signal.) *15 – 6.*
- Say the new problem for the tens. (Signal.) *9 – 2.*
- Say the new problem for the hundreds. (Signal.) *9 – 1.*
- Say the new number for the thousands. (Signal.) *7.*

━━━━━ **TEXTBOOK PRACTICE** ━━━━━

a. Find part 2 in your textbook. ✔
   (Teacher reference:)

For some of these problems, you have to rewrite the first two digits of the top number. For others, you have to rewrite the first three digits of the top number.

b. Copy problem A. Then stop. ✔
- Everybody, read problem A. (Signal.) *8009 – 127.*
- Read the problem for the ones. (Signal.) *9 – 7.*
- Can you work that problem? (Signal.) *Yes.*
- Read the problem for the tens. (Signal.) *0 – 2.*
- Can you work that problem? (Signal.) *No.*
   So you rewrite the first two digits of the top number.
- Work the problem.
   (Observe students and give feedback.)
   (Display:) [64:6G]

$$\begin{array}{r} 7\ 9 \\ \text{a.}\quad \cancel{8\ 0}{}^{1}0\ 9 \\ -\quad 1\ 2\ 7 \\ \hline \boxed{7\ 8\ 8\ 2} \end{array}$$

Here's what you should have.

c. Work the rest of the problems in part 2.
   (Observe students and give feedback.)

d. Check your work.
- Problem B: 8000 – 127. What's the answer? (Signal.) *7873.*
- Problem C: 7003 – 1045. What's the answer? (Signal.) *5958.*
- Problem D: 7003 – 1042. What's the answer? (Signal.) *5961.*
   (Display:) [64:6H]

$$\begin{array}{lll}
\begin{array}{r} 7\ 9\ 9 \\ \text{b.}\quad \cancel{8\ 0\ 0}{}^{1}0 \\ -\ \ 1\ 2\ 7 \\ \hline \boxed{7\ 8\ 7\ 3} \end{array}
&
\begin{array}{r} 6\ 9\ 9 \\ \text{c.}\quad \cancel{7\ 0\ 0}{}^{1}3 \\ -1\ 0\ 4\ 5 \\ \hline \boxed{5\ 9\ 5\ 8} \end{array}
&
\begin{array}{r} 6\ 9 \\ \text{d.}\quad \cancel{7\ 0}{}^{1}0\ 3 \\ -1\ 0\ 4\ 2 \\ \hline \boxed{5\ 9\ 6\ 1} \end{array}
\end{array}$$

Here's what you should have for problems B through D.

EXERCISE 7: EXPONENTS
EVALUATION

[REMEDY]

a. Find part 3 in your textbook. ✔
- Pencils down. ✔
   (Teacher reference:)

These are problems with a base and exponent.

b. Problem A. What's the base? (Signal.) *4.*
- What's the exponent? (Signal.) *3.*
- Say the repeated multiplication for 4 to the third. (Signal.) *4 × 4 × 4.*
c. Problem B. What's the base? (Signal.) *2.*
- What's the exponent? (Signal.) *4.*
- Say the repeated multiplication for 2 to the fourth. (Signal.) *2 × 2 × 2 × 2.*
d. Problem C. What's the base? (Signal.) *3.*
- What's the exponent? (Signal.) *2.*
- Say the repeated multiplication for 3 to the second. (Signal.) *3 × 3.*
e. Problem D. What's the base? (Signal.) *2.*
- What's the exponent? (Signal.) *3.*
- Say the repeated multiplication for 2 to the third. (Signal.) *2 × 2 × 2.*
f. Problem E. What's the base? (Signal.) *5.*
- What's the exponent? (Signal.) *2.*
- Say the repeated multiplication for 5 to the second. (Signal.) *5 × 5.*

g. Go back to problem A. Copy the base and exponent. Show the repeated multiplication and write what 4 to the third equals.
(Observe students and give feedback.)
(Display:)   [64:7A]

> **a.** $4^3 = 4 \times 4 \times 4 = \boxed{64}$

Here's what you should have.
• What does 4 to the third equal? (Signal.) *64.*
h. Work the rest of the items.
(Observe students and give feedback.)
i. Check your work.
• Problem B. Say the repeated multiplication for 2 to the fourth. (Signal.) *2 × 2 × 2 × 2.*
• What does 2 to the fourth equal? (Signal.) *16.*
j. C. Say the repeated multiplication for 3 to the second. (Signal.) *3 × 3.*
• What does 3 to the second equal? (Signal.) *9.*
k. D. Say the repeated multiplication for 2 to the third. (Signal.) *2 × 2 × 2.*
• What does 2 to the third equal? (Signal.) *8.*

l. E. Say the repeated multiplication for 5 to the second. (Signal.) *5 × 5.*
• What does 5 to the second equal?
(Signal.) *25.*
(Display:)   [64:7B]

> **b.** $2^4 = 2 \times 2 \times 2 \times 2 = \boxed{16}$
>
> **c.** $3^2 = 3 \times 3 = \boxed{9}$
>
> **d.** $2^3 = 2 \times 2 \times 2 = \boxed{8}$
>
> **e.** $5^2 = 5 \times 5 = \boxed{25}$

Here's what you should have for items B through E.

Assign Independent Work, Textbook parts 4–9 and Workbook parts 4 and 5.

Optional extra math-fact practice worksheets are available on ConnectED.

Lesson 65

EXERCISE 1: MENTAL MATH
REMAINDERS

a. Open your workbook to Lesson 65 and find part 1. ✔

• Pencils down. ✔
   (Teacher reference:)

a. $3\overline{)16}^{\,R}$ d. $4\overline{)22}^{\,R}$ f. $5\overline{)24}^{\,R}$ h. $3\overline{)29}^{\,R}$

b. $3\overline{)20}^{\,R}$ e. $3\overline{)23}^{\,R}$ g. $3\overline{)25}^{\,R}$ i. $4\overline{)29}^{\,R}$

c. $4\overline{)18}^{\,R}$

You're going to write answers to division problems that show the whole number and the remainder.

• Problem A: 16 ÷ 3. (Pause.) What's the answer? (Signal.) *5 with a remainder of 1.*

• Problem B: 20 ÷ 3. (Pause.) What's the answer? (Signal.) *6 with a remainder of 2.*

b. Write answers to all the problems in part 1. (Observe students and give feedback.)

c. Check your work.

• Problem A. What's 16 ÷ 3? (Signal.) *5 with a remainder of 1.*

• B. What's 20 ÷ 3? (Signal.) *6 with a remainder of 2.*

• C. What's 18 ÷ 4? (Signal.) *4 with a remainder of 2.*

• D. What's 22 ÷ 4? (Signal.) *5 with a remainder of 2.*

• E. What's 23 ÷ 3? (Signal.) *7 with a remainder of 2.*

• F. What's 24 ÷ 5? (Signal.) *4 with a remainder of 4.*

• G. What's 25 ÷ 3? (Signal.) *8 with a remainder of 1.*

• H. What's 29 ÷ 3? (Signal.) *9 with a remainder of 2.*

• I. What's 29 ÷ 4? (Signal.) *7 with a remainder of 1.*

EXERCISE 2: COORDINATE SYSTEM
RATIO AND PROPORTION   REMEDY

a. (Display:)   [65:2A]

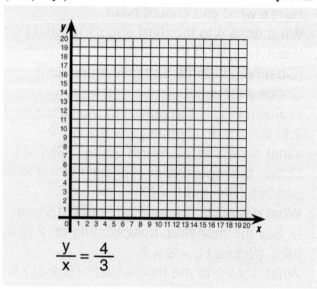

$$\frac{y}{x} = \frac{4}{3}$$

Last time, we saw that straight lines on the coordinate system show equivalent fractions. Each fraction will show Y over X.

• What will each fraction show? (Signal.) *Y over X.*

The equation shows that Y over X = 4/3.

• What's the Y value? (Signal.) *4.*

• What's the X value? (Signal.) *3.*
   (Add to show:)   [65:2B]

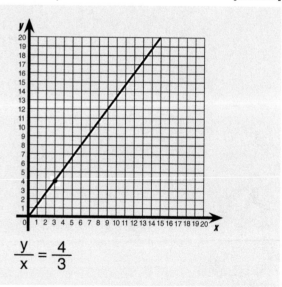

$$\frac{y}{x} = \frac{4}{3}$$

Y is 4. So we'll show points on the line for multiples of 4.

b. What's the Y value for the first point?
(Signal.) *4.*
(Add to show:)   [65:2C]

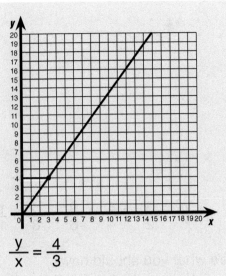

$$\frac{y}{x} = \frac{4}{3}$$

- What's the next Y value? (Signal.) *8.*
- What's the next Y value? (Signal.) *12.*
- What's the next Y value? (Signal.) *16.*
- What's the next Y value? (Signal.) *20.*

(Add to show:)   [65:2D]

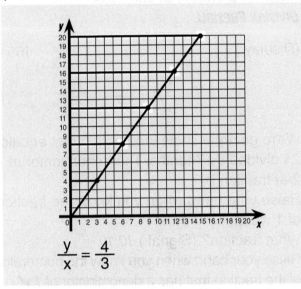

$$\frac{y}{x} = \frac{4}{3}$$

Here are the equations showing the Y values.

(Add to show:)   [65:2E]

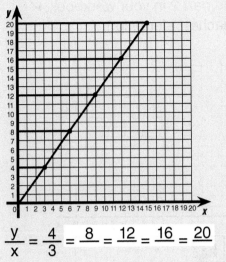

$$\frac{y}{x} = \frac{4}{3} = \frac{8}{} = \frac{12}{} = \frac{16}{} = \frac{20}{}$$

c. You can complete the fractions by writing the X value for each point.

- What's the X value for the first point?
(Signal.) *3.*
So the X values are multiples of 3.
- What's the X value of the next point?
(Signal.) *6.*
- Next point? (Signal.) *9.*
- Next point? (Signal.) *12.*
- Next point? (Signal.) *15.*
So the X values in the equation are 6, 9 ,12, and 15.

(Add to show:)   [65:2F]

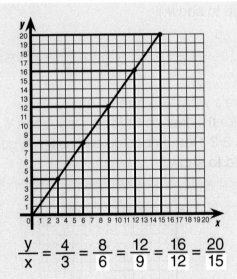

$$\frac{y}{x} = \frac{4}{3} = \frac{8}{6} = \frac{12}{9} = \frac{16}{12} = \frac{20}{15}$$

- Read the equivalent fractions starting with 4/3.
(Signal.) *4/3 = 8/6 = 12/9 = 16/12 = 20/15.*

WORKBOOK PRACTICE

a. Find part 2 in your workbook. ✔
   (Teacher reference:)   R Part D

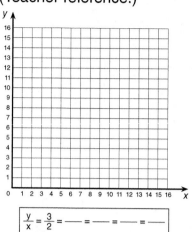

$$\frac{y}{x} = \frac{3}{2} = \text{—} = \text{—} = \text{—} = \text{—}$$

You're going to write fractions for Y over X.

b. The first number for Y is 3.
• What's the next multiple for Y? (Signal.) 6.
• Write the multiples for Y. Then stop. ✔
   (Display:)   [65:2G]

$$\frac{y}{x} = \frac{3}{2} = \frac{6}{} = \frac{9}{} = \frac{12}{} = \frac{15}{}$$

Here's what you should have.

c. Now you'll write the multiples for X.
• What's the first X value? (Signal.) 2.
• Write the other multiples of 2.
   (Observe students and give feedback.)
   (Add to show:)   [65:2H]

$$\frac{y}{x} = \frac{3}{2} = \frac{6}{4} = \frac{9}{6} = \frac{12}{8} = \frac{15}{10}$$

Here's what you should have.

d. Make the first point for the equation Y over
   X = 3/2. Remember, go across for X; go up for Y.
   (Add to show:)   [65:2I]

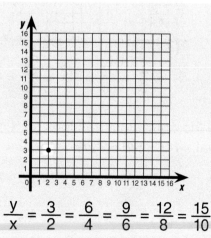

$$\frac{y}{x} = \frac{3}{2} = \frac{6}{4} = \frac{9}{6} = \frac{12}{8} = \frac{15}{10}$$

Here's what you should have.

e. Make the rest of the points.
   (Observe students and give feedback.)
   (Add to show:)   [65:2J]

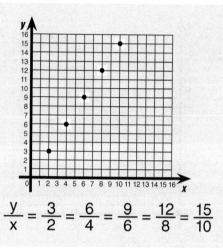

$$\frac{y}{x} = \frac{3}{2} = \frac{6}{4} = \frac{9}{6} = \frac{12}{8} = \frac{15}{10}$$

Here's what you should have.

f. The points show the line for Y/X = 3/2. Make the
   line. If your points are in the right place, the line
   should go through every point and through zero.
   (Observe students and give feedback.)

EXERCISE 3: DECIMAL OPERATIONS
DIVISION PRESKILL

a. (Display:)   [65:3A]

$$\frac{24}{.7} = \frac{}{7}$$

We're going to show the fraction that equals
24 divided by 7 tenths. The denominator of
that fraction is 7.
• Raise your hand when you know the fraction
   of 1 we multiply by. ✔
• What fraction? (Signal.) 10/10.
• Raise your hand when you know the numerator
   of the fraction that has a denominator of 7. ✔
• What's the numerator? (Signal.) 240.
   (Add to show:)   [65:3B]

$$\frac{24}{.7} = \frac{240}{7}$$

The first fraction is 24 divided by 7 tenths.
• What's the first fraction? (Signal.) 24 ÷ 7 tenths.
• What's the other fraction? (Signal.) 240 ÷ 7.

b. (Display:) [65:3C]

$$\frac{4}{.52} = \frac{}{52}$$

We're going to show the fraction that equals 4 divided by 52 hundredths.

- What's the denominator of the fraction we'll complete? (Signal.) *52.*
- Raise your hand when you know the fraction of 1 we multiply by. ✔
- What fraction? (Signal.) *100/100.*
- Raise your hand when you know the numerator of the fraction that has a denominator of 52. ✔
- What's the numerator? (Signal.) *400.*
(Add to show:) [65:3D]

$$\frac{4}{.52} = \frac{400}{52}$$

- Read the first division problem. (Signal.) *4 ÷ 52 hundredths.*
- Read the division problem it equals. (Signal.) *400 ÷ 52.*

c. (Display:) [65:3E]

$$.2\,\overline{)\,8.}$$

- Read this problem. (Signal.) *8 ÷ 2 tenths.* I'll show you the fast way to change the number we divide by into a whole number.
- What does this problem divide by? (Signal.) *2 tenths.*
- So what do we multiply 2 tenths by to get 2? (Signal.) *10.*

d. We multiply both values by 10. I can show that by moving the decimal point of both numbers one place.
(Add to show:) [65:3F]

$$2.\,\overline{)\,8\,0.}$$

We have a zero after the 8.

- Read the new problem. (Signal.) *80 ÷ 2.* When we move the decimal point one place in both numbers, we're multiplying both numbers by 10, so we don't change the value we started with.

e. (Display:) [65:3G]

$$.02\,\overline{)\,8.}$$

New problem: 8 ÷ 2 hundredths.

- What does this problem divide by? (Signal.) *2 hundredths.*
- What do we multiply 2 hundredths by to get 2? (Signal.) *100.*
- How many places do we move both decimal points? (Signal.) *2.*
(Add to show:) [65:3H]

$$02.\,\overline{)\,8\,0\,0.}$$

- Read the new problem. (Signal.) *800 ÷ 2.* We moved the decimal point for 8 two places, so we wrote two zeros.

f. (Display:) [65:3I]

$$.002\,\overline{)\,8.}$$

New problem: 8 ÷ 2 thousandths.

- What does this problem divide by? (Signal.) *2 thousandths.*
- What do we multiply 2 thousandths by to get 2? (Signal.) *1000.*
- How many places do we move both decimal points? (Signal.) *3.*
- How many zeros will we write after 8? (Signal.) *3.*
(Add to show:) [65:3J]

$$002.\,\overline{)\,8\,0\,0\,0.}$$

- Read the new problem. (Signal.) *8000 ÷ 2.*

━━━━━ **WORKBOOK PRACTICE** ━━━━━

a. Find part 3 in your workbook. ✔
(Teacher reference:)

a. $.05\,\overline{)\,14.}$ b. $.5\,\overline{)\,3.}$

You're going to move the decimal points and write zeros for the number under the division sign.

b. Problem A: 14 ÷ 5 hundredths.
- What does the problem divide by? (Signal.) *5 hundredths.*
- Move the decimal point of both numbers so the problem divides by a whole number. ✔
- Everybody, how many places did you move both decimal points? (Signal.) *2.*
- How many zeros did you write after 14? (Signal.) *2.*
- Read the new problem. (Signal.) *1400 ÷ 5.*
  (Display:) [65:3K]

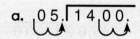

Here's what you should have.

c. Problem B: 3 ÷ 5 tenths.
- Move the decimal point of both numbers so the problem divides by a whole number. ✔
- Everybody, how many places did you move both decimal points? (Signal.) *1.*
- How many zeros did you write after 3? (Signal.) *1.*
- Read the new problem. (Signal.) *30 ÷ 5.*
  (Display:) [65:3L]

Here's what you should have.

EXERCISE 4: ROUNDING   REMEDY

a. You're going to round 9 up.
   (Display:) [65:4A]

349,830
↑

- Read this number. (Signal.) *349,830.*
- Read the part that rounds up. (Signal.) *49.*
- What does 49 round up to? (Signal.) *50.*
  (Add to show:) [65:4B]

349,830
↑
350,000

- Read the rounded number. (Signal.) *350,000.*
b. (Display:) [65:4C]

23,897
↑

- Read this number. (Signal.) *23,897.*

- Read the part that rounds up. (Signal.) *89.*
- What does 89 round up to? (Signal.) *90.*
  (Add to show:) [65:4D]

23,897
↑
23,900

- Read the rounded number. (Signal.) *23,900.*
c. (Display:) [65:4E]

23,894
↑

- Read this number. (Signal.) *23,894.*
- Raise your hand when you know if this number rounds up. ✔
- Does it round up? (Signal.) *No.*
- What's the digit after the arrowed digit? (Signal.) *4.*
  That's not 5 or more, so the number does not round up.
  (Add to show:) [65:4F]

23,894
↑
23,890

- Read the rounded number. (Signal.) *23,890.*

═══ WORKBOOK PRACTICE ═══

a. Find part 4 in your workbook. ✔
   (Teacher reference:)   R Part A

a. 516,982   b. 894,326   c. 30,972
   ↑   ↑   ↑
_____   _____   _____

d. 39,658   e. 194,340
   ↑   ↑
_____   _____

You're going to round nines. For some of these numbers, the 9 rounds up. Remember, look at the digit after the 9.

b. Number A: 516,982.
- Does 9 round up? (Signal.) *Yes.*
- Say the part that rounds up. (Signal.) *69.*
- Write the rounded number below. ✔
- Everybody, read the rounded number. (Signal.) *517,000.*

c. Number B: 894,326.
- Does 9 round up? (Signal.) *No.*
- Write the rounded number below. ✔
- Everybody, read the rounded number. (Signal.) *890,000.*
d. Work the rest of the items in part 4. (Observe students and give feedback.)
e. Check your work.
- Number C. Does 9 round up? (Signal.) *Yes.*
- Read the rounded number. (Signal.) *31,000.*
f. Number D. Does 9 round up? (Signal.) *Yes.*
- Read the rounded number. (Signal.) *40,000.*
g. Number E. Does 9 round up? (Signal.) *No.*
- Read the rounded number. (Signal.) *190,000.*

EXERCISE 5: FRACTION SIMPLIFICATION
GREATEST COMMON FACTOR

a. Open your textbook to Lesson 65 and find part 1. ✔
(Teacher reference:)

a. $\frac{40}{20}$ b. $\frac{15}{18}$ c. $\frac{35}{14}$ d. $\frac{54}{63}$ e. $\frac{48}{40}$

You're going to write simplified fractions. Remember, find the greatest common factor and simplify. If the fraction you write can be simplified again, you didn't find the greatest common factor, so simplify again.
b. Copy and simplify fraction A: 40/20. ✔
The greatest common factor is 20.
- What's the simplified fraction? (Signal.) *2/1.*
Yes, 2 over 1. That equals 2.
c. Copy and simplify fraction B: 15/18. ✔
The greatest common factor is 3.
- What's the simplified fraction? (Signal.) *5/6.*
d. Simplify the rest of the fractions in part 1. (Observe students and give feedback.)
e. Check your work.
- Fraction C: 35/14.
The greatest common factor is 7.
- What's the simplified fraction? (Signal.) *5/2.*
f. Fraction D: 54/63.
The greatest common factor is 9.
- What's the simplified fraction? (Signal.) *6/7.*
g. Fraction E: 48/40.
The greatest common factor is 8.
- What's the simplified fraction? (Signal.) *6/5.*

EXERCISE 6: SUBTRACTION
RENAMING WITH ZERO

a. Find part 2 in your textbook. ✔
(Teacher reference:)

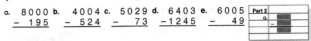

For some of these problems, you have to rewrite **two** digits of the top number. For other problems, you have to rewrite **three** digits of the top number.
b. Problem A: 8000 minus 195.
- Say the problem for the ones. (Signal.) *Zero – 5.*
- Can you work that problem? (Signal.) *No.*
- Raise your hand when you know how many digits of the top number you'll rewrite. ✔
- How many digits? (Signal.) *3.*
Yes, 3 digits.
c. Problem B: 4004 minus 524.
- Say the problem for the ones. (Signal.) *4 – 4.*
- Can you work that problem? (Signal.) *Yes.*
- Say the problem for the tens. (Signal.) *0 – 2.*
- Can you work that problem? (Signal.) *No.*
- Raise your hand when you know how many digits of the top number you'll rewrite. ✔
- How many digits? (Signal.) *2.*
Yes, 2 digits.
d. Copy the problems in part 2 and work them. (Observe students and give feedback.)
e. Check your work.
- Problem A: 8000 – 195. What's the answer? (Signal.) *7805.*
- Problem B. 4004 – 524. What's the answer? (Signal.) *3480.*
- C. 5029 – 73. What's the answer? (Signal.) *4956.*
- D. 6403 – 1245. What's the answer? (Signal.) *5158.*
- E. 6005 – 49. What's the answer? (Signal.) *5956.*
(Display:)   [65:6A]

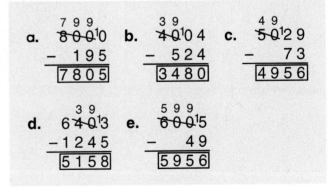

Here's what you should have for each problem.

Exercise 7: Exponents

Evaluation

a. Find part 3 in your textbook. ✔

• Pencils down. ✔

(Teacher reference:)

a. 4^2 b. 1^3 c. 5^3 d. 10^4 e. 2^6

You're going to copy the base and exponent for each problem. You'll show the repeated multiplication and the answer.

b. Read problem A. (Signal.) *4 to the second.*

• Show the repeated multiplication and what 4 to the second equals. ✔

• Everybody, read the multiplication for 4 to the second. (Signal.) *4 × 4.*

• What's the answer? (Signal.) *16.*
(Display:)   [65:7A]

$$\text{a. } 4^2 = 4 \times 4 = \boxed{16}$$

Here's what you should have.

c. Work the rest of the problems in part 3. (Observe students and give feedback.)

d. Check your work.

• Problem B: 1 to the third. Say the multiplication. (Signal.) *1 × 1 × 1.*

• What's the answer? (Signal.) *1.*

e. Problem C: 5 to the third. Say the multiplication. (Signal.) *5 × 5 × 5.*

• What's the answer? (Signal.) *125.*

f. D: 10 to the fourth. Say the multiplication. (Signal.) *10 × 10 × 10 × 10.*

• What's the answer? (Signal.) *10,000.*

g. E: 2 to the sixth. Say the multiplication. (Signal.) *2 × 2 × 2 × 2 × 2 × 2.*

• What's the answer? (Signal.) *64.*

Assign Independent Work, Textbook parts 4–10 and Workbook parts 5 and 6.

Optional extra math-fact practice worksheets are available on ConnectED.

Lessons 66-70 Planning Page

	Lesson 66	Lesson 67	Lesson 68	Lesson 69	Lesson 70
Student Learning Objectives	**Exercises** 1. Solve division problems with remainders using mental math 2. **Plot points for equivalent fractions on a coordinate system and draw the line** 3. Solve division problems that have decimals 4. **Recognize the pattern of numbers with powers of 10** 5. **Learn short division and divide using mental math** 6. Simplify fractions using the greatest common factor 7. Round numbers 8. Complete work independently	**Exercises** 1. Solve division problems with remainders using mental math 2. Solve division problems that have decimals 3. Plot points for equivalent fractions on a coordinate system and draw the line 4. **Solve division problems using short division and mental math** 5. **Round decimals less than 1** 6. Recognize the pattern of numbers with powers of 10 Complete work independently	**Exercises** 1. Find equivalent fractions 2. Solve division problems that have decimals 3. **Find the volume of rectangular prisms using unit cubes and learn the formula for volume** 4. Solve division problems using short division and mental math 5. Round decimals less than 1 6. **Write X/Y equations for ratio sentences** 7. **Write equations with powers of 10** Complete work independently	**Exercises** 1. Find equivalent fractions 2. Solve division problems that have decimals 3. **Write X/Y equations for ratio sentences and plot them on a coordinate system and draw the line** 4. Write equations with powers of 10 5. Solve division problems using short division and mental math 6. Find the volume of rectangular prisms using unit cubes and learn the formula for volume 7. Round decimals less than 1 Complete work independently	**Exercises** 1. Find equivalent fractions 2. Solve division problems using short division and mental math 3. Write equations with powers of 10 4. Write X/Y equations for ratio sentences and plot them on a coordinate system and draw the line 5. **Recognize the division sign ($\div$)** 6. **Find the volume of rectangular prisms using the formula** 7. **Round decimal values that have 9 as a digit** 8. **Set up ratio and fraction-of-a-group word problems** 9. Complete work independently
Common Core State Standards for Mathematics					
5.OA 1	✔	✔	✔	✔	✔
5.NBT 1–2	✔	✔	✔	✔	✔
5.NBT 3		✔	✔	✔	✔
5.NBT 4		✔	✔	✔	✔
5.NBT 5	✔	✔	✔		
5.NBT 6–7	✔	✔	✔	✔	
5.NF 3					✔
5.NF 4–5	✔	✔	✔	✔	✔
5.NF 6	✔		✔		✔
5.MD 3			✔	✔	
5.MD 4			✔	✔	
5.MD 5			✔	✔	✔
5.G 2	✔	✔		✔	✔
Teacher Materials	Presentation Book 1, Board Displays CD or chalk board				
Student Materials	Textbook, Workbook, pencil, lined paper, ruler				
Additional Practice	• Student Practice Software: Block 3: Activity 1 (5.NBT 7), Activity 2, Activity 3 (5.NF 4), Activity 4, Activity 5 (5.NBT 6), Activity 6 (5.NF 4, 5.NF 6, and 5.OA 1) • Provide needed fact practice with Level D or E Math Fact Worksheets.				
Mastery Test					Student Assessment Book (Present Mastery Test 7 following Lesson 70.)

Lesson

EXERCISE 1: MENTAL MATH
REMAINDERS

a. Open your workbook to Lesson 66 and find part 1. ✔
• Pencils down. ✔
(Teacher reference:)

a. $3\overline{\smash)11}\,^R$ b. $5\overline{\smash)18}\,^R$ c. $3\overline{\smash)17}\,^R$ d. $4\overline{\smash)15}\,^R$

e. $2\overline{\smash)15}\,^R$ f. $5\overline{\smash)14}\,^R$ g. $4\overline{\smash)14}\,^R$ h. $3\overline{\smash)14}\,^R$

You're going to write answers to division problems that show the whole number and the remainder.

b. Problem A: 11 ÷ 3. (Pause.) What's the answer? (Signal.) *3 with a remainder of 2.*

c. Problem B: 18 ÷ 5. (Pause.) What's the answer? (Signal.) *3 with a remainder of 3.*

d. Write answers to all the problems in part 1. (Observe students and give feedback.)

e. Check your work.
• Problem A. What's 11 ÷ 3? (Signal.) *3 with a remainder of 2.*
• Problem B. What's 18 ÷ 5? (Signal.) *3 with a remainder of 3.*
• C. What's 17 ÷ 3? (Signal.) *5 with a remainder of 2.*
• D. What's 15 ÷ 4? (Signal.) *3 with a remainder of 3.*
• E. What's 15 ÷ 2? (Signal.) *7 with a remainder of 1.*
• F. What's 14 ÷ 5? (Signal.) *2 with a remainder of 4.*
• G. What's 14 ÷ 4? (Signal.) *3 with a remainder of 2.*
• H. What's 14 ÷ 3? (Signal.) *4 with a remainder of 2.*

EXERCISE 2: COORDINATE SYSTEM
RATIO AND PROPORTION
REMEDY

a. Find part 2 in your workbook. ✔
(Teacher reference:)   R Part E

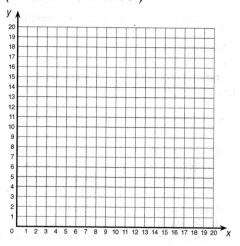

$$\frac{y}{x} = \frac{4}{5} = {-\!-} = {-\!-} = {-\!-}$$

You're going to make a line for the equation Y over X = 4/5.
• Say the equation. (Signal.) *Y over X = 4/5.*
b. What's the first number for Y? (Signal.) *4.*
• What's the next number for Y? (Signal.) *8.*
• Write the multiples for Y. Then write the multiples for X.
(Observe students and give feedback.)
(Display:)   [66:2A]

$$\frac{y}{x} = \frac{4}{5} = \frac{8}{10} = \frac{12}{15} = \frac{16}{20}$$

Here's what you should have.

c. Make the first point for the equation Y over X = 4/5. Remember, go across for X; go up for Y. ✔
(Add to show:) [66:2B]

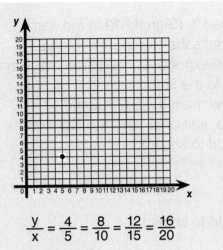

$$\frac{y}{x} = \frac{4}{5} = \frac{8}{10} = \frac{12}{15} = \frac{16}{20}$$

Here's what you should have.
d. Make the rest of the points.
(Observe students and give feedback.)
(Add to show:) [66:2C]

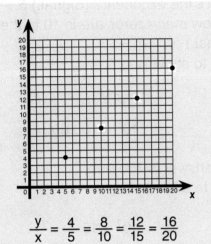

$$\frac{y}{x} = \frac{4}{5} = \frac{8}{10} = \frac{12}{15} = \frac{16}{20}$$

Here's what you should have.
The points show the line for Y over X = 4/5.
• Make the line. If your points are in the right places, the line should go through every point and through zero.
(Observe students and give feedback.)

EXERCISE 3: DECIMAL OPERATIONS
DIVISION

a. Find part 3 in your workbook. ✔
(Teacher reference:)

a. $.002\overline{)6}$ b. $.7\overline{)42}$ c. $.008\overline{)56}$

You're going to move the decimal points and write zeros for the number under the division sign.
All these problems have a whole number under the division sign. Remember, you write the decimal point after the last digit of the whole number.
b. What's the whole number in problem A?
(Signal) 6.
• Where do you write the decimal point?
(Signal) After the 6.
c. What's the whole number in problem B?
(Signal) 42.
• Where do you write the decimal point?
(Signal) After the 2.
d. What's the whole number in problem C?
(Signal) 56.
• Where do you write the decimal point?
(Signal) After the 6.
(Repeat until firm.)
e. Problem A: 6 ÷ 2 thousandths.
• Write the decimal point after the whole number. ✔
• What are you dividing by? (Signal.)
2 thousandths.
• How many places do you move the decimal point to get a whole number? (Signal.) 3.
• Move the decimal point of both numbers three places. Write the zeros you need.
(Observe students and give feedback.)
• How many zeros did you write after 6?
(Signal.) 3.
• Read the new problem. (Signal.) 6000 ÷ 2.
(Display:) [66:3A]

$$002.\overline{)6\,000.}$$

Here's what you should have.

f. Problem B: 42 ÷ 7 tenths.
•   Write the decimal point after the whole
    number. ✔
•   What are you dividing by? (Signal.) *7 tenths.*
•   How many places do you move the decimal
    point to get a whole number? (Signal.) *1.*
•   Move the decimal point of both numbers.
    Write the zeros you need.
    (Observe students and give feedback.)
•   How many zeros did you write after 42?
    (Signal.) *1.*
•   Read the new problem. (Signal.) *420 ÷ 7.*
g. Problem C: 56 ÷ 8 thousandths.
•   Write the decimal point after the whole
    number. ✔
•   Move the decimal points of both numbers.
    Write the zeros you need.
    (Observe students and give feedback.)
•   How many places did you move both decimal
    points? (Signal.) *3.*
•   How many zeros did you write after 6?
    (Signal.) *3.*
•   Read the new problem. (Signal.) *56,000 ÷ 8.*
h. Work all the problems in part 3.
    (Observe students and give feedback.)
i. Check your work.
•   Problem A: 6000 ÷ 2. What's the answer?
    (Signal.) *3000.*
•   Problem B: 420 ÷ 7. What's the answer?
    (Signal.) *60.*
•   Problem C: 56,000 ÷ 8. What's the answer?
    (Signal.) *7000.*
    (Display:)   [66:3B]

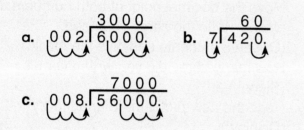

    Here's what you should have.

EXERCISE 4: EXPONENTS
BASE 10

a. You've written repeated multiplication for
    values like 2 to the fifth and 4 to the third.
    When the base is 10, the first digit of the
    answer is 1 followed by zeros. The exponent
    shows how many zeros are in the answer.

b. (Display:)   [66:4A]

$$10^1$$

•   Read it. (Signal.) *10 to the first.*
•   What's the exponent? (Signal.) *1.*
    The exponent shows how many zeros are in
    10 to the first.
•   How many zeros? (Signal.) *1.*
    Yes, just one. 10 to the first equals 10.
    (Add to show:)   [66:4B]

$$10^1 = 10$$

c. (Add to show:)   [66:4C]

$$10^1 = 10$$
$$10^5$$

•   Read it. (Signal.) *10 to the fifth.*
•   What's the exponent? (Signal.) *5.*
•   So how many zeros are in 10 to the fifth?
    (Signal.) *5.*
    (Add to show:)   [66:4D]

$$10^1 = 10$$
$$10^5 = 100,000$$

    The answer is 100,000.
d. (Add to show:)   [66:4E]

$$10^1 = 10$$
$$10^5 = 100,000$$
$$10^3$$

•   Read it. (Signal.) *10 to the third.*
•   What's the exponent? (Signal.) *3.*
•   So how many zeros are in 10 to the third?
    (Signal.) *3.*
    (Add to show:)   [66:4F]

$$10^1 = 10$$
$$10^5 = 100,000$$
$$10^3 = 1000$$

•   What does 10 to the third equal? (Signal.) *1000.*

e. (Add to show:) [66:4G]

$$10^1 = 10$$
$$10^5 = 100,000$$
$$10^3 = 1000$$
$$10^6$$

- Read it. (Signal.) *10 to the sixth.*
- What's the exponent? (Signal.) *6.*
- So how many zeros are in 10 to the sixth? (Signal.) *6.*
  (Add to show:) [66:4H]

$$10^1 = 10$$
$$10^5 = 100,000$$
$$10^3 = 1000$$
$$10^6 = 1,000,000$$

- What does 10 to the sixth equal? (Signal.) *1 million.*

f. (Add to show:) [66:4I]

$$10^1 = 10$$
$$10^5 = 100,000$$
$$10^3 = 1000$$
$$10^6 = 1,000,000$$
$$10^4$$

- Read it. (Signal.) *10 to the fourth.*
- What's the exponent? (Signal.) *4.*
- So how many zeros are in 10 to the fourth? (Signal.) *4.*
  (Add to show:) [66:4J]

$$10^1 = 10$$
$$10^5 = 100,000$$
$$10^3 = 1000$$
$$10^6 = 1,000,000$$
$$10^4 = 10,000$$

- What does 10 to the fourth equal? (Signal.) *10 thousand.*

g. You're going to show that 10 to the fourth equals 10 thousand.
- Say the repeated multiplication for 10 to the fourth. (Signal.) *10 × 10 × 10 × 10.*

(Change to show:) [66:4K]

$$10^4 = 10 \times 10 \times 10 \times 10 = 10,000$$

- What's 10 × 10? (Signal.) *100.*
- What's 100 × 10? (Signal.) *1000.*
- What's 1 thousand times 10? (Signal.) *10,000.*
- So what does 10 to the fourth equal? (Signal.) *10,000.*

h. Remember, the exponents of 10 tell you how many zeros are in the number.

EXERCISE 5: SHORT DIVISION
WITH MENTAL MATH

a. (Display:) [66:5A]

$$4\overline{)256} \qquad 3\overline{)75}$$

We're going to work this problem the fast way without writing a number below. We'll just write the whole-number part in the answer and we'll show the remainder as a **small** number.

b. (Point to $4\overline{)256}$.) Read the problem. (Signal.) *256 ÷ 4.*

c. The first problem you'll work is 25 ÷ 4. What's the whole-number part of the answer? (Signal.) *6.*
- What's the remainder? (Signal.) *1.*
  (Repeat until firm.)

d. Once more. I'll write the numbers this time.
- What's the whole-number part of the answer? (Signal.) *6.*
  (Add to show:) [66:5B]

$$4\overline{)2\overset{6}{5}6} \qquad 3\overline{)75}$$

- What's the remainder? (Signal.) *1.*
  (Add to show:) [66:5C]

$$4\overline{)2\,5_16} \qquad 3\overline{)75}$$

e. The next problem you'll work is 16 divided by 4.
- Say the problem you'll work. (Signal.) *16 ÷ 4.*
- What's the answer? (Signal.) *4.*
  (Add to show:) [66:5D]

$$4\overline{)2\,5_16}^{6\ 4} \qquad 3\overline{)75}$$

- What does 256 ÷ 4 equal? (Signal.) *64.*

f. (Point to $3\overline{)75}$.) Read this problem. (Signal.)
    75 ÷ 3.
• Say the problem for 7. (Signal.) *7 ÷ 3.*
• What's the whole-number part of the answer?
    (Signal.) *2.*
• What's the remainder? (Signal.) *1.*
    (Repeat until firm.)
    (Add to show:)   [66:5E]

$$4\overline{)2\,5{,}6}\;\;\overset{6\ 4}{}\qquad\qquad 3\overline{)7{,}5}\;\;\overset{2}{}$$

g. Say the next problem you'll work. (Signal.)
    15 ÷ 3.
• What's the answer? (Signal.) *5.*
    (Add to show:)   [66:5F]

$$4\overline{)2\,5{,}6}\;\;\overset{6\ 4}{}\qquad\qquad 3\overline{)7{,}5}\;\;\overset{2\ 5}{}$$

• What does 75 ÷ 3 equal? (Signal.) *25.*

========== **WORKBOOK PRACTICE** ==========

a. Find part 4 in your workbook. ✔
    (Teacher reference:)

    a. $5\overline{)345}$ b. $4\overline{)212}$ c. $3\overline{)204}$

b. Read problem A. (Signal.) *345 ÷ 5.*
• Say the first problem you'll work. (Signal.)
    34 ÷ 5.
• What's the whole-number part of the answer?
    (Signal.) *6.*
• Write it above. Then write the remainder in
    front of the 5. ✔
    (Display:)   [66:5G]

$$\text{a. }5\overline{)3\ 4{,}5}\;\;\overset{6}{}$$

    Here's what you should have. The whole
    number is 6, and the remainder is 4.
• Complete the problem. ✔
• Everybody, what does 345 ÷ 5 equal?
    (Signal.) *69.*

c. Read problem B. (Signal.) *212 ÷ 4.*
• Say the first problem you'll work. (Signal.)
    21 ÷ 4.
• What's the whole-number part of the answer?
    (Signal.) *5.*
• Write it above. Then write the remainder in
    front of the 2. ✔

(Display:)   [66:5H]

$$4\overline{)2\ 1{,}2}\;\;\overset{5}{}$$

    Here's what you should have. The whole
    number is 5, and the remainder is 1.
• Complete problem B. ✔
• Everybody, what does 212 ÷ 4 equal?
    (Signal.) *53.*
d. Read problem C. (Signal.) *204 ÷ 3.*
• Say the first problem you'll work. (Signal.)
    20 ÷ 3.
• Work the problem the fast way. Remember,
    write the whole number above and the
    remainder in front of the 4.
    (Observe students and give feedback.)
• Everybody what does 204 ÷ 3 equal?
    (Signal.) *68.*
    (Display:)   [66:5I]

$$\text{c. }3\overline{)2\ 0{,}4}\;\;\overset{6\ 8}{}$$

    Here's what you should have.

EXERCISE 6: FRACTION SIMPLIFICATION
GREATEST COMMON FACTOR

a. Open your textbook to Lesson 66 and find
    part 1. ✔
    (Teacher reference:)

    a. $\frac{20}{50}$ b. $\frac{9}{72}$ c. $\frac{9}{12}$ d. $\frac{80}{30}$ e. $\frac{28}{49}$ f. $\frac{18}{60}$

• Copy and simplify all the fractions in part 1.
    (Observe students and give feedback.)
b. Check your work.
• Fraction A: 20/50.
    The greatest common factor is 10.
• What's the simplified fraction? (Signal.) *2/5.*
c. Fraction B: 9/72.
    The greatest common factor is 9.
• What's the simplified fraction? (Signal.) *1/8.*
d. Fraction C: 9/12.
    The greatest common factor is 3.
• What's the simplified fraction? (Signal.) *3/4.*
e. Fraction D: 80/30.
    The greatest common factor is 10.
• What's the simplified fraction? (Signal.) *8/3.*

f. Fraction E: 28/49.
   The greatest common factor is 7.
   • What's the simplified fraction? (Signal.) *4/7.*
g. Fraction F: 18/60.
   The greatest common factor is 6.
   • What's the simplified fraction? (Signal.) *3/10.*

EXERCISE 7: ROUNDING

REMEDY

a. Find part 2 in your textbook. ✔
   (Teacher reference:)

   You're going to write rounded numbers.
   Remember, if a 9 rounds up, you have to
   change the digit before the 9.
b. Read number A. (Signal.) *596,014.*
   • The arrow points to 9. Does 9 round up?
   (Signal.) *Yes.*
c. Read number B. (Signal.) *309,416.*
   • Does 9 round up? (Signal.) *No.*
d. Write the rounded numbers for items A and
   B. Remember to look at the digit after the
   arrowed digit to see if the number rounds up.
   (Observe students and give feedback.)
e. Check your work.
   • Number A: 596,014. Read the rounded
   number. (Signal.) *600,000.*
   • Number B: 309,416. Read the rounded
   number. (Signal.) *309,000.*
   (Display:)   [66:7A]

   > **a.** 600,000   **b.** 309,000

   Here's what you should have.
f. Write the rounded numbers for the rest of the
   items. Remember to look at the digit after the
   arrowed digit to see if the number rounds up.
   (Observe students and give feedback.)
g. Check your work.
   • Number C: 469,503. Read the rounded
   number. (Signal.) *470,000.*
   • Number D: 519,293. Read the rounded
   number. (Signal.) *519,290.*

EXERCISE 8: INDEPENDENT WORK
WORD PROBLEMS

a. Find part 7 in your textbook. ✔
   (Teacher reference:)

   a. $\frac{4}{10}$ of the bees were flying. There were 60 bees. How many
   were flying?

   b. $\frac{2}{7}$ of the puddles were frozen. There were 30 puddles that
   were frozen. How many puddles were there in all?

   These are word problems that you work by
   multiplying a fraction by a whole number.
   For all problems, you can simplify before you
   multiply.
b. Problem A: 4/10 of the bees were flying. There
   were 60 bees. How many were flying?
   • Write the letter equation. Replace one of the
   letters with a number. Then stop. ✔
   • Everybody, read the equation with a number
   for bees. (Signal.) *4/10 × 60 = f.*
   You can simplify 4/10 times 60.
   • Do it and write the answer with a number and
   a unit name.
   (Observe students and give feedback.)
   • You simplified 60/10 and multiplied 4 × 6. How
   many flying bees were there? (Signal.)
   24 flying bees.
   (Display:)   [66:8A]

   > a. $\frac{4}{10} b = f$
   >
   > $\frac{4}{\cancel{10}} (\cancel{60}^{6}) = f = \boxed{24}$
   >
   > $\boxed{24 \text{ flying bees}}$

   Here's what you should have.
c. You'll work the other problem as part of your
   independent work. Remember to simplify
   before you multiply.

   > Assign Independent Work, Textbook parts 3–9
   > and Workbook part 5.

   > Optional extra math-fact practice worksheets
   > are available on ConnectED.

Lesson 67

EXERCISE 1: MENTAL MATH
REMAINDERS

a. Open your workbook to Lesson 67 and find part 1. ✔
* Pencils down. ✔
  (Teacher reference:)

a. 4$\overline{)11}$ R   b. 5$\overline{)11}$ R   c. 3$\overline{)28}$ R   d. 5$\overline{)28}$ R

e. 4$\overline{)29}$ R   f. 3$\overline{)25}$ R   g. 4$\overline{)25}$ R   h. 6$\overline{)25}$ R

You're going to write answers to division problems that show the whole number and the remainder.

b. Problem A: 11 ÷ 4. (Pause.) What's the answer? (Signal.) *2 with a remainder of 3.*
c. Problem B: 11 ÷ 5. (Pause.) What's the answer? (Signal.) *2 with a remainder of 1.*
d. Write answers to all the problems in part 1. (Observe students and give feedback.)
e. Check your work.
* Problem A. What's 11 ÷ 4? (Signal.) *2 with a remainder of 3.*
* Problem B. What's 11 ÷ 5? (Signal.) *2 with a remainder of 1.*
* C. What's 28 ÷ 3? (Signal.) *9 with a remainder of 1.*
* D. What's 28 ÷ 5? (Signal.) *5 with a remainder of 3.*
* E. What's 29 ÷ 4? (Signal.) *7 with a remainder of 1.*
* F. What's 25 ÷ 3? (Signal.) *8 with a remainder of 1.*
* G. What's 25 ÷ 4? (Signal.) *6 with a remainder of 1.*
* H. What's 25 ÷ 6? (Signal.) *4 with a remainder of 1.*

EXERCISE 2: DECIMAL OPERATIONS
DIVISION

a. Find part 2 in your workbook. ✔
  (Teacher reference:)

a. .04$\overline{)16}$ b. .003$\overline{)9}$

c. .6$\overline{)120}$ d. .02$\overline{)168}$

All these problems divide by a decimal value. You're going to change the problems so they divide by a whole number.

b. Read problem A. (Signal.) *16 ÷ 4 hundredths.*
* Work the problem.
  (Observe students and give feedback.)
* You moved both decimal points. How many places? (Signal.) *2.*
* Read the problem that divides by a whole number. (Signal.) *1600 ÷ 4.*
* What's the answer? (Signal.) *400.*
  (Display:)   [67:2A]

a. .04$\overline{)1600.}$ = 400

Here's what you should have.

c. Work the rest of the problems in part 2. (Observe students and give feedback.)
d. Check your work.
* Problem B: 9 ÷ 3 thousandths.
* How many places did you move both decimal points? (Signal.) *3.*
* Read the problem that divides by a whole number. (Signal.) *9000 ÷ 3.*
* What's the answer? (Signal.) *3000.*
e. Problem C. Read the problem that divides by a whole number. (Signal.) *1200 ÷ 6.*
* What's the answer? (Signal.) *200.*
f. Problem D. Read the problem that divides by a whole number. (Signal.) *16,800 ÷ 2.*
* What's the answer? (Signal.) *8400.*

EXERCISE 3: COORDINATE SYSTEM
RATIO AND PROPORTION

a. Find part 3 in your workbook. ✔
(Teacher reference:)

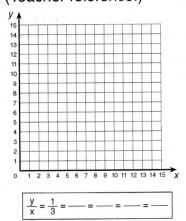

$$\frac{y}{x} = \frac{1}{3} = \underline{\quad} = \underline{\quad} = \underline{\quad} = \underline{\quad}$$

You're going to make a line for the equation Y over X = 1/3.

• Say the equation. (Signal.) *Y over X = 1/3.*

b. What's the first number for Y? (Signal.) *1.*

• What's the next number for Y? (Signal.) *2.*

• Write the multiples for Y. Then write the multiples for X.
(Observe students and give feedback.)
(Display:)   [67:3A]

$$\frac{y}{x} = \frac{1}{3} = \frac{2}{6} = \frac{3}{9} = \frac{4}{12} = \frac{5}{15}$$

Here's what you should have.

c. Make the first point for the equation Y over X = 1/3. Remember, go across for X; go up for Y. ✔
(Add to show:)   [67:3B]

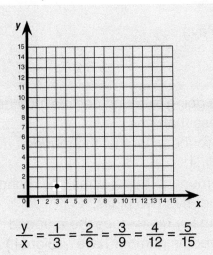

$$\frac{y}{x} = \frac{1}{3} = \frac{2}{6} = \frac{3}{9} = \frac{4}{12} = \frac{5}{15}$$

Here's what you should have.

d. Make the rest of the points.
(Observe students and give feedback.)
(Add to show:)   [67:3C]

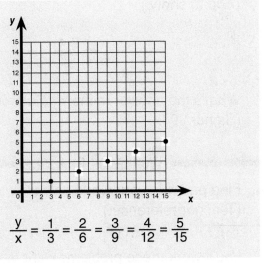

$$\frac{y}{x} = \frac{1}{3} = \frac{2}{6} = \frac{3}{9} = \frac{4}{12} = \frac{5}{15}$$

Here's what you should have.
The points show the line for Y over X = 1/3.

• Make the line. If your points are in the right places, the line should go through every point and through zero.
(Observe students and give feedback.)

EXERCISE 4: SHORT DIVISION
WITH MENTAL MATH   `REMEDY`

a. (Display:)   [67:4A]

$$5\overline{\smash{)}470}$$

You're going to work this problem without writing a number below.

• Read the problem. (Signal.) *470 ÷ 5.*

• Say the first problem you'll work. (Signal.)
47 ÷ 5.

• What's the whole-number part of the answer?
(Signal.) *9.*

• What's the remainder? (Signal.) *2.*
(Add to show:)   [67:4B]

$$5\overline{\smash{)}4\,7_20}^{9}$$

b. Say the next problem you'll work. (Signal.) *20 ÷ 5.*
- What's the answer? (Signal.) *4.*
  (Add to show:)   [67:4C]

$$5\overline{)4\,7_20}^{9\,4}$$

- What's the answer to the whole problem? (Signal.) *94.*

=== **WORKBOOK PRACTICE** ===

a. Find part 4 in your workbook. ✔
   (Teacher reference:)   R Part B

You'll work these problems without writing a number below.

b. Read problem A. (Signal.) *147 ÷ 3.*
- Say the first problem you'll work. (Signal.) *14 ÷ 3.*
- Write the whole number above and the remainder in front of the 7. Then complete the problem.
  (Observe students and give feedback.)
- Everybody, what does 147 ÷ 3 equal? (Signal.) *49.*
  (Display:)   [67:4D]

$$\textbf{a. } 3\overline{)1\,4_27}^{4\,9}$$

Here's what you should have.
c. Read problem B. (Signal.) *112 ÷ 4.*
- Work the problem the fast way.
  (Observe students and give feedback.)
d. Check your work.
- You first worked the problem 11 ÷ 4. What's the answer? (Signal.) *2 with a remainder of 3.*
- Then you worked the problem 32 ÷ 4. What's the answer? (Signal.) *8.*
- What does 112 ÷ 4 equal? (Signal.) *28.*
  (Display:)   [67:4E]

$$\textbf{b. } 4\overline{)1\,1_32}^{2\,8}$$

Here's what you should have.

e. Read problem C. (Signal.) *94 ÷ 2.*
- Work the problem the fast way.
  (Observe students and give feedback.)
f. Check your work.
- You first worked the problem 9 ÷ 2. What's the answer? (Signal.) *4 with a remainder of 1.*
- Then you worked the problem 14 ÷ 2. What's the answer? (Signal.) *7.*
- What does 94 ÷ 2 equal? (Signal.) *47.*
  (Display:)   [67:4F]

$$\textbf{c. } 2\overline{)9_14}^{4\,7}$$

Here's what you should have.
g. Read problem D. (Signal.) *308 ÷ 4.*
- Work the problem the fast way.
  (Observe students and give feedback.)
h. Check your work.
- You first worked the problem 30 ÷ 4. What's the answer? (Signal.) *7 with a remainder of 2.*
- Then you worked the problem 28 ÷ 4. What's the answer? (Signal.) *7.*
- What does 308 ÷ 4 equal? (Signal.) *77.*
  (Display:)   [67:4G]

$$\textbf{d. } 4\overline{)3\,0_28}^{7\,7}$$

Here's what you should have.

EXERCISE 5: ROUNDING
DECIMALS

a. (Display:)   [67:5A]

$$.3\,6\,2\,7$$

We're going to round decimal numbers that are less than one.
If we round to tenths, the rounded number has one digit.
If we round to hundredths, the rounded number has two digits.
- How many digits does the rounded hundredths number have? (Signal.) *2.*
- If we round to thousandths, how many digits does the rounded thousandths number have? (Signal.) *3.*
  (Repeat until firm.)

b. (Add to show:) [67:5B]

.3627
↑

The arrow shows that we round this number to tenths.
- How many digits will be after the decimal point? (Signal.) *1.*
The digit after 3 shows if we round up.
- Is the digit 5 or more? (Signal.) *Yes.*
So we round up.
- What's the rounded value? (Signal.) *4 tenths.*
(Add to show:) [67:5C]

.3627
↑
.4

c. (Change to show:) [67:5D]

.3617
↑

- What place do we round to? (Signal.) *Hundredths.*
- How many digits will be after the decimal point? (Signal.) *2.*
- What's the rounded number? (Signal.) *36 hundredths.*
(Add to show:) [67:5E]

.3617
↑
.36

d. (Change to show:) [67:5F]

.3617
↑

- What place do we round to? (Signal.) *Thousandths.*
- How many digits will be after the decimal point? (Signal.) *3.*
- What's the rounded number? (Signal.) *362 thousandths.*
(Add to show:) [67:5G]

.3617
↑
.362

e. (Display:) [67:5H]

Round .0536 to the tenths place.

We're going to round this number to the tenths place.
The tenths digit is zero. The next digit is 5.
- So does zero round up? (Signal.) *Yes.*
- What's the number rounded to tenths? (Signal.) *1 tenth.*
(Add to show:) [67:5I]

Round .0536 to the tenths place.
.1

f. (Display:) [67:5J]

Round .0536 to the hundredths place.

Now we'll round this number to the hundredths place.
- Raise your hand when you know the rounded number. ✔
- What's the number rounded to hundredths? (Signal.) *5 hundredths.*
(Add to show:) [67:5K]

Round .0536 to the hundredths place.
.05

g. (Display:) [67:5L]

Round .0536 to the thousandths place.

We'll round this number to the thousandths place.
- Raise your hand when you know the rounded number. ✔
- What's the number rounded to thousandths? (Signal.) *54 thousandths.*
(Add to show:) [67:5M]

Round .0536 to the thousandths place.
.054

TEXTBOOK PRACTICE

a. Open your textbook to Lesson 67 and find part 1. ✔
(Teacher reference:)

a. Round .2364 to the tenths place.
b. Round .2364 to the hundredths place.
c. Round .2364 to the thousandths place.

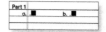

b. Item A: You'll round the number to the tenths place.

• Write the rounded number. ✔
• Everybody, read the rounded number. (Signal.) *2 tenths.*

c. Item B: Round 2364 ten-thousandths to the hundredths place. Remember to look at the digit after the hundredths place to see if you round up.

• Write the rounded number. ✔
• Everybody, read the rounded number. (Signal.) *24 hundredths.*

d. Item C: You'll round the number to the thousandths place.

• Write the rounded number. ✔
• Everybody, read the rounded number. (Signal.) *236 thousandths.*

EXERCISE 6: EXPONENTS

BASE 10
REMEDY

a. You learned a fast way to work with numbers that have a base of 10 and an exponent.
(Display:) [67:6A]

$$10^1$$

• What's the exponent? (Signal.) *1.*
• So how many zeros are in 10 to the first? (Signal.) *1.*
(Add to show:) [67:6B]

$$10^1 = 10$$

• What does 10 to the first equal? (Signal.) *10.*

b. (Add to show:) [67:6C]

$$10^1 = 10$$
$$10^3$$

• What's the exponent? (Signal.) *3.*
• So how many zeros are in 10 to the third? (Signal.) *3.*

(Add to show:) [67:6D]

$$10^1 = 10$$
$$10^3 = 1000$$

• What does 10 to the third equal? (Signal.) *1000.*

c. (Add to show:) [67:6E]

$$10^1 = 10$$
$$10^3 = 1000$$
$$10^6$$

• What's the exponent? (Signal.) *6.*
• So how many zeros are in 10 to the sixth? (Signal.) *6.*
(Add to show:) [67:6F]

$$10^1 = 10$$
$$10^3 = 1000$$
$$10^6 = 1,000,000$$

• What does 10 to the sixth equal? (Signal.) *1 million.*

TEXTBOOK PRACTICE

a. Find part 2 in your textbook. ✔
(Teacher reference:)

a. 10^4 b. 10^2 c. 10^5 d. 10^3

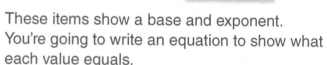

These items show a base and exponent. You're going to write an equation to show what each value equals.

b. Item A. Read it. (Signal.) *10 to the fourth.*
• What's the exponent? (Signal.) *4.*
• So how many zeros are in the answer? (Signal.) *4.*
• Copy the base and exponent. Write the number it equals. ✔
• What does 10 to the fourth equal? (Signal.) *10,000.*

c. Write equations for the rest of the items. (Observe students and give feedback.)

d. Check your work.
• Item B: 10 to the second.
• Read the equation. (Signal.) $10^2 = 100.$

e. C: 10 to the fifth.
- Read the equation. (Signal.) *10^5 = 100,000.*
f. D: 10 to the third.
- Read the equation. (Signal.) *10^3 = 1000.*
g. Find part 3 in your textbook. ✔
   (Teacher reference:)

a. 1000   b. 1,000,000   c. 10,000   d. 100

These items show the numbers. You're going to write the base and exponent.
The base for all of the numbers is 10. The number of zeros shows the number you write for the exponent.
h. Problem A. Read the number. (Signal.) *1 thousand.*
- What's the base? (Signal.) *10.*
- How many zeros are in 1000? (Signal.) *3.*
- What's the exponent? (Signal.) *3.*
   Yes, 1000 equals 10 to the third.

i. Write the equation. Then write equations for the rest of the items.
   (Observe students and give feedback.)
j. Check your work.
- Number B is 1 million.
- How many zeros are in 1 million? (Signal.) *6.*
- What does 1 million equal? (Signal.) *10 to the sixth.*
k. Number C is 10 thousand.
- How many zeros are in 10,000? (Signal.) *4.*
- What does 10,000 equal? (Signal.) *10 to the fourth.*
l. Number D is 1 hundred.
- How many zeros are in 100? (Signal.) *2.*
- What does 100 equal? (Signal.) *10 to the second.*

Assign Independent Work, Textbook parts 4–11 and Workbook parts 5 and 6.

Optional extra math-fact practice worksheets are available on ConnectED.

Lesson  68

EXERCISE 1: EQUIVALENT FRACTIONS

a. (Display:)   [68:1A]

$$\frac{1}{2}$$

You're going to tell me about fractions that equal 1/2.

- What's the denominator of 1/2? (Signal.) *2.*
- How many times bigger is the denominator than the numerator? (Signal.) *2 times.*
  So any fraction that equals 1/2 has a denominator that is 2 times the numerator.
- If the denominator is 2, what's the numerator? (Signal.) *1.*
- If the denominator is 4, what's the numerator? (Signal.) *2.*
- If the denominator is 10, what's the numerator? (Signal.) *5.*
- If the denominator is 8, what's the numerator? (Signal.) *4.*
  (Repeat until firm.)
  Yes, the denominator is 2 times the numerator.

b. (Display:)   [68:1B]

$$\frac{1}{4}$$

Here's 1/4. You're going to tell me about fractions that equal 1/4.

- How many times bigger is the denominator than the numerator? (Signal.) *4 times.*
- If the denominator is 4, what's the numerator? (Signal.) *1.*
- If the denominator is 8, what's the numerator? (Signal.) *2.*
- If the denominator is 20, what's the numerator? (Signal.) *5.*
- If the denominator is 24, what's the numerator? (Signal.) *6.*
  (Repeat until firm.)

c. (Display:)   [68:1C]

$$\frac{1}{3}$$

Here's 1/3. You're going to tell me about fractions that equal 1/3.

- The denominator is how many times the numerator? (Signal.) *3 times.*
- If the denominator is 3, what's the numerator? (Signal.) *1.*
- If the denominator is 9, what's the numerator? (Signal.) *3.*
- If the denominator is 12, what's the numerator? (Signal.) *4.*
- If the denominator is 24, what's the numerator? (Signal.) *8.*
  (Repeat until firm.)

EXERCISE 2: DECIMAL OPERATIONS
DIVISION   `REMEDY`

a. (Display:)   [68:2A]

$$.0\,3\,\overline{\smash{)}1.5\,9\,6}$$

- Read this problem. (Signal.) *1 and 596 thousandths ÷ 3 hundredths.*
  Both numbers have a decimal point.
- What's the number you divide by? (Signal.) *.03.*
- How many places do you move the decimal point? (Signal.) *2.*
  So we move the decimal point of both numbers two places.
  (Add to show:)   [68:2B]

$$0\,3.\,\overline{\smash{)}1\,5\,9.6}$$

b. I show the new decimal point in the answer.
  (Add to show:)   [68:2C]

$$0\,3.\,\overline{\smash{)}1\,5\,9.6}$$

- Read the new problem. (Signal.) *159.6 ÷ 3.*
  (Add to show:)   [68:2D]

$$0\,3.\,\overline{\smash{)}1\,5\,9.6} \qquad .0\,0\,4\,\overline{\smash{)}1\,6.8}$$

c. Read this problem. (Signal.) *16.8 ÷ .004.*
- How many places do you move both decimal points? (Signal.) *3.*
  I show the new decimal point in the answer.
  (Add to show:)   [68:2E]

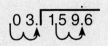 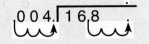

I don't have a digit in every place, so I have to write zeros to fill the spaces at the end.
(Add to show:)   [68:2F]

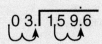 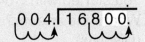

- Read the new problem. (Signal.) *16,800 ÷ 4.*
d. Now we're going to work both problems. (Point to .03) The new problem we work is 159.6 ÷ 3.
- First we work the problem 15 ÷ 3. What's the answer? (Signal.) *5.*
  (Add to show:)   [68:2G]

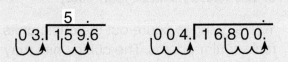

- 9 ÷ 3. What's the answer? (Signal.) *3.*
  (Add to show:)   [68:2H]

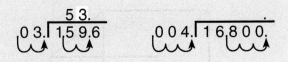

- 6 ÷ 3. What's the answer? (Signal.) *2.*
  (Add to show:)   [68:2I]

- What's the answer to the whole problem? (Signal.) *53.2.*
e. (Point to .004) The new problem we work is 16,800 divided by 4.
- First we work the problem 16 ÷ 4. What's the answer? (Signal.) *4.*
- 8 ÷ 4. What's the answer? (Signal.) *2.*

- What's zero divided by 4? (Signal.) *Zero.*
  So I write two zeros in the answer.
  (Add to show:)   [68:2J]

$$\begin{array}{r} 53.2 \\ 03.\overline{)159.6} \end{array} \qquad \begin{array}{r} 4200. \\ 004.\overline{)16800.} \end{array}$$

- What's the answer to the whole problem? (Signal.) *4200.*
  Remember, write zeros if there are empty places after you move the decimal point.

=== **WORKBOOK PRACTICE** ===

a. Open your workbook to Lesson 68 and find part 1. ✔
   (Teacher reference:)   **R** *Part I*

   a. .08⟌2̄4̄.8̄   b. .5⟌4̄0̄.3̄5̄   c. .03⟌1̄.8̄6̄   d. .2⟌.1̄6̄4̄

b. Problem A: 24.8 divided by .08.
- Move both decimal points and write the decimal point in the answer. Make sure you have a digit in every place. Stop when you've done that much. ✔
  (Display:)   [68:2K]

$$\text{a. } 08.\overline{)2480.}$$

Here's what you should have. Make sure you wrote one zero to show 2480.
- Read the new problem. (Signal.) *2480 ÷ 8.*
c. Problem B: 40.35 ÷ .5.
- Move the decimal points and write the decimal point in the answer. Stop when you've done that much. ✔
  (Display:)   [68:2L]

$$\text{b. } 5.\overline{)403.5}$$

Here's what you should have.
- Read the new problem. (Signal.) *403.5 ÷ 5.*

d. Problem C: 1.86 ÷ .03.
- Move the decimal points and write the decimal point in the answer. Stop when you've done that much. ✔
  (Display:) [68:2M]

c. 0 3.⌐1 8 6.

Here's what you should have.
- Read the new problem. (Signal.) *186 ÷ 3.*

e. Problem D: .164 ÷ .2.
- Move the decimal points and write the decimal point in the answer. Stop when you've done that much. ✔
  (Display:) [68:2N]

d. 2.⌐1 6 4

Here's what you should have.
- Read the new problem. (Signal.) *1.64 ÷ 2.*

f. Work problem A. Pencils down when you're finished.
  (Observe students and give feedback.)

g. Check your work.
  Problem A. You started with 24.8 ÷ .08.
- What's the answer? (Signal.) *310.*
  (Display:) [68:2O]

          3 1 0.
a. 0 8.⌐2 4 8 0.

Here's what you should have.

h. Work problem B.
  (Observe students and give feedback.)

i. Check your work.
  You started with 40.35 divided by .5.
- What's the answer? (Signal.) *80.7*
  (Display:) [68:2P]

          8 0.7
b. 5.⌐4 0 3.5

Here's what you should have.

j. Work problem C.
  (Observe students and give feedback.)

k. Check your work.
  You started with 1.86 ÷ .03.
- What's the answer? (Signal.) *62.*
  (Display:) [68:2Q]

          6 2.
c. 0 3.⌐1 8 6.

Here's what you should have.

l. Work problem D.
  (Observe students and give feedback.)

m. Check your work.
  You started with .164 ÷ .2.
- What's the answer? (Signal.) *.82.*
  (Display:) [68:2R]

          .8 2
d. 2.⌐1 6 4

Here's what you should have.

EXERCISE 3: VOLUME
OF RECTANGULAR PRISMS (UNIT CUBE)

a. You're going to figure out the cubic units for rectangular prisms. The cubic units may be cubic inches, cubic centimeters, cubic feet, or other cubic units.
  (Display:) [68:3A]

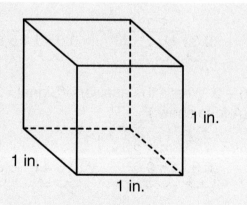
1 in.
1 in.
1 in.

Here's a picture of a **cubic inch.**
A cubic inch has 6 faces—the bottom face, the 4 side faces, and the top face.
- How many faces does a cubic unit have? (Signal.) *6.*
All faces of the cube are the same size.

b. All the faces of a cubic inch are one-inch squares.
All the faces of a cubic foot are one-foot squares.
- What are all faces of a cubic yard? (Signal.) *One-yard squares.*
- How many faces does a cubic yard have? (Signal.) *6.*
- What are all faces of a cubic meter? (Signal.) *One-meter squares.*
- How many faces does a cubic meter have? (Signal.) *6.*
- (Repeat until firm.)

c. (Display:) [68:3B]

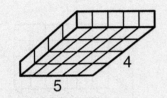

Here's a rectangular prism. You can see the cubic units. The base is 5 by 4. So the area of the base is 5 times 4.
- What's 5 × 4? (Signal.) *20.*
So this rectangular prism has 20 cubic units.

d. (Add to show:) [68:3C]

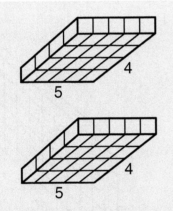

Here's another prism that has 20 cubic units. If we combine the layers, we have 40 cubic units.
(Change to show:) [68:3D]

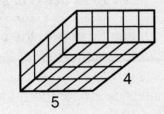

- How many cubic units do we have? (Signal.) *40.*

- If we add another layer, how many cubic units will we have? (Signal.) *60.*
(Add to show:) [68:3E]

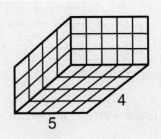

- How many cubic units are in this rectangular prism? (Signal.) *60.*

e. There's a fast way to figure out the number of cubic units.
(Add to show:) [68:3F]

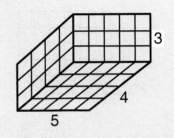

Cubic units = Area of the base × height

- Say the equation. (Signal.) *Cubic units = Area of the base × height.*
- Say the multiplication for the area of the base. (Signal.) *5 × 4.*
- What's the area of the base? (Signal.) *20.*
We multiply 20 × 3.
(Add to show:) [68:3G]

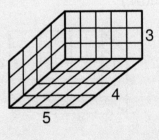

Cubic units = Area of the base × height
         =   20   ×   3

- What's the answer? (Signal.) *60.*

(Add to show:) [68:3H]

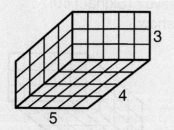

Cubic units = Area of the base × height
$$= \quad 20 \quad × \quad 3$$
$$= \quad \boxed{60}$$

That's the number of cubic units in the whole rectangular prism.

f. (Display:) [68:3I]

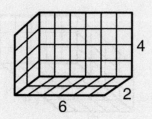

Cubic units = Area of the base × height

Here's a new rectangular prism.

• Say the equation for finding the cubic units.
 (Signal.) *Cubic units = Area of the base × height.*
 (Repeat until firm.)

g. Say the multiplication for the area of the base.
 (Signal.) *6 × 2.*
• What's the answer? (Signal.) *12.*
 (Add to show:) [68:3J]

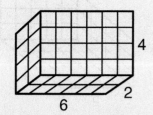

Cubic units = Area of the base × height
$$= \quad 12$$

• What's the height of the prism? (Signal.) *4.*
• So what do we multiply the area of the base by? (Signal.) *4.*
 Yes, 12 × 4.

(Add to show:) [68:3K]

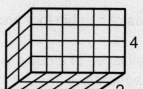

Cubic units = Area of the base × height
$$= \quad 12 \quad × \quad 4$$

h. Raise your hand when you know the number of cubic units. ✔
• What's the number of cubic units? (Signal.) *48.*
 (Add to show:) [68:3L]

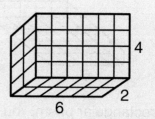

Cubic units = Area of the base × height
$$= \quad 12 \quad × \quad 4$$
$$= \quad \boxed{48}$$

Yes, there are 48 cubic units in the prism.

i. (Display:) [68:3M]

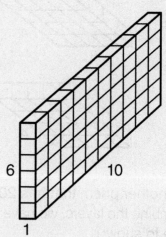

Cubic units = Area of the base × height

• Say the equation for finding the cubic units.
 (Signal.) *Cubic units = Area of the base × height.*
• Say the multiplication for the area of the base.
 (Signal.) *1 × 10.*
• What's the answer? (Signal.) *10.*

(Add to show:)   [68:3N]

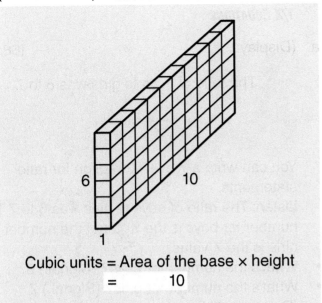

Cubic units = Area of the base × height
        =   10

j. What's the height of the prism? (Signal.) *6.*
   (Add to show:)   [68:3O]

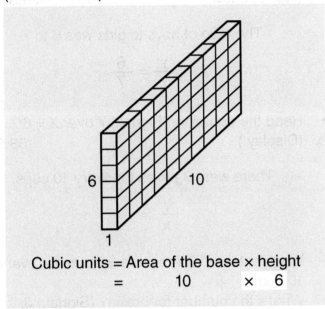

Cubic units = Area of the base × height
        =   10   ×   6

• What's 10 × 6? (Signal.) *60.*

(Add to show:)   [68:3P]

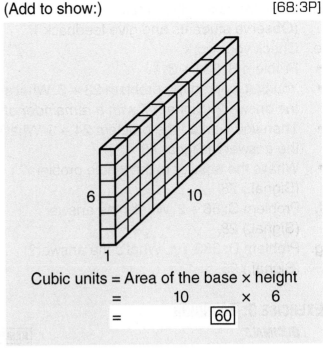

Cubic units = Area of the base × height
             =   10   ×   6
             =   60

That's the number of cubic units in the prism.

EXERCISE 4: SHORT DIVISION
WITH MENTAL MATH   `REMEDY`

a. Find part 2 in your workbook. ✔
   (Teacher reference:)   `R` `Part C`

   a. 5$\overline{)375}$ b. 3$\overline{)234}$ c. 2$\overline{)56}$ d. 4$\overline{)380}$

   You're going to work these problems the fast
   way, without writing numbers below.
b. Read problem A. (Signal.) *375 ÷ 5.*
• Work the problem.
   (Observe students and give feedback.)
c. Check your work.
• You first worked the problem 37 ÷ 5. What's
   the answer? (Signal.) *7 with a remainder of 2.*
• Then you worked the problem 25 ÷ 5. What's
   the answer? (Signal.) *5.*
• What's the answer to the whole problem?
   (Signal.) *75.*
   (Display:)   [68:4A]

$$\begin{array}{r} 7\,5 \\ \text{a. } 5\overline{)3\,7_2 5} \end{array}$$

Here's what you should have.

d. Work the rest of the problems.
   (Observe students and give feedback.)
e. Check your work.
- Problem B: 234 ÷ 3.
- You first worked the problem 23 ÷ 3. What's the answer? (Signal.) *7 with a remainder of 2.*
- Then you worked the problem 24 ÷ 3. What's the answer? (Signal.) *8.*
- What's the answer to the whole problem? (Signal.) *78.*
f. Problem C: 56 ÷ 2. What's the answer? (Signal.) *28.*
g. Problem D: 380 ÷ 4. What's the answer? (Signal.) *95.*

EXERCISE 5: ROUNDING
DECIMALS   REMEDY

a. Open your textbook to Lesson 68 and find part 1. ✔
   (Teacher reference:)

   a. Round .5874 to the hundredths place.
   b. Round .62718 to the thousandths place.
   c. Round .0127 to the hundredths place.
   d. Round .05281 to the tenths place.

   Part 1
   a. ■

   You're going to round these decimal numbers to different places.
- If you round the decimal value to tenths, how many places will be after the decimal point? (Signal.) *1.*
- If you round to hundredths, how many places will be after the decimal point? (Signal.) *2.*
- If you round to thousandths, how many places will be after the decimal point? (Signal.) *3.*
b. Item A. Round the number to the hundredths place. ✔
- Read the rounded number. (Signal.) *59 hundredths.*
c. Item B. Round the number to the thousandths place. ✔
- Read the rounded number. (Signal.) *627 thousandths.*
d. Item C. Round the number to the hundredths place. ✔
- Read the rounded number. (Signal.) *1 hundredth.*
e. Item D. Round the number to the tenths place. ✔
- Read the rounded number. (Signal.) *1 tenth.*

EXERCISE 6: RATIO SENTENCES
Y/X EQUATIONS

a. (Display:)   [68:6A]

   The ratio of boys to girls was 6 to 7.

   $$\frac{y}{x} =$$

   You can write a Y over X fraction for ratio statements.
   Listen: The ratio of boys to girls was 6 to 7. The number for boys is the Y value; the number for girls is the X value.
- What's the number for boys? (Signal.) *6.*
- What's the number for girls? (Signal.) *7.*
  (Repeat until firm.)
  So Y over X equals 6/7.
  (Add to show:)   [68:6B]

   The ratio of boys to girls was 6 to 7.

   $$\frac{y}{x} = \frac{6}{7}$$

- Read the equation. (Signal.) *Y over X = 6/7.*
b. (Display:)   [68:6C]

   There were 3 boxes for every 10 cups.

   $$\frac{y}{x} =$$

   New sentence: There were 3 boxes for every 10 cups.
- What's the number for boxes? (Signal.) *3.*
- What's the number for cups? (Signal.) *10.*
- So what does Y over X equal? (Signal.) *3/10.*
  (Repeat until firm.)
  (Add to show:)   [68:6D]

   There were 3 boxes for every 10 cups.

   $$\frac{y}{x} = \frac{3}{10}$$

- Read the equation. (Signal.) *Y over X = 3/10.*

c. (Display:) [68:6E]

> Every 5 bags weighed 12 pounds.
>
> $$\frac{y}{x} =$$

New sentence: Every 5 bags weighed
12 pounds.
- • What's the number for bags? (Signal.) *5.*
- • What's the number for pounds? (Signal.) *12.*
- • Say the equation for Y over X. (Signal.) *Y over X = 5/12.*
  (Repeat until firm.)
  (Add to show:) [68:6F]

> Every 5 bags weighed 12 pounds.
>
> $$\frac{y}{x} = \frac{5}{12}$$

=== **TEXTBOOK PRACTICE** ===

a. Find part 2 in your textbook. ✔
(Teacher reference:)

> a. The ratio of beans to peas was 9 to 11.
> b. There were 12 phones for every 5 rooms.
> c. Every page had 25 lines.

You're going to write Y over X equations for
each sentence.
b. Read sentence A. (Signal.) *The ratio of beans to peas was 9 to 11.*
- • Write the equation for Y over X. ✔
- • Read the equation. (Signal.) *Y over X = 9/11.*
  (Display:) [68:6G]

> a. $$\frac{y}{x} = \frac{9}{11}$$

Here's what you should have.
c. Write equations for the rest of the sentences.
(Observe students and give feedback.)
d. Check your work.
- • Sentence B: There were 12 phones for every 5 rooms.
- • Read the equation. (Signal.) *Y over X = 12/5.*

e. Sentence C: Every page had 25 lines.
- • Read the equation. (Signal.) *Y over X = 1/25.*
  (Display:) [68:6H]

> b. $$\frac{y}{x} = \frac{12}{5}$$
>
> c. $$\frac{y}{x} = \frac{1}{25}$$

Here's what you should have for B and C.

EXERCISE 7: EXPONENTS
BASE 10   [REMEDY]

a. Find part 3 in your textbook. ✔
(Teacher reference:)

> a. 10,000   b. 10^5 c. 10^3
> d. 1,000,000   e. 10^1 f. 100

You're going to complete the equations so
they show the base and exponent on one side
and the number it equals on the other. The
base number for each item is 10.
b. Item A shows the number. Write the complete
equation with a base and exponent. ✔
- • Everybody, read the equation. (Signal.)
  10,000 = 10^4.
c. Item B shows the base and exponent. Write
the complete equation with the number. ✔
- • Everybody, read the equation. (Signal.)
  10^5 = 100,000.
d. Work the rest of the items in part 3.
(Observe students and give feedback.)
e. Check your work.
- • Item C. Read the equation. (Signal.) *10^3 = 1000.*
- • D. Read the equation. (Signal.) *1 million = 10^6.*
- • E. Read the equation. (Signal.) *10^1 = 10.*
- • F. Read the equation. (Signal.) *100 = 10^2.*

Assign Independent Work, Textbook parts 4–12.

Optional extra math-fact practice worksheets
are available on ConnectED.

Lesson 69

Exercise 1: Equivalent Fractions   [REMEDY]

a. (Display:) [69:1A]

$$\frac{1}{2}$$

You're going to tell me about fractions that equal 1/2.

- What's the denominator of 1/2? (Signal.) *2.*
- How many times bigger is the denominator than the numerator? (Signal.) *2 times.*
- If the denominator is 4, what's the numerator? (Signal.) *2.*
- If the denominator is 6, what's the numerator? (Signal.) *3.*
- If the denominator is 10, what's the numerator? (Signal.) *5.*
- If the denominator is 18, what's the numerator? (Signal.) *9.*
- If the denominator is 20, what's the numerator? (Signal.) *10.*
  (Repeat until firm.)

b. (Display:) [69:1B]

$$\frac{1}{3}$$

- What's the denominator of 1/3? (Signal.) *3.*
- If the denominator is 3, what's the numerator? (Signal.) *1.*
- If the denominator is 6, what's the numerator? (Signal.) *2.*
- If the denominator is 12, what's the numerator? (Signal.) *4.*
- If the denominator is 18, what's the numerator? (Signal.) *6.*
- If the denominator is 30, what's the numerator? (Signal.) *10.*
  (Repeat until firm.)

c. (Display:) [69:1C]

$$\frac{1}{5}$$

- What's the denominator of 1/5? (Signal.) *5.*
- If the denominator is 5, what's the numerator? (Signal.) *1.*
- If the denominator is 10, what's the numerator? (Signal.) *2.*
- If the denominator is 15, what's the numerator? (Signal.) *3.*
- If the denominator is 25, what's the numerator? (Signal.) *5.*
- If the denominator is 30, what's the numerator? (Signal.) *6.*
- If the denominator is 45, what's the numerator? (Signal.) *9.*
- If the denominator is 50, what's the numerator? (Signal.) *10.*
  (Repeat until firm.)

Exercise 2: Decimal Operations
Division

a. Open your workbook to Lesson 69 and find part 1. ✔
   (Teacher reference:)

   a. $.2\overline{)1.084}$ b. $.003\overline{)1.5}$ c. $.06\overline{).0624}$

   Both numbers in each division problem are decimal values.

b. Read problem A. (Signal.) *1.084 ÷ .2.*
- Move the decimals in both numbers so you are dividing by a whole number. Show the new decimal point in the answer. Stop when you've done that much. ✔
- Everybody, read the problem that divides by a whole number. (Signal.) *10.84 ÷ 2.*
  (Display:) [69:2A]

  a. $2.\overline{)1\underset{\curvearrowright}{0.8}4}$

  Here's what you should have.
- Work the problem.
  (Observe students and give feedback.)
- Everybody, what's 1 and 84 thousandths ÷ 2 tenths? (Signal.) *5.42.*

(Add to show:) [69:2B]

$$\begin{array}{r} 5.42 \\ 2.\overline{)10.84} \end{array}$$
a.

Here's what you should have.

c. Read problem B. (Signal.) *1.5 ÷ .003.*
• Move the decimals in both numbers so you are dividing by a whole number. Show the new decimal point in the answer. Make sure you have a digit in every place of the problem. Stop when you've done that much. ✔
• Everybody, read the problem that divides by a whole number. (Signal.) *1500 ÷ 3.*
(Display:) [69:2C]

b. $003.\overline{)1500.}$

Here's what you should have.
• Work the problem.
(Observe students and give feedback.)
• Everybody, what's 1.5 divided by 3 thousandths? (Signal.) *500.*
(Add to show:) [69:2D]

$$\begin{array}{r} 500. \\ 003.\overline{)1500.} \end{array}$$
b.

Here's what you should have.

d. Work problem C. Remember, move the decimal points in both numbers so you divide by a whole number. Show the new decimal point in the answer.
(Observe students and give feedback.)

e. Check your work.
Problem C: 624 ten-thousandths ÷ 6 hundredths.
• Read the problem that divides by a whole number. (Signal.) *6.24 ÷ 6.*
• What's the answer? (Signal.) *1.04.*
(Display:) [69:2E]

$$\begin{array}{r} 1.04 \\ 06.\overline{)06.24} \end{array}$$
c.

Here's what you should have.

EXERCISE 3: COORDINATE SYSTEM
RATIO SENTENCES

a. (Display:) [69:3A]

There were 4 potatoes for every 3 carrots.

Listen: There were 4 potatoes for every 3 carrots. You can write a Y over X equation for this sentence. Y over X equals 4/3.
• Say the Y over X equation for this sentence. (Signal.) *Y over X = 4/3.*

b. (Display:) [69:3B]

The ratio of trees to bushes is 1 to 4.

New sentence: The ratio of trees to bushes is 1 to 4.
• Say the Y over X equation for this sentence. (Signal.) *Y over X = 1/4.*
(Repeat until firm.)

c. (Display:) [69:3C]

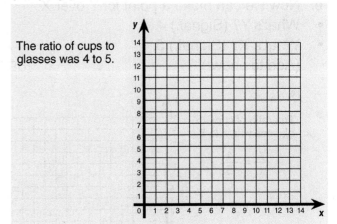

The ratio of cups to glasses was 4 to 5.

(Point to sentence.) We're going to put points for this sentence on the coordinate system.
• Read the sentence. (Signal.) *The ratio of cups to glasses was 4 to 5.*
• Say the Y over X equation. (Signal.) *Y over X = 4/5.*
(Add to show:) [69:3D]

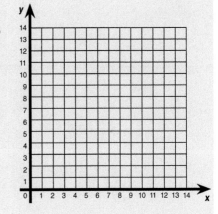

The ratio of cups to glasses was 4 to 5.

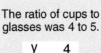
$$\frac{y}{x} = \frac{4}{5}$$

d. We need to show the names for Y and X on the coordinate system. The ratio of cups to glasses was 4 to 5.
• What's the name for Y? (Signal.) *Cups.*
• What's the name for X? (Signal.) *Glasses.*
(Repeat until firm.)
(Add to show:) [69:3E]

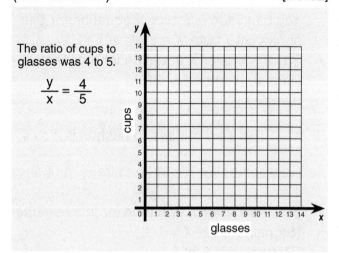
The ratio of cups to glasses was 4 to 5.

$$\frac{y}{x} = \frac{4}{5}$$

Here are the names on the coordinate system.
e. Now we can make a point for Y over X.
• What's Y? (Signal.) *4.*
• What's X? (Signal.) *5.*
(Add to show:) [69:3F]

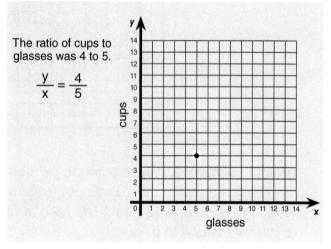
The ratio of cups to glasses was 4 to 5.

$$\frac{y}{x} = \frac{4}{5}$$

Here's the point for Y over X = 4/5.
The point shows that there were 4 cups for every 5 glasses.

f. (Display:) [69:3G]

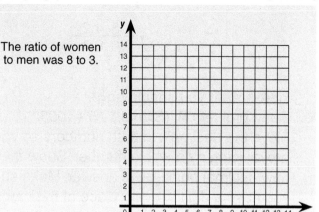
The ratio of women to men was 8 to 3.

• New sentence. Read the sentence. (Signal.) *The ratio of women to men was 8 to 3.*
• Say the Y over X equation. (Signal.) *Y over X = 8/3.*
(Add to show:) [69:3H]

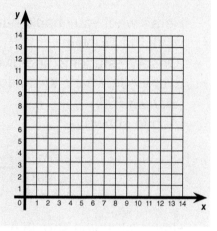
The ratio of women to men was 8 to 3.

$$\frac{y}{x} = \frac{8}{3}$$

g. We need to show the names for Y and X on the coordinate system.
• What's the name for Y? (Signal.) *Women.*
• What's the name for X? (Signal.) *Men.*
(Repeat until firm.)
(Add to show:) [69:3I]

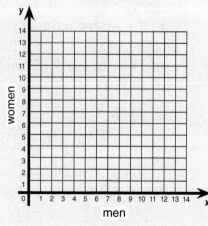
The ratio of women to men was 8 to 3.

$$\frac{y}{x} = \frac{8}{3}$$

Here are the names on the coordinate system.

h. Now we can make a point for Y over X.
- What's Y? (Signal.) *8.*
- What's X? (Signal.) *3.*

(Add to show:)   [69:3J]

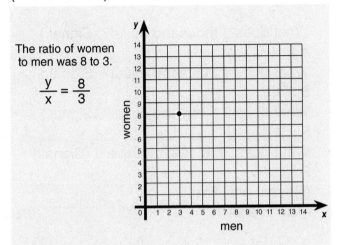

The ratio of women to men was 8 to 3.

$$\frac{y}{x} = \frac{8}{3}$$

Here's the point for Y over X equals 8/3. The point shows that there are 8 women for every 3 men.

━━━━━━━ **WORKBOOK PRACTICE** ━━━━━━━

a. Find part 2 in your workbook. ✔
(Teacher reference:)

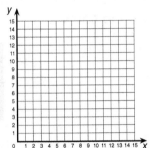

a. The ratio of books to magazines was 7 to 5.

$$\frac{y}{x} =$$

- There were 14 books and 10 magazines.

b. Item A. Read the sentence. (Signal.) *The ratio of books to magazines was 7 to 5.*
You're going to write the names for Y and X.
- What's the name for Y? (Signal.) *Books.*
- What's the name for X? (Signal.) *Magazines.*
- Write the names in the blanks on the coordinate system. ✔

(Display:)   [69:3K]

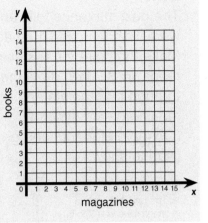

a. The ratio of books to magazines was 7 to 5.

$$\frac{y}{x} =$$

- There were 14 books and 10 magazines.

Here's what you should have.

c. Now complete the equation for Y over X. ✔
(Add to show:)   [69:3L]

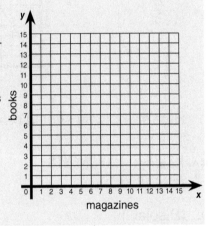

a. The ratio of books to magazines was 7 to 5.

$$\frac{y}{x} = \frac{7}{5}$$

- There were 14 books and 10 magazines.

Here's what you should have.

d. Now make a point for Y over X = 7/5. ✔
(Add to show:)   [69:3M]

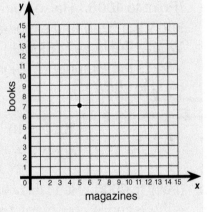

a. The ratio of books to magazines was 7 to 5.

$$\frac{y}{x} \quad \frac{7}{5}$$

- There were 14 books and 10 magazines.

Here's what you should have.

e. You're going to make another point and draw a line.

The next sentence tells about the point: There were 14 books and 10 magazines.

• Make the point for 14 books and 10 magazines. Then draw the line. The line goes through both points and goes through zero on the coordinate system.

(Observe students and give feedback.)

(Add to show:) [69:3N]

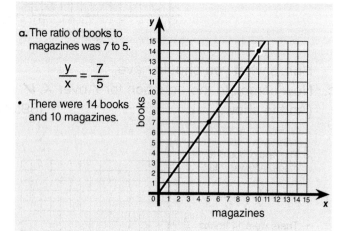

a. The ratio of books to magazines was 7 to 5.

$$\frac{y}{x} = \frac{7}{5}$$

• There were 14 books and 10 magazines.

Here's what you should have.

EXERCISE 4: EXPONENTS
BASE 10

a. (Display:) [69:4A]

$$1000$$
$$4000$$

• (Point to **1000**.) Raise your hand when you know the base and exponent for 1000. ✔
• What's the base? (Signal.) *10.*
• What's the exponent? (Signal.) *3.*

(Add to show:) [69:4B]

$$1000 = 10^3$$
$$4000 =$$

b. (Point to **4000**.) Read this number. (Signal.) *4000.*
• 4 thousand is how many times 1 thousand? (Signal.) *4.*

Yes, 4 thousand is 4 times 1 thousand, so it is 4 times 10 to the third.

4000 equals 4 times 10 to the third.

• What does 1 thousand equal? (Signal.) *10^3.*
• What does 4 thousand equal? (Signal.) *4×10^3.*

(Repeat until firm.)

(Add to show:) [69:4C]

$$1000 = 10^3$$
$$4000 = 4 \times 10^3$$

• What does 4 thousand equal? (Signal.) *4×10^3.*
• What does 5 thousand equal? (Signal.) *5×10^3.*
• What does 9 thousand equal? (Signal.) *9×10^3.*
• What does 26 thousand equal? (Signal.) *26×10^3.*

(Repeat until firm.)

c. (Add to show:) [69:4D]

$$1000 = 10^3$$
$$4000 = 4 \times 10^3$$
$$400 =$$

4 hundred does not equal 4 times 10 to the third.

• How many zeros does 400 have? (Signal.) *2.* So it equals 4 times 10 to the second.

(Add to show:) [69:4E]

$$1000 = 10^3$$
$$4000 = 4 \times 10^3$$
$$400 = 4 \times 10^2$$

• What does 400 equal? (Signal.) *4×10^2.*
• What does 200 equal? (Signal.) *2×10^2.*
• What does 9 hundred equal? (Signal.) *9×10^2.*
• What does 9 thousand equal? (Signal.) *9×10^3.*
• What does 6 hundred equal? (Signal.) *6×10^2.*
• What does 6 thousand equal? (Signal.) *6×10^3.*

(Repeat until firm.)

d. Remember, if the number has 3 zeros, it's the first part times 10 to the third.

WORKBOOK PRACTICE

a. Find part 3 in your workbook. ✔
(Teacher reference:)

a. 8000 = _____   d. 13,000 = _____

b. 200 = _____   e. 700 = _____

c. 9000 = _____   f. 500 = _____

These are hundreds numbers and thousands numbers.

b. Number A is 8 thousand. Write what it equals. ✔
(Display:)   [69:4F]

> **a.** $8000 = 8 \times 10^3$

Here's what you should have.

c. Write what the other numbers equal.
(Observe students and give feedback.)

d. Check your work.

• Number B is 200. What does it equal?
(Signal.) 2×10^2.

• C is 9000. What does it equal? (Signal.)
9×10^3.

• D is 13 thousand. What does it equal?
(Signal.) 13×10^3.

• E is 700. What does it equal? (Signal.) 7×10^2.

• F is 500. What does it equal? (Signal.) 5×10^2.
(Add to show:)   [69:4G]

a. $8000 = 8 \times 10^3$	**d.** $13,000 = 13 \times 10^3$
b. $200 = 2 \times 10^2$	**e.** $700 = 7 \times 10^2$
c. $9000 = 9 \times 10^3$	**f.** $500 = 5 \times 10^2$

Here's what you should have.

EXERCISE 5: SHORT DIVISION
WITH MENTAL MATH

a. Find part 4 in your workbook. ✔
(Teacher reference:)

a. $3\overline{)135}$ b. $5\overline{)95}$ c. $4\overline{)152}$ d. $2\overline{)74}$

• Work all the problems.
(Observe students and give feedback.)

b. Check your work. Read each problem and the answer.

• Problem A. (Signal.) $135 \div 3 = 45$.

• B. (Signal.) $95 \div 5 = 19$.

• C. (Signal.) $152 \div 4 = 38$.

• D. (Signal.) $74 \div 2 = 37$.
(Display:)   [69:5A]

a. $3\overline{)13_15}$ $\quad\dfrac{45}{}$ b. $5\overline{)9_45}$ $\quad\dfrac{19}{}$

c. $4\overline{)15_32}$ $\quad\dfrac{38}{}$ d. $2\overline{)7_14}$ $\quad\dfrac{37}{}$

Here's what you should have for each problem.

EXERCISE 6: VOLUME
OF RECTANGULAR PRISMS (UNIT CUBE)

a. (Display:)   [69:6A]

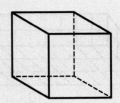

• You learned about cubic units. How many faces does a cubic unit have? (Signal.) 6.
All the faces are the same size.

b. (Display:)   [69:6B]

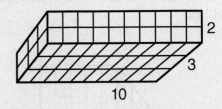

Cubic units = Area of the base × height

Here's a rectangular prism that has layers of cubic units. Cubic units = Area of the base × height.

• Say the equation for cubic units. (Signal.)
Cubic units = Area of the base × height.
(Repeat until firm.)

c. Say the multiplication for the area of the base.
    (Signal.) *10 × 3.*
•   What's the area of the base? (Signal.) *30.*
    (Add to show:)   [69:6C]

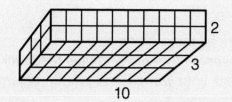

Cubic units = Area of the base × height
          =   30

•   What's the height of the prism? (Signal.) *2.*
•   So what do you multiply 30 by? (Signal.) *2.*
    (Add to show:)   [69:6D]

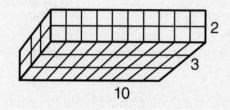

Cubic units = Area of the base × height
          =   30   × 2

•   What's 30 × 2? (Signal.) *60.*
    (Add to show:)   [69:6E]

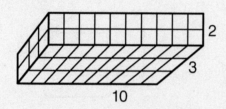

Cubic units = Area of the base × height
          =   30   × 2
          =   60

•   So how many cubic units are in the prism?
    (Signal.) *60.*
d. (Display:)   [69:6F]

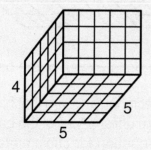

Cubic units = Area of the base × height

Here's a new rectangular prism.

┌─• Say the equation for finding the cubic units.
│   (Signal.) *Cubic units = Area of the base × height.*
└─   (Repeat until firm.)
e. Say the multiplication for the area of the base.
    (Signal.) *5 × 5.*
•   What's the area of the base? (Signal.) *25.*
    (Add to show:)   [69:6G]

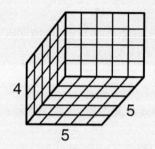

Cubic units = Area of the base × height
          =   25

•   What's the height of the prism? (Signal.) *4.*
•   Say the multiplication for the area of the base
    times the height. (Signal.) *25 × 4.*
    (Add to show:)   [69:6H]

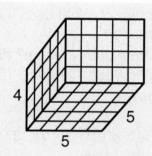

Cubic units = Area of the base × height
          =   25   × 4

Yes, 25 × 4.
•   What's the number of cubic units? (Signal.) *100.*
    (Add to show:)   [69:6I]

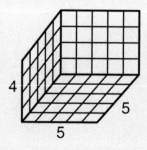

Cubic units = Area of the base × height
          =   25   × 4
          =   100

Yes, there are 100 cubic units in the prism.

f. (Display:) [69:6J]

> Cubic units = (A_b) h

Here's the equation for cubic units.
(Point to **(A_b) h.**)
Ab × H. That's area of the base times height.
- Say the equation for cubic units. (Signal.)
Cubic units = Area of the base × height.

=========== **WORKBOOK PRACTICE** ===========

a. Find part 5 in your workbook. ✔
(Teacher reference:)

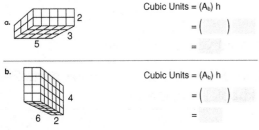

Cubic Units = (A_b) h

= ( )

=

Cubic Units = (A_b) h

= ( )

=

You can see the equation for cubic units.
b. Problem A. You'll write the area of the base under Ab.
- Say the multiplication for the area of the base. (Signal.) *5 × 3.*
- What's the answer? (Signal.) *15.*
- Write 15 for area of the base. ✔
- What's the height of the prism? (Signal.) *2.*
- Write 2 for the height. ✔
- Say the multiplication for area of the base times height. (Signal.) *15 × 2.*
- Show the multiplication and write the answer below.
(Observe students and give feedback.)
- Everybody, what's the number of cubic units in the prism? (Signal.) *30.*
(Add to show:) [69:6K]

> **a.** Cubic units = (A_b) h
>
> = (15) 2
>
> = 30

Here's what you should have.

c. Problem B. You'll write the area of the base under Ab.
- Say the multiplication for the area of the base. (Signal.) *6 × 2.*
- What's the answer? (Signal.) *12.*
- Write 12 for the area of the base. ✔
- Everybody, what's the height of the prism? (Signal.) *4.*
- Write 4 for the height. ✔
- Say the multiplication for area of the base times height. (Signal.) *12 × 4.*
- Show the multiplication and write the answer below.
(Observe students and give feedback.)
- Everybody, what's the number of cubic units in the prism? (Signal.) *48.*
(Display:) [69:6L]

> **b.** Cubic units = (A_b) h
>
> = (12) 4
>
> = 48

Here's what you should have.

EXERCISE 7: ROUNDING
DECIMALS REMEDY

a. (Display:) [69:7A]

> Round .0396 to the tenths place.

We'll round this number to the tenths place.
- Does zero round up? (Signal.) *No.*
- So what do we write in the tenths place? (Signal.) *Zero.*
(Add to show:) [69:7B]

> Round .0396 to the tenths place.
> .0

- Read the rounded value. (Signal.) *Zero tenths.*

b. (Display:) [69:7C]

> Round .2048 to the hundredths place.

We'll round this number to the hundredths place.

- Does zero round up? (Signal.) *No.*
  So the rounded value is 20 hundredths.
  (Add to show:) [69:7D]

> Round .2048 to the hundredths place.
> .20

- Read the rounded value. (Signal.) *20 hundredths.*

═══ TEXTBOOK PRACTICE ═══

a. Open your textbook to Lesson 69 and find part 1. ✔
   (Teacher reference:)

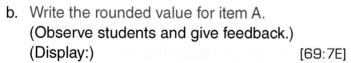

a. Round .0048 to the hundredths place.
b. Round .0048 to the tenths place.
c. Round .0048 to the thousandths place.

b. Write the rounded value for item A.
   (Observe students and give feedback.)
   (Display:) [69:7E]

> a. .00

Here's what you should have, zero hundredths. That's two zeros after the decimal point.

c. Write the rounded value for item B. ✔
- Everybody, read the rounded value. (Signal.) *Zero tenths.*
  (Display:) [69:7F]

> b. .0

Yes, zero tenths. That's one zero after the decimal point.

d. Write the rounded value for item C. ✔
- Everybody, read the rounded value. (Signal.) *5 thousandths.*
  (Display:) [69:7G]

> c. .005

Here's what you should have.

> Assign Independent Work, Textbook parts 2–8 and Workbook parts 6 and 7.

Optional extra math-fact practice worksheets are available on ConnectED.

Lesson 70

EXERCISE 1: EQUIVALENT FRACTIONS `REMEDY`

a. (Display:)   [70:1A]

$$\frac{1}{2} = \frac{}{10}$$

• (Point to $\frac{1}{2}$.) How many times bigger is the denominator than the numerator? (Signal.) *2 times.*
• So how many tenths equal 1/2? (Signal.) *5 tenths.*
  (Add to show:)   [70:1B]

$$\frac{1}{2} = \frac{5}{10}$$

Yes, 5/10 = 1/2.

b. (Display:)   [70:1C]

$$\frac{1}{2} = \frac{}{12}$$

• Raise your hand when you know how many 12ths equal 1/2. ✔
• How many 12ths equal 1/2? (Signal.) *6/12.*
  (Add to show:)   [70:1D]

$$\frac{1}{2} = \frac{6}{12}$$

Yes, 6/12 = 1/2.

c. (Display:)   [70:1E]

$$\frac{1}{2} = \frac{}{16}$$

• Raise your hand when you know how many 16ths equal 1/2. ✔
• How many 16ths equal 1/2? (Signal.) *8/16.*
  (Add to show:)   [70:1F]

$$\frac{1}{2} = \frac{8}{16}$$

Yes, 8/16 = 1/2.

d. (Display:)   [70:1G]

$$\frac{1}{5} = \frac{}{30}$$

• Raise your hand when you know how many 30ths equal 1/5. ✔
• How many 30ths equal 1/5? (Signal.) *6/30.*
  (Add to show:)   [70:1H]

$$\frac{1}{5} = \frac{6}{30}$$

Yes, 6/30 = 1/5.

e. (Display:)   [70:1I]

$$\frac{1}{3} = \frac{}{18}$$

• Raise your hand when you know how many 18ths equal 1/3. ✔
• How many 18ths equal 1/3? (Signal.) *6/18.*
  (Add to show:)   [70:1J]

$$\frac{1}{3} = \frac{6}{18}$$

Yes, 6/18 = 1/3.

f. (Display:)   [70:1K]

$$\frac{1}{9} = \frac{}{36}$$

• Raise your hand when you know how many 36ths equal 1/9. ✔
• How many 36ths equal 1/9? (Signal.) *4/36.*
  (Add to show:)   [70:1L]

$$\frac{1}{9} = \frac{4}{36}$$

Yes, 4/36 = 1/9.

g. (Display:)   [70:1M]

$$\frac{1}{4} = \frac{}{24}$$

• Raise your hand when you know how many 24ths equal 1/4. ✔
• How many 24ths equal 1/4? (Signal.) *6/24.*
  (Add to show:)   [70:1N]

$$\frac{1}{4} = \frac{6}{24}$$

Yes, 6/24 = 1/4.

===== **WORKBOOK PRACTICE** =====

a. Open your workbook to Lesson 70 and find part 1. ✔
   (Teacher reference:)   R Part J

  a. $\frac{1}{4} = \frac{}{12}$ b. $\frac{1}{9} = \frac{}{27}$ c. $\frac{1}{6} = \frac{}{48}$

  d. $\frac{1}{6} = \frac{}{18}$ e. $\frac{1}{10} = \frac{}{60}$ f. $\frac{1}{5} = \frac{}{20}$

b. Problem A: 1/4 = how many 12ths?
• Complete the fraction. ✔
• Everybody, read the fraction equation. (Signal.) *1/4 = 3/12.*
c. Complete the rest of the equations in part 1. (Observe students and give feedback.)
d. Check your work. Read each equation.
• Equation B. (Signal.) *1/9 = 3/27.*
• C. (Signal.) *1/6 = 8/48.*
• D. (Signal.) *1/6 = 3/18.*
• E. (Signal.) *1/10 = 6/60.*
• F. (Signal.) *1/5 = 4/20.*
   (Display:)   [70:10]

  a. $\frac{1}{4} = \frac{3}{12}$ **b.** $\frac{1}{9} = \frac{3}{27}$ **c.** $\frac{1}{6} = \frac{8}{48}$

  d. $\frac{1}{6} = \frac{3}{18}$ **e.** $\frac{1}{10} = \frac{6}{60}$ **f.** $\frac{1}{5} = \frac{4}{20}$

Here's what you should have.

EXERCISE 2: SHORT DIVISION
WITH MENTAL MATH

a. (Display:)   [70:2A]

$$3\overline{)7\,3\,5}$$

Here's a problem that will have three digits in the answer.
• Read the problem. (Signal.) *735 ÷ 3.*
   We'll work it the fast way.
b. Say the first problem we'll work. (Signal.) *7 ÷ 3.*
• What's the answer? (Signal.) *2 with a remainder of 1.*
   (Add to show:)   [70:2B]

$$3\overline{)7_13\,5}^{\,2}$$

c. Say the next problem we'll work. (Signal.) *13 ÷ 3.*
• What's the answer? (Signal.) *4 with a remainder of 1.*
   (Add to show:)   [70:2C]

$$3\overline{)7_13_15}^{\,2\,4}$$

d. Say the last problem. (Signal.) *15 ÷ 3.*
• What's the answer? (Signal.) *5.*
   (Add to show:)   [70:2D]

$$3\overline{)7_13_15}^{\,2\,4\,5}$$

• What does 735 ÷ 3 equal? (Signal.) *245.*

===== **WORKBOOK PRACTICE** =====

a. Find part 2 in your workbook. ✔
   (Teacher reference:)

  a. $5\overline{)8\,2\,5}$ b. $3\overline{)7\,3\,2}$ c. $4\overline{)3\,5\,7\,6}$

b. Read problem A. (Signal.) *825 ÷ 5.*
• Say the first problem you'll work. (Signal.) *8 ÷ 5.*
• What's the answer? (Signal.) *1 with a remainder of 3.*
• Write 1 and the remainder. Then complete the problem. (Observe students and give feedback.)
c. Check your work.
• What's the answer to the whole problem? (Signal.) *165.*
   (Display:)   [70:2E]

$$\textbf{a.}\ \ 5\overline{)8_32_25}^{\,1\,6\,5}$$

Here's what you should have.
The first problem you worked is 8 ÷ 5.
The next problem you worked is 32 ÷ 5.
The last problem you worked is 25 ÷ 5.
d. Read problem B. (Signal.) *732 ÷ 3.*
• Work the whole problem. (Observe students and give feedback.)

e. Check your work.
- What's the answer to the whole problem? (Signal.) *244.*
  (Display:) [70:2F]

$$3 \overline{)7_13_12} \quad \begin{array}{c} 2\ 4\ 4 \end{array}$$
b.

Here's what you should have.
- Say the first problem you worked. (Signal.) *7 ÷ 3.*
- Say the next problem you worked. (Signal.)
  13 ÷ 3.
- Say the last problem you worked. (Signal.)
  12 ÷ 3.
f. Read problem C. (Signal.) *3576 ÷ 4.*
- Work the problem.
  (Observe students and give feedback.)
g. Check your work.
- What's the answer to the problem? (Signal.) *894.*
  (Display:) [70:2G]

$$4 \overline{)3\ 5_37_16} \quad \begin{array}{c} 8\ 9\ 4 \end{array}$$
c.

Here's what you should have.
- Say the first problem you worked. (Signal.)
  35 ÷ 4.
- Say the next problem you worked. (Signal.)
  37 ÷ 4.
- Say the last problem you worked. (Signal.)
  16 ÷ 4.

EXERCISE 3: EXPONENTS
BASE 10

a. (Display:) [70:3A]

7000

Last time, you learned that numbers like 7000 can be written as a number times 10 with an exponent. 7 thousand has 3 zeros. So it equals 7 times 10 to the third.
- What does 7 thousand equal? (Signal.)
  7×10^3.
b. (Display:) [70:3B]

700

- How many zeros does 7 hundred have? (Signal.) *2.*

- So what does 700 equal? (Signal.) 7×10^2.
  Yes, 700 equals 7 times 10 to the second.
c. (Display:) [70:3C]

700,000

- Read this number. (Signal.) *700,000.*
- Raise your hand when you know how many zeros it has. ✔
- How many zeros are in 700,000? (Signal.) *5.*
- So what does 700,000 equal? (Signal.)
  7×10^5.
d. (Display:) [70:3D]

70,000

- Read this number. (Signal.) *70,000.*
- How many zeros does it have? (Signal.) *4.*
- So what does 70 thousand equal? (Signal.)
  7×10^4.
e. (Display:) [70:3E]

45,000

- Read this number. (Signal.) *45,000.*
- How many zeros does it have? (Signal.) *3.*
- So what does 45,000 equal? (Signal.)
  45×10^3.
f. (Display:) [70:3F]

2100

- Read this number. (Signal.) *2100.*
- How many zeros does it have? (Signal.) *2.*
- So what does 2100 equal? (Signal.) 21×10^2.
g. (Display:) [70:3G]

360

- Read this number. (Signal.) *360.*
- How many zeros does it have? (Signal.) *1.*
- So what does 360 equal? (Signal.) 36×10^1.

=== **WORKBOOK PRACTICE** ===

a. Find part 3 in your workbook. ✔
  (Teacher reference:)

  a. 60,000 = _____   d. 8,100,000 = _____

  b. 17,000 = _____   e.   470 = _____

  c. 12,300 = _____   f. 3,000,000 = _____

  You're going to write what each number equals.

b. Read number A. (Signal.) *60,000.*
- How many zeros does it have? (Signal.) *4.*
- So what does it equal? (Signal.) *6 × 10⁴.*
- Write what 60,000 equals. ✔
  (Display:) [70:3H]

  a. $60,000 = 6 \times 10^4$

Here's what you should have.
c. Read number B. (Signal.) *17,000.*
- How many zeros does it have? (Signal.) *3.*
- Write what it equals. ✔
- Everybody, what does 17,000 equal? (Signal.) *17 × 10³.*
d. Complete the rest of the equations.
  (Observe students and give feedback.)
e. Check your work.
- Number C: 12,300. What does it equal? (Signal.) *123 × 10².*
- D: 8,100,000. What does it equal? (Signal.) *81 × 10⁵.*
- E: 470. What does it equal? (Signal.) *47 × 10¹.*
- F: 3,000,000. What does it equal? (Signal.) *3 × 10⁶.*

EXERCISE 4: COORDINATE SYSTEM
RATIO SENTENCES

a. Find part 4 in your workbook. ✔
  (Teacher reference:)

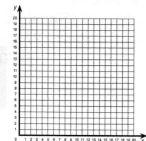
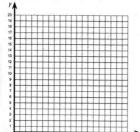

a. There were 2 cakes for every 9 forks.
- There were 4 cakes and 18 forks.

b. The ratio of chickens to roosters is 3 to 2.
- There were 12 chickens and 8 roosters.

Each item tells about two points on a line.
b. Item A: There were 2 cakes for every 9 forks.
- Say the Y over X equation. (Signal.) *Y over X = 2/9.*
  (Repeat until firm.)
- What's the name for Y? (Signal.) *Cakes.*
- What's the name for X? (Signal.) *Forks.*
- Write the names. Then make the point for Y over X equals 2/9.
  (Observe students and give feedback.)

(Display:) [70:4A]

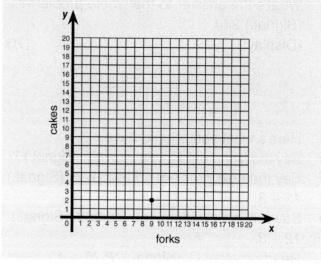

Here's what you should have.
c. The next sentence tells about another point on the line: There were 4 cakes and 18 forks.
- Make the point for 4 cakes and 18 forks. Then draw the line. The line goes through both points and goes through zero.
  (Observe students and give feedback.)
  (Add to show:) [70:4B]

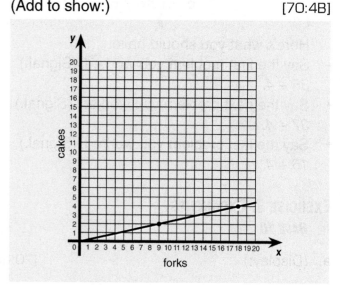

Here's what you should have.
d. Touch item B. ✔
  Item B says: The ratio of chickens to roosters is 3 to 2.
- Say the Y over X equation. (Signal.) *Y over X = 3/2.*
- What's the name for Y? (Signal.) *Chickens.*
- What's the name for X? (Signal.) *Roosters.*
- Write the names. Then make the point for Y over X equals 3/2.
  (Observe students and give feedback.)

(Display:) [70:4C]

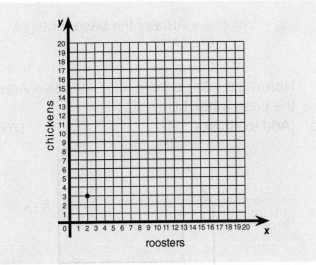

Here's what you should have.

e. The next sentence says: There were 12 chickens and 8 roosters.

• Make the point for 12 chickens and 8 roosters. Then draw the line. Remember, the line goes through both points and goes through zero.
(Observe students and give feedback.)
(Add to show:) [70:4D]

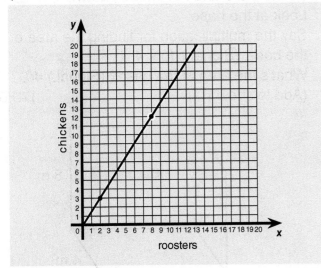

Here's what you should have.

EXERCISE 5: DIVISION SIGN

a. (Display:) [70:5A]

$$\frac{15}{3}$$

• Read this problem. (Signal.) *15/3.*
• Read the fraction as a division problem. (Signal.) *15 ÷ 3.*

b. (Display:) [70:5B]

$$15 \div 3$$

Here's the same problem with a new divide-by sign. It looks like a fraction with a dot on top and a dot on the bottom.
I'll read this problem: 15 divided by 3.
• Your turn: Read the problem. (Signal.) *15 ÷ 3.*
• What's the answer? (Signal.) *5.*
(Add to show:) [70:5C]

$$15 \div 3 = 5$$

• Read the equation. (Signal.) *15 divided by 3 = 5.*

c. (Display:) [70:5D]

$$32 \div 8$$

• Read the problem. (Signal.) *32 ÷ 8.*
• What's the answer? (Signal.) *4.*
(Add to show:) [70:5E]

$$32 \div 8 = 4$$

• Read the equation. (Signal.) *32 ÷ 8 = 4.*

d. (Display:) [70:5F]

$$96 \div 12$$

• Read the problem. (Signal.) *96 ÷ 12.*
The answer is 8.
(Add to show:) [70:5G]

$$96 \div 12 = 8$$

• Read the equation. (Signal.) *96 ÷ 12 = 8.*

e. (Display:) [70:5H]

$$\frac{3}{4} \div \frac{2}{3}$$

• Read the problem. (Signal.) *3/4 divided by 2/3.*
The answer is 9/8.
(Add to show:) [70:5I]

$$\frac{3}{4} \div \frac{2}{3} = \frac{9}{8}$$

f. In a few lessons, you're going to learn how to work problems that divide by a fraction.

a. Open your textbook to Lesson 70 and find part 1. ✔
* Pencils down. ✔
  (Teacher reference:)

a. 56 ÷ 7	b. 27 ÷ 9	c. 45 ÷ 5	d. 90 ÷ 9
e. 64 ÷ 8	f. 36 ÷ 4	g. 72 ÷ 9	h. 28 ÷ 4

Part 1
a. ■ ÷ ■ = ■

b. Read problem A. (Signal.) *56 divided by 7.*
* What's the answer? (Signal.) *8.*
c. Read problem B. (Signal.) *27 divided by 9.*
* What's the answer? (Signal.) *3.*
d. Read problem C. (Signal.) *45 divided by 5.*
* What's the answer? (Signal.) *9.*
  (Repeat until firm.)
e. Copy all the problems in part 1 and write the answers.
  (Observe students and give feedback.)
f. Check your work. Read each equation.
* Problem A. (Signal.) *56 ÷ 7 = 8.*
* B. (Signal.) *27 ÷ 9 = 3.*
* C. (Signal.) *45 ÷ 5 = 9.*
* D. (Signal.) *90 ÷ 9 = 10.*
* E. (Signal.) *64 ÷ 8 = 8.*
* F. (Signal.) *36 ÷ 4 = 9.*
* G. (Signal.) *72 ÷ 9 = 8.*
* H. (Signal.) *28 ÷ 4 = 7.*
g. Remember, this division sign looks like a fraction with a dot on top and a dot on the bottom.

EXERCISE 6: VOLUME
OF RECTANGULAR PRISMS (EQUATION) [REMEDY]

a. (Display:) [70:6A]

> Volume

* Here's a new word: Volume. Say **volume.** (Signal.) *Volume.*
  When you find the number of cubic units in a rectangular prism, you find the volume.
* What do you find? (Signal.) *The volume.*
  (Add to show:) [70:6B]

> Volume = Area of the base × height

Here's the equation for finding the volume of a rectangular prism: Volume = Area of the base × height.

* Say the equation. (Signal.) *Volume = Area of the base × height.*

(Add to show:) [70:6C]

> Volume = Area of the base × height
> $V = (A_b) h$

Here's the letter equation for Volume = Area of the base × height.

b. (Add to show:) [70:6D]

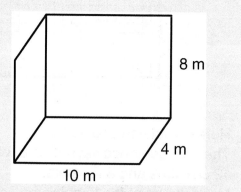

Volume = Area of the base × height
$V = (A_b) h$

You can't see the cubic units, but the numbers for the base and height are shown.
Look at the base.
* Say the multiplication for finding the area of the base. (Signal.) *10 × 4.*
* What's the area of the base? (Signal.) *40.*
  (Add to show:) [70:6E]

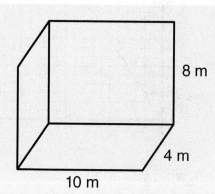

Volume = Area of the base × height
$V = (A_b) h$
$V = (40)$

c. Now we multiply the area of the base times the height to find the cubic units.
- The area of the base is 40. What's the height? (Signal.) *8.*
  (Add to show:) [70:6F]

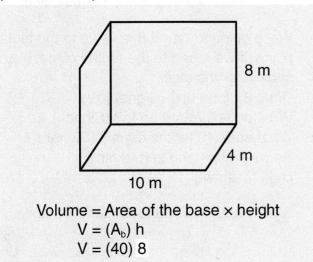

Volume = Area of the base × height
$V = (A_b) h$
$V = (40) 8$

- Say the multiplication for finding the volume. (Signal.) *40 × 8.*
- Raise your hand when you know the answer. ✔
- What's 40 × 8? (Signal.) *320.*
  So the volume of the prism is 320 cubic meters.
  (Add to show:) [70:6G]

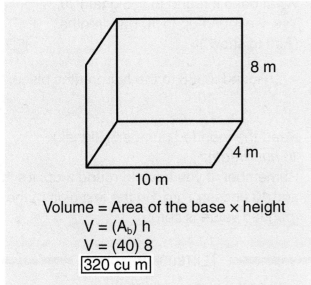

Volume = Area of the base × height
$V = (A_b) h$
$V = (40) 8$
$\boxed{320 \text{ cu m}}$

- What's the volume of the prism? (Signal.) *320 cubic meters.*

a. Find part 2 in your textbook. ✔
   (Teacher reference:)

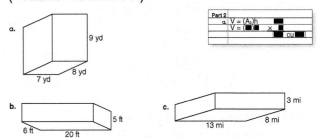

You'll find the volume of each rectangular prism. Listen: If the units are yards, the units for the **volume** of the prism are cubic yards.
- If the units are miles, what are the units for volume? (Signal.) *Cubic miles.*
- If the units are feet, what are the units for volume? (Signal.) *Cubic feet.*
- Say the equation for volume. (Signal.) *Volume = Area of the base × height.*
  (Repeat until firm.)

b. Figure A. Copy the letter equation for volume. Below, put in numbers for the area of the base and the height.
- Raise your hand when you have the equation with a number for area of the base and a number for height. ✔
- You multiplied 7 × 8 to find the area of the base. What's the number for the area of the base? (Signal.) *56.*
- What's the height? (Signal.) *9.*
  (Display:) [70:6H]

a. $V = (A_b) h$
   $V = (56) 9$

Here's what you should have.

c. Figure out the volume and write the answer with the unit name **cubic yards.**
   (Observe students and give feedback.)
- Everybody, what's the volume of figure A? (Signal.) *504 cubic yards.*
  (Add to show:) [70:6I]

a. $V = (A_b) h$
   $V = (56) 9$

$$\begin{array}{r} \overset{5}{5}\,6 \\ \times\quad 9 \\ \hline \boxed{5\,0\,4 \text{ cu yd}} \end{array}$$

Here's what you should have.

d. Figure B. Copy the letter equation. Below, put in numbers for the area of the base and the height.
   Stop when you've done that much. ✔

e. Check your work.
- For the base, you multiplied 20 × 6. What's the number for the area of the base? (Signal.) *120.*
- What's the height? (Signal.) *5.*
(Display:) [70:6J]

> **b.** $V = (A_b)\ h$
> $V = (120)\ 5$

Here's what you should have.

f. Find the volume of figure B.
(Observe students and give feedback.)

g. Check your work.
- What's 120 × 5? (Signal.) *600.*
- What's the unit name in the answer? (Signal.) *Cubic feet.*
(Add to show:) [70:6K]

> **b.** $V = (A_b)\ h$ $\begin{array}{r} \overset{1}{1}\ 2\ 0 \\ \times \quad 5 \\ \hline \boxed{6\ 0\ 0}\ \text{cu ft} \end{array}$
> $V = (120)\ 5$

Here's what you should have.

h. Figure C. Copy the letter equation. Below, put in numbers for the area of the base and the height.
Stop when you've done that much. ✔

i. Check your work.
- What's the area of the base? (Signal.) *104.*
- What's the height? (Signal.) *3.*
(Display:) [70:6L]

> **c.** $V = (A_b)\ h$
> $V = (104)\ 3$

Here's what you should have.

j. Find the volume of figure C.
(Observe students and give feedback.)

k. Check your work.
- What's 104 × 3? (Signal.) *312.*
- What's the unit name in the answer? (Signal.) *Cubic miles.*
(Add to show:) [70:6M]

> **c.** $V = (A_b)\ h$ $\begin{array}{r} \overset{1}{1}\ 0\ 4 \\ \times \quad 3 \\ \hline \boxed{3\ 1\ 2}\ \text{cu mi} \end{array}$
> $V = (104)\ 3$

Here's what you should have.

EXERCISE 7: ROUNDING
DECIMALS WITH 9

a. (Display:) [70:7A]

> Round .9638 to the tenths place.

We're going to round this value to the tenths place. If we round to the tenths place, we must show a tenths digit.
- Does 9 round up? (Signal.) *Yes.*
- What does it round up to? (Signal.) *10.*
- What does 10 tenths equal? (Signal.) *1.*
So we write 1 and zero tenths.
(Add to show:) [70:7B]

> Round .9638 to the tenths place.
> 1.0

- Read the rounded number. (Signal.) *1 and zero tenths.*

b. (Display:) [70:7C]

> Round .0988 to the hundredths place.

- What's the hundredths digit? (Signal.) *9.*
- Do we round 9 up? (Signal.) *Yes.*
- What does it round to? (Signal.) *10.*
Yes, we round up to 10 hundredths.
(Add to show:) [70:7D]

> Round .0988 to the hundredths place.
> .10

- Read the rounded number. (Signal.) *10 hundredths.*

c. Remember, if you have to round 9 up, it's 10, and 10 ends in zero. So the last digit of the rounded value is zero.

━━━━━━━━━ **TEXTBOOK PRACTICE** ━━━━━━━━━

a. Find part 3 in your textbook. ✔
(Teacher reference:)

> **a.** Round .9876 to the tenths place.
> **b.** Round .0095 to the thousandths place.
> **c.** Round .0962 to the hundredths place.

For all these items, the 9 rounds up.

b. Write the rounded value for item A. Remember, the last digit is zero. (Observe students and give feedback.) (Display:) [70:7E]

> **a. 1.0**

Here's what you should have.
- Read the rounded value. (Signal.) *1 and zero tenths.*

c. Item B. Write the rounded value. Remember, the last digit is zero. ✔ (Display:) [70:7F]

> **b. .010**

Here's what you should have.
- Read the rounded value. (Signal.) *10 thousandths.*

d. Item C. Write the rounded value. Remember, the last digit is zero. ✔ (Display:) [70:7G]

> **c. .10**

Here's what you should have.
- Read the rounded value. (Signal.) *10 hundredths.*

EXERCISE 8: WORD PROBLEMS
RATIO/FRACTION OF A GROUP

a. For some word problems, you write ratio equations. For other problems, you write a fraction times something.

b. (Display:) [70:8A]

> $\frac{3}{5}$ of the girls were laughing.

- Does this sentence tell about a fraction times something? (Signal.) *Yes.*
  Yes, 3/5 of the girls is 3/5 times the girls.
- Say the equation for this sentence. (Signal.) *3/5 G = L.*

c. (Display:) [70:8B]

> $\frac{2}{3}$ of the ducks are swimming.

- Does this sentence tell about a fraction times something? (Signal.) *Yes.*
  Yes, 2/3 of the ducks is 2/3 times the ducks.
- Say the equation. (Signal.) *2/3 D = S.*

d. (Display:) [70:8C]

> There were 2 ducks for every 5 fish.

- Does this sentence tell about a fraction times something? (Signal.) *No.*
  Right, this sentence tells about a ratio.

e. (Display:) [70:8D]

> There were 9 boys for every 2 dogs.

- Does this sentence tell about a fraction times something? (Signal.) *No.*
  Right. It tells about a ratio.
- What are the letters for the ratio? (Signal.) *B and D.*
  (Add to show:) [70:8E]

> There were 9 boys for every 2 dogs.
>
> $\frac{b}{d}$

- What are the numbers for the ratio? (Signal.) *9 and 2.*
  (Add to show:) [70:8F]

> There were 9 boys for every 2 dogs.
>
> $\frac{b}{d}$ $\frac{9}{2}$

Here's the ratio.

f. (Display:) [70:8G]

> There were 6 birds for every 9 geese.

- Does this sentence tell about a fraction times something? (Signal.) *No.*
- Write the ratio fraction with letters and with numbers in part A. ✔
  (Add to show:) [70:8H]

> There were 6 birds for every 9 geese.
>
> $\frac{b}{g}$ $\frac{6}{9}$

Here's what you should have.

TEXTBOOK PRACTICE

a. Find part 4 in your textbook. ✔
(Teacher reference:)

a. There were 4 black bricks for every 3 red bricks.
b. $\frac{7}{10}$ of the animals were hungry.
c. There were 6 pumpkins for every 11 napkins.
d. There were 5 girls for every 9 children.
e. $\frac{3}{8}$ of the children were girls.

Some sentences tell about a fraction times something. Some tell about a ratio.

b. Sentence A: There were 4 black bricks for every 3 red bricks.

• Does that sentence tell about a fraction times something or about a ratio fraction? (Signal.)
A ratio fraction.

• Write the fraction with letters and the fraction with numbers. ✔
(Display:) [70:8I]

$$\text{a. } \frac{b}{r} \quad \frac{4}{3}$$

Here's what you should have.

c. Sentence B: 7/10 of the animals were hungry.

• Does that sentence tell about a fraction times something or about a ratio fraction? (Signal.)
A fraction times something.

• Write the equation. ✔
(Display:) [70:8J]

$$\text{b. } \frac{7}{10}a = h$$

Here's what you should have.

d. Write ratio fractions or fraction equations for the rest of the items.
(Observe students and give feedback.)

e. Check your work.

• Sentence C: There were 6 pumpkins for every 11 napkins.
(Display:) [70:8K]

$$\text{c. } \frac{p}{n} \quad \frac{6}{11}$$

Here's what you should have.

• Sentence D: There were 5 girls for every 9 children.
(Display:) [70:8L]

$$\text{d. } \frac{g}{c} \quad \frac{5}{9}$$

Here's what you should have.

• Sentence E: 3/8 of the children were girls.
(Display:) [70:8M]

$$\text{e. } \frac{3}{8}c = g$$

Here's what you should have.

EXERCISE 9: INDEPENDENT WORK
DECIMAL OPERATIONS—DIVISION

a. Find part 5 in your workbook. ✔
(Teacher reference:)

a. $.03\overline{)27.9}$ b. $.8\overline{)4.88}$ c. $.09\overline{)8109}$

b. Read problem A. (Signal.) *27.9 ÷ .03.*
Remember how to work these problems. Move the decimal points in both numbers so you are dividing by a whole number. Show the new decimal point in the answer. Make sure you have a digit in every place of the problem. Stop when you've done that much. ✔

• Everybody, read the problem that divides by a whole number. (Signal.) *2790 ÷ 3.*
(Display:) [70:9A]

$$\text{a. } 03.\overline{)27\,90.}$$

Here's what you should have.
You'll work the problem as part of your independent work.

c. Read problem B. (Signal.) *4.88 ÷ .8.*

• Move the decimal points in both numbers so you are dividing by a whole number. Show the new decimal point in the answer. Make sure you have a digit in every place of the problem. Stop when you've done that much. ✔

• Everybody, read the problem that divides by a whole number. (Signal.) *48.8 ÷ 8.*
(Display:) [70:9B]

$$\text{b. } 8.\overline{)4\,8.8}$$

Here's what you should have.

Assign Independent Work, Workbook parts 5 and 6 and Textbook parts 5–12.

Optional extra math-fact practice worksheets are available on ConnectED.

Mastery Test 7

Teacher Presentation

a. Find Test 7 in your test booklet. ✔
   You'll work the test by yourself. Read the directions carefully and do your best work. Put your pencil down when you've finished the test. (Observe but do not give feedback.)

Scoring Notes

a. Collect test booklets. Use the *Answer Key* and Passing Criteria Table to score the tests.

Passing Criteria Table—Mastery Test 7			
Part	Score	Possible Score	Passing Score
1	1 for each item	4	3
2	1 for each digit in the answer	6	5
3	1 for each digit in the answer	10	8
4	1 for each item	3	3
5	1 for each item	3	3
6	1 for each fraction 1 for each point 1 for the line	10	8
7	1 for each item	4	4
8	1 for each item	3	3
9	1 for each fraction	4	3
10	2 for each item (correct digits, correct decimal point (optional for c))	6	5
11	1 for each item	3	3
12	1 for each item (repeated multiplication, answer)	6	5
	Total	62	

b. Complete the Mastery Test 7 Remedy Summary Sheet to determine whether group remedies are needed. Reproducible Remedy Summary Sheets are at the back of the *Answer Key* and the back of the *Teacher's Guide.*

• If ¼ or more of the students did not pass a test part, present the remedy for that part before beginning Lesson 71. The Remedy Table follows and also appears at the end of the Mastery Test 7 *Answer Key.* Remedies worksheets follow Mastery Test 7 in the *Student Assessment Book.*

Part	Test Items	Remedy Lesson	Remedy Ex.	Remedies Worksheet	Textbook
		65	4	Part A	—
1	Rounding	66	7	—	Part 2
		67	—	—	Part 10
2	Short Division (With Mental Math)	67	4	Part B	—
		68	4	Part C	—
3	Subtraction (Renaming with Zeros)	63	5	—	Part 2
		64	6	—	Part 2
4	Fraction Multiplication (Cross Cancellation)	61	4	—	Part 1
		62	2	—	Part 1
5	Exponents for Repeated Multiplication	61	5	—	Part 2
		62	3	—	Part 2
6	Coordinate System (Ratio and Proportion)	65	2	Part D	—
		66	2	Part E	—
7	Exponents (Base 10)	67	6	—	Part 2 Part 3*
		68	7	—	Part 3
8	Rounding (Decimals)	68	5	—	Part 1
		69	7	—	Part 1
9	Equivalent Fractions	61	3	Part F	—
		62	4	Part G	—
10	Decimal Division	62	5	Part H	—
		68	2	Part I	—
11	Equivalent Fractions	69	1	—	—
		70	1	Part J	—
12	Exponents (Evaluation)	63	7	—	Part 3
		64	7	—	Part 3

*For Lesson 67, Exercise 6, there are two Textbook parts for one exercise.

Retest
Retest individual students on any part failed.

	Lesson 71	Lesson 72	Lesson 73	Lesson 74	Lesson 75
Student Learning Objectives	**Exercises** 1. **Compare fractions** 2. Solve division problems using short division and mental math 3. **Solve ratio word problems, plot them on a coordinate system, and draw the line** 4. Recognize the division sign (÷) 5. Write equations with powers of 10 6. Find the volume of rectangular prisms using the formula 7. Round decimals that have 9 as a digit 8. Set up ratio and fraction-of-a-group word problems Complete work independently	**Exercises** 1. Compare fractions 2. Solve ratio word problems, plot them on a coordinate system, and draw the line 3. **Find the area of circles** 4. Solve division problems using short division and mental math 5. **Solve multiplication problems that include powers of 10** 6. Round decimals that have 9 as a digit 7. **Solve ratio and fraction-of-a-group word problems** Complete work independently	**Exercises** 1. Compare fractions 2. Multiply and divide using mental math 3. **Divide a whole number by a fraction** 4. Solve ratio word problems, plot them on a coordinate system, and draw the line 5. **Change fractions to mixed numbers** 6. Find the area of circles 7. Round decimals that have 9 as a digit 8. Complete work independently	**Exercises** 1. Compare fractions 2. Multiply and divide using mental math 3. **Divide a fraction by a fraction** 4. Solve ratio word problems, plot them on a coordinate system, and draw the line 5. **Change fractions to mixed numbers** 6. Find the area of circles 7. Round decimals that have 9 as a digit 8. Complete work independently	**Exercises** 1. **Simply fractions with a common factor of 10** 2. Compare fractions 3. **Say expanded notations for decimal place value** 4. Divide a fraction by a fraction 5. Change fractions to mixed numbers 6. Find the area of circles 7. Round decimals that have 9 as a digit 8. Complete work independently
Common Core State Standards for Mathematics					
5.OA 1	✔	✔	✔	✔	✔
5.NBT 1–2	✔	✔	✔		✔
5.NBT 3		✔		✔	✔
5.NBT 4	✔	✔	✔	✔	✔
5.NBT 5	✔		✔	✔	✔
5.NBT 6			✔		✔
5.NBT 7	✔	✔	✔	✔	✔
5.NF 3					✔
5.NF 4–6	✔	✔	✔	✔	✔
5.NF 7			✔		✔
5.MD 5	✔	✔	✔	✔	
5.G 2	✔	✔	✔	✔	✔
Teacher Materials	Presentation Book 2, Board Displays CD or chalk board				
Student Materials	Textbook, Workbook, pencil, lined paper, ruler				
Additional Practice	• Student Practice Software: Block 3: Activity 1 (5.NBT 7), Activity 2, Activity 3 (5.NF 4), Activity 4, Activity 5 (5.NBT 6), Activity 6 (5.NF 4, 5.NF 6, and 5.OA 1) • Provide needed fact practice with Level D or E Math Fact Worksheets.				
Mastery Test					

Lesson 71

EXERCISE 1: FRACTION ANALYSIS
COMPARISON

a. (Display:)   [71:1A]

a. $\frac{1}{4}$	$\frac{5}{12}$	b. $\frac{1}{4}$	$\frac{3}{20}$
c. $\frac{1}{2}$	$\frac{5}{8}$	d. $\frac{1}{5}$	$\frac{7}{40}$

Each item shows two fractions. We're going to figure out which fraction is more.

b. (Point to **A.**) Read these fractions. (Signal.) *1/4, 5/12.*
- Which fraction has the larger denominator? (Signal.) *5/12.*
  So you ask: How many 12ths is 1/4?
- What do you ask? (Signal.) *How many 12ths is 1/4?*

c. (Point to **B.**) Read these fractions. (Signal.) *1/4, 3/20.*
- Which fraction has the larger denominator? (Signal.) *3/20.*
  So you ask: How many 20ths is 1/4?
- What do you ask? (Signal.) *How many 20ths is 1/4?*

d. (Point to **A.**) What do you ask for these fractions? (Signal.) *How many 12ths is 1/4?*
- (Point to **B.**) What do you ask for these fractions? (Signal.) *How many 20ths is 1/4?*
  (Repeat until firm.)

e. (Point to **C.**) Which fraction has the larger denominator? (Signal.) *5/8.*
- So what do you ask? (Signal.) *How many 8ths is 1/2?*

f. (Point to **D.**) Which fraction has the larger denominator? (Signal.) *7/40.*
- So what do you ask? (Signal.) *How many 40ths is 1/5?*
  (Repeat until firm.)

g. (Point to **A.**) What do you ask? (Signal.) *How many 12ths is 1/4?*
- Raise your hand when you know the answer. ✔
- What's the answer? (Signal.) *3/12.*
- Which is more—3/12 or 5/12? (Signal.) *5/12.*
- So which is more—1/4 or 5/12? (Signal.) *5/12.*
  (Repeat until firm.)
  I make the sign to show 5/12 is more.
  (Add to show:)   [71:1B]

a. $\frac{1}{4} < \frac{5}{12}$		b. $\frac{1}{4}$	$\frac{3}{20}$
c. $\frac{1}{2}$	$\frac{5}{8}$	d. $\frac{1}{5}$	$\frac{7}{40}$

h. (Point to **B.**) What do you ask? (Signal.) *How many 20ths is 1/4?*
- Raise your hand when you know the answer. ✔
- What's the answer? (Signal.) *5/20.*
- Which is more—5/20 or 3/20? (Signal.) *5/20.*
- So which is more—1/4 or 3/20? (Signal.) *1/4.*
  (Repeat until firm.)
  I make the sign to show 1/4 is more.
  (Add to show:)   [71:1C]

a. $\frac{1}{4} < \frac{5}{12}$		b. $\frac{1}{4} > \frac{3}{20}$	
c. $\frac{1}{2}$	$\frac{5}{8}$	d. $\frac{1}{5}$	$\frac{7}{40}$

i. (Point to **C.**) What do you ask? (Signal.) *How many 8ths is 1/2?*
- Raise your hand when you know the answer. ✔
- What's the answer? (Signal.) *4/8.*
- Which is more—4/8 or 5/8? (Signal.) *5/8.*
- So which is more—1/2 or 5/8? (Signal.) *5/8.*
  (Repeat until firm.)
  I make the sign to show 5/8 is more.
  (Add to show:)   [71:1D]

a. $\frac{1}{4} < \frac{5}{12}$		b. $\frac{1}{4} > \frac{3}{20}$	
c. $\frac{1}{2} < \frac{5}{8}$		d. $\frac{1}{5}$	$\frac{7}{40}$

j. (Point to **D**.) What do you ask? (Signal.) *How many 40ths is 1/5?*
- Raise your hand when you know the answer. ✔
- What's the answer? (Signal.) *8/40.*
- Which is more—8/40 or 7/40? (Signal.) *8/40.*
- So which is more—1/5 or 7/40? (Signal.) *1/5.*
  (Repeat until firm.)
  I make the sign to show 1/5 is more.
  (Add to show:)   [71:1E]

a. $\dfrac{1}{4} < \dfrac{5}{12}$	b. $\dfrac{1}{4} > \dfrac{3}{20}$
c. $\dfrac{1}{2} < \dfrac{5}{8}$	d. $\dfrac{1}{5} > \dfrac{7}{40}$

EXERCISE 2: SHORT DIVISION
WITH MENTAL MATH
REMEDY

a. (Display:)   [71:2A]

$$3\overline{)478}$$

- Read this problem. (Signal.) *478 ÷ 3.*
  We'll work it the fast way.
b. Say the first problem you work. (Signal.) *4 ÷ 3.*
- What's the answer? (Signal.) *1 with a remainder of 1.*
  (Add to show:)   [71:2B]

$$3\overline{)4_17\,8}^{\,1}$$

c. Say the next problem you work. (Signal.) *17 ÷ 3.*
- What's the answer? (Signal.) *5 with a remainder of 2.*
  (Add to show:)   [71:2C]

$$3\overline{)4_17_28}^{\,1\,5}$$

d. Say the last problem you work. (Signal.) *28 ÷ 3.*
- What's the answer? (Signal.) *9 with a remainder of 1.*
- We write the remainder as a fraction. What fraction? (Signal.) *1/3.*
  (Add to show:)   [71:2D]

$$3\overline{)4_17_28}^{\,1\,5\,9\frac{1}{3}}$$

a. Open your workbook to Lesson 71 and find part 1. ✔
  (Teacher reference:)   **R** **Part D**

  a. $5\overline{)638}$ b. $4\overline{)225}$ c. $3\overline{)704}$ d. $2\overline{)97}$

  All these problems have remainders. You'll write the remainders as fractions.
b. Problem A: 638 ÷ 5.
- Work the problem the fast way. Remember, write the last remainder as a fraction.
  (Observe students and give feedback.)
c. Check your work.
- Say the first problem you worked. (Signal.) *6 ÷ 5.*
- What's the answer? (Signal.) *1 with a remainder of 1.*
- Say the next problem you worked. (Signal.) *13 ÷ 5.*
- What's the answer? (Signal.) *2 with a remainder of 3.*
- Say the last problem you worked. (Signal.) *38 ÷ 5.*
- What's the answer? (Signal.) *7 with a remainder of 3.*
  You showed the remainder as 3/5.
  (Display:)   [71:2E]

  a. $5\overline{)6_13_38}^{\,1\,2\,7\frac{3}{5}}$

  Here's what you should have.
- What's 638 ÷ 5? (Signal.) *127 and 3/5.*
d. Work problem B.
  (Observe students and give feedback.)
e. Check your work.
- Say the first problem you worked. (Signal.) *22 ÷ 4.*
- What's the answer? (Signal.) *5 with a remainder of 2.*
- Say the last problem you worked. (Signal.) *25 ÷ 4.*
- What's the answer? (Signal.) *6 with a remainder of 1.*
- What's 225 ÷ 4? (Signal.) *56 and 1/4.*
  (Display:)   [71:2F]

  b. $4\overline{)2\,2_25}^{\,5\,6\frac{1}{4}}$

  Here's what you should have.

f. Work the rest of the problems in part 1.
   (Observe students and give feedback.)
g. Check your work.
• Problem C: 704 ÷ 3.
• What's the answer? (Signal.) *234 and 2/3.*
   (Display:)   [71:2G]

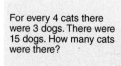

   c. $3\overline{)7_10_14}$ $234\frac{2}{3}$

   Here's what you should have.
h. Problem D: 97 ÷ 2.
• What's the answer? (Signal.) *48 and 1/2.*
   (Display:)   [71:2H]

   d. $2\overline{)9_17}$ $48\frac{1}{2}$

   Here's what you should have.

EXERCISE 3: COORDINATE SYSTEM
WORD PROBLEMS

a. (Display:)   [71:3A]

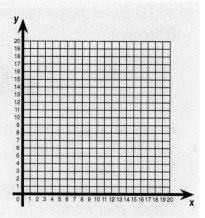

   Here's a problem: For every 4 cats there were
   3 dogs. There were 15 dogs. How many cats
   were there?
   First, we'll solve this problem. Then we'll make
   a line and answer other questions about the
   cats and dogs.
   For every 4 cats there were 3 dogs. There
   were 15 dogs. How many cats were there?

• Work the problem in part A.
   (Observe students and give feedback.)
• Everybody, how many cats were there?
   (Signal.) *20 cats.*
   (Add to show:)   [71:3B]

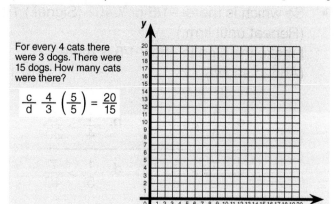

b. Here's the problem you worked. This equation
   has two equivalent fractions—4/3 and 20/15.
   First, we put the names for Y and X on the
   coordinate system.
• What's the name for Y? (Signal.) *Cats.*
• What's the name for X? (Signal.) *Dogs.*
   (Add to show:)   [71:3C]

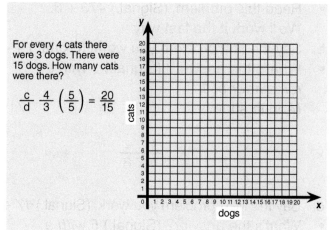

c. Now we make points for both fractions. The
   first fraction is 4/3.
• What's the Y value? (Signal.) *4.*
• What's the X value? (Signal.) *3.*
d. The other fraction is 20/15.
• What's the Y value? (Signal.) *20.*
• What's the X value? (Signal.) *15.*
   Here are the two points and a line that
   connects them. The line goes through zero.

(Add to show:) [71:3D]

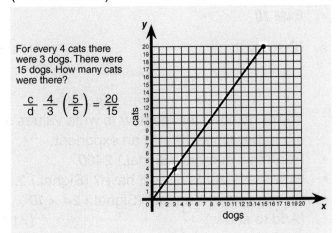

For every 4 cats there were 3 dogs. There were 15 dogs. How many cats were there?

$$\frac{c}{d} \quad \frac{4}{3}\left(\frac{5}{5}\right) = \frac{20}{15}$$

e. We can use the line to answer other questions about the cats and dogs.

(Add to show:) [71:3E]

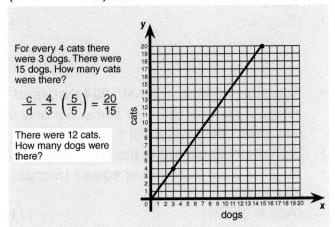

For every 4 cats there were 3 dogs. There were 15 dogs. How many cats were there?

$$\frac{c}{d} \quad \frac{4}{3}\left(\frac{5}{5}\right) = \frac{20}{15}$$

There were 12 cats. How many dogs were there?

Here's a question: There were 12 cats. How many dogs were there?

We just make a point on the line for 12 cats. The point shows the number of dogs.

(Add to show:) [71:3F]

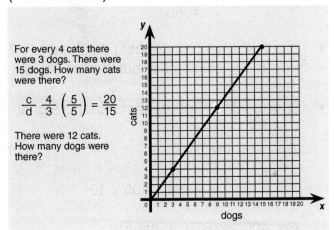

For every 4 cats there were 3 dogs. There were 15 dogs. How many cats were there?

$$\frac{c}{d} \quad \frac{4}{3}\left(\frac{5}{5}\right) = \frac{20}{15}$$

There were 12 cats. How many dogs were there?

f. Raise your hand when you know the number of dogs. ✔

• What's the number of dogs? (Signal.) 9.
  Yes, if there were 12 cats, there were 9 dogs.

(Change to show:) [71:3G]

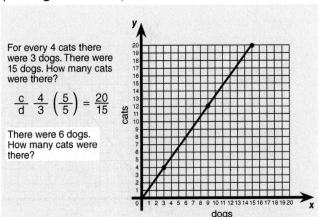

For every 4 cats there were 3 dogs. There were 15 dogs. How many cats were there?

$$\frac{c}{d} \quad \frac{4}{3}\left(\frac{5}{5}\right) = \frac{20}{15}$$

There were 6 dogs. How many cats were there?

New question: There were 6 dogs. How many cats were there?

We make a point for 6 dogs. The point shows the number of cats.

(Add to show:) [71:3H]

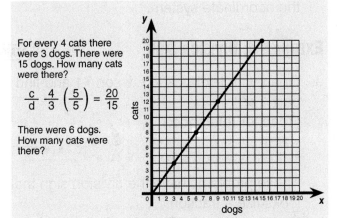

For every 4 cats there were 3 dogs. There were 15 dogs. How many cats were there?

$$\frac{c}{d} \quad \frac{4}{3}\left(\frac{5}{5}\right) = \frac{20}{15}$$

There were 6 dogs. How many cats were there?

• Raise your hand when you know the number of cats. ✔

• What's the number of cats? (Signal.) 8.
  Yes, if there were 6 dogs, there were 8 cats.

g. **(Change to show:)** [71:3I]

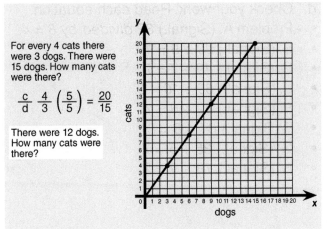

For every 4 cats there were 3 dogs. There were 15 dogs. How many cats were there?

$$\frac{c}{d} \quad \frac{4}{3}\left(\frac{5}{5}\right) = \frac{20}{15}$$

There were 12 dogs. How many cats were there?

New question: There were 12 dogs. How many cats were there?

We make the point for 12 dogs.

(Add to show:) [71:3J]

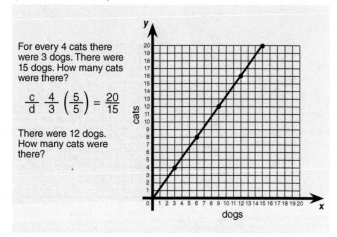

For every 4 cats there were 3 dogs. There were 15 dogs. How many cats were there?

$$\frac{c}{d} \quad \frac{4}{3} \left(\frac{5}{5}\right) = \frac{20}{15}$$

There were 12 dogs. How many cats were there?

- Raise your hand when you know the number of cats. ✔
- What's the number of cats? (Signal.) *16.*
  Yes, if there were 12 dogs, there were 16 cats.
h. Remember, you can work ratio problems on the coordinate system.

EXERCISE 4: DIVISION SIGN
REMEDY

a. Open your textbook to Lesson 71 and find part 1. ✔
   (Teacher reference:)

| a. 32 ÷ 8 | b. 90 ÷ 9 | c. 49 ÷ 7 | d. 60 ÷ 10 |
| e. 42 ÷ 6 | f. 64 ÷ 8 | g. 35 ÷ 5 | h. 18 ÷ 9 |

Part 1
a. ■ ÷ ■ = ■

These problems have the division sign that looks like a fraction.
b. Read problem A. (Signal.) *32 divided by 8.*
- Read problem B. (Signal.) *90 ÷ 9.*
- Read problem C. (Signal.) *49 ÷ 7.*
c. Copy all the problems in part 1 and write the answers.
   (Observe students and give feedback.)
d. Check your work. Read each equation.
- Problem A. (Signal.) *32 divided by 8 = 4.*
- B. (Signal.) *90 ÷ 9 = 10.*
- C. (Signal.) *49 ÷ 7 = 7.*
- D. (Signal.) *60 ÷ 10 = 6.*
- E. (Signal.) *42 ÷ 6 = 7.*
- F. (Signal.) *64 ÷ 8 = 8.*
- G. (Signal.) *35 ÷ 5 = 7.*
- H. (Signal.) *18 ÷ 9 = 2.*

EXERCISE 5: EXPONENTS
BASE 10

a. (Display:) [71:5A]

2400

You're going to tell me how to write values as a number times 10 with an exponent.
- Read this number. (Signal.) *2400.*
- How many zeros does it have? (Signal.) *2.*
- So what does it equal? (Signal.) *24 × 10².*
  (Add to show:) [71:5B]

$$2400 = 24 \times 10^2$$

Yes, 24 times 10 to the second.
b. (Display:) [71:5C]

204,000

- Read this number. (Signal.) *204 thousand.*
  Here's a number with four zeros, but only the last three zeros tell about the exponent.
- How many zeros are at the end? (Signal.) *3.*
- So what does the number equal? (Signal.) *204 × 10³.*
  (Add to show:) [71:5D]

$$204,000 = 204 \times 10^3$$

Yes, 204 times 10 to the third.
c. (Display:) [71:5E]

4050

- Read this number. (Signal.) *4 thousand 50.*
- How many zeros tell about the exponent? (Signal.) *1.*
- So what does this number equal? (Signal.) *405 × 10¹.*
  (Add to show:) [71:5F]

$$4050 = 405 \times 10^1$$

Yes, 405 times 10 to the first.

d. (Display:) [71:5G]

> 101,000

- Read this number. (Signal.) *101 thousand.*
- How many zeros tell about the exponent? (Signal.) *3.*
- So what does this number equal? (Signal.) *101 × 10³.*
  (Add to show:) [71:5H]

> $101,000 = 101 × 10^3$

Yes, 101 times 10 to the third.

TEXTBOOK PRACTICE

a. Find part 2 in your textbook. ✔
   (Teacher reference:)

   a. 4090   b. 300,200   c. 375,000
   d. 190,200   e. 163,000

b. Number A: 4 thousand 90.
- Copy the number. Write what it equals. ✔
- Everybody, read equation A. (Signal.) *4090 = 409 × 10¹.*
c. Write equations for the rest of the numbers.
   (Observe students and give feedback.)
d. Check your work.
- Number B: 300,200. What does it equal? (Signal.) *3002 × 10².*
- C: 375,000. What does it equal? (Signal.) *375 × 10³.*
- D: 190,200. What does it equal? (Signal.) *1902 × 10².*
- E: 163,000. What does it equal? (Signal.) *163 × 10³.*

EXERCISE 6: VOLUME
OF RECTANGULAR PRISMS   [REMEDY]

a. Find part 3 in your textbook. ✔
   (Teacher reference:)

   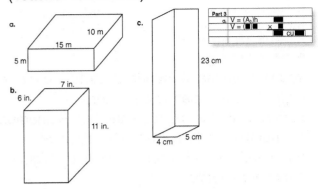

You're going to find the volume of these rectangular prisms.
b. Prism A. It shows the area of the base at the top of the prism.
- Raise your hand when you can say the multiplication for the area of the base. ✔
- Say the multiplication for the area of the base. (Signal.) *15 × 10.*
  (Repeat until firm.)
c. You multiply the area of the base times the height.
- What's the height? (Signal.) *5 meters.*
- Start with the letter equation, figure out the base, and then multiply by the height.
  (Observe students and give feedback.)
d. Check your work.
   You multiplied 15 × 10 for the base.
- What's the answer? (Signal.) *150.*
  Then you multiplied 150 × 5.
- What's the volume of the rectangular prism? (Signal.) *750 cubic meters.*
e. Work problem B.
   (Observe students and give feedback.)
f. Check your work.
   You worked the problem 6 × 7 for the base.
- What's 6 × 7? (Signal.) *42.*
  Then you multiplied 42 × 11.
- What's the volume of the rectangular prism? (Signal.) *462 cubic inches.*
g. Work problem C.
   (Observe students and give feedback.)
h. Check your work.
   You worked the problem 4 × 5 for the base.
- What's 4 × 5? (Signal.) *20.*
- What's 20 × 23? (Signal.) *460.*
- What's the volume of the rectangular prism? (Signal.) *460 cubic centimeters.*

EXERCISE 7: ROUNDING

DECIMALS WITH 9

a. Find part 4 in your textbook. ✔
   (Teacher reference:)

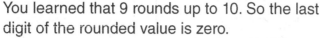

   a. Round .9842 to the tenths place.
   b. Round .0094 to the thousandths place.
   c. Round .0991 to the hundredths place.
   d. Round .9395 to the tenths place.

   You learned that 9 rounds up to 10. So the last digit of the rounded value is zero.

b. Item A. Write the rounded number. Remember, the number ends in the tenths place. ✔
   • Everybody, read the rounded value. (Signal.)
     1 and zero tenths.
     (Display:)   [71:7A]

   a. 1.0

   Here's what you should have.

c. Write the rounded value for B. Check to see if it rounds up. ✔
   • Everybody, the digit after 9 is 4. So does the number round up? (Signal.) *No.*
   • What's the rounded value? (Signal.)
     9 thousandths.
     (Display:)   [71:7B]

   b. .009

   Here's what you should have.

d. Write the rounded value for the rest of the items.
   (Observe students and give feedback.)

e. Check your work.
   • Item C: Read the rounded value. (Signal.)
     10 hundredths.
     (Display:)   [71:7C]

   c. .10

   Here's what you should have.

f. Item D: Read the rounded value. (Signal.)
   9 tenths.
   (Display:)   [71:7D]

   d. .9

   Here's what you should have.

g. Rounding 9 up works the same way for other values.
   (Display:)   [71:7E]

   .39   .29

   .59   .89

h. (Point to .39.) Read this number. (Signal.)
   39 hundredths.
   • If you round 39 up, what do you get?
     (Signal.) *40.*
     So, 39 hundredths rounds up to 40 hundredths.
   • What's the last digit of 40 hundredths?
     (Signal.) *Zero.*

i. (Point to .59.) Read this number. (Signal.)
   59 hundredths.
   • If you round 59 up, what do you get?
     (Signal.) *60.*
   • What's the last digit of 60 hundredths?
     (Signal.) *Zero.*

j. If you round 29 up, what do you get?
   (Signal.) *30.*
   • If you round 89 up, what do you get? (Signal.) *90.*
     (Add to show:)   [71:7F]

   .396   .296

   .596   .896

   If we round these numbers to hundredths, all of them will round up.
   You can show how many hundredths there are by underlining the first two digits.
   (Add to show:)   [71:7G]

   .39̲6   .29̲6

   .59̲6   .89̲6

k. (Point to .396.) The underlined part shows the hundredths.
- How many hundredths are in this number? (Signal.) *39.*
- So if you round 39 up, what do you get? (Signal.) *40.*
(Add to show:) [71:7H]

.<u>39</u>6   .<u>29</u>6
.40

.<u>59</u>6   .<u>89</u>6

Yes, the rounded number is 40 hundredths.

l. (Point to .596.) How many hundredths are in this number? (Signal.) *59.*
- So if you round 59 up, what do you get? (Signal.) *60.*
(Add to show:) [71:7I]

.<u>39</u>6   .<u>29</u>6
.40

.<u>59</u>6   .<u>89</u>6
.60

Yes, 60 hundredths.

m. (Point to .296.) How many hundredths are in this number? (Signal.) *29.*
- So if you round 29 up, what do you get? (Signal.) *30.*
(Add to show:) [71:7J]

.<u>39</u>6   .<u>29</u>6
.40   .30

.<u>59</u>6   .<u>89</u>6
.60

Yes, 30 hundredths.

n. (Point to .896.) How many hundredths are in this number? (Signal.) *89.*
- So if you round 89 up, what do you get? (Signal.) *90.*
(Add to show:) [71:7K]

.<u>39</u>6   .<u>29</u>6
.40   .30

.<u>59</u>6   .<u>89</u>6
.60   .90

Yes, 90 hundredths.

o. Find part 5 in your textbook. ✔
(Teacher reference:)

a. .497   b. .297   c. .697

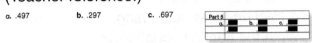

For all these numbers, the hundredths digit is 9.
- Copy these numbers in a row. Underline the first two digits. ✔
p. Read number A. (Signal.) *497 thousandths.*
- How many hundredths are there? (Signal.) *49.* The 9 rounds up.
- Write the rounded number for hundredths below. ✔
- Everybody, read the rounded number. (Signal.) *50 hundredths.*
q. Round the rest of the numbers to hundredths. ✔
- Read the rounded number for B. (Signal.) *30 hundredths.*
- Read the rounded number for C. (Signal.) *70 hundredths.*

EXERCISE 8: WORD PROBLEMS
RATIO/FRACTION OF A GROUP

a. Find part 6 in your textbook.
(Teacher reference:)

    a. $\frac{9}{10}$ of the bricks were red.

    b. There were 7 cats for every 9 squirrels.

    c. There were 9 bikes for every 4 motorcycles.

    d. $\frac{1}{6}$ of the children were sick.

    e. $\frac{6}{11}$ of the books were new.

    f. Every 2 boxes weigh 25 pounds.

Some sentences tell about a fraction times
something. Some tell about a ratio.

b. Sentence A: 9/10 of the bricks were red.

•   Does that sentence tell about a fraction times
something or about a ratio fraction? (Signal.)
A fraction times something.

•   Write the fraction equation. ✔
(Display:)   [71:8A]

$$\text{a. } \frac{9}{10} \, b = r$$

Here's what you should have.

c. Sentence B: There were 7 cats for every
9 squirrels. ·

•   Does that sentence tell about a fraction times
something or about a ratio fraction? (Signal.)
A ratio fraction.

•   Write the letter fraction and the number
fraction for the ratio. ✔
(Display:)   [71:8B]

$$\text{b. } \frac{c}{s} \quad \frac{7}{9}$$

Here's what you should have.

d. Sentence C: There were 9 bikes for every
4 motorcycles.

•   Write the equation or the ratio fractions. ✔
(Display:)   [71:8C]

$$\text{c. } \frac{b}{m} \quad \frac{9}{4}$$

Here's what you should have.

e. Sentence D: 1/6 of the children were sick.

•   Write the equation or the ratio fractions. ✔
(Display:)   [71:8D]

$$\text{d. } \frac{1}{6} \, C = S$$

Here's what you should have.

f. Sentence E: 6/11 of the books were new.

•   Write the equation or the ratio fractions. ✔
(Display:)   [71:8E]

$$\text{e. } \frac{6}{11} \, b = n$$

Here's what you should have.

g. Sentence F: Every 2 boxes weigh 25 pounds.

•   Write the equation or the ratio fractions. ✔
(Display:)   [71:8F]

$$\text{f. } \frac{b}{p} \quad \frac{2}{25}$$

Here's what you should have.

Assign Independent Work, Textbook parts 7–11
and Workbook parts 2–4.

Optional extra math-fact practice worksheets
are available on ConnectED.

Lesson 72

EXERCISE 1: FRACTION ANALYSIS
COMPARISON

a. (Display:) [72:1A]

a. $\frac{1}{3}$	$\frac{6}{15}$		**b.** $\frac{1}{3}$	$\frac{9}{24}$
c. $\frac{1}{5}$	$\frac{4}{30}$		**d.** $\frac{1}{2}$	$\frac{7}{18}$

We're going to write signs to show which fraction is more.

b. Read the fractions for A. (Signal.) *1/3, 6/15.*
• Which fraction has the larger denominator? (Signal.) *6/15.*
  So you ask: How many 15ths is 1/3?
• What do you ask? (Signal.) *How many 15ths is 1/3?*
• Raise your hand when you know the answer. ✔
• What's the answer? (Signal.) *5/15.*
c. Read the fractions for B. (Signal.) *1/3, 9/24.*
• Which fraction has the larger denominator? (Signal.) *9/24.*
• So what do you ask? (Signal.) *How many 24ths is 1/3?*
• Raise your hand when you know the answer. ✔
• What's the answer? (Signal.) *8/24.*
d. Read the fractions for C. (Signal.) *1/5, 4/30.*
• Which fraction has the larger denominator? (Signal.) *4/30.*
• So what do you ask? (Signal.) *How many 30ths is 1/5?*
• Raise your hand when you know the answer. ✔
• What's the answer? (Signal.) *6/30.*
e. Read the fractions for D. (Signal.) *1/2, 7/18.*
• Which fraction has the larger denominator? (Signal.) *7/18.*
• So what do you ask? (Signal.) *How many 18ths is 1/2?*
• Raise your hand when you know the answer. ✔
• What's the answer? (Signal.) *9/18.*

f. This time, we're going to write the signs.
• (Point to **A.**) Say the question about 15ths. (Signal.) *How many 15ths is 1/3?*
• What's the answer? (Signal.) *5/15.*
• Which is more—5/15 or 6/15? (Signal.) *6/15.*
• So which is more—1/3 or 6/15? (Signal.) *6/15.*
  (Repeat until firm.)
  (Add to show:) [72:1B]

a. $\frac{1}{3} <$	$\frac{6}{15}$		**b.** $\frac{1}{3}$	$\frac{9}{24}$
c. $\frac{1}{5}$	$\frac{4}{30}$		**d.** $\frac{1}{2}$	$\frac{7}{18}$

g. (Point to **B.**) Say the question about 24ths. (Signal.) *How many 24ths is 1/3?*
• What's the answer? (Signal.) *8/24.*
• Which is more—8/24 or 9/24? (Signal.) *9/24.*
• So which is more—1/3 or 9/24? (Signal.) *9/24.*
  (Repeat until firm.)
  (Add to show:) [72:1C]

a. $\frac{1}{3} <$	$\frac{6}{15}$		**b.** $\frac{1}{3} <$	$\frac{9}{24}$
c. $\frac{1}{5}$	$\frac{4}{30}$		**d.** $\frac{1}{2}$	$\frac{7}{18}$

h. (Point to **C.**) Say the question about 30ths. (Signal.) *How many 30ths is 1/5?*
• What's the answer? (Signal.) *6/30.*
• Which is more—6/30 or 4/30? (Signal.) *6/30.*
• So which is more—1/5 or 4/30? (Signal.) *1/5.*
  (Repeat until firm.)
  (Add to show:) [72:1D]

a. $\frac{1}{3} <$	$\frac{6}{15}$		**b.** $\frac{1}{3} <$	$\frac{9}{24}$
c. $\frac{1}{5} >$	$\frac{4}{30}$		**d.** $\frac{1}{2}$	$\frac{7}{18}$

i. (Point to **D.**) Say the question about 18ths. (Signal.) *How many 18ths is 1/2?*
• What's the answer? (Signal.) *9/18.*
• Which is more—9/18 or 7/18? (Signal.) *9/18.*
• So which is more—1/2 or 7/18? (Signal.) *1/2.*
  (Repeat until firm.)

(Add to show:) [72:1E]

a. $\frac{1}{3} < \frac{6}{15}$	b. $\frac{1}{3} < \frac{9}{24}$
c. $\frac{1}{5} > \frac{4}{30}$	d. $\frac{1}{2} > \frac{7}{18}$

===== WORKBOOK PRACTICE =====

a. Open your workbook to Lesson 72 and find part 1.
(Teacher reference:)

a. $\frac{1}{2}$ $\frac{7}{12}$ b. $\frac{1}{3}$ $\frac{3}{15}$ c. $\frac{1}{5}$ $\frac{6}{20}$

You're going to make signs to show which fraction is more.

b. Read the fractions in A. (Signal.) *1/2, 7/12.*
• Which fraction has the larger denominator? (Signal.) *7/12.*
• So what do you ask? (Signal.) *How many 12ths is 1/2?*

c. Read the fractions in B. (Signal.) *1/3, 3/15.*
• Which fraction has the larger denominator? (Signal.) *3/15.*
• So what do you ask? (Signal.) *How many 15ths is 1/3?*

d. Read the fractions in C. (Signal.) *1/5, 6/20.*
• Which fraction has the larger denominator? (Signal.) *6/20.*
• So what do you ask? (Signal.) *How many 20ths is 1/5?*

e. Go back to problem A. ✔
• Say the question you ask. (Signal.) *How many 12ths is 1/2?*
• Raise your hand when you know the answer. ✔
• What's the answer? (Signal.) *6/12.*
• Write the sign to show which fraction is more. ✔
(Display:) [72:1F]

a. $\frac{1}{2} < \frac{7}{12}$

Here's what you should have.
1/2 is 6/12. 7/12 is more than 6/12.

f. Problem B. Say the question you ask. (Signal.) *How many 15ths is 1/3?*
• Raise your hand when you know the answer. ✔
• What's the answer? (Signal.) *5/15.*
• Write the sign to show which fraction is more. ✔
(Display:) [72:1G]

b. $\frac{1}{3} > \frac{3}{15}$

Here's what you should have.
1/3 is 5/15. 5/15 is more than 3/15.

g. Problem C. Say the question you ask. (Signal.) *How many 20ths is 1/5?*
• Raise your hand when you know the answer. ✔
• What's the answer? (Signal.) *4/20.*
• Write the sign to show which fraction is more. ✔
(Display:) [72:1H]

c. $\frac{1}{5} < \frac{6}{20}$

Here's what you should have.
1/5 is 4/20. 6/20 is more than 4/20.

EXERCISE 2: COORDINATE SYSTEM
WORD PROBLEMS

a. Find part 2 in your workbook. ✔
(Teacher reference:)

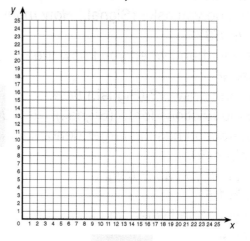

• There were 2 stamps for every 5 coins.
a. There were 15 coins. How many stamps were there? _____
$$\left(\quad \right) =$$
b. If there were 8 stamps, how many coins were there? _____
c. If there were 10 coins, how many stamps were there? _____
d. If there were 10 stamps, how many coins were there? _____

You're going to work ratio problems by making points and a line on the coordinate system.

b. There were 2 stamps for every 5 coins.
The first sentence tells the names for Y and X.
It also gives the numbers for the first point on
the coordinate system.

• Write the names and make the first point on
the coordinate system.
(Observe students and give feedback.)
(Display:) [72:2A]

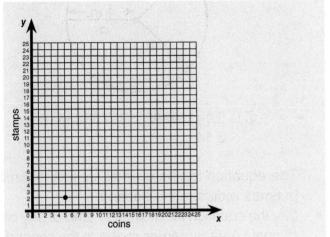

Here's what you should have.

c. Now you're going to work a problem to figure
out another point on the coordinate system.
Problem A: There were 15 coins. How many
stamps were there?

• Work the problem. Write the ratio equation and
answer the question.
(Observe students and give feedback.)

• Everybody, how many stamps were there?
(Signal.) 6 stamps.
(Add to show:) [72:2B]

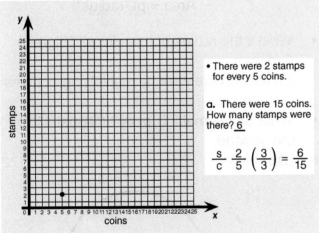

Here's what you should have.

d. The fraction 6/15 tells about another point on
the coordinate system. Make that point. Then
make a line that connects the points and goes
through zero.
(Observe students and give feedback.)

(Add to show:) [72:2C]

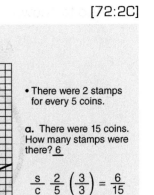

Here's what you should have.

e. Now you can use the line to answer the rest of
the questions.

f. Read question B. (Signal.) *If there were
8 stamps, how many coins were there?*

• Make the point for 8 stamps, and write the
answer to the question.
(Observe students and give feedback.)

• Everybody, if there were 8 stamps, how many
coins were there? (Signal.) *20.*

g. Read question C. (Signal.) *If there were
10 coins, how many stamps were there?*

• Make the point for 10 coins, and write the
answer to the question. ✔

• Everybody, if there were 10 coins, how many
stamps were there? (Signal.) *4.*

h. Read question D. (Signal.) *If there were
10 stamps, how many coins were there?*

• Make the point and write the answer. ✔

• Everybody, if there were 10 stamps, how
many coins were there? (Signal.) *25.*

EXERCISE 3: AREA
OF CIRCLES

a. (Display:) [72:3A]

$$Area = pi \ (radius^2)$$

You're going to figure out the area of circles.
Here's the equation you use: Area = pi times
radius to the second. The radius is the
distance from the center of a circle to the edge
of a circle.

b. (Add to show:) [72:3B]

Area = pi (radius²)

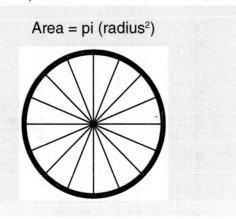

A radius is like a spoke in a wheel. It goes
from the center of the circle to the edge of the
circle. Every radius is the same length.
(Change to show:) [72:3C]

Area = pi (radius²)

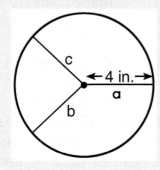

Here's a diagram that shows you more than
one radius. Radius A is 4 inches.
• So the length of radius B is also 4 inches.
• What's the length of radius B? (Signal.) *4 inches.*
• What's the length of radius C? (Signal.) *4 inches.*
Yes, every radius in a circle is the same length.
The area of any circle is pi times the radius to
the second.

c. Pi is a strange number. Its decimal places go
on forever. Here is pi shown with 20 decimal
places.
(Add to show:) [72:3D]

Area = pi (radius²)

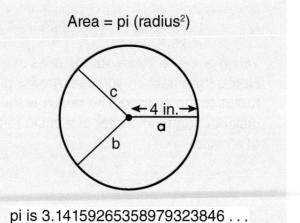

pi is 3.14159265358979323846 . . .

We will round pi to 3 and 14 hundredths.

• What will we round pi to? (Signal.) *3 and
14 hundredths.*
(Add to show:) [72:3E]

Area = pi (radius²)

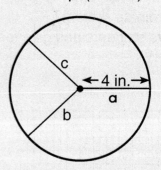

pi is 3.14159265358979323846 . . .
3.14

The equation for the area of a circle is: Area =
pi times radius to the second.
• Say the equation for finding the area of any circle.
(Signal.) *Area = pi times radius to the second.*
(Repeat until firm.)

d. (Display:) [72:3F]

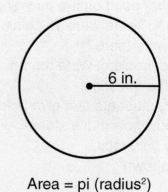

Area = pi (radius²)

• What's the radius of this circle? (Signal.)
6 inches.
So we find the area by multiplying pi times 6 to
the second.
• What do we multiply pi by? (Signal.) *6 to
the second.*

e. (Display:) [72:3G]

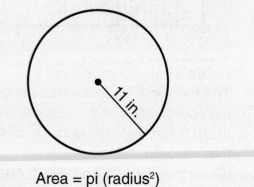

Area = pi (radius²)

New circle.

- What's the radius? (Signal.) *11 inches.*
- So what do we multiply pi by to find the area? (Signal.) *11 to the second.*

f. (Display:) [72:3H]

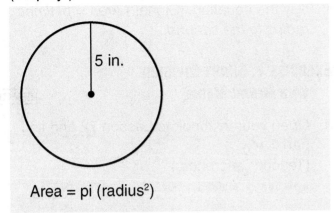

Area = pi (radius²)

New circle.
- What's the radius? (Signal.) *5 inches.*
- So what do we multiply pi by to find the area? (Signal.) *5 to the second.*

g. Let's work the problem.
- What number do we use for pi? (Signal.)
  3 and 14 hundredths.
  (Add to show:) [72:3I]

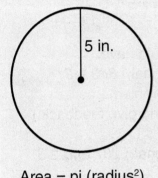

Area = pi (radius²)
     = 3.14

So we work the problem 3 and 14 hundredths times 5 to the second.
- What's 5 to the second? (Signal.) *25.*
  (Add to show:) [72:3J]

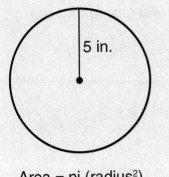

Area = pi (radius²)
     = 3.14 (25)

So we work the problem 3 and 14 hundredths times 25.

- Say the problem. (Signal.) *3 and 14 hundredths × 25.*

h. Figure out the area in part A. The unit name in the answer is square inches.
   (Observe students and give feedback.)
- Everybody, what's the area of the circle? (Signal.) *78 and 50 hundredths square inches.*
  (Add to show:) [72:3K]

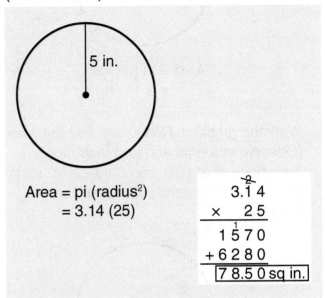

Area = pi (radius²)
     = 3.14 (25)

```
      2
    3.1 4
  ×   2 5
    1 5 7 0
  + 6 2 8 0
  ⎣7 8.5 0 sq in.⎦
```

Here's what you should have.

i. (Display:) [72:3L]

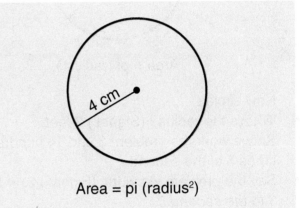

Area = pi (radius²)

New circle.
- What's the radius? (Signal.) *4 centimeters.*
  So we work the problem 3 and 14 hundredths times 4 to the second.
- Say the problem we work. (Signal.) *3 and 14 hundredths × 4 to the second.*
  (Repeat until firm.)

- What's 4 to the second? (Signal.) *16.*
  (Add to show:)   [72:3M]

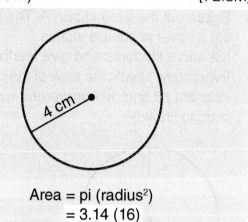

Area = pi (radius²)
= 3.14 (16)

- Work the problem. Remember the unit name.
  (Observe students and give feedback.)
- Everybody, what's the area of the circle? (Signal.)
  50 and 24 hundredths square centimeters.

j. (Display:)   [72:3N]

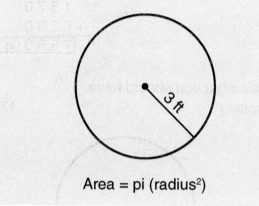

Area = pi (radius²)

New circle.
- What's the radius? (Signal.) *3 feet.*
  So we work the problem 3 and 14 hundredths
  times 3 to the second.
- Say the problem we work. (Signal.) *3.14 times
  3 to the second.*
- What's 3 to the second? (Signal.) *9.*
  (Add to show:)   [72:3O]

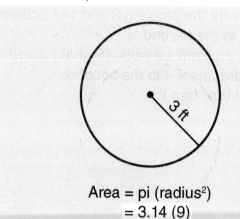

Area = pi (radius²)
= 3.14 (9)

- Work the problem.
  (Observe students and give feedback.)

- Everybody, what's the area of the circle?
  (Signal.) *28 and 26 hundredths square feet.*
k. Remember the equation for finding the area of
   a circle.
- Say the equation. (Signal.) *Area = pi times
  radius to the second.*

EXERCISE 4: SHORT DIVISION
WITH MENTAL MATH   REMEDY

a. Open your textbook to Lesson 72 and find
   part 1. ✔
   (Teacher reference:)

a. $7\overline{)1065}$ b. $3\overline{)863}$ c. $2\overline{)77}$ d. $5\overline{)728}$

These problems have a fraction remainder.
b. Read problem A. (Signal.) *1065 ÷ 7.*
- Copy and work the problem.
  (Observe students and give feedback.)
c. Check your work.
- What's 1065 ÷ 7? (Signal.) *152 and 1/7.*
  (Display:)   [72:4A]

$$\text{a.} \quad 7\overline{)1\,0_36_15}\;{}^{152\frac{1}{7}}$$

Here's what you should have.
d. Read problem B. (Signal.) *863 ÷ 3.*
- Copy and work the problem.
  (Observe students and give feedback.)
e. Check your work.
- What's 863 ÷ 3? (Signal.) *287 and 2/3.*
  (Display:)   [72:4B]

$$\text{b.} \quad 3\overline{)8_26_23}\;{}^{287\frac{2}{3}}$$

Here's what you should have.
f. Read problem C. (Signal.) *77 ÷ 2.*
- Copy and work the problem.
  (Observe students and give feedback.)
g. Check your work.
- What's 77 ÷ 2? (Signal.) *38 and 1/2.*
  (Display:)   [72:4C]

$$\text{c.} \quad 2\overline{)7_17}\;{}^{38\frac{1}{2}}$$

Here's what you should have.

h. Read problem D. (Signal.) *728 ÷ 5.*
- Copy and work the problem.
  (Observe students and give feedback.)
i. Check your work.
- What's 728 ÷ 5? (Signal.) *145 and 3/5.*
  (Display:) [72:4D]

$$\text{d. } 5\overline{)7_2 2_2 8} \; 1\,4\,5\tfrac{3}{5}$$

Here's what you should have.

EXERCISE 5: EXPONENTS
BASE 10

a. (Display:) [72:5A]

$$351 \times 10^2 =$$

- Read this problem. (Signal.) *351 × 10 to the second.*
  We're going to write what 351 × 10 to the second equals.
- What are we multiplying by? (Signal.) *10 to the second.*
- So how many zeros do we write after 351? (Signal.) *2.*
  (Add to show:) [72:5B]

$$351 \times 10^2 = 35{,}100$$

- Read the equation. (Signal.) *351 × 10² = 35,100.*

b. (Display:) [72:5C]

$$24 \times 10^4 =$$

- Read this problem. (Signal.) *24 × 10⁴.*
- What are we multiplying by? (Signal.) *10⁴.*
- So how many zeros do we write after 24? (Signal.) *4.*
  (Add to show:) [72:5D]

$$24 \times 10^4 = 240{,}000$$

- Everybody, read the equation. (Signal.)
  24 × 10⁴ = 240,000.

a. Find part 2 in your textbook. ✔
  (Teacher reference:)

  a. $135 \times 10^3 =$ ▬ d. $90 \times 10^3 =$ ▬
  b. $70{,}200 =$ ▬ $\times$ ▪▪ e. $1{,}080{,}000 =$ ▬ $\times$ ▪▪
  c. $28 \times 10^5 =$ ▬

Part 2		
a.	▪ × ▪▪ =	▬
b.	▬ = ▪ × ▪▪	

b. Read problem A. (Signal.) *135 × 10³.*
- Copy the problem and complete the equation.
  (Observe students and give feedback.)
- Everybody, read the equation. (Signal.)
  135 × 10³ = 135,000.
c. Read number B. (Signal.) *70,200.*
  You'll write a number times 10 with an exponent.
- How many zeros tell about the exponent? (Signal.) *2.*
- Write what the number equals. ✔
- Everybody, read the equation. (Signal.)
  70,200 = 702 × 10².
d. Work the rest of the problems in part 2. You'll either figure out the number or figure out the exponent.
  (Observe students and give feedback.)
e. Check your work.
- Read the equation for C. (Signal.)
  28 × 10⁵ = 2,800,000.
- Read the equation for D. (Signal.)
  90 × 10³ = 90,000.
- Read the equation for E. (Signal.)
  1,080,000 = 108 × 10⁴.

EXERCISE 6: ROUNDING
DECIMALS WITH 9

a. You learned that if you round 9 up, you get a tens number.
- If you round 29 thousandths up, what do you get? (Signal.) *30 thousandths.*
- If you round 69 hundredths up, what do you get? (Signal.) *70 hundredths.*
- If you round 9 tenths up, what do you get? (Signal.) *10 tenths.*
  (Repeat until firm.)
  Yes, 10 tenths, which is one and zero tenths.

b. Listen: If you round 269 thousandths up, what do you get? (Signal.) *270 thousandths.*
• If you round 819 thousandths up, what do you get? (Signal.) *820 thousandths.*
• If you **don't** round 289 thousandths up, what do you get? (Signal.) *289 thousandths.*
(Repeat until firm.)

═══ TEXTBOOK PRACTICE ═══

a. Find part 3 in your textbook. ✔
(Teacher reference:)

a. Round .2696 to the thousandths place.
b. Round .4093 to the thousandths place.
c. Round .5982 to the hundredths place.
d. Round .1297 to the thousandths place.
e. Round .946 to the tenths place.

Part 3
a. ▮ b. ▮
.▮ .▮

• Copy the numbers in a row. ✔
b. Item A. The directions tell you to round the number to thousandths.
• Underline the digits through thousandths. ✔
• How many thousandths does the number have? (Signal.) *269.*
• Look at the digit after the 9 and write the rounded number below. ✔
• Read the rounded number. (Signal.) *270 thousandths.*
(Display:)   [72:6A]

a. .2 6 9 6
‾‾‾‾‾
.2 7 0

Here's what you should have.

c. Item B tells you to round the number to thousandths. Do it. Underline the number through thousandths. Write the rounded number below. ✔
• Read the rounded number. (Signal.) *409 thousandths.*
(Display:)   [72:6B]

b. .4 0 9 3
‾‾‾‾‾
.4 0 9

Here's what you should have.

d. Item C tells you to round the number to hundredths. Do it. Underline the number through hundredths. Write the rounded number below. ✔
• Read the rounded number. (Signal.) *60 hundredths.*
(Display:)   [72:6C]

c. .5 9 8 2
‾‾‾
.6 0

Here's what you should have.

e. Item D tells you to round the number to thousandths. Do it. Underline the number through thousandths. Write the rounded number below. ✔
• Read the rounded number. (Signal.) *130 thousandths.*
(Display:)   [72:6D]

d. .1 2 9 7
‾‾‾‾‾
.1 3 0

Here's what you should have.

f. Item E tells you to round the number to tenths. Do it. Underline the number through tenths. Write the rounded number below. ✔
• Read the rounded number. (Signal.) *9 tenths.*
(Display:)   [72:6E]

e. .9 4 6
‾
.9

Here's what you should have.

EXERCISE 7: WORD PROBLEMS
RATIO/FRACTION OF A GROUP

a. Find part 4 in your textbook.
(Teacher reference:)

a. $\frac{4}{10}$ of the windows were dirty. There were 60 windows. How many dirty windows were there?
b. There were 2 dirty windows for every 7 windows. There were 32 dirty windows. How many windows were there?
c. $\frac{3}{5}$ of the ducks were swimming. There were 45 ducks in all. How many were swimming?
d. There were 2 open books for every 3 books. There were 16 open books. How many books were there in all?

For some of these problems, you'll write a fraction equation. For others, you'll write a ratio equation.
The first sentence in all the problems tells what kind of equation you write.

b. Problem A: 4/10 of the windows were dirty. There were 60 windows. How many dirty windows were there? Listen to the first sentence again: 4/10 of the windows were dirty.

- To work that problem, do you write a fraction equation or a ratio equation? (Signal.) *A fraction equation.*

c. Problem B: There were 2 dirty windows for every 7 windows. There were 32 dirty windows. How many windows were there? Listen to the first sentence again: There were 2 dirty windows for every 7 windows.

- To work that problem, do you write a fraction equation or a ratio equation? (Signal.) *A ratio equation.*
(Repeat until firm.)

d. Problem C: 3/5 of the ducks were swimming. There were 45 ducks in all. How many were swimming? Listen to the first sentence again: 3/5 of the ducks were swimming.

- To work that problem, do you write a fraction equation or a ratio equation? (Signal.) *A fraction equation.*

e. Problem D: There were 2 open books for every 3 books. There were 16 open books. How many books were there in all? Listen to the first sentence again: There were 2 open books for every 3 books.

- To work that problem, do you write a fraction equation or a ratio equation? (Signal.) *A ratio equation.*
(Repeat until firm.)

f. Go back to problem A. Work the whole problem.
(Observe students and give feedback.)
Check your work.

- Read the fraction equation you started with. (Signal.) *4/10 W = D.*
- How many dirty windows were there? (Signal.) *24 dirty windows.*
(Display:)   [72:7A]

$$\text{a. } \frac{4}{10} \, w = d$$
$$\frac{4}{\cancel{10}} \, (\cancel{60})^6 = d = 24$$
$$\boxed{24 \text{ dirty windows}}$$

Here's what you should have.

g. Work problem B.
(Observe students and give feedback.)
Check your work.

- Did you start with a fraction equation or a ratio equation? (Signal.) *A ratio equation.*
- How many windows were there? (Signal.) *112 windows.*
(Display:)   [72:7B]

$$\text{b. } \frac{d}{w} \quad \frac{2}{7} \left(\frac{16}{16}\right) = \frac{32}{\boxed{112 \text{ windows}}}$$

Here's what you should have.

h. Work problem C.
(Observe students and give feedback.)
Check your work.

- Did you start with a fraction equation or a ratio equation? (Signal.) *A fraction equation.*
- Read the fraction equation. (Signal.) *3/5 D = S.*
- How many ducks were swimming? (Signal.) *27 swimming ducks.*
(Display:)   [72:7C]

$$\text{c. } \frac{3}{5} \, d = s$$
$$\frac{3}{\cancel{5}} \, (\cancel{45})^9 = s = 27$$
$$\boxed{27 \text{ swimming ducks}}$$

Here's what you should have.

i. Work problem D.
(Observe students and give feedback.)
Check your work.

- Did you start with a fraction equation or a ratio equation? (Signal.) *A ratio equation.*
- How many books were there in all? (Signal.) *24 books.*
(Display:)   [72:7D]

$$\text{d. } \frac{o}{b} \quad \frac{2}{3} \left(\frac{8}{8}\right) = \frac{16}{\boxed{24 \text{ books}}}$$

Here's what you should have.

Assign Independent Work, Textbook parts 5–9 and Workbook parts 3–7.

Optional extra math-fact practice worksheets are available on ConnectED.

Lesson 73

EXERCISE 1: FRACTION ANALYSIS

COMPARISON REMEDY

a. Open your workbook to Lesson 73 and find part 1.
(Teacher reference:) R **Part I**

a. $\frac{1}{4}$ $\frac{2}{12}$ b. $\frac{1}{3}$ $\frac{9}{24}$ c. $\frac{1}{5}$ $\frac{7}{20}$ d. $\frac{1}{10}$ $\frac{3}{40}$

You're going to make signs to show which fraction is more.

b. Read the fractions in A. (Signal.) *1/4, 2/12.*
• Which fraction has the larger denominator? (Signal.) *2/12.*
• So what do you ask? (Signal.) *How many 12ths is 1/4?*

c. Read the fractions in B. (Signal.) *1/3, 9/24.*
• Which fraction has the larger denominator? (Signal.) *9/24.*
• So what do you ask? (Signal.) *How many 24ths is 1/3?*

d. Read the fractions in C. (Signal.) *1/5, 7/20.*
• Which fraction has the larger denominator? (Signal.) *7/20.*
• So what do you ask? (Signal.) *How many 20ths is 1/5?*

e. Go back to problem A. ✔
• Say the question you ask? (Signal.) *How many 12ths is 1/4?*
• Raise your hand when you know the answer. ✔
• How many 12ths is 1/4? (Signal.) *3/12.*
• Write the sign to show which fraction is more. ✔
(Display:)   [73:1A]

> **a.** $\frac{1}{4} > \frac{2}{12}$

Here's what you should have.
1/4 is 3/12. 3/12 is more than 2/12.

f. Problem B.
• Say the question you ask. (Signal.) *How many 24ths is 1/3?*
• Raise your hand when you know the answer. ✔
• How many 24ths is 1/3? (Signal.) *8/24.*
• Write the sign to show which fraction is more. ✔
(Display:)   [73:1B]

> **b.** $\frac{1}{3} < \frac{9}{24}$

Here's what you should have.
1/3 is 8/24. 9/24 is more than 8/24.

g. Problem C.
• Say the question you ask. (Signal.) *How many 20ths is 1/5?*
• Raise your hand when you know the answer. ✔
• How many 20ths is 1/5? (Signal.) *4/20.*
• Write the sign to show which fraction is more. ✔
(Display:)   [73:1C]

> **c.** $\frac{1}{5} < \frac{7}{20}$

Here's what you should have.
1/5 is 4/20. 7/20 is more than 4/20.

h. Problem D.
• Say the question you ask. (Signal.) *How many 40ths is 1/10?*
• Raise your hand when you know the answer. ✔
• How many 40ths is 1/10? (Signal.) *4/40.*
• Write the sign to show which fraction is more. ✔
(Display:)   [73:1D]

> **d.** $\frac{1}{10} > \frac{3}{40}$

Here's what you should have.
1/10 is 4/40. 4/40 is more than 3/40.

EXERCISE 2: MENTAL MATH

a. Time for some mental math.
- Listen: What's 40 divided by 10? (Signal.) *4.*
- What's 90 ÷ 10? (Signal.) *9.*
- What's 900 ÷ 10? (Signal.) *90.*

b. Listen: 950 divided by 10 is 95.
- What's 850 ÷ 10? (Signal.) *85.*
- What's 830 ÷ 10? (Signal.) *83.*
  (Repeat until firm.)

c. Listen: What's 80 ÷ 10? (Signal.) *8.*
- 80 ÷ 4. (Pause.) (Signal.) *20.*
- 200 ÷ 4. (Pause.) (Signal.) *50.*
- 200 ÷ 5. (Pause.) (Signal.) *40.*
- 200 ÷ 10. (Pause.) (Signal.) *20.*
- 200 ÷ 2. (Pause.) (Signal.) *100.*
  (Repeat until firm.)

d. New problem type.
- Listen: 40 times 2. (Pause.) (Signal.) *80.*
- 40 times 3. (Pause.) (Signal.) *120.*
- 40 times 5. (Pause.) (Signal.) *200.*
  (Repeat until firm.)

EXERCISE 3: FRACTION OPERATIONS
DIVISION

a. Dividing by a fraction is the same as multiplying by the reciprocal of the fraction.
- Dividing by 2/5 is the same as multiplying by 5/2.
- What's the same as dividing by 2/5? (Signal.) *Multiplying by 5/2.*
- What's the same as dividing by 4/5? (Signal.) *Multiplying by 5/4.*
- What's the same as dividing by 1/5? (Signal.) *Multiplying by 5 over 1.*
- What's the same as dividing by 3/10? (Signal.) *Multiplying by 10/3.*
- What's the same as dividing by 4/9? (Signal.) *Multiplying by 9/4.*
- What's the same as dividing by 8/3? (Signal.) *Multiplying by 3/8.*
  (Repeat until firm.)

b. (Display:) [73:3A]

$$20 \div \frac{3}{5} = 20 \times \frac{\square}{\square}$$

Here's 20 divided by 3/5.
- That's the same as multiplying 20 by what fraction? (Signal.) *5/3.*

c. (Change to show:) [71:3B]

$$20 \div \frac{2}{10} = 20 \times \frac{\square}{\square}$$

Here's 20 divided by 2/10.
- That's the same as multiplying 20 by what fraction? (Signal.) *10/2.*

d. (Display:) [73:3C]

$$80 \div \frac{4}{5} = 80 \times \frac{\square}{\square}$$

Here's 80 divided by 4/5.
- That's the same as multiplying 80 by what fraction? (Signal.) *5/4.*

e. (Display:) [73:3D]

$$10 \div \frac{6}{3} = 10 \times \frac{\square}{\square}$$

Here's 10 divided by 6/3.
- That's the same as multiplying 10 by what fraction? (Signal.) *3/6*
  (Add to show:) [73:3E]

$$10 \div \frac{6}{3} = 10 \times \frac{3}{6}$$

f. Let's see if that's true.
- 6/3 equals a whole number. What number? (Signal.) *2.*
  So 10 divided by 6/3 is the same as 10 divided by 2.
  (Add to show:) [73:3F]

$$10 \div \frac{6}{3} = 10 \times \frac{3}{6}$$
$$10 \div 2$$

- What's 10 divided by 2? (Signal.) *5.*
  (Point left.) So on this side of the equation we have 5.
  (Add to show:) [73:3G]

$$10 \div \frac{6}{3} = 10 \times \frac{3}{6}$$
$$10 \div 2$$
$$5$$

g. On the other side, we have 10 times 3/6.
- Say the multiplication for the numerators. (Signal.) *10 × 3.*
- What's the answer? (Signal.) *30.*
- What's the denominator? (Signal.) *6.* (Add to show:) [73:3H]

$$10 \div \frac{6}{3} = 10 \times \frac{3}{6}$$
$$10 \div 2 = \frac{30}{6}$$
$$5$$

- What whole number does 30/6 equal? (Signal.) *5.* (Add to show:) [73:3I]

$$10 \div \frac{6}{3} = 10 \times \frac{3}{6}$$
$$10 \div 2 = \frac{30}{6}$$
$$5 = 5$$

h. We have 5 on each side of the equation. So dividing by a fraction is the same as multiplying by the reciprocal of that fraction.

=========== **WORKBOOK PRACTICE** ===========

a. Find part 2 in your workbook. ✔
- Pencils down. ✔ (Teacher reference:)

a. $30 \div \frac{4}{5} = 30 \times$ d. $100 \div \frac{17}{12} = 100 \times$

b. $13 \div \frac{3}{2} = 13 \times$ e. $3 \div \frac{5}{8} = 3 \times$

c. $34 \div \frac{7}{9} = 34 \times$

Each equation shows what you divide by. You're going to complete each equation to show the fraction you multiply by.

b. Item A: 30 divided by 4/5 = 30 times a fraction.
- What fraction? (Signal.) *5/4.*
- Say the equation. (Signal.) *30 ÷ 4/5 = 30 × 5/4.*
c. Item B: 13 divided by 3/2 = 13 times a fraction.
- What fraction? (Signal.) *2/3.*
- Say the equation. (Signal.) *13 ÷ 3/2 = 13 × 2/3.*

d. Item C: 34 divided by 7/9 = 34 times a fraction.
- What fraction? (Signal.) *9/7.*
- Say the equation. (Signal.) *34 ÷ 7/9 = 34 × 9/7.*
e. Complete each equation to show the fraction you multiply by. (Observe students and give feedback.)
f. Check your work.
- Item A: 30 ÷ 4/5 = 30 times what fraction? (Signal.) *5/4.*
- B: 13 ÷ 3/2 = 13 times what fraction? (Signal.) *2/3.*
- C: 34 ÷ 7/9 = 34 times what fraction? (Signal.) *9/7.*
- D: 100 ÷ 17/12 = 100 times what fraction? (Signal.) *12/17.*
- E: 3 ÷ 5/8 = 3 times what fraction? (Signal.) *8/5.*

EXERCISE 4: COORDINATE SYSTEM
WORD PROBLEMS

a. Find part 3 in your workbook. ✔
(Teacher reference:)   R Part A

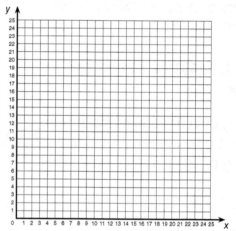

- The ratio of chairs to people is 4 to 3.

a. If there are 12 people, how many chairs are there? _____

$$\left(\ \ \right) =$$

b. If there are 8 chairs, how many people are there? _____

c. If there are 20 chairs, how many people are there? _____

d. If there are 9 people, how many chairs are there? _____

e. If there are 18 people, how many chairs are there? _____

You're going to work ratio problems by making points and a line on the coordinate system.

b. Listen: The ratio of chairs to people is 4 to 3. The first sentence tells the names for Y and X. It also gives the numbers for the first point on the coordinate system.

• Write the names and make the first point on the coordinate system.
(Observe students and give feedback.)
(Display:)   [73:4A]

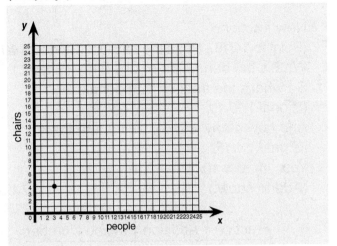

Here's what you should have.

c. Now you're going to work a problem to figure out another point on the coordinate system. Question A: If there are 12 people, how many chairs are there?

• Work the problem. Write the ratio equation and the answer.
(Observe students and give feedback.)

• Everybody, how many chairs are there? (Signal.) *16 chairs.*
(Add to show:)   [73:4B]

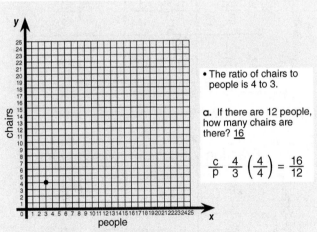

Here's what you should have.

d. The fraction 16/12 tells about another point on the coordinate system. Make that point. Then make a line that connects the points and goes through zero.
(Add to show:)   [73:4C]

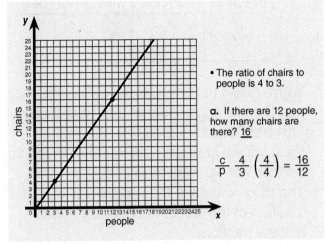

Here's what you should have.
Now you can use the line to answer the rest of the questions.

e. Read question B. (Signal.) *If there are 8 chairs, how many people are there?*

• Make the point for 8 chairs, and write the answer to the question. ✔

• Everybody, if there are 8 chairs, how many people are there? (Signal.) *6.*

f. Read question C. (Signal.) *If there are 20 chairs, how many people are there?*

• Make the point for 20 chairs and write the answer to the question. ✔

• Everybody, if there are 20 chairs, how many people are there? (Signal.) *15.*

g. Read question D. (Signal.) *If there are 9 people, how many chairs are there?*

• Make the point for 9 people, and write the answer to the question. ✔

• Everybody, if there are 9 people, how many chairs are there? (Signal.) *12.*

h. Read question E. (Signal.) *If there are 18 people, how many chairs are there?*

• Make the point for 18 people, and write the answer to the question. ✔

• Everybody, if there are 18 people, how many chairs are there? (Signal.) *24.*

EXERCISE 5: FRACTIONS
AS MIXED NUMBERS

a. (Display:)   [73:5A]

Fraction	Addition	Mixed Number
$\frac{8}{5}$		

You're going to learn a fast way to write mixed numbers for fractions that are more than one. The first column in the table shows the fractions we start with. All fractions in the first column will be more than one.

The next column in the table will show the addition of the whole number and the leftover fraction.

The last column will show the mixed number.

b. (Point to **8/5**.) Read the fraction. (Signal.) *8/5.*
We write 8/5 as 1 plus the leftover fraction.
- What's the denominator of 8/5? (Signal.) *5.*
So the fraction that equals 1 is 5/5.
(Add to show:)   [73:5B]

Fraction	Addition	Mixed Number
$\frac{8}{5}$	$\frac{5}{5} +$	

- 8/5 = 5/5 plus some fraction. What fraction? (Signal.) *3/5.*
(Add to show:)   [73:5C]

Fraction	Addition	Mixed Number
$\frac{8}{5}$	$\frac{5}{5} + \frac{3}{5}$	

5/5 is 1. So 8/5 = 1 and 3/5.
(Add to show:)   [73:5D]

Fraction	Addition	Mixed Number
$\frac{8}{5}$	$\frac{5}{5} + \frac{3}{5}$	$1\frac{3}{5}$

- Read the mixed number that equals 8/5. (Signal.) *1 and 3/5.*

c. (Add to show:)   [73:5E]

Fraction	Addition	Mixed Number
$\frac{8}{5}$	$\frac{5}{5} + \frac{3}{5}$	$1\frac{3}{5}$
$\frac{14}{10}$		

New fraction.
- (Point to **14/10**.) Read the fraction. (Signal.) *14/10.*
- What's the denominator of 14/10? (Signal.) *10.*
- So what's the fraction that equals 1? (Signal.) *10/10.*
- And how many leftover tenths are there? (Signal.) *4/10.*
Yes, 14/10 = 10/10 + 4/10.
(Add to show:)   [73:5F]

Fraction	Addition	Mixed Number
$\frac{8}{5}$	$\frac{5}{5} + \frac{3}{5}$	$1\frac{3}{5}$
$\frac{14}{10}$	$\frac{10}{10} + \frac{4}{10}$	

10/10 is 1. So 14/10 equals what mixed number? (Signal.) *1 and 4/10.*
(Add to show:)   [73:5G]

Fraction	Addition	Mixed Number
$\frac{8}{5}$	$\frac{5}{5} + \frac{3}{5}$	$1\frac{3}{5}$
$\frac{14}{10}$	$\frac{10}{10} + \frac{4}{10}$	$1\frac{4}{10}$

d. (Add to show:)   [73:5H]

Fraction	Addition	Mixed Number
$\frac{8}{5}$	$\frac{5}{5} + \frac{3}{5}$	$1\frac{3}{5}$
$\frac{14}{10}$	$\frac{10}{10} + \frac{4}{10}$	$1\frac{4}{10}$
$\frac{16}{9}$		

New fraction.
- (Point to **16/9**.) Read the fraction. (Signal.) *16/9.*
This fraction equals 1 plus leftover parts.
- What's the fraction that equals 1? (Signal.) *9/9.*

- What's the fraction for the leftover ninths?
  (Signal.) *7/9.*
  (Add to show:)   [73:5I]

Fraction	Addition	Mixed Number
$\frac{8}{5}$	$\frac{5}{5} + \frac{3}{5}$	$1\frac{3}{5}$
$\frac{14}{10}$	$\frac{10}{10} + \frac{4}{10}$	$1\frac{4}{10}$
$\frac{16}{9}$	$\frac{9}{9} + \frac{7}{9}$	

- What's the mixed number that equals 16/9?
  (Signal.) *1 and 7/9.*
  (Add to show:)   [73:5J]

Fraction	Addition	Mixed Number
$\frac{8}{5}$	$\frac{5}{5} + \frac{3}{5}$	$1\frac{3}{5}$
$\frac{14}{10}$	$\frac{10}{10} + \frac{4}{10}$	$1\frac{4}{10}$
$\frac{16}{9}$	$\frac{9}{9} + \frac{7}{9}$	$1\frac{7}{9}$

══════ **WORKBOOK PRACTICE** ══════

a. Find part 4 in your workbook. ✔
   (Teacher reference:)

	Fraction	Addition	Mixed Number
a.	$\frac{17}{12}$	+	
b.	$\frac{13}{8}$	+	
c.	$\frac{18}{11}$	+	
d.	$\frac{17}{9}$	+	

All the fractions are shown in the first column.
You're going to write the addition and the
mixed number each fraction equals.
b. Read fraction A. (Signal.) *17/12.*
- What's the fraction that equals 1? (Signal.) *12/12.*
- Write the fraction that equals 1 plus the
  leftover twelfths. Then write the mixed number
  that equals 17/12.
  (Observe students and give feedback.)
  Check your work.
- 17/12 = 12/12 + 5/12. What mixed number is
  that? (Signal.) *1 and 5/12.*

(Display:)   [73:5K]

	Fraction	Addition	Mixed Number
a.	$\frac{17}{12}$	$\frac{12}{12} + \frac{5}{12}$	$1\frac{5}{12}$

Here's what you should have.
c. Work the rest of the items. Show the fraction
   that equals 1 plus the leftovers. Then write the
   mixed number.
   (Observe students and give feedback.)
d. Check your work.
- Item B. 13/8 = 8/8 + 5/8. What mixed number
  is that? (Signal.) *1 and 5/8.*
- C. 18/11 = 11/11 + 7/11. What mixed number is
  that? (Signal.) *1 and 7/11.*
- D. 17/9 = 9/9 + 8/9. What mixed number is
  that? (Signal.) *1 and 8/9.*

EXERCISE 6: AREA
OF CIRCLES   `REMEDY`

a. You learned an equation for finding the area of
   a circle. The equation is area = pi times radius
   to the second.
- Say the equation. (Signal.) *Area = pi times
  radius to the second.*
  The number for pi is 3 and 14 hundredths.
- What's pi? (Signal.) *3.14.*
b. (Display:)   [73:6A]

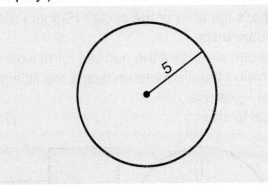

The radius is the distance from the center of a
circle to the edge.
- What's the radius of this circle? (Signal.) *5.*
  Every radius we show for this circle will be the
  same length as the radius shown.
- What's the length? (Signal.) *5.*
  So the multiplication for finding the area is
  3.14 × 5 to the second.
- Say the multiplication for finding the area.
  (Signal.) *3.14 × 5 to the second.*

c. (Display:) [73:6B]

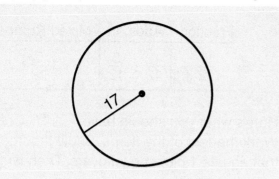

- What's the radius of this circle? (Signal.) *17.*
- Say the multiplication for finding the area. (Signal.) *3.14 × 17 to the second.*
  (Repeat until firm.)

d. (Display:) [73:6C]

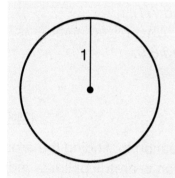

- What's the radius of this circle? (Signal.) *1.*
- Say the multiplication for finding the area. (Signal.) *3.14 × 1 to the second.*
  (Repeat until firm.)

e. What's 1 to the second? (Signal.) *1.*
   So the area of the circle is 3.14 square units.
- What's the area of the circle? (Signal.) *3.14 square units.*

f. We can show that the number for pi works by showing the 3 square units and the little part that is left over.
   (Add to show:) [73:6D]

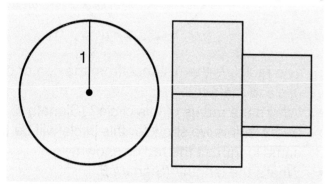

Here's 3 and 14 hundredths square units. You can see that the side of each unit is the same length as the radius. If we fit these parts in a circle that has a radius of 1, here's what we get.

(Change to show:) [73:6E]

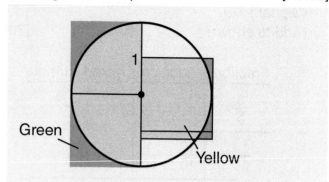

You can see that the 3 square units fill most of the circle. Those units are yellow.
Now we put the green parts inside the circle.
(Change to show:) [73:6F]

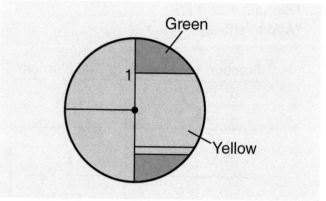

The circle is filled.

━━━━━ **TEXTBOOK PRACTICE** ━━━━━

a. Open your textbook to Lesson 73 and find part 1. ✔
(Teacher reference:)

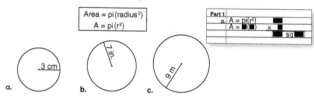

- Read the equation with words. (Signal.) *Area = pi times radius to the second.*
  Below is the equation with letters. It says the same thing: Area = pi times radius to the second.
- Say the equation. (Signal.) *Area = pi times radius to the second.*

b. Problem A. Write the letter equation for the area of a circle. ✔
- What's the radius of A? (Signal.) *3 centimeters.*

- Write the equation below with a number for pi and a number for the radius to the second. ✔
  (Display:)   [73:6G]

  > **a.** $A = pi (r^2)$
  > $A = 3.14 (9)$

  Here's what you should have: $A = 3.14 \times 9$.
- Work the problem and write the answer. Remember the unit name.
  (Observe students and give feedback.)
- Check your work. What's the area of circle A? (Signal.) *28 and 26 hundredths square centimeters.*
  (Display:)   [73:6H]

  > **a.** $A = pi (r^2)$
  > $A = 3.14 (9)$
  >
  > $$\begin{array}{r} \overset{1}{3}.\overset{3}{1}\,4 \\ \times \quad 9 \\ \hline \boxed{2\,8.2\,6 \text{ sq cm}} \end{array}$$

  Here's what you should have.
c. Problem B. Write the letter equation for the area of any circle. Below, write the equation with numbers for circle B. Stop when you've done that much.
  (Observe students and give feedback.)
- Read the equation with numbers. (Signal.) *A = 3.14 (49).*
- Work the problem and write the answer with the unit name.
  (Observe students and give feedback.)
- Check your work. What's the area of circle B? (Signal.) *153.86 square inches.*
d. Work problem C. Write the equation with letters. Below, write the equation with numbers for circle C. Figure out the area.
  (Observe students and give feedback.)
- Check your work. What's the area of circle C? (Signal.) *254.34 square meters.*

Exercise 7: Rounding
Decimals with 9

a. Find part 2 in your textbook. ✔
  (Teacher reference:)

  > a. Round .391 to the hundredths place.
  > b. Round .9862 to the tenths place.
  > c. Round .4959 to the hundredths place.
  > d. Round .72988 to the thousandths place.

- Copy the numbers in a row. ✔
  You're going to round the numbers to the place each item tells. See if you can work the problems without underlining the digits. If you have trouble, underline the digits.
b. Item A tells you to round the number to hundredths. Do it. ✔
- Read the rounded number. (Signal.)
  39 hundredths.
c. Item B tells you to round the number to tenths. Do it. ✔
- Read the rounded number. (Signal.) *1 and zero tenths.*
d. Item C tells you to round the number to hundredths. Do it. ✔
- Read the rounded number. (Signal.)
  50 hundredths.
e. Item D tells you to round the number to thousandths. Do it. ✔
- Read the rounded number. (Signal.)
  730 thousandths.

Exercise 8: Independent Work
Word Problems—Ratio/Fraction of a Group

a. Find part 3 in your textbook.
  (Teacher reference:)

  > a. There were 5 red bugs for every 7 green bugs. There were 28 green bugs. How many red bugs were there?
  > b. $\frac{5}{6}$ of the ants were workers. There were 60 worker ants. How many ants were there in all?
  > c. Rachel walks 3 miles every 8 days. How many days will it take her to walk 15 miles?
  > d. $\frac{4}{9}$ of the pictures had frames. There were 72 pictures. How many pictures were framed?

  For some of these problems, you'll write a fraction equation. For others, you'll write a ratio equation.

b. Problem A: There were 5 red bugs for every 7 green bugs. There were 28 green bugs. How many red bugs were there?
- To work that problem, do you write a fraction equation or a ratio equation? (Signal.) *A ratio equation.*

c. Problem B: 5/6 of the ants were workers. There were 60 worker ants. How many ants were there in all?
- To work that problem, do you write a fraction equation or a ratio equation? (Signal.) *A fraction equation.*
(Repeat until firm.)

d. Problem C: Rachel walks 3 miles every 8 days. How many days will it take her to walk 15 miles?
- To work that problem, do you write a fraction equation or a ratio equation? (Signal.) *A ratio equation.*

e. Problem D: 4/9 of the pictures had frames. There were 72 pictures. How many pictures were framed?
- To work that problem, do you write a fraction equation or a ratio equation? (Signal.) *A fraction equation.*
(Repeat until firm.)

f. Go back to problem A. Work the problem. (Observe students and give feedback.) Check your work.
- Did you start with a fraction equation or a ratio equation? (Signal.) *A ratio equation.*

- How many red bugs were there? (Signal.) *20 red bugs.*
(Display:) [73:8A]

$$\text{a. } \frac{r}{g} \quad \frac{5}{7} \left(\frac{4}{4} \right) = \frac{\boxed{20 \text{ red bugs}}}{28}$$

Here's what you should have.

g. Work problem B. (Observe students and give feedback.) Check your work.
- Did you start with a fraction equation or a ratio equation? (Signal.) *A fraction equation.*
- Read the fraction equation. (Signal.) *5/6 A = W.*
- How many ants were there in all? (Signal.) *72 ants.*
(Display:) [73:8B]

$$\text{b. } \frac{5}{6} a = w$$
$$\left(\frac{6}{5} \right) \frac{5}{6} a = (\cancel{60})^{12} \left(\frac{6}{5} \right)$$
$$a = 72$$
$$\boxed{72 \text{ ants}}$$

Here's what you should have.

h. You'll work the rest of the problems as part of your independent work.

Assign Independent Work, the rest of Textbook part 3 and parts 4–9 and Workbook part 5.

Optional extra math-fact practice worksheets are available on ConnectED.

Lesson 74

EXERCISE 1: FRACTION ANALYSIS
COMPARISON

REMEDY

a. Open your workbook to Lesson 74 and find part 1.
   (Teacher reference:)

   R Part J

   a. $\frac{1}{2}$ ▮ $\frac{5}{8}$ b. $\frac{1}{6}$ ▮ $\frac{2}{18}$ c. $\frac{1}{4}$ ▮ $\frac{6}{20}$ d. $\frac{1}{8}$ ▮ $\frac{5}{56}$

   You're going to ask questions to figure out which fraction is more.

b. Problem A. Read the fractions. (Signal.) *1/2, 5/8.*
   - Say the question about how many eighths. (Signal.) *How many 8ths is 1/2?*
c. Problem B. Read the fractions. (Signal.) *1/6, 2/18.*
   - Say the question about how many 18ths. (Signal.) *How many 18ths is 1/6?*
d. Problem C. Read the fractions. (Signal.) *1/4, 6/20.*
   - Say the question. (Signal.) *How many 20ths is 1/4?*
   (Repeat until firm.)
e. Problem D. Read the fractions. (Signal.) *1/8, 5/56.*
   - Say the question. (Signal.) *How many 56ths is 1/8?*
   (Repeat until firm.)
f. Work all the problems. Write the sign to show which fraction is more.
   (Observe students and give feedback.)
g. Check your work.
   - Problem A. How many 8ths is 1/2? (Signal.) *4/8.*
   - So which is more—1/2 or 5/8? (Signal.) *5/8.*
h. Problem B. How many 18ths is 1/6? (Signal.) *3/18.*
   - So which is more—1/6 or 2/18? (Signal.) *1/6.*
i. Problem C. How many 20ths is 1/4? (Signal.) *5/20.*
   - So which is more—1/4 or 6/20? (Signal.) *6/20.*
j. Problem D. How many 56ths is 1/8? (Signal.) *7/56.*
   - So which is more—1/8 or 5/56? (Signal.) *1/8.*

(Display:)   [74:1A]

a. $\frac{1}{2} < \frac{5}{8}$ b. $\frac{1}{6} > \frac{2}{18}$

c. $\frac{1}{4} < \frac{6}{20}$ d. $\frac{1}{8} > \frac{5}{56}$

Here's what you should have for each item.

EXERCISE 2: MENTAL MATH

a. Time for some mental math.
   - Listen: What's 60 divided by 10? (Signal.) *6.*
   - 60 ÷ 3. (Pause.) (Signal.) *20.*
   - 60 ÷ 2. (Pause.) (Signal.) *30.*
   - 60 ÷ 6. (Pause.) (Signal.) *10.*
   (Repeat until firm.)
b. New problem type.
   - Listen: 600 divided by 6. (Pause.) (Signal.) *100.*
   - 600 ÷ 3. (Pause.) (Signal.) *200.*
   - 600 ÷ 2. (Pause.) (Signal.) *300.*
   - 800 ÷ 2. (Pause.) (Signal.) *400.*
   - 800 ÷ 10. (Pause.) (Signal.) *80.*
   - 800 ÷ 4. (Pause.) (Signal.) *200.*
   (Repeat until firm.)
c. New problem type.
   - Listen: 2 times 30. (Pause.) (Signal.) *60.*
   - 2 × 200. (Pause.) (Signal.) *400.*
   - 4 × 200. (Pause.) (Signal.) *800.*
   - 5 × 200. (Pause.) (Signal.) *1000.*
   - 9 × 200. (Pause.) (Signal.) *1800.*
   - 7 × 200. (Pause.) (Signal.) *1400.*
   (Repeat until firm.)

EXERCISE 3: FRACTION OPERATIONS
DIVISION

a. You learned that dividing by a fraction is the same as multiplying by the reciprocal of the fraction.
   - 8 divided by 2/5 equals 8 times what fraction? (Signal.) *5/2.*
   - 10 ÷ 3/2 equals 10 times what fraction? (Signal.) *2/3.*

b. (Display:) [74:3A]

$$\frac{2}{9} \div \frac{3}{4}$$

Here's a fraction divided by a fraction.
- Read the division problem. (Signal.) *2/9 ÷ 3/4.*
  It equals 2/9 times 4/3.
- What does it equal? (Signal.) *2/9 × 4/3.*
  (Add to show:) [74:3B]

$$\frac{2}{9} \div \frac{3}{4} = \frac{2}{9} \times \frac{4}{3}$$

- Read the equation. (Signal.) *2/9 ÷ 3/4 =
  2/9 × 4/3.*

c. (Display:) [74:3C]

$$\frac{3}{10} \div \frac{5}{4} = \frac{3}{10} \times \frac{\square}{\square}$$

- 3/10 ÷ 5/4 = 3/10 times what fraction?
  (Signal.) *4/5.*
  (Add to show:) [72:3D]

$$\frac{3}{10} \div \frac{5}{4} = \frac{3}{10} \times \frac{4}{5}$$

- Read the equation. (Signal.) *3/10 ÷ 5/4 =
  3/10 × 4/5.*

=========== WORKBOOK PRACTICE ===========

a. Find part 2 in your workbook. ✔
- Pencils down. ✔
  (Teacher reference:)

a. $\frac{7}{3} \div \frac{2}{5} = \frac{7}{3} \times$ d. $\frac{7}{12} \div \frac{3}{8} = \frac{7}{12} \times$

b. $\frac{3}{4} \div \frac{8}{7} = \frac{3}{4} \times$ e. $\frac{4}{3} \div \frac{5}{9} = \frac{4}{3} \times$

c. $\frac{9}{10} \div \frac{4}{3} = \frac{9}{10} \times$ f. $\frac{6}{5} \div \frac{9}{2} = \frac{6}{5} \times$

b. Problem A says 7/3 ÷ 2/5 = 7/3 times a
   fraction.
- What fraction? (Signal.) *5/2.*
- Say the whole equation. (Signal.) *7/3 ÷ 2/5 =
  7/3 × 5/2.*
c. Problem B says 3/4 ÷ 8/7 = 3/4 times a
   fraction.
- What fraction? (Signal.) *7/8.*
- Say the equation. (Signal.) *3/4 ÷ 8/7 = 3/4 × 7/8.*
d. Problem C says 9/10 ÷ 4/3 = 9/10 times a
   fraction.
- What fraction? (Signal.) *3/4.*
- Say the equation. (Signal.) *9/10 ÷ 4/3 =
  9/10 × 3/4.*

e. Complete all the equations.
   (Observe students and give feedback.)
f. Check your work.
- Problem A: 7/3 ÷ 2/5 = 7/3 times what
  fraction? (Signal.) *5/2.*
- Say the equation. (Signal.) *7/3 ÷ 2/5 = 7/3 × 5/2.*
g. B: 3/4 ÷ 8/7 = 3/4 times what fraction?
   (Signal.) *7/8.*
- Say the equation. (Signal.) *3/4 ÷ 8/7 = 3/4 × 7/8.*
h. C: 9/10 ÷ 4/3 = 9/10 times what fraction?
   (Signal.) *3/4.*
- Say the equation. (Signal.) *9/10 ÷ 4/3 =
  9/10 × 3/4.*
i. D: 7/12 ÷ 3/8 = 7/12 times what fraction?
   (Signal.) *8/3.*
- Say the equation. (Signal.) *7/12 ÷ 3/8 =
  7/12 × 8/3.*
j. E: 4/3 ÷ 5/9 = 4/3 times what fraction?
   (Signal.) *9/5.*
- Say the equation. (Signal.) *4/3 ÷ 5/9 = 4/3 × 9/5.*
k. F: 6/5 ÷ 9/2 = 6/5 times what fraction?
   (Signal.) *2/9.*
- Say the equation. (Signal.) *6/5 ÷ 9/2 = 6/5 × 2/9.*

EXERCISE 4: COORDINATE SYSTEM
WORD PROBLEMS   `REMEDY`

a. Find part 3 in your workbook. ✔
   (Teacher reference:)   `R` `Part B`

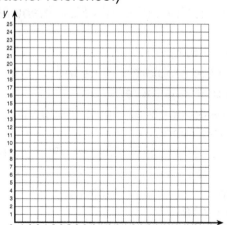

- There are 2 bananas for every 4 apples.
a. If there are 20 apples, how many bananas are there? _____

$$\left(\quad\right) =$$

b. If there are 6 bananas, how many apples are there? _____
c. If there are 2 apples, how many bananas are there? _____
d. If there are 8 apples, how many bananas are there? _____
e. If there are 8 bananas, how many apples are there? _____

You're going to work ratio problems by making
points and a line on the coordinate system.

b. There are 2 bananas for every 4 apples.
Question A: If there are 20 apples, how many bananas are there?

• Write the names and make the first point for the first sentence. Then figure out the answer to question A and make a second point.
(Observe students and give feedback.)

c. Check your work.
(Display:) [74:4A]

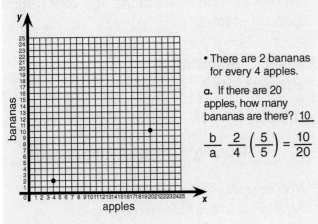

• There are 2 bananas for every 4 apples.

a. If there are 20 apples, how many bananas are there? 10

$$\frac{b}{a} \frac{2}{4} \left(\frac{5}{5} \right) = \frac{10}{20}$$

Here's what you should have.

d. Make the line. ✔
Now you're going to show other points on the line and answer more questions.

e. Read question B. (Signal.) *If there are 6 bananas, how many apples are there?*

• Read question C. (Signal.) *If there are 2 apples, how many bananas are there?*

• Read question D. (Signal.) *If there are 8 apples, how many bananas are there?*

• Read question E. (Signal.) *If there are 8 bananas, how many apples are there?*

f. Show the points for each question and write the answer.
(Observe students and give feedback.)

g. Check your work.

• Question B: If there are 6 bananas, how many apples are there? (Signal.) *12.*

• C: If there are 2 apples, how many bananas are there? (Signal.) *1.*

• D: If there are 8 apples, how many bananas are there? (Signal.) *4.*

• E: If there are 8 bananas, how many apples are there? (Signal.) *16.*

As Mixed Numbers

a. Find part 4 in your workbook. ✔
(Teacher reference:)

	Fraction	Addition	Mixed Number
a.	$\frac{19}{10}$	+	
b.	$\frac{7}{6}$	+	
c.	$\frac{17}{14}$	+	
d.	$\frac{25}{20}$	+	

You're going to write the mixed numbers for these fractions. First, you figure out the fraction that equals 1. Then you add the fraction for the leftovers.

b. Read fraction A. (Signal.) *19/10.*

• What's the fraction that equals 1? (Signal.) *10/10.*

• What's the fraction for the leftovers? (Signal.) *9/10.*
So 19/10 = 10/10 + 9/10.

• Write the fraction addition and then the mixed number that equals 19/10. ✔

• Everybody, what mixed number equals 19/10? (Signal.) *1 and 9/10.*

c. Read fraction B. (Signal.) *7/6.*

• Write the addition and then the mixed number that equals 7/6. ✔
Check your work.

• 7/6 = 6/6 + 1/6. What's the mixed number? (Signal.) *1 and 1/6.*

d. Work the rest of the problems in part 4.
(Observe students and give feedback.)

e. Check your work.

• Fraction C. What mixed number does 17/14 equal? (Signal.) *1 and 3/14.*

• Fraction D. What mixed number does 25/20 equal? (Signal.) *1 and 5/20.*

EXERCISE 6: AREA
OF CIRCLES
REMEDY

a. Open your textbook to Lesson 74 and find part 1. ✔
(Teacher reference:)

- Say the word equation for the area of a circle. (Signal.) *Area = pi times radius to the second.*
- What does pi equal? (Signal.) *3.14.*

b. Circle A. What's the radius? (Signal.) *4 feet.*
- What's 4 to the second? (Signal.) *16.*
- Write the equation with letters. Below, write the equation with numbers. Stop when you've done that much.
(Observe students and give feedback.)
(Display:) [74:6A]

$$A = pi\,(r^2)$$
$$A = 3.14\,(16)$$

Here's what you should have.
- Work the problem and write the answer. Remember, the unit name is square feet.
(Observe students and give feedback.)

c. Check your work.
- Read the problem you worked. (Signal.) *3.14 × 16.*
- What's the area of circle A? (Signal.) *50.24 square feet.*

d. Work problem B. Write the equation with letters. Below, write the equation with numbers and figure out the area of the circle.
(Observe students and give feedback.)

e. Check your work. The equation with numbers is A = 3.14 × 100.
- What's the area? (Signal.) *314 square inches.*

f. Work the rest of the problems in part 1. Remember, first write the equation with letters. Then write the equation with numbers below.
(Observe students and give feedback.)

g. Check your work.
- Circle C. What's the radius? (Signal.) *8 yards.*
- What's 8 to the second? (Signal.) *64.*
- What's 3.14 × 64? (Signal.) *200.96.*
- What's the unit name in the answer? (Signal.) *Square yards.*

h. Circle D. What's the radius? (Signal.) *5 meters.*
- What's 5 to the second? (Signal.) *25.*
- What's 3.14 × 25? (Signal.) *78.50.*
- What's the unit name in the answer? (Signal.) *Square meters.*

EXERCISE 7: ROUNDING
DECIMALS WITH 9

a. (Display:) [74:7A]

> Round 6.95 to the tenths place.

You're going to work problems that round tenths up to a whole number.
(Add to show:) [74:7B]

> Round 6.95 to the tenths place.

- There are 69 tenths. How many tenths? (Signal.) *69 tenths.*
- What does that round up to? (Signal.) *70 tenths.* So the rounded number is 70 tenths. That's 7 and zero tenths.
(Add to show:) [74:7C]

> Round 6.95 to the tenths place.
> 7.0

- Read the rounded number. (Signal.) *7 and zero tenths.*

b. (Display:) [74:7D]

> 3 4.9 5

We'll round this number to the tenths place.
(Add to show:) [74:7E]

> 3 4.9 5

- How many tenths are there? (Signal.) *349.*

- If we round up, how many tenths will we have? (Signal.) *350.*
(Add to show:)   [74:7F]

$$\underline{34.9}5$$
$$35.0$$

Here's the rounded number.
- Read the rounded number. (Signal.) *35 and zero tenths.*

c. (Display:)   [74:7G]

$$80.96$$

We'll round this number to the tenths place.
(Add to show:)   [74:7H]

$$\underline{80.9}6$$

- How many tenths are there? (Signal.) *809.*
- If we round up, how many tenths will we have? (Signal.) *810.*
(Add to show:)   [74:7I]

$$\underline{80.9}6$$
$$81.0$$

Here's the rounded number.
- Read the rounded number. (Signal.) *81 and zero tenths.*

=== TEXTBOOK PRACTICE ===

a. Find part 2 in your textbook. ✔
(Teacher reference:)

a. 4.98   b. 21.97   c. 1.954   d. 10.96

All these numbers have a 9 in the tenths place that rounds up. You're going to write the rounded numbers to the tenths place.
- Write the numbers in a row. Underline each number through the tenths place. Stop when you've done that much. ✔

b. Number A. You'll round 4.98 to the tenths place.
- Everybody, how many tenths are there? (Signal.) *49.*
- What does 49 round up to? (Signal.) *50.*
- Write the number rounded to tenths. ✔
- Everybody, read the rounded number. (Signal.) *5 and zero tenths.*

c. Number B: 21 and 97 hundredths.
- Everybody, how many tenths are there? (Signal.) *219.*
- Write the rounded number. ✔
- Everybody, read the rounded number. (Signal.) *22 and zero tenths.*
d. Number C: 1 and 954 thousandths.
- Everybody, how many tenths are there? (Signal.) *19.*
- Write the rounded number. ✔
- Everybody, read the rounded number. (Signal.) *2 and zero tenths.*
e. Number D: 10 and 96 hundredths.
- Everybody, how many tenths are there? (Signal.) *109.*
- Write the rounded number. ✔
- Everybody, read the rounded number. (Signal.) *11 and zero tenths.*

EXERCISE 8: INDEPENDENT WORK
EQUIVALENT FRACTIONS

a. Find part 3 in your textbook. ✔
(Teacher reference:)

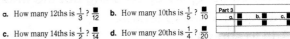

a. How many 12ths is $\frac{1}{3}$? $\frac{\blacksquare}{12}$ b. How many 10ths is $\frac{1}{5}$? $\frac{\blacksquare}{10}$
c. How many 14ths is $\frac{1}{2}$? $\frac{\blacksquare}{14}$ d. How many 20ths is $\frac{1}{4}$? $\frac{\blacksquare}{20}$

Each item asks a question and shows the denominator of the answer.
b. Item A: How many twelfths is 1/3?
- Write the fraction that has a denominator of 12. ✔
- Everybody, how many twelfths is 1/3? (Signal.) *4/12.*
c. You'll work the rest of the problems as part of your independent work.

Assign Independent Work, the rest of Textbook part 3 and parts 4–8 and Workbook parts 5–9.

Optional extra math-fact practice worksheets are available on ConnectED.

Lesson

EXERCISE 1: FRACTION SIMPLIFICATION
COMMON FACTOR OF 10

a. (Display:)   [75:1A]

$$\frac{60}{30} \qquad \frac{80}{20}$$

$$\frac{100}{20} \qquad \frac{120}{30}$$

Both parts of these fractions end in zero.

b. (Point to **60**.) 60 equals 6 times 10.
(Point to **30**.) 30 equals 3 times 10.
(Add to show:)   [75:1B]

$$\frac{60}{30} = \frac{6 \times 10}{3 \times 10} \qquad \frac{80}{20}$$

$$\frac{100}{20} \qquad \frac{120}{30}$$

So we can simplify by crossing out the fraction that equals one: 10 tenths.
(Add to show:)   [75:1C]

$$\frac{60}{30} = \frac{6 \times \cancel{10}}{3 \times \cancel{10}} \qquad \frac{80}{20}$$

$$\frac{100}{20} \qquad \frac{120}{30}$$

So 60/30 equals 6 thirds.
• And 6/3 equals a whole number. What number? (Signal.) *2.*
(Add to show:)   [75:1D]

$$\frac{60}{30} = \frac{6 \times \cancel{10}}{3 \times \cancel{10}} = 2 \qquad \frac{80}{20}$$

$$\frac{100}{20} \qquad \frac{120}{30}$$

c. (Point to **80/20**.) Read this fraction. (Signal.) *80 twentieths.*
It equals 8 times 10 over 2 times 10. So I can just cross out the zeros.
(Add to show:)   [75:1E]

$$\frac{60}{30} = \frac{6 \times \cancel{10}}{3 \times \cancel{10}} = 2 \qquad \frac{8\cancel{0}}{2\cancel{0}}$$

$$\frac{100}{20} \qquad \frac{120}{30}$$

• What fraction does 80/20 equal? (Signal.)
8 halves.
• And what whole number does 8/2 equal?
(Signal.) *4.*
(Add to show:)   [75:1F]

$$\frac{60}{30} = \frac{6 \times \cancel{10}}{3 \times \cancel{10}} = 2 \qquad \frac{8\cancel{0}}{2\cancel{0}} = 4$$

$$\frac{100}{20} \qquad \frac{120}{30}$$

d. (Point to **100/20**.) Read this fraction. (Signal.)
100 twentieths.
Both numbers end in zero. So I cross out the zeros.
(Add to show:)   [75:1G]

$$\frac{60}{30} = \frac{6 \times \cancel{10}}{3 \times \cancel{10}} = 2 \qquad \frac{8\cancel{0}}{2\cancel{0}} = 4$$

$$\frac{10\cancel{0}}{2\cancel{0}} \qquad \frac{120}{30}$$

• What does 10/2 equal? (Signal.) *5.*
(Add to show:)   [75:1H]

$$\frac{60}{30} = \frac{6 \times \cancel{10}}{3 \times \cancel{10}} = 2 \qquad \frac{8\cancel{0}}{2\cancel{0}} = 4$$

$$\frac{10\cancel{0}}{2\cancel{0}} = 5 \qquad \frac{120}{30}$$

e. (Point to **120/30**.) Read this fraction. (Signal.)
120 thirtieths.
Both numbers end in zero. So I cross out the zeros.
(Add to show:)   [75:1I]

$$\frac{60}{30} = \frac{6 \times \cancel{10}}{3 \times \cancel{10}} = 2 \qquad \frac{8\cancel{0}}{2\cancel{0}} = 4$$

$$\frac{10\cancel{0}}{2\cancel{0}} = 5 \qquad \frac{12\cancel{0}}{3\cancel{0}}$$

• What does 12/3 equal? (Signal.) *4.*
(Add to show:)   [75:1J]

$$\frac{60}{30} = \frac{6 \times \cancel{10}}{3 \times \cancel{10}} = 2 \qquad \frac{8\cancel{0}}{2\cancel{0}} = 4$$

$$\frac{10\cancel{0}}{2\cancel{0}} = 5 \qquad \frac{12\cancel{0}}{3\cancel{0}} = 4$$

WORKBOOK PRACTICE

a. Open your workbook to Lesson 75 and find part 1. ✔
(Teacher reference:)

a. $\frac{80}{40} =$ b. $\frac{180}{30} =$ c. $\frac{200}{50} =$

d. $\frac{120}{20} =$ e. $\frac{100}{50} =$

Both parts of these fractions end in zero. You're going to cross out the zeros and write the whole number each fraction equals.

b. Work problem A. ✔
* Everybody, what does 80/40 equal?(Signal.) *2.*
(Display:)   [75:1K]

a. $\frac{8\cancel{0}}{4\cancel{0}} = 2$

Here's what you should have.

c. Work the rest of the problems in part 1.
(Observe students and give feedback.)

d. Check your work.
* Problem B: 180/30.
   What does it equal? (Signal.) *6.*
* C: 200/50.
   What does it equal? (Signal.) *4.*
* D: 120/20.
   What does it equal? (Signal.) *6.*
* E: 100/50.
   What does it equal? (Signal.) *2.*

EXERCISE 2: FRACTION ANALYSIS
COMPARISON

a. Find part 2 in your workbook.
(Teacher reference:)

a. $\frac{1}{3}$ $\frac{1}{6}$ b. $\frac{1}{5}$ $\frac{3}{20}$ c. $\frac{1}{2}$ $\frac{7}{10}$ d. $\frac{1}{4}$ $\frac{5}{24}$

You're going to ask questions to figure out which fraction is more.

b. Problem A. Read the fractions. (Signal.) *1/3, 1/6.*
* Say the question. (Signal.) *How many 6ths is 1/3?*
c. Problem B. Read the fractions. (Signal.) *1/5, 3/20.*
* Say the question. (Signal.) *How many 20ths is 1/5?*
d. Problem C. Read the fractions. (Signal.) *1/2, 7/10.*
* Say the question. (Signal.) *How many 10ths is 1/2?*
e. Problem D. Read the fractions. (Signal.) *1/4, 5/24.*
* Say the question. (Signal.) *How many 24ths is 1/4?*
(Repeat until firm.)
f. Work all the problems. Write the sign to show which fraction is more.
(Observe students and give feedback.)
g. Check your work.
* Problem A. How many 6ths is 1/3? (Signal.) *2/6.*
* So which is more—1/3 or 1/6? (Signal.) *1/3.*
h. Problem B. How many 20ths is 1/5?
(Signal.) *4/20.*
* So which is more—1/5 or 3/20? (Signal.) *1/5.*
i. Problem C. How many 10ths is 1/2?
(Signal.) *5/10.*
* So which is more—1/2 or 7/10? (Signal.) *7/10.*
j. Problem D. How many 24ths is 1/4?
(Signal.) *6/24.*
* So which is more—1/4 or 5/24? (Signal.) *1/4.*
(Display:)   [75:2A]

a. $\frac{1}{3} > \frac{1}{6}$ b. $\frac{1}{5} > \frac{3}{20}$

c. $\frac{1}{2} < \frac{7}{10}$ d. $\frac{1}{4} > \frac{5}{24}$

Here's what you should have for each problem.

EXERCISE 3: DECIMAL PLACE VALUE
EXPANDED NOTATION

a. (Display:) [75:3A]

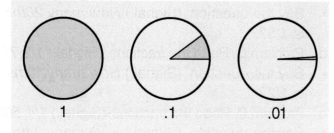

| 1 | .1 | .01 |

Here are circles that show 1, 1 tenth, one hundredth.

b. (Point to **1.**) What does this show? (Signal.) *1.* Yes, one whole.

• (Point to **.1**) What does this circle show? (Signal.) *1 tenth.*

• How many of these shaded parts would you need to fill one whole circle? (Signal.) *10.* Yes, you would need 10 parts.

• (Point to **.01**) What does this circle show? (Signal.) *1 hundredth.*

• How many of these shaded parts would you need to fill one whole circle? (Signal.) *100.* Yes, you would need 100 parts.

• If we had a circle with 1 thousandth shaded, how many parts would you need to fill one whole circle? (Signal.) *1000.*

c. Which is greater—1 or 1/10? (Signal.) *1.*

• How many times greater? (Signal.) *10 times.*

• Which is greater—1/10 or 1/100? (Signal.) *1/10.*

• How many times greater? (Signal.) *10 times.*

• Which is greater—1/100 or 1/1000? (Signal.) *1 hundredth.*

• How many times greater? (Signal.) *10 times.* (Repeat until firm.)

d. (Display:) [75:3B]

.456	.01
.196	.302
.409	

• (Point to **.456.**) Read this number. (Signal.) *456 thousandths.* My turn to say the place-value addition for 456 thousandths: 4 tenths plus 5 hundredths plus 6 thousandths.

• Your turn: Say the place-value addition for 456 thousandths. (Signal.) *4 tenths + 5 hundredths + 6 thousandths.* (Repeat until firm.)

e. (Point to **.196.**) Read this number. (Signal.) *196 thousandths.*

• Say the place-value addition for .196. (Signal.) *1 tenth + 9 hundredths + 6 thousandths.*

f. (Point to **.409.**) Read this number. (Signal.) *409 thousandths.*

• Say the place-value addition for .409. (Signal.) *4 tenths + zero hundredths + 9 thousandths.*

g. (Point to **.01.**) Read this number. (Signal.) *1 hundredth.*

• Say the place-value addition. (Signal.) *Zero tenths + 1 hundredth.*

h. (Point to **.302.**) Read this number. (Signal.) *302 thousandths.*

• Say the place-value addition. (Signal.) *3 tenths + zero hundredths + 2 thousandths.* (Repeat until firm.)

EXERCISE 4: FRACTION OPERATIONS
DIVISION

a. Find part 3 in your workbook. ✔

• Pencils down. ✔ (Teacher reference:)

a. $\frac{3}{5} \div \frac{2}{3} = \frac{3}{5} \times$ d. $\frac{6}{7} \div \frac{3}{2} = \frac{6}{7} \times$

b. $\frac{1}{3} \div \frac{1}{5} = \frac{1}{3} \times$ e. $\frac{4}{3} \div \frac{2}{5} = \frac{4}{3} \times$

c. $\frac{3}{8} \div 7 = \frac{3}{8} \times$

These problems have a fraction divided by a fraction. You're going to complete each equation to show the **first** fraction multiplied by a fraction.

b. Problem A: 3/5 ÷ 2/3 = 3/5 times a fraction. What fraction? (Signal.) *3/2.*

• Problem B: 1/3 ÷ 1/5 = 1/3 times a fraction. What fraction? (Signal.) *5/1.* Or you could just write 5.

c. Problem C: 3/8 ÷ 7 = 3/8 times a fraction. What fraction? (Signal.) *1/7.*

• Problem D: 6/7 ÷ 3/2. What does 6/7 divided by 3/2 equal? (Signal.) *6/7 × 2/3.*

• Problem E: 4/3 ÷ 2/5. What does 4/3 divided by 2/5 equal? (Signal.) *4/3 × 5/2.* (Repeat until firm.)

d. Complete all the equations. (Observe students and give feedback.)

e. Check your work. Read each equation.
- A. (Signal.) 3/5 ÷ 2/3 = 3/5 × 3/2.
- B. (Signal.) 1/3 ÷ 1/5 = 1/3 × 5/1.
- C. (Signal.) 3/8 ÷ 7 = 3/8 × 1/7.
- D. (Signal.) 6/7 ÷ 3/2 = 6/7 × 2/3.
- E. (Signal.) 4/3 ÷ 2/5 = 4/3 × 5/2.

EXERCISE 5: FRACTIONS
As Mixed Numbers   REMEDY

a. (Display:)   [75:5A]

$$\frac{15}{12} = 1\frac{\square}{\square}$$

$$\frac{9}{5} = 1\frac{\square}{\square}$$

$$\frac{17}{9} = 1\frac{\square}{\square}$$

These fractions equal 1 and leftovers.
You're going to figure out the mixed numbers.

b. (Point to **15/12**.) 15/12 = 1 and leftover
twelfths.
- Raise your hand when you know the fraction
for the leftovers. ✔
- Say the fraction for the leftovers. (Signal.) 3/12.
- Say the mixed number that equals 15/12.
(Signal.) 1 and 3/12.
(Add to show:)   [75:5B]

$$\frac{15}{12} = 1\frac{3}{12}$$

$$\frac{9}{5} = 1\frac{\square}{\square}$$

$$\frac{17}{9} = 1\frac{\square}{\square}$$

c. (Point to **9/5**.) 9/5 = 1 and leftover fifths.
- Raise your hand when you know the fraction
for the leftovers. ✔
- Say the fraction for the leftovers. (Signal.) 4/5.
- Say the mixed number that equals 9/5.
(Signal.) 1 and 4/5.
(Add to show:)   [75:5C]

$$\frac{15}{12} = 1\frac{3}{12}$$

$$\frac{9}{5} = 1\frac{4}{5}$$

$$\frac{17}{9} = 1\frac{\square}{\square}$$

d. (Point to **17/9**.) 17/9 = 1 and leftover ninths.
- Raise your hand when you know the fraction
for the leftovers. ✔
- Say the fraction for the leftovers. (Signal.) 8/9.
- Say the mixed number that equals 17/9.
(Signal.) 1 and 8/9.
(Add to show:)   [75:5D]

$$\frac{15}{12} = 1\frac{3}{12}$$

$$\frac{9}{5} = 1\frac{4}{5}$$

$$\frac{17}{9} = 1\frac{8}{9}$$

━━━━━━━ **WORKBOOK PRACTICE** ━━━━━━━

a. Find part 4 in your workbook. ✔
(Teacher reference:)   R Part C

a. $\frac{10}{7} =$ b. $\frac{11}{9} =$

c. $\frac{29}{20} =$ d. $\frac{17}{14} =$

You're going to write the mixed number that
equals each fraction.
b. Problem A. What's the fraction? (Signal.) 10/7.
- Write the mixed number that equals 10/7. ✔
- Everybody, what mixed number equals 10/7?
(Signal.) 1 and 3/7.
c. Problem B. What's the fraction? (Signal.) 11/9.
- Write the mixed number that equals 11/9. ✔
- Everybody, what mixed number equals 11/9?
(Signal.) 1 and 2/9.
d. Work the rest of the items.
(Observe students and give feedback.)
e. Check your work.
- Fraction C is 29/20. What's the mixed
number? (Signal.) 1 and 9/20.
- Fraction D is 17/14. What's the mixed number?
(Signal.) 1 and 3/14.

EXERCISE 6: AREA
OF CIRCLES

a. Open your textbook to Lesson 75 and find part 1. ✔
(Teacher reference:)

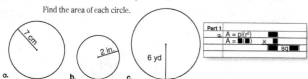

You're going to find the area of circles.
- Say the equation for the area of a circle. (Signal.) *Area = pi times radius to the second.*

b. Work problem A. Write the letter equation for area. Below, write the equation with numbers. Stop when you've done that much. ✔
(Display:) [75:6A]

$$A = pi\,(r^2)$$
$$A = 3.14\,(49)$$

Here's what you should have.
The equation with numbers is $A = 3.14 \times 49$.
- Figure out the area of the circle.
(Observe students and give feedback.)
- Check your work. What's the area? (Signal.) *153.86 square centimeters.*
(Add to show:) [75:6B]

$$A = pi\,(r^2)$$
$$A = 3.14\,(49)$$

$$
\begin{array}{r}
3.1\,4 \\
\times \quad 4\,9 \\
\hline
2\,8\,2\,6 \\
+\,1\,2\,5\,6\,0 \\
\hline
1\,5\,3.8\,6\ \text{sq cm}
\end{array}
$$

Here's what you should have.
c. Work the rest of the problems in part 1. Remember, first write the equation with letters. Then write the equation below with numbers.
(Observe students and give feedback.)
d. Check your work.
- Circle B. What's the number for the radius? (Signal.) *2.*
- What's the radius to the second? (Signal.) *4.*
- What's 3.14 × 4? (Signal.) *12.56.*
- What's the unit name in the answer? (Signal.) *Square inches.*

e. Circle C. What's the number for the radius? (Signal.) *6.*
- What's the radius to the second? (Signal.) *36.*
- What's 3.14 × 36? (Signal.) *113.04.*
- What's the unit name in the answer? (Signal.) *Square yards.*

EXERCISE 7: ROUNDING
DECIMALS WITH 9

a. Find part 2 in your textbook. ✔
(Teacher reference:)

a. 42.95   b. 0.93   c. 163.96   d. 44.94   e. 8.983

You'll round these numbers to the tenths place. See if you can do it without underlining the number through the tenths. If you have trouble, you can underline the number through the tenths.
- Write the numbers in a row. ✔
b. Read number A. (Signal.) *42 and 95 hundredths.*
- Below, write the rounded number to the tenths place. ✔
- Everybody, read the rounded number. (Signal.) *43 and zero tenths.*
(Display:) [75:7A]

a. 4 2.9 5
4 3.0

Here's what you should have.
c. Write the rest of the rounded numbers.
(Observe students and give feedback.)
d. Check your work.
Read each rounded number.
- B. (Signal.) *9 tenths.*
- C. (Signal.) *164 and zero tenths.*
- D. (Signal.) *44 and 9 tenths.*
- E. (Signal.) *9 and zero tenths.*
(Display:) [75:7B]

b. 0.9 3   c. 1 6 3.9 6
0.9   1 6 4.0

d. 4 4.9 4   e. 8.9 8 3
4 4.9   9.0

Here's what you should have.

EXERCISE 8: INDEPENDENT WORK
COORDINATE SYSTEM—POINTS ON A LINE

a. Find part 5 in your workbook. ✔
   (Teacher reference:)

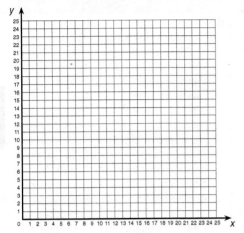

• The ratio of plates to cups is 5 to 3.

a. If there are 12 cups, how many plates are there? _____

$$\left(\quad\right) =$$

b. If there are 9 cups, how many plates are there? _____

c. If there are 10 plates, how many cups are there? _____

d. If there are 15 cups, how many plates are there? _____

You'll work the first problem and show two points on the coordinate system. Then you'll make the line and answer the rest of the questions.

Assign Independent Work, Workbook parts 5–8 and Textbook parts 3–10.

Optional extra math-fact practice worksheets are available on ConnectED.

Lessons 76-80 Planning Page

	Lesson 76	Lesson 77	Lesson 78	Lesson 79	Lesson 80
Student Learning Objectives	**Exercises** 1. Simplify fractions with a common factor of 10 2. **Write expanded notations for decimal place value** 3. Divide a fraction by a fraction 4. **Find the volume of cylinders** 5. **Add and subtract mixed numbers** 6. Round decimals 7. Change fractions to mixed numbers 8. Complete work independently	**Exercises** 1. Multiply and divide using mental math 2. Write expanded notations for decimal place value 3. **Solve parts-of-a-group word problems** 4. Divide by a fraction 5. Find the volume of cylinders 6. Add and subtract mixed numbers 7. Change fractions to mixed numbers Complete work independently	**Exercises** 1. Multiply and divide using mental math 2. **Compare decimals** 3. **Rewrite addition problems with a fraction more than 1** 4. Solve parts-of-a-group word problems 5. **Solve fraction division problems** 6. Find the volume of cylinders Complete work independently	**Exercises** 1. Compare fractions 2. Multiply and divide using mental math 3. Rewrite addition problems with a fraction more than 1 4. Compare decimals 5. Solve fraction division problems 6. Solve parts-of-a-group word problems 7. Find the volume of cylinders 8. Simplify fractions 9. Complete work independently	**Exercises** 1. Write expanded notations for decimal place value 2. Compare fractions 3. Multiply and divide using mental math 4. Compare decimals 5. **Show division with fractions as a diagram** 6. Solve parts-of-a-group word problems 7. Find the volume of cylinders and rectangular prisms 8. Rewrite addition problems with a fraction more than 1 9. Simplify fractions Complete work independently
Common Core State Standards for Mathematics					
5.OA 1	✔	✔	✔	✔	
5.NBT 1–2	✔				
5.NBT 3	✔	✔	✔	✔	✔
5.NBT 4	✔			✔	
5.NBT 5		✔	✔	✔	✔
5.NBT 6		✔			
5.NBT 7	✔	✔	✔	✔	✔
5.NF 1–2			✔		✔
5.NF 3	✔				
5.NF 4–7	✔	✔	✔	✔	✔
5.MD 5	✔	✔			✔
5.G 2	✔	✔		✔	
Teacher Materials	Presentation Book 2, Board Displays CD or chalk board				
Student Materials	Textbook, Workbook, pencil, lined paper, ruler				
Additional Practice	• Student Practice Software: Block 3: Activity 1 (5.NBT 7), Activity 2, Activity 3 (5.NF 4), Activity 4, Activity 5 (5.NBT 6), Activity 6 (5.NF 4, 5.NF 6, and 5.OA 1) • Provide needed fact practice with Level D or E Math Fact Worksheets.				
Mastery Test					Student Assessment Book (Present Mastery Test 8 following Lesson 80.)

Lesson

EXERCISE 1: FRACTION SIMPLIFICATION
COMMON FACTOR OF 10

a. Open your workbook to Lesson 76 and find part 1. ✔
(Teacher reference:)

  a. $\frac{150}{50} =$ b. $\frac{350}{70} =$ c. $\frac{90}{30} =$

  d. $\frac{200}{20} =$ e. $\frac{240}{60} =$

Both parts of these fractions end in zero. You learned a fast way to simplify them. You cross out both zeros because 10 tenths equals one.

b. Read fraction A. (Signal.) *150/50.*
• Cross out the zeros and write the whole number this fraction equals. ✔
• Everybody, what does 150/50 equal? (Signal.) *3.*
(Display:)   [76:1A]

> a. $\frac{15\cancel{0}}{5\cancel{0}} = 3$

Here's what you should have.

c. Cross out the zeros in the rest of the problems and write the numbers the fractions equal.
(Observe students and give feedback.)
d. Check your work.
• Fraction B: 350/70.
What does it equal? (Signal.) *5.*
• C: 90/30.
What does it equal? (Signal.) *3.*
• D: 200/20.
What does it equal? (Signal.) *10.*
• E: 240/60.
What does it equal? (Signal.) *4.*

EXERCISE 2: DECIMAL PLACE VALUE
EXPANDED NOTATION   REMEDY

a. (Display:)   [76:2A]

> .873
>
> .051
>
> 2.13
>
> 15.02

b. (Point to .873.) Read this number. (Signal.)
873 thousandths.
• Say the place-value addition for .873. (Signal.)
8 tenths + 7 hundredths + 3 thousandths.
(Add to show:)   [76:2B]

> .873 = .8 + .07 + .003
>
> .051
>
> 2.13
>
> 15.02

Here's the place-value addition for 8 tenths + 7 hundredths + 3 thousandths.

c. (Point to .051.) Read this number. (Signal.)
51 thousandths.
• Say the place-value addition for .051. (Signal.)
Zero tenths + 5 hundredths + 1 thousandth.
(Add to show:)   [76:2C]

> .873 = .8 + .07 + .003
>
> .051 = .0 + .05 + .001
>
> 2.13
>
> 15.02

Here's the place-value addition for
zero tenths + 5 hundredths + 1 thousandth.

d. (Point to **2.13**.) Read this number. (Signal.) *2 and 13 hundredths.*
My turn to say the place-value addition: 2 + 1 tenth + 3 hundredths.
• Say the place-value addition for 2.13. (Signal.) *2 + 1 tenth + 3 hundredths.*
(Add to show:) [76:2D]

$$.873 = .8 + .07 + .003$$
$$.051 = .0 + .05 + .001$$
$$2.13 \;= 2 + .1 + .03$$
$$15.02$$

Here's the place-value addition.

e. (Point to **15.02**.) Read this number. (Signal.) *15 and 2 hundredths.*
• Say the place-value addition. (Signal.) *10 + 5 + zero tenths + 2 hundredths.*
(Repeat until firm.)
(Add to show:) [76:2E]

$$.873 = .8 + .07 + .003$$
$$.051 = .0 + .05 + .001$$
$$2.13 \;= 2 + .1 + .03$$
$$15.02 \;= 10 + 5 + .0 + .02$$

Here's the place-value addition.

==================== **WORKBOOK PRACTICE** ====================

a. Find part 2 in your workbook. ✔
(Teacher reference:) R Part E

a. .62 = _____

b. 8.023 = _____

c. 65.31 = _____

d. .076 = _____

You'll complete the place-value equation for each decimal number.

b. Read number A. (Signal.) *62 hundredths.*
• Write what 62 hundredths equals. ✔
• Everybody, say the place-value addition for .62. (Signal.) *6 tenths + 2 hundredths.*

c. Read number B. (Signal.) *8 and 23 thousandths.*
• Complete the place-value equation. ✔
• Everybody, say the place-value addition for 8 and 23 thousandths. (Signal.) *8 + zero tenths + 2 hundredths + 3 thousandths.*

d. Read number C. (Signal.) *65 and 31 hundredths.*
• Complete the place-value equation. ✔
• Everybody, say the place-value addition for 65 and 31 hundredths. (Signal.) *60 + 5 + 3 tenths + 1 hundredth.*

e. Read number D. (Signal.) *76 thousandths.*
• Complete the place-value equation. ✔
• Everybody, say the place-value addition for 76 thousandths. (Signal.) *Zero tenths + 7 hundredths + 6 thousandths.*

EXERCISE 3: FRACTION OPERATIONS
DIVISION

a. Find part 3 in your workbook. ✔
• Pencils down. ✔
(Teacher reference:)

a. $\frac{3}{4} \div \frac{1}{2} =$ d. $\frac{9}{10} \div \frac{3}{8} =$

b. $6 \div \frac{4}{5} =$ e. $\frac{30}{5} \div \frac{6}{3} =$

c. $\frac{1}{3} \div 4 =$

You're going to write fraction division problems as multiplication problems.

b. Problem A: 3/4 divided by 1/2.
Say the multiplication problem. (Signal.) *3/4 × 2/1.*
• B: 6 divided by 4/5.
Say the multiplication problem. (Signal.) *6 × 5/4.*
• C: 1/3 divided by 4.
Say the multiplication problem. (Signal.) *1/3 × 1/4.*

c. Complete all the equations.
(Observe students and give feedback.)

d. Check your work. Read each equation.
• A. (Signal.) *3/4 ÷ 1/2 = 3/4 × 2/1.*
• B. (Signal.) *6 ÷ 4/5 = 6 × 5/4.*
• C. (Signal.) *1/3 ÷ 4 = 1/3 × 1/4.*
• D. (Signal.) *9/10 ÷ 3/8 = 9/10 × 8/3.*
• E. (Signal.) *30/5 ÷ 6/3 = 30/5 × 3/6.*

e. (Display:) [76:3A]

$$e.\ \frac{30}{5} \div \frac{6}{3} = \frac{30}{5} \times \frac{3}{6}$$

Let's see if our equations are correct.
(Point to first **30/5**.) The fractions on this side
equal whole numbers.
- What does 30/5 equal? (Signal.) *6.*
- What does 6/3 equal? (Signal.) *2.*
(Add to show:) [76:3B]

$$e.\ \frac{30}{5} \div \frac{6}{3} = \frac{30}{5} \times \frac{3}{6}$$
$$6 \div 2$$

- What's 6 divided by 2? (Signal.) *3.*
(Add to show:) [76:3C]

$$e.\ \frac{30}{5} \div \frac{6}{3} = \frac{30}{5} \times \frac{3}{6}$$
$$6 \div 2$$
$$3$$

f. We have 3 on one side. On the other side, we
have 30/5 times 3/6.
- What does 30 × 3 equal? (Signal.) *90.*
- What does 5 × 6 equal? (Signal.) *30.*
(Add to show:) [76:3D]

$$e.\ \frac{30}{5} \div \frac{6}{3} = \frac{30}{5} \times \frac{3}{6}$$
$$6 \div 2 = \frac{90}{30}$$
$$3$$

- Raise your hand when you know the whole
number 90/30 equals. ✔
- What does 90/30 equal? (Signal.) *3.*
(Add to show:) [74:3E]

$$e.\ \frac{30}{5} \div \frac{6}{3} = \frac{30}{5} \times \frac{3}{6}$$
$$6 \div 2 = \frac{90}{30}$$
$$3 \quad = 3$$

So we have 3 on each side of the equation.
The equations are correct.

EXERCISE 4: VOLUME
OF CYLINDERS

a. You know how to find the volume of a
rectangular prism.
(Display:) [76:4A]

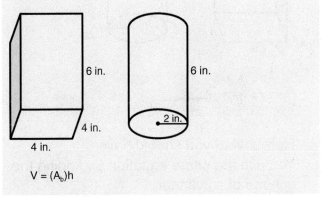

V = (A_b)h

The equation is: Volume equals Area of the
base times height.
- Say the equation. (Signal.) *Volume = Area of
the base × height.*
First you figure out the area of the base.
Then you multiply by the height.
- Raise your hand when you know the area of
the rectangular prism's base. ✔
- What's the area of the base? (Signal.)
16 square inches.
- What's the height? (Signal.) *6 inches.*
(Add to show:) [76:4B]

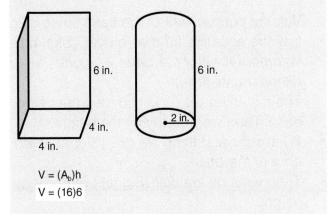

$$V = (A_b)h$$
$$V = (16)6$$

b. In part A, figure out the volume. Remember,
the unit name is cubic inches. ✔
- You multiplied 16 times 6. What's the volume
of the prism? (Signal.) *96 cubic inches.*

(Add to show:) [76:4C]

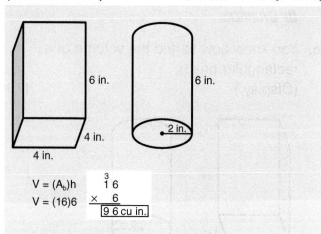

$V = (A_b)h$
$V = (16)6$
$\begin{array}{r} \overset{3}{1}6 \\ \times\ \ 6 \\ \hline \boxed{96 \text{ cu in.}} \end{array}$

Here's what you should have.

c. You use the same equation for finding the volume of a cylinder.

(Change to show:) [76:4D]

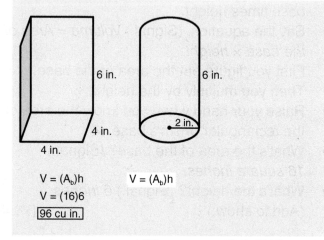

$V = (A_b)h$ $V = (A_b)h$
$V = (16)6$
$\boxed{96 \text{ cu in.}}$

Volume equals Area of the base times height.
- Say the equation for the volume. (Signal.) *Volume = Area of the base × height.*
  (Repeat until firm.)

  The first thing we do is find the area of the base. Then we multiply by the height.
- What's the first thing we do? (Signal.) *Find the area of the base.*
- Then what do we do? (Signal.) *Multiply by the height.*
  (Repeat until firm.)

d. The base of the cylinder is a circle.
  To find the area of the base, we multiply pi times radius to the second.
- What do we multiply? (Signal.) *Pi times radius to the second.*
  (Repeat until firm.)

- What's the radius of the base? (Signal.) *2 inches.*
- Raise your hand when you know the numbers you'll multiply to find the area of the base. ✔
- Say the numbers you'll multiply to find the area of the base. (Signal.) *3.14 × 4.*
  (Repeat until firm.)

e. Figure out the area. Pencils down when you've done that much. ✔
- Everybody, what's 3.14 times 4? (Signal.) *12.56.* So the area of the base is 12.56 square inches.

(Add to show:) [76:4E]

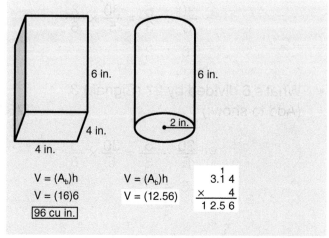

$V = (A_b)h$ $V = (A_b)h$ $\begin{array}{r} \overset{1}{3.1}4 \\ \times\ \ \ 4 \\ \hline 12.56 \end{array}$
$V = (16)6$ $V = (12.56)$
$\boxed{96 \text{ cu in.}}$

Here's the problem you worked for the area of the base.

f. To find the volume, we multiply the area of the base by the height.
- What's the height? (Signal.) *6 inches.*
- Say the multiplication for the area of the base times the height. (Signal.) *12.56 × 6.*
  (Repeat until firm.)

(Add to show:) [76:4F]

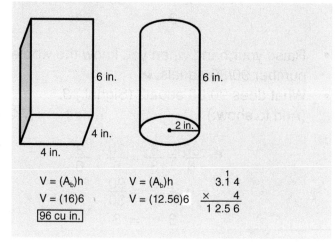

$V = (A_b)h$ $V = (A_b)h$ $\begin{array}{r} \overset{1}{3.1}4 \\ \times\ \ \ 4 \\ \hline 12.56 \end{array}$
$V = (16)6$ $V = (12.56)6$
$\boxed{96 \text{ cu in.}}$

- Figure out the volume. Remember, the unit name is cubic inches. ✔
- Everybody, what's the volume of the cylinder? (Signal.) *75.36 cubic inches.*

(Add to show:) [76:4G]

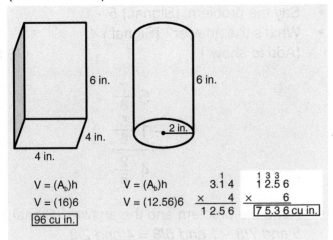

V = (A_b)h   V = (A_b)h
V = (16)6   V = (12.56)6
96 cu in.

$$
\begin{array}{r}
1\\
3.1\,4\\
\times4\\
\hline
1\,2.5\,6
\end{array}
\qquad
\begin{array}{r}
1\,3\,3\\
1\,2.5\,6\\
\times6\\
\hline
7\,5.3\,6\text{ cu in.}
\end{array}
$$

Here's the problem you worked.

g. Raise your hand when you know which has a greater volume—the rectangular prism or the cylinder. ✔
• Which has the greater volume? (Signal.) *The rectangular prism.*
  Yes, if the rectangular prism is a box, the cylinder would just fit inside the box.
  (Display:) [76:4H]

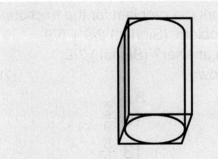

Here's how it would look.
h. (Display:) [76:4I]

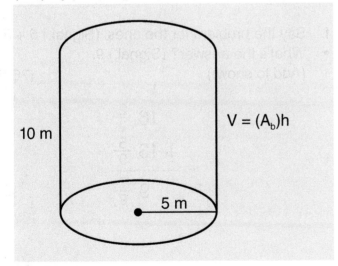

New cylinder.
• What's the equation for finding the volume of a cylinder? (Signal.) *Volume = Area of the base × height.*
  (Repeat until firm.)

• What's the radius of the base? (Signal.) *5 meters.*
• Say the multiplication for the area of the base. (Signal.) *3.14 × 25.*
• Figure out the area of the base. Then stop. ✔
• You worked the problem 3.14 × 25. What's the answer? (Signal.) *78.50.*
  The height is 10.
• Say the problem you'll work to find the volume. (Signal.) *78.50 × 10.*
  (Add to show:) [76:4J]

V = (A_b)h
V = (78.50)10

• Figure out the volume of the cylinder. Write the answer with the unit name. ✔
• You multiplied 78.50 by 10. What's the volume of the cylinder? (Signal.) *785 cubic meters.*
i. (Display:) [76:4K]

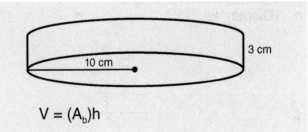

V = (A_b)h

New cylinder.
• What's the equation for finding the volume? (Signal.) *Volume = Area of the base × height.*
• The base is a circle, so you find the area of that circle. What's the radius of the base? (Signal.) *10 centimeters.*
• Say the multiplication for the area of the base. (Signal.) *3.14 × 100.*
• Figure out the area of the base. Then stop. ✔
• Everybody, what's 3.14 × 100? (Signal.) *314.*
  The height is 3.
• Say the problem you'll work to find the volume. (Signal.) *314 × 3.*

(Add to show:) [76:4L]

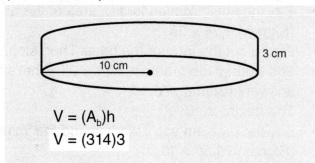

$V = (A_b)h$
$V = (314)3$

- Figure out the area of the base times the height. (Observe students and give feedback.)
- Everybody, what's the volume of the cylinder? (Signal.) *942 cubic centimeters.*
(Add to show:) [76:4M]

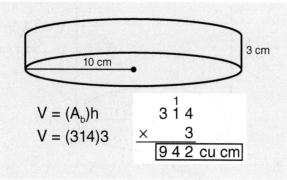

$V = (A_b)h$
$V = (314)3$

$$\begin{array}{r} \overset{1}{3}\,1\,4 \\ \times \quad 3 \\ \hline \boxed{9\,4\,2 \text{ cu cm}} \end{array}$$

Here's the problem you worked.

EXERCISE 5: MIXED-NUMBER OPERATIONS
ADDITION/SUBTRACTION

REMEDY

a. (Display:) [76:5A]

$$\begin{array}{r} 5\dfrac{7}{8} \\ -1\dfrac{5}{8} \\ \hline \end{array}$$

Here's a problem that has mixed numbers.
- Read the problem. (Signal.) *5 and 7/8 – 1 and 5/8.*
b. To work the problem, we start with the fractions, 7/8 minus 5/8.
- Say the problem for the fractions. (Signal.) *7/8 – 5/8.*
- What's the answer? (Signal.) *2/8.*
(Add to show:) [76:5B]

$$\begin{array}{r} 5\dfrac{7}{8} \\ -1\dfrac{5}{8} \\ \hline \dfrac{2}{8} \end{array}$$

c. Now we work the problem for the ones.
- Say the problem. (Signal.) *5 – 1.*
- What's the answer? (Signal.) *4.*
(Add to show:) [76:5C]

$$\begin{array}{r} 5\dfrac{7}{8} \\ -1\dfrac{5}{8} \\ \hline 4\dfrac{2}{8} \end{array}$$

- Read the problem and the answer. (Signal.) *5 and 7/8 – 1 and 5/8 = 4 and 2/8.*
d. (Display:) [76:5D]

$$\begin{array}{r} 16\dfrac{2}{9} \\ +13\dfrac{5}{9} \\ \hline \end{array}$$

New problem.
- Read the problem. (Signal.) *16 and 2/9 + 13 and 5/9.*
e. We first work the problem for the fractions.
- Say the problem. (Signal.) *2/9 + 5/9.*
- What's the answer? (Signal.) *7/9.*
(Add to show:) [76:5E]

$$\begin{array}{r} 16\dfrac{2}{9} \\ +13\dfrac{5}{9} \\ \hline \dfrac{7}{9} \end{array}$$

f. Say the problem for the ones. (Signal.) *6 + 3.*
- What's the answer? (Signal.) *9.*
(Add to show:) [76:5F]

$$\begin{array}{r} 16\dfrac{2}{9} \\ +13\dfrac{5}{9} \\ \hline 9\dfrac{7}{9} \end{array}$$

g. Say the problem for the tens. (Signal.) *1 + 1.*
• What's the answer? (Signal.) *2.*
(Add to show:)   [76:5G]

$$16 \frac{2}{9}$$
$$+ 13 \frac{5}{9}$$
$$\overline{29 \frac{7}{9}}$$

• Read the problem and the answer. (Signal.)
16 and 2/9 + 13 and 5/9 = 29 and 7/9.

========= **TEXTBOOK PRACTICE** =========

a. Open your textbook to Lesson 76 and find
part 1. ✔
(Teacher reference:)

a. $13\frac{7}{10}$ b. $6\frac{1}{12}$ c. $21\frac{3}{9}$ d. $9\frac{5}{6}$
   $-11\frac{2}{10}$ $+4\frac{10}{12}$ $+ 8\frac{4}{9}$ $-7\frac{3}{6}$

b. Read problem A. (Signal.) *13 and 7/10 – 11
and 2/10.*
• Copy problem A and work it. Remember, work
the problem for the fractions first.
(Observe students and give feedback.)
• Check your work. What's 13 and 7/10 minus
11 and 2/10? (Signal.) *2 and 5/10.*
c. Copy problem B and work it.
(Observe students and give feedback.)
• Check your work. What's 6 and 1/12 plus 4
and 10/12? (Signal.) *10 and 11/12.*
d. Copy problem C and work it. ✔
• Check your work. What's 21 and 3/9 plus 8
and 4/9? (Signal.) *29 and 7/9.*
e. Copy problem D and work it. ✔
• Check your work. What's 9 and 5/6 minus 7
and 3/6? (Signal.) *2 and 2/6.*

EXERCISE 6: ROUNDING
DECIMALS

a. Find part 2 in your textbook. ✔
(Teacher reference:)

a. Round 30.698 to the hundredths place.
b. Round 4.925 to the tenths place.
c. Round 15.32 to the tenths place.
d. Round 1.099 to the hundredths place.
e. Round 145.087 to the tenths place.
f. Round 0.65 to the tenths place.
g. Round 406.958 to the hundredths place.
h. Round 73.925 to the tenths place.
i. Round 0.36951 to the thousandths place.

You're going to write rounded numbers. Some
numbers have nines that round up. Some
numbers don't round up. You don't have to
copy the numbers. Just write the rounded
numbers for A through I.
b. Write rounded number A. ✔
• Everybody, read the rounded number.
(Signal.) *30 and 70 hundredths.*
c. Write rounded numbers for the rest of the
items.
(Observe students and give feedback.)
d. Check your work. Read each rounded number.
• Item B. (Signal.) *4 and 9 tenths.*
• C. (Signal.) *15 and 3 tenths.*
• D. (Signal.) *1 and 10 hundredths.*
• E. (Signal.) *145 and 1 tenth.*
• F. (Signal.) *7 tenths.*
• G. (Signal.) *406 and 96 hundredths.*
• H. (Signal.) *73 and 9 tenths.*
• I. (Signal.) *370 thousandths.*

EXERCISE 7: FRACTIONS
AS MIXED NUMBERS

REMEDY

a. Find part 3 in your textbook. ✔
(Teacher reference:)

a. $\frac{15}{11}$ b. $\frac{37}{30}$ c. $\frac{13}{7}$ d. $\frac{8}{5}$

You're going to copy each fraction and write the mixed number it equals.

b. Fraction A: 15/11.
• Write the equation that shows 15/11 and the mixed number it equals.
(Observe students and give feedback.)
• Everybody, read the equation for fraction A. (Signal.) *15/11 = 1 and 4/11.*

c. Write equations for the rest of the fractions in part 3.
(Observe students and give feedback.)

d. Check your work.
• Fraction B: 37/30. Read the equation. (Signal.) *37/30 = 1 and 7/30.*
• C: 13/7. Read the equation. (Signal.) *13/7 = 1 and 6/7.*
• D: 8/5. Read the equation. (Signal.) *8/5 = 1 and 3/5.*

EXERCISE 8: INDEPENDENT WORK
WORD PROBLEMS—RATIO/FRACTION OF A GROUP

a. Find part 4 in your textbook. ✔
(Teacher reference:)

a. In a swimming race $\frac{5}{8}$ of the swimmers are women. There are 120 women swimmers. How many swimmers are there in all? How many swimmers are men?

b. The ratio of old dogs to young dogs was 2 to 3. There were 150 dogs. How many were young? How many were old?

c. The ratio of hens to roosters was 3 to 5. There were 75 roosters. How many hens were there?

d. $\frac{3}{7}$ of the windows were clean. There were 42 windows. How many clean windows were there?

We'll work the first problem together. Then you'll do the rest of the items as part of your independent work.

b. Problem A: In a swimming race 5/8 of the swimmers are women. There are 120 women swimmers. How many swimmers are there in all? How many swimmers are men?
The problem has two questions. You'll figure out the answer to the first question and then add or subtract to answer the second question.

c. The first question is: How many swimmers are there in all?
• Write the equation with letters. Stop when you've done that much. ✔
• Everybody, read the equation. (Signal.) *5/8 S = W.*
• Figure out how many swimmers there are in all. ✔
• Everybody, how many swimmers are there in all? (Signal.) *192 swimmers.*

d. Add or subtract to find the answer to the second question. ✔
• Everybody, did you add or subtract? (Signal.) *Subtract.*
• Read the subtraction problem and the answer. (Signal.) *192 – 120 = 72.*
• How many swimmers are men? (Signal.) *72 men.*

Assign Independent Work, the rest of Textbook part 4 and parts 5–9 and Workbook parts 4–7.

Optional extra math-fact practice worksheets are available on ConnectED.

Lesson 77

EXERCISE 1: MENTAL MATH

a. Time for some mental math.
- Listen: What's 3 times 6? (Signal.) *18.*
- So what's 30 × 6? (Signal.) *180.*
- What's 30 × 7? (Signal.) *210.*
- What's 30 × 9? (Signal.) *270.*
  (Repeat until firm.)
b. Listen: 4 times 200. (Pause.) (Signal.) *800.*
- 5 × 200. (Pause.) (Signal.) *1000.*
- 9 × 200. (Pause.) (Signal.) *1800.*
  (Repeat until firm.)
c. 9 times 300. (Pause.) (Signal.) *2700.*
- 9 × 500. (Pause.) (Signal.) *4500.*
- 9 × 900. (Pause.) (Signal.) *8100.*
  (Repeat until firm.)
d. Listen: 200 divided by 10. (Pause.)
  (Signal.) *20.*
- 200 ÷ 4. (Pause.) (Signal.) *50.*
- 200 ÷ 2. (Pause.) (Signal.) *100.*
  (Repeat until firm.)
e. 200 divided by 5. (Pause.) (Signal.) *40.*
- 300 ÷ 5. (Pause.) (Signal.) *60.*
- 400 ÷ 5. (Pause.) (Signal.) *80.*
- 450 ÷ 5. (Pause.) (Signal.) *90.*
  (Repeat until firm.)

EXERCISE 2: DECIMAL PLACE VALUE
EXPANDED NOTATION   REMEDY

a. Open your workbook to Lesson 77 and find
   part 1. ✔
   (Teacher reference:)   R Part F
   a. 39.41 = _____
   b. .003 = _____
   c. 2.47 = _____
   d. .603 = _____

b. Complete the place-value equation for each
   number.
   (Observe students and give feedback.)
c. Check your work.
- Item A. Read the place-value addition for
  39 and 41 hundredths. (Signal.) *30 + 9 +
  4 tenths + 1 hundredths.*

- B. Read the place-value addition for
  3 thousandths. (Signal.) *Zero tenths +
  zero hundredths + 3 thousandths.*
- C. Read the place-value addition for 2 and
  47 hundredths. (Signal.) *2 + 4 tenths +
  7 hundredths.*
- D. Read the place-value addition for
  603 thousandths. (Signal.) *6 tenths +
  zero hundredths + 3 thousandths.*

EXERCISE 3: WORD PROBLEMS
PARTS OF A GROUP

a. (Display:)   [77:3A]

> $\frac{2}{3}$ of the boats were red. There were 60
> boats. How many of the boats were red?
> How many boats were not red?

Here's a new kind of problem. It asks two
questions: 2/3 of the boats were red. There
were 60 boats. How many boats were red?
How many boats were not red?
b. To work this problem, you write a letter
   equation and solve it. That answers one of
   the questions. 2/3 of the boats were red. There
   were 60 boats.
   (Add to show:)   [77:3B]

> $\frac{2}{3}$ of the boats were red. There were 60
> boats. How many of the boats were red?
> How many boats were not red?
>
> $\frac{2}{3} b = r$
>
> $\frac{2}{3} (60) = r$

- Read the letter equation. (Signal.) *2/3 B = R.*
- Read the equation with a number for boats.
  (Signal.) *2/3 × 60 = R.*
  We solve this equation.
- Say the problem we work. (Signal.) *2/3 × 60.*
  The answer is 40 red boats.

(Add to show:) [77:3C]

$\dfrac{2}{3}$ of the boats were red. There were 60 boats. How many of the boats were red? How many boats were not red?

$\dfrac{2}{3}$ b = r

$\dfrac{2}{3}$ (̶6̶0̶) = r = 40
    20

$\boxed{\text{40 red boats}}$

That's the answer to one of the questions.
- Read that question. (Signal.) *How many of the boats were red?*
- Say the answer. (Signal.) *40 red boats.*
c. We figure out the answer to the other question by subtracting.
- Read that question. (Signal.) *How many boats were not red?*
  We know the number of all the boats and the number of red boats. So we can subtract to find the number of boats that were not red.
- Raise your hand when you can say the subtraction problem. ✔
- Say the subtraction problem. (Signal.) *60 – 40.*
- What's the answer? (Signal.) *20.*
  Yes, 20 boats were not red.
(Add to show:) [77:3D]

$\dfrac{2}{3}$ of the boats were red. There were 60 boats. How many of the boats were red? How many boats were not red?

$\dfrac{2}{3}$ b = r

$\dfrac{2}{3}$ (̶6̶0̶) = r = 40
    20

$\boxed{\text{40 red boats}}$

$\begin{array}{r} 6\ 0 \\ -\ 4\ 0 \\ \hline \boxed{2\ 0 \text{ not red boats}} \end{array}$

d. Remember how we work the problem. We write the letter equation and solve it. Then we subtract to find the answer to the other question.
e. (Display:) [77:3E]

$\dfrac{3}{5}$ of the plants were alive. 15 plants were alive. How many plants were dead? How many plants were there?

Here's another problem that asks two questions: 3/5 of the plants were alive. 15 plants were alive. How many plants were dead? How many plants were there?

f. First we write the letter equation. Raise your hand when you can say the equation. ✔
- Say the letter equation. (Signal.) *3/5 p = A.*
- We write a second equation with a number for alive. What's that number? (Signal.) *15.*
(Add to show:) [77:3F]

$\dfrac{3}{5}$ of the plants were alive. 15 plants were alive. How many plants were dead? How many plants were there?

$\dfrac{3}{5}$ P = a

$\dfrac{3}{5}$ P = 15

When we solve this equation, we'll answer one of the questions.
- Raise your hand when you know that question. ✔
- Which question will this equation answer? (Signal.) *How many plants were there?*
g. Work the problem in part A. Remember to simplify before you multiply. Raise your hand when you know the answer.
(Observe students and give feedback.)
- You worked the problem 15 times 5/3. How many plants were there? (Signal.) *25.*
  That answers one of the questions.
h. The other question is: How many plants were dead? We find the answer to that question by subtracting.
- Raise your hand when you can say the subtraction problem. ✔
- Say the subtraction problem. (Signal.) *25 – 15.*
(Add to show:) [77:3G]

$\dfrac{3}{5}$ of the plants were alive. 15 plants were alive. How many plants were dead? How many plants were there?

$\dfrac{3}{5}$ P = a

$\left(\dfrac{5}{3}\right) \dfrac{3}{5}$ P = ̶1̶5̶ $\left(\dfrac{5}{3}\right)$
                                5

P = 25

$\begin{array}{r} 2\ 5 \\ -\ 1\ 5 \end{array}$

$\boxed{\text{25 plants}}$

- What's the answer? (Signal.) *10.*
- So how many plants were dead? (Signal.) *10 dead plants.*

(Add to show:) [77:3H]

$\frac{3}{5}$ of the plants were alive. 15 plants were alive. How many plants were dead? How many plants were there?

$$\frac{3}{5} P = a$$

$$\left(\frac{5}{3}\right) \frac{3}{5} P = \overset{5}{\cancel{15}} \left(\frac{5}{\cancel{3}}\right)$$

$$P = 25$$

$$\begin{array}{r} 25 \\ -15 \\ \hline 10 \text{ dead plants} \end{array}$$

25 plants

i. Remember, solve the letter equation. Then subtract to find the answer to the other question.

EXERCISE 4: FRACTON OPERATIONS

DIVISION REMEDY

a. (Display:) [77:4A]

$$\frac{4}{5} \div \frac{9}{5}$$

We're going to work this problem.
First, we rewrite it as a multiplication problem. Then we work the multiplication problem and write the answer.

• Say the multiplication problem that equals 4/5 divided by 9/5. (Signal.) *4/5 × 5/9.*
(Repeat until firm.)
(Add to show:) [77:4B]

$$\frac{4}{5} \div \frac{9}{5} = \frac{4}{5} \times \frac{5}{9}$$

b. Before we multiply, we can simplify. We can cross out 5/5.
(Add to show:) [77:4C]

$$\frac{4}{5} \div \frac{9}{5} = \frac{4}{\cancel{5}} \times \frac{\cancel{5}}{9}$$

• What's left in the numerator? (Signal.) *4.*
• What's in the denominator? (Signal.) *9.*
• So the answer is 4/9.
(Add to show:) [77:4D]

$$\frac{4}{5} \div \frac{9}{5} = \frac{4}{\cancel{5}} \times \frac{\cancel{5}}{9} = \frac{4}{9}$$

• What does 4/5 divided by 9/5 equal?
(Signal.) *4/9.*

━━━━━ **TEXTBOOK PRACTICE** ━━━━━

a. Open your textbook to Lesson 77 and find part 1. ✔
(Teacher reference:)

a. $\frac{2}{7} \div \frac{10}{14}$ b. $\frac{9}{10} \div \frac{6}{5}$ c. $\frac{20}{3} \div 4$ d. $\frac{1}{3} \div \frac{9}{4}$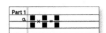

These are like the problem we just worked.

b. Read problem A. (Signal.) *2/7 ÷ 10/14.*
• Say the multiplication problem you work. (Signal.) *2/7 × 14/10.*
• Write the problem, simplify if you can, and write the answer.
(Observe students and give feedback.)
• Check your work. What does 2/7 divided by 10/14 equal? (Signal.) *2/5.*
(Display:) [77:4E]

a. $\dfrac{\overset{1}{\cancel{2}}}{\underset{}{\cancel{7}}} \times \dfrac{\overset{2}{\cancel{14}}}{\underset{5}{\cancel{10}}} = \boxed{\dfrac{2}{5}}$

Here's what you should have.
14/7 equals 2. 2/10 equals 1/5.

c. Work problem B. ✔
• Check your work. What does 9/10 divided by 6/5 equal? (Signal.) *3/4.*
(Display:) [77:4F]

b. $\dfrac{\overset{3}{\cancel{9}}}{\underset{2}{\cancel{10}}} \times \dfrac{\overset{1}{\cancel{5}}}{\underset{2}{\cancel{6}}} = \boxed{\dfrac{3}{4}}$

Here's what you should have.

d. Work problem C. ✔
• Check your work. What does 20/3 divided by 4 equal? (Signal.) *5/3.*
(Display:) [77:4G]

c. $\dfrac{\overset{5}{\cancel{20}}}{3} \times \dfrac{1}{\cancel{4}} = \boxed{\dfrac{5}{3}}$

Here's what you should have.

e. Work problem D. ✔
• Check your work. What does 1/3 divided by 9/4 equal? (Signal.) *4/27.*
(Display:) [77:4H]

d. $\dfrac{1}{3} \times \dfrac{4}{9} = \boxed{\dfrac{4}{27}}$

Here's what you should have.

EXERCISE 5: VOLUME
OF CYLINDERS

REMEDY

a. (Display:) [77:5A]

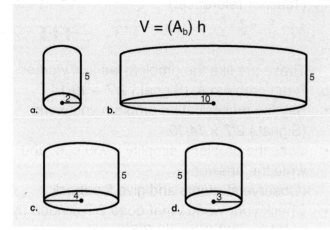

$$V = (A_b) h$$

We're going to find the volume of these cylinders. Remember: Volume equals Area of the base times the height.
You find the area of the base by multiplying 3.14 times radius to the second.

b. (Point to **A.**) What's the radius of this cylinder? (Signal.) *2.*

- Say the multiplication for the area of the base. (Signal.) *3.14 × 4.*
- Then what do you multiply the answer by to find the volume? (Signal.) *5.*

c. (Point to **B.**) What's the radius of this cylinder? (Signal.) *10.*

- Say the multiplication for the area of the base. (Signal.) *3.14 × 100.*
- Then what do you multiply the answer by to find the volume? (Signal.) *5.*

d. (Point to **C.**) What's the radius of this cylinder? (Signal.) *4.*

- Say the multiplication for the area of the base. (Signal.) *3.14 × 16.*
- Then what do you multiply the answer by to find the volume? (Signal.) *5.*

e. (Point to **D.**) What's the radius of this cylinder? (Signal.) *3.*

- Say the multiplication for the area of the base. (Signal.) *3.14 × 9.*
- Then what do you multiply the answer by to find the volume? (Signal.) *5.*
(Repeat until firm.)

f. (Display:) [77:5B]

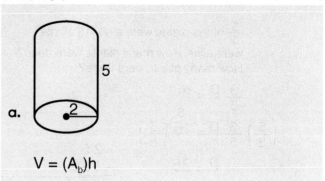

$$V = (A_b)h$$

Cylinder A.

- Copy the letter equation in part B. Find the area of the base and then stop.
(Observe students and give feedback.)
- Everybody, what's the area of the base? (Signal.) *12.56.*
(Add to show:) [77:5C]

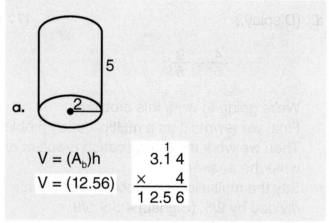

$$V = (A_b)h$$
$$V = (12.56)$$

$$\begin{array}{r} \overset{1}{3.1\,4} \\ \times \quad 4 \\ \hline 1\,2.5\,6 \end{array}$$

g. Now we'll multiply the area of the base times the height.

- Say the multiplication for the area of the base times the height. (Signal.) *12.56 × 5.*
- Below the letter equation, write the equation with two numbers. Then stop. ✔
(Add to show:) [77:5D]

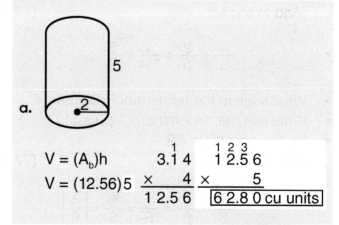

$$V = (A_b)h$$
$$V = (12.56)5$$

$$\begin{array}{r} \overset{1}{3.1\,4} \\ \times \quad 4 \\ \hline 1\,2.5\,6 \end{array}$$ $$\begin{array}{r} \overset{1\;2\;3}{1\,2.5\,6} \\ \times \quad 5 \\ \hline \boxed{6\,2.8\,0\ \text{cu units}} \end{array}$$

Here's the multiplication and the answer.

- What's the volume of the cylinder? (Signal.) *62.80 cubic units.*

h. (Display:)　　　　　　　　　　　　[77:5E]

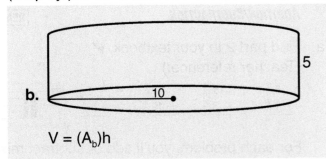

V = (A_b)h

- Cylinder B. Copy the letter equation. Find the area of the base and then stop.
  (Observe students and give feedback.)
- Everybody, what's the area of the base?
  (Signal.) *314.*
  (Add to show:)　　　　　　　　　[77:5F]

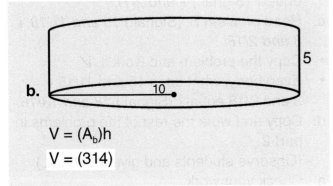

V = (A_b)h
V = (314)

i. Now we'll multiply the area of the base times the height.
- Say the multiplication for the area of the base times the height. (Signal.) *314 × 5.*
- Below the letter equation, write the equation with two numbers. Then stop. ✔
  (Add to show:)　　　　　　　　　[77:5G]

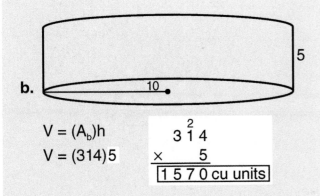

V = (A_b)h
V = (314)5

$$\begin{array}{r} 3\overset{2}{1}4 \\ \times\quad 5 \\ \hline \boxed{1\,5\,7\,0\text{ cu units}} \end{array}$$

Here's the multiplication and the answer.
- What's the volume of the cylinder? (Signal.)
  1570 cubic units.

j. (Display:)　　　　　　　　　　　　[77:5H]

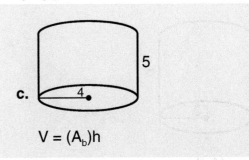

V = (A_b)h

- Cylinder C. Copy the letter equation. Find the area of the base and then stop.
  (Observe students and give feedback.)
- Everybody, what's the area of the base?
  (Signal.) *50.24.*
  (Add to show:)　　　　　　　　　[77:5I]

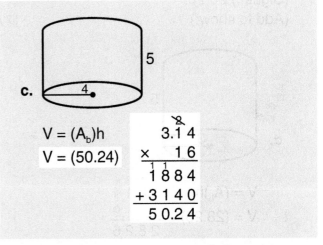

V = (A_b)h
V = (50.24)

$$\begin{array}{r} 3.\overset{\cancel{2}}{1}4 \\ \times\quad 1\,6 \\ \hline \overset{1\,1}{1}\,8\,8\,4 \\ +\,3\,1\,4\,0 \\ \hline 5\,0.2\,4 \end{array}$$

k. Now we'll multiply the area of the base times the height.
- Say the multiplication for the area of the base times the height. (Signal.) *50.24 × 5.*
- Write the equation with two numbers. Then stop. ✔
  (Add to show:)　　　　　　　　　[77:5J]

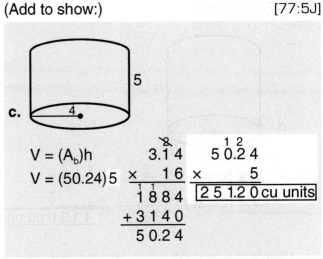

V = (A_b)h
V = (50.24)5

$$\begin{array}{r} 3.\overset{\cancel{2}}{1}4 \\ \times\quad 1\,6 \\ \hline \overset{1\,1}{1}\,8\,8\,4 \\ +\,3\,1\,4\,0 \\ \hline 5\,0.2\,4 \end{array} \qquad \begin{array}{r} \overset{1\,2}{5}\,0.2\,4 \\ \times\quad\quad 5 \\ \hline \boxed{2\,5\,1.2\,0\text{ cu units}} \end{array}$$

Here's the multiplication and the answer.
- What's the volume of the cylinder? (Signal.)
  251.20 cubic units.

l. (Display:) [77:5K]

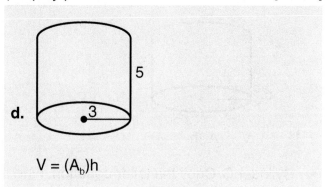

V = (A_b)h

- Cylinder D. Copy the letter equation. Find the area of the base and then stop.
  (Observe students and give feedback.)
- Everybody, what's the area of the base? (Signal.) 28.26.
  (Add to show:) [77:5L]

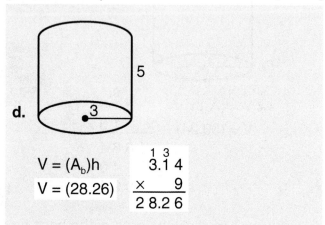

$$V = (A_b)h$$
$$V = (28.26)$$

$$\begin{array}{r} {\scriptstyle 1\ 3} \\ 3.1\,4 \\ \times\qquad 9 \\ \hline 2\,8.2\,6 \end{array}$$

m. Now we'll multiply the area of the base times the height.

- Say the multiplication for the area of the base times the height. (Signal.) 28.26 × 5.
- Write the equation with two numbers. ✔
  (Add to show:) [77:5M]

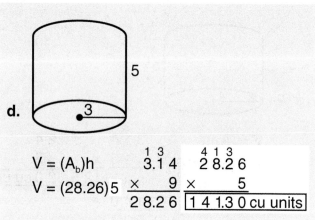

$$V = (A_b)h$$
$$V = (28.26)5$$

$$\begin{array}{r} {\scriptstyle 1\ 3} \\ 3.1\,4 \\ \times\qquad 9 \\ \hline 2\,8.2\,6 \end{array}\qquad \begin{array}{r} {\scriptstyle 4\ 1\ 3} \\ 2\,8.2\,6 \\ \times\qquad 5 \\ \hline \boxed{1\,4\,1.3\,0\ \text{cu units}} \end{array}$$

Here's the multiplication and the answer.

- What's the volume of the cylinder? (Signal.) 141.30 cubic units.

EXERCISE 6: MIXED-NUMBER OPERATIONS
ADDITION/SUBTRACTION

a. Find part 2 in your textbook. ✔
   (Teacher reference:)

a.	b.	c.	d.	
$8\frac{9}{11}$	$15\frac{11}{18}$	$42\frac{7}{10}$	$8\frac{1}{9}$	Part 2
$-1\frac{7}{11}$	$+3\frac{2}{18}$	$-19\frac{3}{10}$	$+25\frac{6}{9}$	

For each problem, you'll add or subtract mixed numbers. You'll write the answer as a mixed number.

b. Read problem A. (Signal.) *8 and 9/11 – 1 and 7/11.*
- Copy the problem and work it. ✔
- Everybody, what does 8 and 9/11 – 1 and 7/11 equal? (Signal.) *7 and 2/11.*

c. Read problem B. (Signal.) *15 and 11/18 + 3 and 2/18.*
- Copy the problem and work it. ✔
- Everybody, what does 15 and 11/18 + 3 and 2/18 equal? (Signal.) *18 and 13/18.*

d. Copy and work the rest of the problems in part 2.
   (Observe students and give feedback.)

e. Check your work.
- Problem C. You worked the problem 42 and 7/10 – 19 and 3/10.
- What's the answer? (Signal.) *23 and 4/10.*
- Problem D. You worked the problem 8 and 1/9 + 25 and 6/9.
- What's the answer? (Signal.) *33 and 7/9.*

EXERCISE 7: FRACTIONS

AS MIXED NUMBERS

a. Find part 3 in your textbook. ✔
   (Teacher reference:)

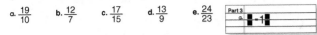

a. $\frac{19}{10}$ b. $\frac{12}{7}$ c. $\frac{17}{15}$ d. $\frac{13}{9}$ e. $\frac{24}{23}$

   You're going to copy each fraction and write the mixed number it equals.

b. Fraction A: 19/10.
- Write the fraction that shows 19/10 and the mixed number it equals.
   (Observe students and give feedback.)
- Everybody, read the equation for fraction A. (Signal.) *19/10 = 1 and 9/10.*

c. Write equations for the rest of the fractions in part 3.
   (Observe students and give feedback.)

d. Check your work.
- Fraction B: 12/7. Read the equation. (Signal.) *12/7 = 1 and 5/7.*
- C: 17/15. Read the equation. (Signal.) *17/15 = 1 and 2/15.*
- D: 13/9. Read the equation. (Signal.) *13/9 = 1 and 4/9.*
- E: 24/23. Read the equation. (Signal.) *24/23 = 1 and 1/23.*

Assign Independent Work, Textbook parts 4–10 and Workbook parts 2–5.

Optional extra math-fact practice worksheets are available on ConnectED.

Lesson 78

EXERCISE 1: MENTAL MATH

a. Time for some mental math.
- Listen: 360 divided by 10. (Pause.) (Signal.) *36.*
- 350 ÷ 10. (Pause.) (Signal.) *35.*
- 350 ÷ 5. (Pause.) (Signal.) *70.*
- 350 ÷ 7. (Pause.) (Signal.) *50.*
- 450 ÷ 5. (Pause.) (Signal.) *90.*
- (Repeat until firm.)

b. Listen: 9 times 200. (Pause.) (Signal.) *1800.*
- 9 × 500. (Pause.) (Signal.) *4500.*
- 9 × 800. (Pause.) (Signal.) *7200.*
- 9 × 300. (Pause.) (Signal.) *2700.*
- (Repeat until firm.)

c. 60 divided by 3. (Pause.) (Signal.) *20.*
- 80 ÷ 4. (Pause.) (Signal.) *20.*
- 120 ÷ 3. (Pause.) (Signal.) *40.*
- 160 ÷ 4. (Pause.) (Signal.) *40.*
- (Repeat until firm.)

d. 180 divided by 9. (Pause.) (Signal.) *20.*
- 450 ÷ 9. (Pause.) (Signal.) *50.*
- 360 ÷ 9. (Pause.) (Signal.) *40.*
- (Repeat until firm.)

EXERCISE 2: DECIMAL PLACE VALUE
COMPARISON

[REMEDY]

a. (Display:) [78:2A]

> .3627   .35

We're going to write the sign that shows which number is greater.

b. We figure out which number is greater by comparing the digits.
- What's the tenths digit for both numbers? (Signal.) *3.*
- What's the hundredths digit for the first number? (Signal.) *6.*

- What's the hundredths digit for the other number? (Signal.) *5.*
So the first number is greater.
(Add to show:) [78:2B]

> .3627 > .35

c. (Display:) [78:2C]

> .5062   .507

- What's the tenths digit for both numbers? (Signal.) *5.*
- What's the hundredths digit for both numbers? (Signal.) *Zero.*
- What's the thousandths digit for the first number? (Signal.) *6.*
- What's the thousandths digit for the other number? (Signal.) *7.*
- So which number is greater—the first or the second? (Signal.) *The second.*
(Add to show:) [78:2D]

> .5062 < .507

d. (Display:) [78:2E]

> .860   .8600

- What's the tenths digit for both numbers? (Signal.) *8.*
- What's the hundredths digit for both numbers? (Signal.) *6.*
- What's the thousandths digit for both numbers? (Signal.) *Zero.*
The zeros don't change the value, so the numbers are equal.
(Add to show:) [78:2F]

> .860 = .8600

a. Open your workbook to Lesson 78 and find part 1. ✔

(Teacher reference:)   R Part G

a. .920 ▨ .92   d. .0601 ▨ .601

b. .3466 ▨ .3486   e. .0032 ▨ .003200

c. .102 ▨ .1012

b. Read the numbers for item A. (Signal.)
920 thousandths, 92 hundredths.

• Write the sign that compares the two numbers.
(Display:)   [78:2G]

a. .920 = .92

Here's what you should have. The values are equal.

c. Write signs for the rest of the items in part 1.
(Observe students and give feedback.)

d. Check your work.

• Item B. Is one of the numbers greater than the other? (Signal.) *Yes.*

• Which number is greater—the first number or the second number? (Signal.) *The second number.*
(Display:)   [78:2H]

b. .3466 < .3486

Here's what you should have.
3486 ten thousandths is greater.

e. Item C. Is one of the numbers greater than the other? (Signal.) *Yes.*

• Which number is greater? (Signal.)
102 thousandths.

f. Item D. Is one of the numbers greater than the other? (Signal.) *Yes.*

• Which number is greater? (Signal.)
601 thousandths.

g. Item E. Is one of the numbers greater than the other? (Signal.) *No.*

• What sign did you write? (Signal.) *Equals.*

EXERCISE 3: FRACTIONS
AS MIXED NUMBERS

a. (Display:)   [78:3A]

$$4 + \frac{8}{5}$$

We're going to rewrite addition problems that have a fraction that is more than one.

b. (Point to **4 + 8/5**.) Read the problem. (Signal.)
4 + 8/5.
The fraction is more than one. So I'll rewrite the fraction.

• Raise your hand when you know the mixed number that equals 8/5. ✔

• What's the mixed number? (Signal.) *1 and 3/5.*
We write the fraction below and put the 1 with the other whole number.
(Add to show:)   [78:3B]

$$4 + \frac{\overset{1}{8}}{5}$$
$$\frac{3}{5}$$

c. Now we add whole numbers 1 + 4.

• What's the answer? (Signal.) *5.*
(Add to show:)   [78:3C]

$$4 + \frac{\overset{1}{8}}{5}$$
$$5 + \frac{3}{5}$$

So the answer is 5 plus 3/5.

d. (Display:)   [78:3D]

$$3 + \frac{12}{7}$$

• 12/7 is more than one. Raise your hand when you know the mixed number that equals 12/7. ✔

• What's the mixed number? (Signal.) *1 and 5/7.*

We write the fraction below and put the 1 with the other whole number.
(Add to show:) [78:3E]

$$3 + \dfrac{12}{7}$$

$$\dfrac{5}{7}$$

e. Now we add the whole numbers.
• What's 1 plus 3? (Signal.) *4.*
(Add to show:) [78:3F]

$$3 + \dfrac{12}{7}$$

$$4 + \dfrac{5}{7}$$

• Read the answer. (Signal.) *4 + 5/7.*

━━━━ **WORKBOOK PRACTICE** ━━━━

a. Find part 2 in your workbook. ✔
(Teacher reference:)

a. $6 + \dfrac{14}{10}$ b. $2 + \dfrac{9}{5}$ c. $12 + \dfrac{17}{14}$

These are like the problems we just worked.
b. Read problem A. (Signal.) *6 + 14/10.*
You're going to write 14/10 as a mixed number.
• Write 1 above the 6 and the fraction below 14/10. ✔
• What fraction did you write? (Signal.) *4/10.*
• Now add the whole numbers and write the answer below the 6. ✔
• Everybody, what does 6 + 14/10 equal? (Signal.) *7 + 4/10.*
c. Read problem B. (Signal.) *2 + 9/5.*
You're going to write 9/5 as a mixed number.
• Write the 1 above the other whole number and the fraction below 9/5. Then add the whole numbers. ✔
• Everybody, what does 2 + 9/5 equal? (Signal.) *3 + 4/5.*
d. Read problem C. (Signal.) *12 + 17/14.*
• Write 17/14 as a mixed number. Then add the whole numbers. ✔
• Everybody, what does 12 + 17/14 equal? (Signal.) *13 + 3/14.*

EXERCISE 4: WORD PROBLEMS
PARTS OF A GROUP | REMEDY |

a. (Display:) [78:4A]

$\dfrac{5}{8}$ of the windows were dirty. There were 40 windows. How many clean windows were there? How many dirty windows were there?

This problem asks two questions: 5/8 of the windows were dirty. There were 40 windows. How many clean windows were there? How many dirty windows were there?
We write the letter equation for one of the sentences. We replace one of the letters with a number. That equation answers one of the questions. We find the answer to the other question by subtracting.

b. Write the letter equation in part A. Put in the number the problem gives. Then stop. ✔
• Everybody, read the letter equation. (Signal.) *5/8 W = D.*
• Read the equation with numbers. (Signal.) *5/8 (40) = D.*
(Add to show:) [78:4B]

$\dfrac{5}{8}$ of the windows were dirty. There were 40 windows. How many clean windows were there? How many dirty windows were there?

$$\dfrac{5}{8} \, w = d$$

$$\dfrac{5}{8} \, (40) = d$$

• Solve the equation and write the answer to one of the questions. Pencils down when you've done that much.
(Observe students and give feedback.)
c. Check your work. You wrote the answer to one question.
• Read that question. (Signal.) *How many dirty windows were there?*
• What's the answer? (Signal.) *25 dirty windows.*

d. Read the other question. (Signal.) *How many clean windows were there?*
You figure out that answer by subtracting.
- Raise your hand when you can say the subtraction problem. ✔
- Say the subtraction problem. (Signal.) *40 – 25.*
e. Work the problem and write the answer. ✔
- Everybody, read the problem you worked. (Signal.) *40 – 25.*
- How many windows were clean? (Signal.) *15 clean windows.*

───────────── **TEXTBOOK PRACTICE** ─────────────

a. Open your textbook to Lesson 78 and find part 1. ✔
(Teacher reference:)

a. $\frac{2}{7}$ of the dogs were sleeping. 8 dogs were sleeping. How many dogs were awake? How many dogs were there?

b. There are 45 students. $\frac{4}{9}$ of the students are girls. How many girls are there? How many boys are there?

These are like the problem we just worked.
b. Problem A: 2/7 of the dogs were sleeping. 8 dogs were sleeping. How many dogs were awake? How many dogs were there?
- Write the letter equation. Replace one of the letters with a number. Stop when you've done that much. ✔
- Everybody, read the letter equation. (Signal.) *2/7 D = S.*
- Read the equation with a number. (Signal.) *2/7 D = 8.*
(Display:)   [78:4C]

> a. $\frac{2}{7}$ d = s
>
> $\frac{2}{7}$ d = 8

Here's what you should have.
- Work the problem and write the unit name in the answer. Remember to simplify before you multiply. (Observe students and give feedback.)

c. Check your work.
- What does D equal? (Signal.) *28.*
- Which question does that answer? (Signal.) *How many dogs were there?*
(Add to show:)   [78:4D]

> a. $\frac{2}{7}$ d = s
>
> $\left(\frac{7}{2}\right) \frac{2}{7}$ d = $\overset{4}{\cancel{8}}\left(\frac{7}{\cancel{2}}\right)$
>
>   d = 28
>
>   28 dogs

Here's what you should have.
d. Now figure the answer to the other question. ✔
The other question is: How many dogs were awake?
- Everybody, say the subtraction problem you worked. (Signal.) *28 – 8.*
- What's the answer? (Signal.) *20.*
(Display:)   [78:4E]

> a. $\frac{2}{7}$ d = s
>
> $\left(\frac{7}{2}\right) \frac{2}{7}$ d = $\overset{4}{\cancel{8}}\left(\frac{7}{\cancel{2}}\right)$
>
>   d = 28   2 8
>   –   8
>   28 dogs   2 0 awake dogs

Here's what you should have.
e. Work problem B. Answer both questions. Remember to simplify before you multiply. (Observe students and give feedback.)
f. Check your work.
You wrote the letter equation 4/9 S = G. The equation with a number for S is 4/9 times 45 = G.
- What does G equal? (Signal.) *20.*
- Which question does that answer? (Signal.) *How many girls are there?*

g. The other question is: How many boys are there?
- Say the subtraction problem for answering that question. (Signal.) *45 – 20.*
- What's the answer? (Signal.) *25.*
(Display:) [78:4F]

$$\textbf{b.} \ \frac{4}{9} \, s = g$$

$$\frac{4}{9} \, (\cancel{45}^{\,5}) = g = 20$$

20 girls

$$\begin{array}{r} 4\,5 \\ -2\,0 \\ \hline 2\,5 \ \text{boys} \end{array}$$

Here's what you should have.

EXERCISE 5: FRACTION OPERATIONS
DIVISION   REMEDY

a. Find part 2 in your textbook. ✔
(Teacher reference:)

a. $\frac{1}{4} \div \frac{7}{8}$ b. $\frac{8}{5} \div \frac{1}{10}$ c. $\frac{9}{2} \div \frac{7}{3}$ d. $\frac{8}{9} \div 4$

These problems divide fractions. You'll write the multiplication problem, simplify if you can, and figure out the answer.
b. Problem A: 1/4 ÷ 7/8.
- Work the multiplication problem.
(Observe students and give feedback.)
Check your work.
- Read the multiplication problem you worked. (Signal.) *1/4 × 8/7.*
- What's the answer? (Signal.) *2/7.*
(Display:) [78:5A]

$$\textbf{a.} \ \frac{1}{\cancel{4}} \times \frac{\cancel{8}^{\,2}}{7} = \boxed{\frac{2}{7}}$$

Here's what you should have.
c. Work the rest of the problems in part 2.
(Observe students and give feedback.)
d. Check your work.
- Problem B: 8/5 ÷ 1/10.
- Read the multiplication problem you worked. (Signal.) *8/5 × 10/1.*
- What's the answer? (Signal.) *16.*

(Display:) [78:5B]

$$\textbf{b.} \ \frac{8}{\cancel{5}} \times \frac{\cancel{10}^{\,2}}{1} = \boxed{16}$$

Here's what you should have.
e. Problem C: 9/2 ÷ 7/3.
- Read the multiplication problem you worked. (Signal.) *9/2 × 3/7.*
- What's the answer? (Signal.) *27/14.*
(Display:) [78:5C]

$$\textbf{c.} \ \frac{9}{2} \times \frac{3}{7} = \boxed{\frac{27}{14}}$$

Here's what you should have.
f. Problem D: 8/9 ÷ 4.
- Read the multiplication problem you worked. (Signal.) *8/9 × 1/4.*
- What's the answer? (Signal.) *2/9.*
(Display:) [78:5D]

$$\textbf{d.} \ \frac{\cancel{8}^{\,2}}{9} \times \frac{1}{\cancel{4}} = \boxed{\frac{2}{9}}$$

Here's what you should have.

EXERCISE 6: VOLUME
OF CYLINDERS   REMEDY

a. (Display:) [78:6A]

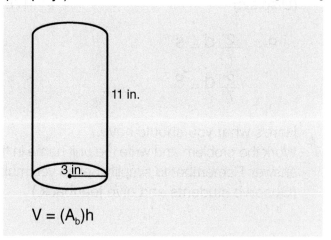

$$V = (A_b)h$$

You're going to figure out the volume of this cylinder.
- What's the equation for finding the volume of a rectangular prism? (Signal.) *Volume = Area of the base × height.*
- What's the equation for finding the volume of a cylinder? (Signal.) *Volume = Area of the base × height.*

b. In part B, figure out the area of the base. Then stop.
(Observe students and give feedback.)
You worked the problem 3.14 times 9.
- What's the number for the area of the base? (Signal.) *28 and 26 hundredths.*
- What's the height? (Signal.) *11.*
(Add to show:) [78:6B]

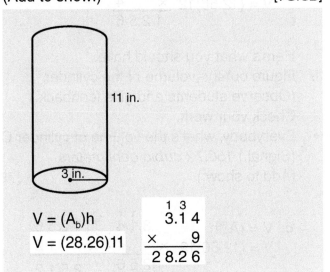

$V = (A_b)h$
$V = (28.26)11$

$$\begin{array}{r} \overset{1\ 3}{3.1\ 4} \\ \times\quad 9 \\ \hline 2\ 8.2\ 6 \end{array}$$

c. Copy the equation with numbers. Then figure out the volume of the cylinder.
(Observe students and give feedback.)
You worked the problem 28.26 times 11.
- What's the answer? (Signal.) *310.86.*
- What's the volume of the cylinder? (Signal.) *310.86 cubic inches.*
(Add to show:) [78:6C]

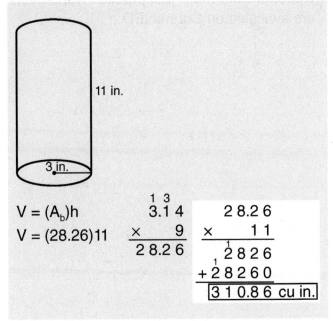

$V = (A_b)h$
$V = (28.26)11$

$$\begin{array}{r} \overset{1\ 3}{3.1\ 4} \\ \times\quad 9 \\ \hline 2\ 8.2\ 6 \end{array}$$

$$\begin{array}{r} 2\ 8.2\ 6 \\ \times\quad 1\ 1 \\ \hline \overset{1}{2}\ 8\ 2\ 6 \\ +\ \underset{1}{2}\ 8\ 2\ 6\ 0 \\ \hline \boxed{3\ 1\ 0.8\ 6 \text{ cu in.}} \end{array}$$

Here's what you should have.

---

a. Find part 3 in your textbook. ✔
(Teacher reference:)

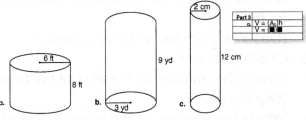

- Cylinder A. The number for the base is shown on top. What's the radius of the base? (Signal.) *6 feet.*
- Copy the equation for volume. First figure out the area of the base. Raise your hand when you have an equation with a number for the area of the base and a number for the height.
(Observe students and give feedback.)

b. Check your work.
- What's the number for the area of the base? (Signal.) *113.04.*
- What's the number for the height? (Signal.) *8.*
(Display:) [78:6D]

a. $V = (A_b)h$
$V = (113.04)8$

$$\begin{array}{r} \overset{1}{\overset{\cancel{2}}{3}}.1\ 4 \\ \times\quad 3\ 6 \\ \hline \overset{1}{\underset{1}{1}}\ 8\ 8\ 4 \\ +\ 9\ 4\ 2\ 0 \\ \hline 1\ 1\ 3.0\ 4 \end{array}$$

Here's what you should have.

c. Figure out the volume of the cylinder.
(Observe students and give feedback.)
Check your work.
- Everybody, what's the volume of cylinder A? (Signal.) *904.32 cubic feet.*
(Add to show:) [78:6E]

a. $V = (A_b)h$
$V = (113.04)8$

$$\begin{array}{r} \overset{1}{\overset{\cancel{2}}{3}}.1\ 4 \\ \times\quad 3\ 6 \\ \hline \overset{1}{\underset{1}{1}}\ 8\ 8\ 4 \\ +\ 9\ 4\ 2\ 0 \\ \hline 1\ 1\ 3.0\ 4 \end{array}$$

$$\begin{array}{r} \overset{1\ 2\quad 3}{1\ 1\ 3.0\ 4} \\ \times\quad\quad 8 \\ \hline \boxed{9\ 0\ 4.3\ 2 \text{ cu ft}} \end{array}$$

Here's what you should have.

---

d. Cylinder B. Copy the equation for volume. First figure out the area of the base. Raise your hand when you have an equation with a number for the area of the base and a number for the height.
(Observe students and give feedback.)

e. Check your work.
- What's the number for the area of the base? (Signal.) *28.26.*
- What's the number for the height? (Signal.) *9.* (Display:) [78:6F]

$$\textbf{b.} \quad V = (A_b)h \qquad \overset{1\ 3}{3.1\ 4}$$
$$V = (28.26)9 \qquad \underset{2\,8.2\,6}{\times \qquad 9}$$

Here's what you should have.

f. Figure out the volume of the cylinder.
(Observe students and give feedback.)
Check your work.
- Everybody, what's the volume of cylinder B? (Signal.) *254.34 cubic yards.*
(Add to show:) [78:6G]

$$\textbf{b.} \quad V = (A_b)h \qquad \overset{1\ 3}{3.1\ 4} \qquad \overset{7\ 2\ 5}{2\ 8.2\ 6}$$
$$V = (28.26)9 \qquad \underset{2\,8.2\,6}{\times \qquad 9} \times \underset{\boxed{2\,5\,4.3\,4\ \text{cu yd}}}{\qquad 9}$$

Here's what you should have.

g. Cylinder C. Copy the equation for volume. First figure out the area of the base. Raise your hand when you have an equation with a number for the area of the base and a number for the height.
(Observe students and give feedback.)

h. Check your work.
- What's the number for the area of the base? (Signal.) *12.56.*
- What's the number for the height? (Signal.) *12.* (Display:) [78:6H]

$$\textbf{c.} \quad V = (A_b)h \qquad \overset{1}{3.1\ 4}$$
$$V = (12.56)12 \qquad \underset{1\,2.5\,6}{\times \qquad 4}$$

Here's what you should have.

i. Figure out the volume of the cylinder.
(Observe students and give feedback.)
Check your work.
- Everybody, what's the volume of cylinder C? (Signal.) *150.72 cubic centimeters.*
(Add to show:) [78:6I]

$$\textbf{c.} \quad V = (A_b)h \qquad \overset{1}{3.1\ 4} \qquad \overset{1\ 1}{1\ 2.5\ 6}$$
$$V = (12.56)12 \qquad \underset{1\,2.5\,6}{\times \qquad 4} \times \qquad 1\,2$$
$$\underset{\boxed{1\,5\,0.7\,2\ \text{cu cm}}}{\begin{array}{r} \overset{1}{2\,5\,1\,2} \\ +\,1\,2\,5\,6\,0 \end{array}}$$

Here's what you should have.

Assign Independent Work, Textbook parts 4–9 and Workbook parts 3–6.

Optional extra math-fact practice worksheets are available on ConnectED.

Lesson 79

EXERCISE 1: FRACTION ANALYSIS
COMPARISON

a. Open your workbook to Lesson 79 and find part 1.
(Teacher reference:)

a. $\frac{4}{15}$ $\frac{1}{5}$ b. $\frac{9}{16}$ $\frac{1}{2}$ c. $\frac{1}{3}$ $\frac{2}{12}$

d. $\frac{3}{20}$ $\frac{1}{4}$ e. $\frac{1}{9}$ $\frac{4}{27}$

You're going to write the signs to show which fraction is more. For some problems, the fraction with the larger denominator is first.

b. Read the fractions for problem A. (Signal.)
4/15, 1/5.
• Which fraction has the larger denominator? (Signal.) *4/15.*
• Say the question about 15ths. (Signal.)
How many 15ths is 1/5?
• What's the answer? (Signal.) *3/15.*
(Repeat until firm.)

c. Read the fractions for problem B. (Signal.)
9/16, 1/2.
• Which fraction has the larger denominator? (Signal.) *9/16.*
• Say the question about 16ths. (Signal.)
How many 16ths is 1/2?
• What's the answer? (Signal.) *8/16.*
(Repeat until firm.)

d. Write signs for all the problems.
(Observe students and give feedback.)

e. Check your work.
• Problem A. Which is more—4/15 or 1/5? (Signal.) *4/15.*
• Problem B. Which is more—9/16 or 1/2? (Signal.) *9/16.*
• C. Which is more—1/3 or 2/12? (Signal.) *1/3.*
• D. Which is more—3/20 or 1/4? (Signal.) *1/4.*
• E. Which is more—1/9 or 4/27? (Signal.) *4/27.*

(Display:)   [79:1A]

a. $\frac{4}{15} > \frac{1}{5}$ b. $\frac{9}{16} > \frac{1}{2}$ c. $\frac{1}{3} > \frac{2}{12}$

d. $\frac{3}{20} < \frac{1}{4}$ e. $\frac{1}{9} < \frac{4}{27}$

Here's what you should have.

EXERCISE 2: MENTAL MATH

a. Time for some mental math.
• Listen: 360 divided by 9. (Pause.) (Signal.) *40.*
• Listen: 540 ÷ 9. (Pause.) (Signal.) *60.*
• 720 ÷ 9. (Pause.) (Signal.) *80.*
• 810 ÷ 9. (Pause.) (Signal.) *90.*
(Repeat until firm.)

b. Listen: 450 divided by 9. (Pause.) (Signal.) *50.*
• 480 ÷ 8. (Pause.) (Signal.) *60.*
• 640 ÷ 8. (Pause.) (Signal.) *80.*
(Repeat until firm.)

c. Listen: 6 times 9. (Pause.) (Signal.) *54.*
• Listen: 6 × 90. (Pause.) (Signal.) *540.*
• 4 × 90. (Pause.) (Signal.) *360.*
• 4 × 50. (Pause.) (Signal.) *200.*
(Repeat until firm.)

d. Listen: 4 times 80. (Pause.) (Signal.) *320.*
• 6 × 80. (Pause.) (Signal.) *480.*
• 3 × 80. (Pause.) (Signal.) *240.*
• 5 × 80. (Pause.) (Signal.) *400.*
(Repeat until firm.)

e. Listen: 360 divided by 9. (Pause.) (Signal.) *40.*
• 360 ÷ 4. (Pause.) (Signal.) *90.*
• 90 times 4. (Pause.) (Signal.) *360.*
• 90 × 2. (Pause.) (Signal.) *180.*
• 90 × 3. (Pause.) (Signal.) *270.*
(Repeat until firm.)

EXERCISE 3: FRACTIONS
AS MIXED NUMBERS

a. (Display:)   [79:3A]

$$6 + \frac{7}{5}$$

- Read the problem. (Signal.) *6 + 7/5.*
  We're going to rewrite 7/5 as a mixed number.
- Raise your hand when you know the mixed number. ✔
- What's the mixed number? (Signal.) *1 and 2/5.*
  We write the fraction below and put the 1 with the other whole number.
  (Add to show:)   [79:3B]

$$\overset{1}{6} + \frac{7}{5}$$
$$\frac{2}{5}$$

- What's the new whole number? (Signal.) *7.*
  (Add to show:)   [79:3C]

$$\overset{1}{6} + \frac{7}{5}$$
$$7 + \frac{2}{5}$$

- What does 6 + 7/5 equal? (Signal.) *7 + 2/5.*

━━━━ WORKBOOK PRACTICE ━━━━

a. Find part 2 in your workbook. ✔
  (Teacher reference:)

   a. $3 + \frac{7}{6}$ b. $16 + \frac{12}{9}$ c. $9 + \frac{25}{20}$ d. $75 + \frac{9}{8}$

These are like the problem we just worked.
b. Read problem A. (Signal.) *3 + 7/6.*
- Write 7/6 as a mixed number and show the new addition for 3 + 7/6. ✔
- Everybody, what does 3 + 7/6 equal? (Signal.) *4 + 1/6.*
c. Work the rest of the problems in part 2.
  (Observe students and give feedback.)

d. Check your work.
- Problem B: 16 + 12/9.
  Read the new addition. (Signal.) *17 + 3/9.*
- C: 9 + 25/20.
  Read the new addition. (Signal.) *10 + 5/20.*
- D: 75 + 9/8.
  Read the new addition. (Signal.) *76 + 1/8.*

EXERCISE 4: DECIMAL PLACE VALUE
COMPARISON   [REMEDY]

a. (Display:)   [79:4A]

3.27   3.027

You're going to compare decimal numbers. Remember, you start with the first digit and compare the numbers. If the digits are the same then you do the same thing with the next digits until you find digits that are not the same.
- Read these numbers. (Signal.) *3 and 27 hundredths, 3 and 27 thousandths.*
b. Look at the digits for the whole number. Are those digits the same? (Signal.) *Yes.*
- Look at the tenths. Are those digits the same? (Signal.) *No.*
- Which digit is more? (Signal.) *2.*
- So which number is greater? (Signal.) *3.27.*
  (Add to show:)   [79:4B]

3.27 > 3.027

Here's the sign.
c. (Display:)   [79:4C]

6.29   5.291

- Read the numbers. (Signal.) *6 and 29 hundredths, 5 and 291 thousandths.*
- Are the digits for the whole numbers the same? (Signal.) *No.*
- Which digit is more? (Signal.) *6.*
- So which number is greater? (Signal.) *6.29.*
  (Add to show:)   [79:4D]

6.29 > 5.291

Here's the sign.

d. (Display:) [79:4E]

$$.734 \qquad 7.34$$

- Read the numbers. (Signal.) *734 thousandths, 7 and 34 hundredths.*
- Are the digits for the whole numbers the same? (Signal.) *No.*
- Which number is greater? (Signal.) *7.34.*
(Add to show:) [79:4F]

$$.734 < 7.34$$

Yes, 734 thousandths is the same as zero and 734 thousandths.

e. (Display:) [79:4G]

$$.529 \qquad 5.29$$

- Read the numbers. (Signal.) *529 thousandths, 5 and 29 hundredths.*
- Are the digits for the whole numbers the same? (Signal.) *No.*
- Which number is greater? (Signal.) *5.29.*
(Add to show:) [79:4H]

$$.529 < 5.29$$

Yes, 529 thousandths is the same as zero and 529 thousandths.

━━━━━━ **WORKBOOK PRACTICE** ━━━━━━

a. Find part 3 in your workbook. ✔
(Teacher reference:)   [R] [Part H]

a. 7.052   7.502   b. 16.130   16.13   c. .432   4

d. .036   .035   e. 12.2   1.22

b. Read the numbers for item A. (Signal.) *7 and 52 thousandths, 7 and 502 thousandths.*
- Write the sign to show which number is greater or if the numbers are equal. ✔
- Which number is greater? (Signal.) *7.502.*
c. Write signs for the rest of the number pairs. (Observe students and give feedback.)

d. Check your work.
- Item B: The numbers are 16 and 130 thousandths and 16 and 13 hundredths.
- Is one number greater than the other? (Signal.) *No.*
- What sign did you write? (Signal.) *Equals.*
e. C. The numbers are 432 thousandths and 4.
- Which number is greater? (Signal.) *4.*
f. D. The numbers are 36 thousandths and 35 thousandths.
- Which number is greater? (Signal.) *36 thousandths.*
g. E. The numbers are 12 and 2 tenths and 1 and 22 hundredths.
- Which number is greater? (Signal.) *12 and 2 tenths.*

EXERCISE 5: FRACTION OPERATIONS
DIVISION

a. (Display:) [79:5A]

$$\frac{1}{3} \div 2 =$$

We're going to complete a diagram to show the answer to a problem that divides a fraction by a whole number. 1/3 divided by 2.
- Say the multiplication. (Signal.) *1/3 × 1/2.*
(Add to show:) [79:5B]

$$\frac{1}{3} \div 2 = \frac{1}{3} \times \frac{1}{2}$$

- What does 1/3 times 1/2 equal? (Signal.) *1/6.*
(Add to show:) [79:5C]

$$\frac{1}{3} \div 2 = \frac{1}{3} \times \frac{1}{2} = \boxed{\frac{1}{6}}$$

Yes, 1/3 ÷ 2 = 1/6.

b. (Change to show:)   [79:5D]

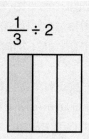

$$\frac{1}{3} \div 2$$

Let's make a diagram of the problem and see if the answer is correct. Here's a diagram that shows one third. The problem we started with is 1/3 divided by 2. So we divide the 1/3 into two parts.

(Change to show:)   [79:5E]

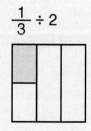

$$\frac{1}{3} \div 2$$

• Raise your hand when you know what fraction of the whole unit is shaded. ✔
• What's the fraction? (Signal.) *1/6.*
  So 1/3 divided by 2 equals 1/6.
  (Add to show:)   [79:5F]

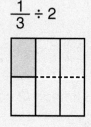

$$\frac{1}{3} \div 2$$

You can see that the shaded part is 1/6 of the whole unit. That's the same answer we got when we worked the problem.

WORKBOOK PRACTICE

a. Find part 4 in your workbook. ✔
    (Teacher reference:)

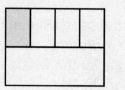

a. $\frac{1}{2} \div 4 =$ =

b. Problem A: 1/2 ÷ 4.
• Say the multiplication. (Signal.) *1/2 × 1/4.*
• Work the multiplication problem. Figure out what 1/2 ÷ 4 equals. ✔
• Everybody, what does 1/2 times 1/4 equal? (Signal.) *1/8.*
• Divide one of the halves into 4 parts. Then shade one of those parts.
   (Observe students and give feedback.)
   (Display:)   [79:5G]

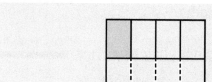

Here's one way to show the division.
• Everybody, what fraction of the whole unit did you shade? (Signal.) *1/8.*
   Yes, 1/2 divided by 4 equals 1/8.
   (Add to show:)   [79:5H]

EXERCISE 6: WORD PROBLEMS
PARTS OF A GROUP

REMEDY

a. Open your textbook to Lesson 79 and find part 1. ✔
   (Teacher reference:)

   a. There were 100 doors. $\frac{2}{5}$ of the doors were closed. How many open doors were there? How many closed doors were there?

   b. $\frac{1}{6}$ of the birds were robins. There were 12 robins. How many birds were there? How many birds were not robins?

   Each problem asks two questions. You work the problems by writing the equation with letters and solving that equation. That answers one of the questions. Then you subtract to answer the other question.

b. Problem A: There were 100 doors. 2/5 of the doors were closed. How many open doors were there? How many closed doors were there?

• Work the problem. Raise your hand when you have an answer to one of the questions.
   (Observe students and give feedback.)

• You solved the equation 2/5 × 100 = C. What does C equal? (Signal.) *40.*

• Is that the number for closed doors or open doors? (Signal.) *Closed doors.*

c. Figure out the answer to the other question.
   (Observe students and give feedback.)

• Everybody, say the subtraction problem you worked. (Signal.) *100 – 40.*

• How many open doors were there? (Signal.) *60 open doors.*

d. Problem B: 1/6 of the birds were robins. There were 12 robins. How many birds were there? How many birds were not robins?

• Work that problem. Answer both questions.
   (Observe students and give feedback.)

e. Check your work.
   You solved the equation 1/6 B = 12.

• What does B equal? (Signal.) *72.*

• Is that the number for robins or for birds? (Signal.) *Birds.*

• Say the subtraction problem you worked. (Signal.) *72 – 12.*

• What's the answer? (Signal.) *60.*

• Is that the number for robins or for not robins? (Signal.) *Not robins.*

EXERCISE 7: VOLUME
OF CYLINDERS

a. Find part 2 in your textbook. ✔
   (Teacher reference:)

   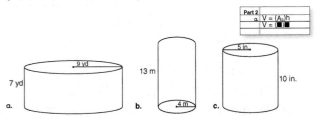

   You're going to find the volume of cylinders. The equation you use is: Volume = Area of the base times height.

b. What shape is the base of figure A? (Signal.) *A circle.*

• Say the multiplication for the area of the base. (Signal.) *3.14 × 81.*

c. Cylinder B. Say the multiplication for the area of the base. (Signal.) *3.14 × 16.*

• Cylinder C. Say the multiplication for the area of the base. (Signal.) *3.14 × 25.*

d. Cylinder A. Copy the letter equation. Figure out the area of the base and write the equation with a number for the area of the base and a number for the height. Stop when you've done that much.
   (Observe students and give feedback.)
   (Display:) [79:7A]

```
a. V = (A_b)h              1 3
   V = (254.34)7          3.1 4
                        ×    8 1
                          3 1 4
                      + 2 5 1 2 0
                        2 5 4.3 4
```

   Here's what you should have.

e. Figure out the volume. Remember, the unit name for the answer is cubic yards.
   (Observe students and give feedback.)

f. Check your work.
   You multiplied 254.34 by 7.

• What's the volume of the cylinder? (Signal.) *1780.38 cubic yards.*

(Add to show:) [79:7B]

a. $V = (A_b)h$
$V = (254.34)7$

$$\begin{array}{r} \overset{1\ 3}{3.1\,4} \\ \times\quad 8\,1 \\ \hline 3\,1\,4 \\ +2\,5\,1\,2\,0 \\ \hline 2\,5\,4.3\,4 \end{array}$$

$$\begin{array}{r} \overset{3\ 3\ 2\ 2}{2\,5\,4.3\,4} \\ \times\qquad 7 \\ \hline \boxed{1\,7\,8\,0.3\,8\ \text{cu yd}} \end{array}$$

Here's what you should have.

g. Find the volume of cylinder B. Start with the letter equation. Figure out the area of the base and multiply by the height. Remember, the unit name for the answer is cubic meters.
(Observe students and give feedback.)

h. Check your work. Cylinder B. The radius is 4.
• Say the multiplication for the area of the base. (Signal.) *3.14 times 16.*
• What's the number for area of the base? (Signal.) *50.24.*
You multiplied 50.24 by 13.
• What's the volume of the cylinder? (Signal.) *653.12 cubic meters.*
(Display:) [79:7C]

b. $V = (A_b)h$
$V = (50.24)13$

$$\begin{array}{r} \overset{\not{8}}{3.1\,4} \\ \times\quad 1\,6 \\ \hline \overset{1\ 1}{1\,8\,8\,4} \\ +3\,1\,4\,0 \\ \hline 5\,0.2\,4 \end{array}$$

$$\begin{array}{r} \overset{\not{7}}{5\,0.2\,4} \\ \times\quad 1\,3 \\ \hline \overset{1}{1\,5\,0\,7\,2} \\ 5\,0\,2\,4\,0 \\ \hline \boxed{6\,5\,3.1\,2\ \text{cu m}} \end{array}$$

Here's what you should have.

i. Find the volume of cylinder C. Figure out the area of the base and multiply by the height. Remember, the unit name in the answer is cubic inches.
(Observe students and give feedback.)

j. Check your work. The radius is 5.
• Say the multiplication for the area of the base. (Signal.) *3.14 times 25.*
• What's the answer? (Signal.) *78.50.*
You multiplied 78.50 by 10.
• What's the volume of the cylinder? (Signal.) *785 cubic inches.*
(Display:) [79:7D]

c. $V = (A_b)h$
$V = (78.50)10$
$\boxed{785\ \text{cu in.}}$

$$\begin{array}{r} \overset{\not{8}}{3.1\,4} \\ \times\quad 2\,5 \\ \hline \overset{1}{1\,5\,7\,0} \\ +6\,2\,8\,0 \\ \hline 7\,8.5\,0 \end{array}$$

Here's what you should have.

EXERCISE 8: FRACTION SIMPLIFICATION

ANSWERS

a. (Display:) [79:8A]

$$\frac{8}{12}$$

$$\frac{75}{9}$$

From now on, you're going to make sure that all answers are simplified.
If an answer is a fraction, you show it simplified.
If a fraction is more than 1, you show it as a mixed number.
If the fraction part of a mixed number can be simplified, you show it simplified.

b. (Point to $\frac{8}{12}$.) Read this fraction. (Signal.) *8/12.*
• Can it be simplified? (Signal.) *Yes.*
• Raise your hand when you know the simplified fraction that equals 8/12. ✔
• What's the simplified fraction? (Signal.) *2/3.*
(Add to show:) [79:8B]

$$\frac{8}{12} = \boxed{\frac{2}{3}}$$

$$\frac{75}{9}$$

• Is the fraction more than 1? (Signal.) *No.*
• So will we write it as a mixed number? (Signal.) *No.*

c. (Point to $\frac{75}{9}$.) Read this fraction. (Signal.) *75/9.*
The fraction can be simplified.
• Raise your hand when you know the common factor. ✔
• What's the common factor? (Signal.) *3.*
• Copy the fraction in part A. Show the simplified fraction. Then show the mixed number it equals. ✔

d. Check your work.
• What's the simplified fraction? (Signal.) *25/3.*
• What's the simplified mixed number? (Signal.) *8 and 1/3.*
(Add to show:) [79:8C]

$$\frac{8}{12} = \boxed{\frac{2}{3}}$$

$$\frac{75}{9} = \frac{25}{3} = \boxed{8\frac{1}{3}}$$

Here's what you should have.
Remember, simplify the fraction first.

TEXTBOOK PRACTICE

a. Find part 3 in your textbook. ✔
(Teacher reference:)

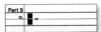

- Copy each fraction and write the simplified
fraction. If the simplified fraction is more than
1, write it as a mixed number.
(Observe students and give feedback.)

b. Check your work.
- Fraction A: 38/6. What's the simplified
fraction? (Signal.) *19/3.*
- What's the simplified mixed number?
(Signal.) *6 and 1/3.*

c. Fraction B: 50/12. What's the simplified
fraction? (Signal.) *25/6.*
- What's the simplified mixed number?
(Signal.) *4 and 1/6.*

d. Fraction C: 4/20. What's the simplified
fraction? (Signal.) *1/5.*

e. Fraction D: 56/16. What's the simplified
fraction? (Signal.) *7/2.*
- What's the simplified mixed number?
(Signal.) *3 and 1/2.*

f. Find part 4 in your textbook. ✔
(Teacher reference:)

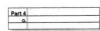

(Display:)   [79:8D]

> Work each problem. S

(Point to S.) You'll see this symbol after some
directions in your independent work. The S
reminds you to simplify the answers to the
problems in that part.

EXERCISE 9: INDEPENDENT WORK
AREA OF A CIRCLE

a. Find part 9 in your textbook. ✔
(Teacher reference:)

Find the area of the circle. Start with the letter equation.

I'll read the directions: **Find the area of the
circle. Start with the letter equation.**

b. Remember, Area = Pi times radius to the
second.
- Say the equation. (Signal.) *Area = Pi times
radius to the second.*
Start with the letter equation when you work
the problem.

Assign Independent Work, Textbook parts 4–9
and Workbook parts 5–8.

Optional extra math-fact practice worksheets
are available on ConnectED.

Lesson

EXERCISE 1: DECIMAL PLACE VALUE
EXPANDED NOTATION

a. (Display:)   [80:1A]

> .283

- Read the number. (Signal.) *283 thousandths.*
- Say the place-value addition. (Signal.)
  2 tenths + 8 hundredths + 3 thousandths.
  (Add to show:)   [80:1B]

> .283 = .2 + .08 + .003

b. We can show the place-value addition works
   by writing it in a column.
   We'll add 2 tenths + 8 hundredths +
   3 thousandths.
   (Add to show:)   [80:1C]

> .283 = .2 + .08 + .003
>
>   .2
>   .0 8
>   + .0 0 3

c. How many are in the thousandths column?
   (Signal.) *3.*
- How many are in the hundredths column?
  (Signal.) *8.*
- How many are in the tenths column? (Signal.) *2.*
  (Add to show:)   [80:1D]

> .283 = .2 + .08 + .003
>
>   .2
>   .0 8
>   + .0 0 3
>   ─────────
>   .2 8 3

- What number did we end up with? (Signal.)
  283 thousandths.
  That's the number we started with. So the
  place-value addition works.

━━━━━━━━━━ WORKBOOK PRACTICE ━━━━━━━━━━

a. Open your workbook to Lesson 80 and find
   part 1. ✔
   (Teacher reference:)

> a. .412   b. 2.03

You're going to write the place-value addition
in a column. Remember, you'll need zeros
after the decimal point for some numbers.

b. Number A: 412 thousandths.
- Say the place-value addition. (Signal.)
  4 tenths + 1 hundredth + 2 thousandths.
- Write the place-value addition in columns
  starting with 4 tenths. Then show the answer.
  (Observe students and give feedback.)
  (Display:)   [80:1E]

> a.   .4
>   .0 1
>   + .0 0 2
>   ─────────
>   .4 1 2

Here's what you should have.
- Everybody, what's the answer? (Signal.)
  412 thousandths.
  That's the number we started with. So the
  place-value addition works.
c. Number B: 2 and 3 hundredths.
- Write the place-value addition in a column,
  starting with 2. Then show the answer.
  (Observe students and give feedback.)
  (Display:)   [80:1F]

> b.   2
>   .0
>   + .0 3
>   ─────────
>   2.0 3

Here's what you should have.
- Everybody, what's the answer? (Signal.) *2 and
  3 hundredths.*
  That's the number we started with. So the
  place-value addition works.

EXERCISE 2: FRACTION ANALYSIS
COMPARISON

a. Find part 2 in your workbook.
   (Teacher reference:)

   a. $\frac{3}{10}$ $\frac{1}{5}$ b. $\frac{4}{18}$ $\frac{1}{3}$ c. $\frac{1}{2}$ $\frac{4}{10}$

   d. $\frac{4}{60}$ $\frac{1}{10}$ e. $\frac{1}{4}$ $\frac{7}{20}$

   You're going to write signs to show which fraction is more.

b. Read the fractions for problem A. (Signal.) *3/10, 1/5.*
 • Which fraction has the larger denominator? (Signal.) *3/10.*
 • Say the question about 10ths. (Signal.) *How many 10ths is 1/5?*
 • What's the answer? (Signal.) *2/10.*
   (Repeat until firm.)
c. Write signs for all the problems.
   (Observe students and give feedback.)
d. Check your work.
 • Problem A. Which is more—3/10 or 1/5? (Signal.) *3/10.*
 • Problem B. Which is more—4/18 or 1/3? (Signal.) *1/3.*
 • C. Which is more—1/2 or 4/10? (Signal.) *1/2.*
 • D. Which is more—4/60 or 1/10? (Signal.) *1/10.*
 • E. Which is more—1/4 or 7/20? (Signal.) *7/20.*
   (Display:)   [80:2A]

   a. $\frac{3}{10} > \frac{1}{5}$ b. $\frac{4}{18} < \frac{1}{3}$ c. $\frac{1}{2} > \frac{4}{10}$

   d. $\frac{4}{60} < \frac{1}{10}$ e. $\frac{1}{4} < \frac{7}{20}$

   Here's what you should have.

EXERCISE 3: MENTAL MATH

a. Time for some mental math.
 • Listen: 360 divided by 9. (Pause.) (Signal.) *40.*
 • 360 ÷ 4. (Pause.) (Signal.) *90.*
 • 270 ÷ 9. (Pause.) (Signal.) *30.*
 • 270 ÷ 3. (Pause.) (Signal.) *90.*
   (Repeat until firm.)

b. Listen: 90 × 4. (Pause.) (Signal.) *360.*
 • 90 × 6. (Pause.) (Signal.) *540.*
 • 90 × 3. (Pause.) (Signal.) *270.*
   (Repeat until firm.)
c. Listen: 4 × 80. (Pause.) (Signal.) *320.*
 • 6 × 80. (Pause.) (Signal.) *480.*
 • 10 × 80. (Pause.) (Signal.) *800.*
 • 5 × 80. (Pause.) (Signal.) *400.*
   (Repeat until firm.)
d. Listen: 5 × 40. (Pause.) (Signal.) *200.*
 • 8 × 40. (Pause.) (Signal.) *320.*
 • 3 × 40. (Pause.) (Signal.) *120.*
   (Repeat until firm.)
e. Listen: 5 × 60. (Pause.) (Signal.) *300.*
 • 2 × 60. (Pause.) (Signal.) *120.*
 • 10 × 60. (Pause.) (Signal.) *600.*
   (Repeat until firm.)

EXERCISE 4: DECIMAL PLACE VALUE
COMPARISON

a. Find part 3 in your workbook. ✔
   (Teacher reference:)

   a. .006 .06   e. .560 5.6

   b. 10.1 1.01   f. 3.400 3.4

   c. 7.0399 7.0339   g. .55 .5501

   d. .034 .03

   You're going to compare the pairs of numbers and make the signs.
b. Item A. Read the numbers. (Signal.)
   6 thousandths, 6 hundredths.
 • Write the sign. Then work the rest of the items in part 3.
   (Observe students and give feedback.)
c. Check your work.
   (Display:)   [80:4A]

a. .006 < .06	e. .560 < 5.6
b. 10.1 > 1.01	f. 3.400 = 3.4
c. 7.0399 > 7.0339	g. .55 < .5501
d. .034 > .03	

   Here's what you should have for each item.
   Check each sign that you made.

EXERCISE 5: FRACTIONS OPERATIONS
DIVISION

a. Find part 4 in your workbook. ✔
   (Teacher reference:)

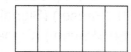

You're going to make a diagram of the problem 1/5 divided by 2.
- Say the problem. (Signal.) *1/5 ÷ 2.*
- Work the multiplication problem. Figure out what 1/5 divided by 2 equals. ✔
- Everybody, what does 1/5 divided by 2 equal? (Signal.) *1/10.*

b. The diagram shows fifths. Divide one of the fifths into 2 parts and shade one of those parts. ✔
- Everybody, what fraction of the whole unit did you shade? (Signal.) *1/10.*
   Yes, 1/5 divided by 2 equals 1/10.
   (Display:)   [80:5A]

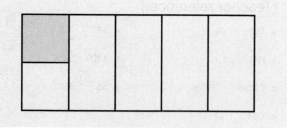

Here's one way to show the division.

c. We can show that 1/5 divided by 2 is 1/10 by dividing the rest of the parts by 2.
   (Add to show:)   [80:5B]

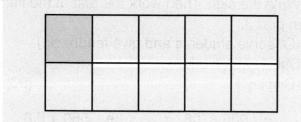

- What fraction of the diagram is shaded? (Signal.) *1/10.*

EXERCISE 6: WORD PROBLEMS
PARTS OF A GROUP

a. Open your textbook to Lesson 80 and find part 1. ✔
   (Teacher reference:)

a. $\frac{2}{5}$ of the pies were baked. There were 14 baked pies. How many pies were not baked? How many pies were there in all?

b. The pie shop baked apple pies and blackberry pies. There were 60 pies in all. $\frac{3}{4}$ of the pies were apple pies. How many were blackberry pies? How many were apple pies?

b. Problem A: 2/5 of the pies were baked. There were 14 baked pies. How many pies were not baked? How many pies were there in all?
- Write the equation with letters and solve that equation. Make sure you know which question it answers. Stop when you have an answer to one of the questions.
   (Observe students and give feedback.)

c. Check your work.
   You solved the equation 2/5 P = 14.
- What does P equal? (Signal.) *35.*
- Is that the number for all pies or pies that are not baked? (Signal.) *All pies.*
   Yes, P tells about all the pies.

d. Figure out the answer to the other question. ✔
- Everybody, say the subtraction problem you worked. (Signal.) *35 – 14.*
- What's the whole answer? (Signal.) *21 pies not baked.*
   Yes, 21 pies were not baked.

e. Work problem B. Remember to answer both questions.
   (Observe students and give feedback.)

f. Check your work.
   You solved the equation 3/4 × 60 = A.
- What does A equal? (Signal.) *45.*
- What's the unit name? (Signal.) *Apple pies.*
   Yes, there were 45 apple pies.
- Say the subtraction problem you worked. (Signal.) *60 – 45.*
- What's the whole answer? (Signal.) *15 blackberry pies.*
   Yes, there were 15 blackberry pies.

EXERCISE 7: VOLUME
OF CYLINDERS/RECTANGULAR PRISMS

a. Find part 2 in your textbook. ✔
    (Teacher reference:)

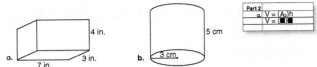

You're going to find the volume of these figures. You use the same equation for both volumes.

*   Say the equation. (Signal.) *Volume = Area of the base × height.*
b. Figure A. What kind of figure is it? (Signal.) *Rectangular prism.*
*   Start with the letter equation. Figure out the area of the base and multiply by the height. Write the unit name in the answer.
    (Observe students and give feedback.)
c. Check your work.
*   You figured out the area of the base. What's 7 times 3? (Signal.) *21.*
*   You multiplied 21 by 4. What's the volume of figure A? (Signal.) *84 cubic inches.*
    (Display:)   [80:7A]

> a. $V = (A_b)h$
> $V = (21)4$
>
> $\begin{array}{r} 2\,1 \\ \times\quad 4 \\ \hline \boxed{8\,4 \text{ cu in.}} \end{array}$

Here's what you should have.
d. Figure B. What kind of figure is it? (Signal.) *Cylinder.*
*   Start with the letter equation. Figure out the area of the base and multiply by the height.
    (Observe students and give feedback.)
e. Check your work.
*   You figured out the area of the base. What's the number for the area of the base? (Signal.) *28.26.*
*   What's the number for the height? (Signal.) *5.*
*   You multiplied 28.26 by 5. What's the volume of the cylinder? (Signal.) *141.3 cubic centimeters.*

(Display:)   [80:7B]

> b. $V = (A_b)h$
> $V = (28.26)5$
>
> $\begin{array}{r} {}^{1\,3} \\ 3.1\,4 \\ \times\quad 9 \\ \hline 2\,8.2\,6 \end{array}$
> $\begin{array}{r} {}^{4\,1\,3} \\ 2\,8.2\,6 \\ \times\quad 5 \\ \hline \boxed{1\,4\,1.3\,0 \text{ cu cm}} \end{array}$

Here's what you should have.

EXERCISE 8: FRACTIONS
AS MIXED NUMBERS

a. Find part 3 in your textbook. ✔
    (Teacher reference:)

a. $13 + \frac{15}{11}$ b. $9 + \frac{23}{20}$ c. $7 + \frac{19}{10}$ d. $22 + \frac{18}{17}$

You're going to copy each problem. Below, you'll show the addition with a fraction that is less than one.
b. Read problem A. (Signal.) *13 + 15/11.*
*   Write the problem and show the new addition below.
    (Observe students and give feedback.)
c. Check your work.
    (Display:)   [80:8A]

> a. $13 + \overset{1}{\frac{15}{11}}$
>
> $\boxed{14 + \frac{4}{11}}$

Here's what you should have. 13 + 15/11 is 14 + 4/11.
d. Work the rest of the problems in part 3.
    (Observe students and give feedback.)
e. Check your work.
*   Problem B. What does 9 + 23/20 equal? (Signal.) *10 + 3/20.*
*   C. What does 7 + 19/10 equal? (Signal.) *8 + 9/10.*
*   D. What does 22 + 18/17 equal? (Signal.) *23 + 1/17.*

EXERCISE 9: FRACTION SIMPLIFICATION

ANSWERS

a. Find part 4 in your textbook. ✔
   (Teacher reference:)

a. $\frac{18}{54}$ b. $\frac{57}{9}$ c. $\frac{37}{10}$ d. $\frac{75}{100}$

Remember, from now on, simplify fractions first. Then you'll show all of your answers with a simplified fraction or a simplified mixed number.

b. Fraction A: 18/54.
• Copy the fraction and show the simplified fraction it equals.
  (Observe students and give feedback.)
• Check your work.
  You may have written the fraction 2/6. That fraction can be simplified.
• What's the simplified fraction for 18/54? (Signal.) *1/3.*

c. Fraction B: 57/9.
• Write the simplified fraction. ✔
• Everybody, what's the simplified fraction? (Signal.) *19/3.*
• Now write the simplified mixed number. ✔
• Everybody, what's the simplified mixed number? (Signal.) *6 and 1/3.*

d. Fraction C: 37/10
• Raise your hand when you know if the fraction can be simplified. ✔
• Can 37/10 be simplified? (Signal.) *No.*
• So just write it as a mixed number. ✔
• Everybody, what does 37/10 equal? (Signal.) *3 and 7/10.*

e. Fraction D: 75/100.
• Write the simplified value. ✔
• Everybody, what's the simplified value? (Signal.) *3/4.*

f. Find part 6 in your textbook. ✔
   (Teacher reference:)

Copy and work each problem. Ⓢ

a. $24\frac{3}{5}$ b. $2\frac{8}{15}$ c. $21\frac{13}{14}$
   $-\ 8\frac{3}{5}$ $+\ 7\frac{6}{15}$ $-\ 10\frac{11}{14}$

Remember, the S written after directions tells you to simplify answers if you can.

Assign Independent Work, Textbook parts 5–11 and Workbook parts 5 and 6.

Optional extra math-fact practice worksheets are available on ConnectED.

Mastery Test 8

Note: Mastery Tests are administered to the entire group. Each student will need a pencil and the *Student Assessment Book.* Try to arrange students so they cannot look at other students' responses.

Note: Students will need a ruler for part 1 of the test.

Teacher Presentation

a. Find Test 8 in your test booklet. ✔
You'll work the test by yourself. Read the directions carefully and do your best work. Put your pencil down when you've finished the test.
(Observe but do not give feedback.)

Scoring Notes

a. Collect test booklets. Use the *Answer Key* and Passing Criteria Table to score the tests.

Part	Score	Possible Score	Passing Score
Passing Criteria Table—Mastery Test 8			
1	1 for each answer, each axis label, each point, line	12	10
2	1 for each item	3	3
3	3 for each item (letter equation, number equation, answer with unit name)	6	5
4	1 for each item	4	3
5	1 for each digit in answer 1 for each remainder	10	8
6	1 for each addend	9	8
7	2 for each item (whole number, fraction)	4	3
8	3 for each item (letter equation, answers)	6	5
9	3 for each item (letter equation, multiplication, answer with unit name)	6	5
10	1 for each item	4	3
11	2 for each item (multiplication equation, answer)	6	5
12	4 for each item (letter equation, area of base, number answer, unit name)	8	7
13	1 for each item	3	3
	Total	81	

b. Complete the Mastery Test 8 Remedy Summary Sheet to determine whether group remedies are needed. Reproducible Remedy Summary Sheets are at the back of the *Answer Key* and the back of the *Teacher's Guide.*

• If ¼ or more of the students did not pass a test part, present the remedy for that part before beginning Lesson 81. The Remedy Table follows and also appears at the end of the Mastery Test 8 *Answer Key.* Remedies worksheets follow Mastery Test 8 in the *Student Assessment Book.*

Part	Test Items	Remedy Lesson	Remedy Ex.	Remedies Worksheet	Textbook
1	Coordinate System (Word Problems)	73	4	Part A	—
		74	4	Part B	—
2	Division Sign	70	5	—	Part 1
		71	4	—	Part 1
3	Volume (Of Rectangular Prisms)	70	6	—	Part 2
		71	6	—	Part 3
4	Fractions (As Mixed Numbers)	75	5	Part C	—
		76	7	—	Part 3
5	Short Division (With Mental Math)	71	2	Part D	—
		72	4	—	Part 1
6	Decimal Place Value (Expanded Notation)	76	2	Part E	—
		77	2	Part F	—
7	Mixed-Number Operations (Addition/Subtraction)	76	5	—	Part 1
		77	6	—	Part 2
8	Word Problems (Parts of a Group)	78	4	—	Part 1
		79	6	—	Part 1
9	Area (Of Circles)	73	6	—	Part 1
		74	6	—	Part 1
10	Decimal Place Value (Comparison)	78	2	Part G	—
		79	4	Part H	—
11	Fraction Operations (Division)	77	4	—	Part 1
		78	5	—	Part 2
12	Volume (Of Cylinders)	77	5	—	—
		78	6	—	Part 3
13	Fraction Analysis (Comparison)	73	1	Part I	—
		74	1	Part J	—

Remedy Table—Mastery Test 8

Retest
Retest individual students on any part failed.

	Lesson 81	Lesson 82	Lesson 83	Lesson 84	Lesson 85
Student Learning Objectives	**Exercises** 1. Multiply and divide using mental math 2. Solve multiplication problems that include powers of 10 3. Solve fraction division problems 4. **Solve word problems that have mixed numbers** 5. **Understand temperatures** 6. **Simplify mixed numbers with a fraction more than 1** 7. Complete work independently	**Exercises** 1. Multiply and divide using mental math 2. Solve multiplication problems that include powers of 10 3. Show division with fractions as a diagram 4. Solve word problems that have mixed numbers 5. **Find the average temperature** 6. Simplify mixed numbers with a fraction more than 1 7. **Write fractions for parts-of-a-group sentences** Complete work independently	**Exercises** 1. Multiply and divide using mental math 2. **Add mixed numbers** 3. **Solve division problems that divide by powers of 10** 4. **Add 1 more to a fraction** 5. **Evaluate expressions that have parentheses** 6. **Solve fraction word problems with division** 7. **Find the average of values in a table** 8. Write fractions for parts-of-a-group sentences Complete work independently	**Exercises** 1. Multiply and divide using mental math 2. **Add and subtract mixed numbers** 3. Solve division problems that divide by powers of 10 4. Add 1 more to a fraction 5. Evaluate expressions that have parentheses 6. Solve fraction word problems with division 7. **Find the average of values in a bar graph** 8. Say fractions for parts-of-a-group sentences 9. Complete work independently	**Exercises** 1. Multiply and divide using mental math 2. Add mixed numbers 3. Solve division problems that divide by powers of 10 4. **Evaluate expressions that have parentheses and a letter** 5. Solve fraction word problems with division 6. Find the average of values in a bar graph 7. Write equations for parts-of-a-group sentences 8. **Rewrite mixed numbers as a preskill for borrowing** Complete work independently
Common Core State Standards for Mathematics					
5.OA 1			✔	✔	✔
5.NBT 1–3	✔	✔	✔	✔	✔
5.NBT 4	✔		✔		✔
5.NBT 5–6	✔	✔	✔	✔	✔
5.NBT 7	✔	✔		✔	✔
5.NF 1				✔	
5.NF 2	✔	✔	✔	✔	✔
5.NF 4–7	✔	✔	✔	✔	✔
5.MD 5		✔		✔	
5.G 2		✔	✔	✔	
Teacher Materials	Presentation Book 2, Board Displays CD or chalk board				
Student Materials	Textbook, Workbook, pencil, lined paper, ruler				
Additional Practice	• Student Practice Software: Block 4: Activity 1 (5.NF 5), Activity 2, Activity 3 (5.NBT 4), Activity 4 (5.NBT 3), Activity 5 (5.NBT 7), Activity 6 (5.MD 3 and 5.MD 4) • Provide needed fact practice with Level D or E Math Fact Worksheets.				
Mastery Test					

Lesson

EXERCISE 1: MENTAL MATH

a. Time for some mental math.
- Listen: 60 × 2. (Pause.) (Signal.) *120.*
- 40 × 2. (Pause.) (Signal.) *80.*
- 40 × 4. (Pause.) (Signal.) *160.*
- 60 × 4. (Pause.) (Signal.) *240.*
- 90 × 4. (Pause.) (Signal.) *360.*
- (Repeat until firm.)

b. New problem type.
- Listen: 270 divided by 3. (Pause.) (Signal.) *90.*
- 360 ÷ 9. (Pause.) (Signal.) *40.*
- 360 ÷ 4. (Pause.) (Signal.) *90.*
- 360 ÷ 6. (Pause.) (Signal.) *60.*
- 360 ÷ 2. (Pause.) (Signal.) *180.*
- (Repeat until firm.)

c. Listen: 270 divided by 3. (Pause.) (Signal.) *90.*
- 180 ÷ 2. (Pause.) (Signal.) *90.*
- 800 ÷ 10. (Pause.) (Signal.) *80.*
- 600 ÷ 3. (Pause.) (Signal.) *200.*
- 800 ÷ 4. (Pause.) (Signal.) *200.*
- (Repeat until firm.)

d. New problem type.
- Listen: 4 × 30. (Pause.) (Signal.) *120.*
- 4 × 80. (Pause.) (Signal.) *320.*
- 4 × 200. (Pause.) (Signal.) *800.*
- 6 × 60. (Pause.) (Signal.) *360.*
- 3 × 30. (Pause.) (Signal.) *90.*
- (Repeat until firm.)

EXERCISE 2: EXPONENTS
MULTIPLICATION BY BASE 10

REMEDY

a. You're going to multiply decimal numbers by ten with an exponent.
(Display:)   [81:2A]

$$\underline{\quad\quad} \times 10^2$$
$$\underline{\quad\quad} \times 10^3$$
$$\underline{\quad\quad} \times 10^6$$

The exponent tells you how many places to move the decimal point.

- (Point to 10^2.) If the exponent is 2, you move the decimal point two places.
- (Point to 10^3.) If the exponent is 3, how many places do you move the decimal point? (Signal.) *3 places.*
- (Point to 10^6.) If the exponent is 6, how many places do you move the decimal point? (Signal.) *6 places.*
- (Repeat until firm.)

b. (Display:)   [81:2B]

$$13.4 \times 10^3$$

- Read the problem. (Signal.) *13 and 4 tenths* $\times 10^3$
- What does 10 to the third equal? (Signal.) *1000.*
- How many places do we move the decimal point? (Signal.) *3 places.*
(Add to show:)   [81:2C]

$$13.4 \times 10^3 = 13{,}400.$$

- What does 13.4×10^3 equal? (Signal.) *13,400.*

c. (Display:)   [81:2D]

$$.015 \times 10^2$$

- Read the problem. (Signal.) *15 thousandths* $\times 10^2$.
- What does 10^2 equal? (Signal.) *100.*
- How many places do we move the decimal point? (Signal.) *2 places.*
(Add to show:)   [81:2E]

$$.015 \times 10^2 = 01.5$$

The answer is more than 1, so we don't show a zero in front of the number.
(Change to show:)   [81:2F]

$$.015 \times 10^2 = 1.5$$

- What does $.015 \times 10^2$ equal? (Signal.) *1.5.*

d. (Display:) [81:2G]

$$.203 \times 10^5$$

- Read the problem. (Signal.) *203 thousandths × 10⁵.*
- How many places do we move the decimal point? (Signal.) *5 places.*
  (Add to show:) [81:2H]

$$.203 \times 10^5 = 20{,}300.$$

- What does .203 × 10⁵ equal? (Signal.) *20,300.*

================= WORKBOOK PRACTICE =================

a. Open your workbook to Lesson 81 and find part 1. ✔
  (Teacher reference:)   R Part D

  a. 4.56 × 10² =

  b. .034 × 10³ =

  c. .02 × 10⁴ =

b. Read problem A. (Signal.)
  4 and 56 hundredths × 10².
- How many places will you move the decimal point? (Signal.) *2 places.*
- Complete the equation. ✔
- Everybody, what does 4.56 × 10² equal?
  (Signal.) *456.*
  (Display:) [81:2I]

$$\textbf{a. } 4.56 \times 10^2 = 4{,}56.$$

Here's what you should have. You moved the decimal point two places. You don't need a decimal point after 456, but it's not wrong if you write one.

c. Read problem B. (Signal.) *34 thousandths × 10³.*
- Complete the equation. Don't write a zero before the whole number. ✔
- Everybody, what does .034 × 10³ equal? (Signal.) *34.*
  (Display:) [81:2J]

$$\textbf{b. } .034 \times 10^3 = 34.$$

Here's what you should have. You moved the decimal point three places. You don't need a decimal point after 34, but it's not wrong if you write one.

d. Read problem C. (Signal.) *2 hundredths × 10⁴.*
- Complete the equation. ✔
- Everybody, what does .02 × 10⁴ equal? (Signal.) *200.*
  (Display:) [81:2K]

$$\textbf{c. } .02 \times 10^4 = 200.$$

Here's what you should have. You moved the decimal point four places.

e. Remember, the exponent tells how many places to move the decimal point.

EXERCISE 3: FRACTION OPERATIONS
DIVISION

a. (Display:) [81:3A]

$$3 \div \frac{1}{2}$$

- Read the problem. (Signal.) *3 ÷ 1/2.*
- Raise your hand when you know the answer. ✔
- What's 3 divided by 1/2? (Signal.) *6.*
  Yes, if you divide 3 into halves, you end up with 6 parts.

b. (Add to show:) [81:3B]

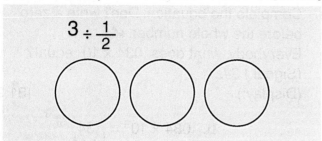

$$3 \div \frac{1}{2}$$

Here are 3 units.
We divide each unit into halves.
(Add to show:) [81:3C]

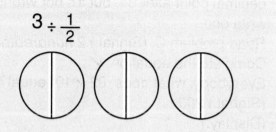

$$3 \div \frac{1}{2}$$

- How many parts did we end up with?
  (Signal.) 6.

================ WORKBOOK PRACTICE ================

a. Find part 2 in your workbook. ✔
   (Teacher reference:)

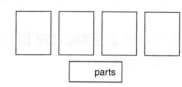

a. $4 \div \frac{1}{3} =$ ☐ = ☐ ☐ ☐ ☐

parts

Here's a problem of the same type.
4 divided by 1/3.

- Work the multiplication problem and write the answer. ✔
- Everybody, what's 4 divided by 1/3? (Signal.) 12.

b. You're going to divide the 4 units into thirds.

- So how many parts will each unit have? (Signal.) 3.
- Divide the units into thirds and write the total number of parts in the box.
  (Observe students and give feedback.)
- Everybody, how many parts is 4 divided by 1/3? (Signal.) 12 parts.

EXERCISE 4: MIXED-NUMBER OPERATIONS
WORD PROBLEMS
REMEDY

a. Open your textbook to Lesson 81 and find part 1. ✔
   (Teacher reference:)

a. A builder had a board that was $8\frac{3}{4}$ feet long. He cut it into two boards. One was $2\frac{1}{4}$ feet long. How long was the other board?

b. Ginger ran $7\frac{1}{8}$ miles on Monday and $3\frac{5}{8}$ miles on Tuesday. How far did she run on both days?

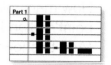

b. Problem A: A builder had a board that was 8 and 3/4 feet long. He cut it into two boards. One was 2 and 1/4 feet long. How long was the other board?

- Write the problem with two mixed numbers. Stop when you've done that much. ✔
- Everybody, read the problem you wrote. (Signal.) 8 and 3/4 – 2 and 1/4.
  (Display:) [81:4A]

$$\begin{array}{r} a. \quad 8\frac{3}{4} \\ -\ 2\frac{1}{4} \\ \hline \end{array}$$

Here's what you should have.

c. Work the subtraction problem. ✔

- Everybody, what's 8 and 3/4 minus 2 and 1/4? (Signal.) 6 and 2/4.
- Can you simplify the fraction part of the mixed number? (Signal.) Yes.
- Do it. Write the simplified mixed number and the unit name. ✔
- Everybody, one board was 2 and 1/4 feet long. How long was the other board? (Signal.) 6 and 1/2 feet.
  (Add to show:) [81:4B]

$$\begin{array}{r} a. \quad 8\frac{3}{4} \\ -\ 2\frac{1}{4} \\ \hline 6\frac{2}{4} = \boxed{6\frac{1}{2} \text{ feet}} \end{array}$$

Here's what you should have.

d. Problem B: Ginger ran 7 and 1/8 miles on Monday and 3 and 5/8 miles on Tuesday. How far did she run on both days?
• Work the problem. Write the answer as a simplified mixed number and a unit name. (Observe students and give feedback.)
e. Check your work.
• Everybody, read the mixed-number problem you worked. (Signal.) *7 and 1/8 + 3 and 5/8.*
(Display:)   [81:4C]

$$
\begin{array}{r}
\textbf{b.} \quad 7\frac{1}{8} \\
+ \ 3\frac{5}{8} \\
\hline
10\frac{6}{8}
\end{array}
$$

Here's the answer to that problem.
You simplified 10 and 6/8.
• How far did Ginger run on both days? (Signal.) *10 and 3/4 miles.*
(Add to show:)   [81:4D]

$$
\begin{array}{r}
\textbf{b.} \quad 7\frac{1}{8} \\
+ \ 3\frac{5}{8} \\
\hline
10\frac{6}{8} = \boxed{10\frac{3}{4}} \text{ miles}
\end{array}
$$

Here's what you should have.

EXERCISE 5: TEMPERATURE

a. We measure angles in degrees. We use the little circle to show degrees.
(Display:)   [81:5A]

$$46°$$

• What does this say? (Signal.) *46 degrees.*
b. Listen: We measure **temperature** in degrees. We use the same symbol to show temperature degrees.
(Display:)   [81:5B]

$$81°$$
$$14°$$

• (Point to **81°**.) What's this temperature? (Signal.) *81 degrees.*
• (Point to **14°**.) What's this temperature? (Signal.) *14 degrees.*

a. Find part 2 in your textbook. ✔
(Teacher reference:)

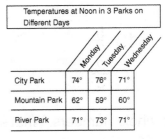

Temperatures at Noon in 3 Parks on Different Days	Monday	Tuesday	Wednesday
City Park	74°	76°	71°
Mountain Park	62°	59°	60°
River Park	71°	73°	71°

a. What was the lowest temperature on Monday?
b. Which park had the highest temperature on Tuesday?
c. How much colder was Wednesday than Tuesday at City Park?
d. What was the coldest temperature of the three-day period?

You're going to answer questions about this table.
The table shows the temperatures at noon in three parks on three days.
b. I'll read the questions.
• Question A: What was the lowest temperature on Monday?
• Question B: Which park had the highest temperature on Tuesday?
• Question C: How much colder was Wednesday than Tuesday at City Park?
• Question D: What was the coldest temperature of the three-day period?
• Write answers to the questions. Remember to show the degree symbol in your answers. ✔
c. Check your work.
• Question A: What was the lowest temperature on Monday? (Signal.) *62 degrees.*
• B: Which park had the highest temperature on Tuesday? (Signal.) *City Park.*
• C: How much colder was Wednesday than Tuesday at City Park? (Signal.) *5 degrees.*
• D: What was the coldest temperature of the three-day period? (Signal.) *59 degrees.*

EXERCISE 6: FRACTIONS
AS MIXED NUMBERS   REMEDY

a. (Display:)   [81:6A]

$$13\frac{5}{3}$$

Here's 13 and 5/3. That's the same as 13 **plus** 5/3.
The fraction is more than 1. So we rewrite the mixed number to show a fraction that is less than 1.

- 5/3 equals 1 and how many thirds? (Signal.) *2/3.*
  We write 2/3 below and write 1 above 13.
  (Add to show:) [81:6B]

$$13 \overset{1}{\frac{5}{3}}$$
$$\frac{2}{3}$$

b. Now we add the whole numbers. What's
   1 + 13? (Signal.) *14.*
   (Add to show:) [81:6C]

$$13 \overset{1}{\frac{5}{3}}$$
$$\boxed{14 \frac{2}{3}}$$

13 and 5/3 equals 14 and 2/3.

c. (Display:) [81:6D]

$$2 \frac{12}{7}$$

- Read the mixed number. (Signal.) *2 and 12/7.*
  The fraction is more than 1, so we have to
  rewrite the fraction.
- Raise your hand when you know what 12/7
  equals. ✔
- What does 12/7 equal? (Signal.) *1 and 5/7.*
  I write the fraction below and the 1 above the
  other whole number.
  (Add to show:) [81:6E]

$$2 \overset{1}{\frac{12}{7}}$$
$$\frac{5}{7}$$

- Everybody, what's the new whole number?
  (Signal.) *3.*
  (Add to show:) [81:6F]

$$2 \overset{1}{\frac{12}{7}}$$
$$\boxed{3 \frac{5}{7}}$$

Yes, 2 and 12/7 equals 3 and 5/7.

a. Find part 3 in your textbook. ✔
   (Teacher reference:)

   a. $15\frac{11}{8}$ b. $4\frac{19}{15}$ c. $10\frac{14}{9}$ d. $26\frac{13}{12}$

b. Read mixed number A. (Signal.) *15 and 11/8.*
- Copy the number. Below, write the mixed
  number with a fraction that is less than 1. ✔
- Everybody, what does 15 and 11/8 equal?
  (Signal.) *16 and 3/8.*
c. Copy and write mixed number B. ✔
- Everybody, what does 4 and 19/15 equal?
  (Signal.) *5 and 4/15.*
d. Copy and rewrite the rest of the mixed
   numbers.
   (Observe students and give feedback.)
e. Check your work.
- Item C. What does 10 and 14/9 equal?
  (Signal.) *11 and 5/9.*
- Item D. What does 26 and 13/12 equal?
  (Signal.) *27 and 1/12.*

EXERCISE 7: INDEPENDENT WORK
PARTS OF A GROUP

a. Find part 4 in your textbook. ✔
   (Teacher reference:)

   a. There were 90 students. $\frac{4}{5}$ of the students passed the test. How
   many students passed the test? How many failed the test?

You're going to work this problem as part of
your independent work. The problem asks two
questions. Remember to answer both of them.

Assign Independent Work, Textbook parts 4–8
and Workbook parts 3–8.

Optional extra math-fact practice worksheets
are available on ConnectED.

Lesson

EXERCISE 1: MENTAL MATH

a. Time for some mental math.
- Listen: 60 × 6. (Pause.) (Signal.) *360.*
- 90 × 6. (Pause.) (Signal.) *540.*
- 30 × 4. (Pause.) (Signal.) *120.*
- 70 × 4. (Pause.) (Signal.) *280.*
- 90 × 4. (Pause.) (Signal.) *360.*
- (Repeat until firm.)

b. Listen: 80 × 4. (Pause.) (Signal.) *320.*
- 20 × 5. (Pause.) (Signal.) *100.*
- 100 × 5. (Pause.) (Signal.) *500.*
- 200 × 4. (Pause.) (Signal.) *800.*
- (Repeat until firm.)

c. Listen: 360 divided by 2. (Pause.) (Signal.) *180.*
- 180 ÷ 2. (Pause.) (Signal.) *90.*
- 180 ÷ 3. (Pause.) (Signal.) *60.*
- 240 ÷ 3. (Pause.) (Signal.) *80.*
- 240 ÷ 4. (Pause.) (Signal.) *60.*
- (Repeat until firm.)

d. Listen: 270 divided by 3. (Pause.) (Signal.) *90.*
- 360 ÷ 9. (Pause.) (Signal.) *40.*
- 360 ÷ 4. (Pause.) (Signal.) *90.*
- 72 ÷ 8. (Pause.) (Signal.) *9.*
- 720 ÷ 8. (Pause.) (Signal.) *90.*
- (Repeat until firm.)

EXERCISE 2: EXPONENTS
MULTIPLICATION BY BASE 10

a. Open your workbook to Lesson 82 and find part 1. ✔
    (Teacher reference:)

a. $1.052 \times 10^2 =$ d. $2.91 \times 10^4 =$

b. $.42 \times 10^3 =$ e. $.37 \times 10^1 =$

c. $.002 \times 10^2 =$

You're going to complete each equation.

- If you multiply by 10 to the first, how many places do you move the decimal point? (Signal.) *1 place.*
- If you multiply by 10 to the third, how many places do you move the decimal point? (Signal.) *3 places.*

b. Read problem A. (Signal.) *1 and 52 thousandths × 10².*
- Complete the equation. ✔
- Everybody, what does 1.052×10^2 equal? (Signal.) *105.2.*

c. Complete the rest of the equations in part 1. (Observe students and give feedback.)

d. Check your work.
- Problem B. What does $.42 \times 10^3$ equal? (Signal.) *420.*
- C. What does $.002 \times 10^2$ equal? (Signal.) *2 tenths.*
- D. What does 2.91×10^4 equal? (Signal.) *29,100.*
- E. What does $.37 \times 10^1$ equal? (Signal.) *3.7*

EXERCISE 3: FRACTION OPERATIONS
DIVISION

a. Find part 2 in your workbook. ✔
    (Teacher reference:)

a. $2 \div \frac{1}{4} =$ $=$ ◯ ◯   [parts]

You're going to divide 2 by 1/4.
- Work the problem and write the answer. ✔
- Everybody, what's 2 divided by 1/4? (Signal.) *8.*

b. Divide each unit into fourths and write the number of parts in the box. ✔
- Everybody, how many parts is 2 divided by 1/4? (Signal.) *8 parts.*

EXERCISE 4: MIXED-NUMBER OPERATIONS
WORD PROBLEMS
REMEDY

a. Open your textbook to Lesson 82 and find part 1. ✔
(Teacher reference:)

a. Tom picked 13⅖ pounds of apples. Al picked 7⅖ pounds of apples. How many pounds of apples did the men have together?

b. The dog weighed 18 11/16 pounds. The dog lost 2 10/16 pounds. How much did the dog weigh then?

b. Problem A: Tom picked 13 and 2/5 pounds of apples. Al picked 7 and 2/5 pounds of apples. How many pounds of apples did the men have together?
• Work the problem. Show the answer as a mixed number with a unit name.
(Observe students and give feedback.)
c. Check your work.
• How many pounds of apples did the men have together? (Signal.) *20 and 4/5 pounds.*
(Display:) [82:4A]

$$\begin{array}{r} \overset{1}{13}\frac{2}{5} \\ + \ 7\frac{2}{5} \\ \hline \boxed{20\frac{4}{5}} \text{ pounds} \end{array}$$

Here's what you should have.
d. Problem B: The dog weighed 18 and 11/16 pounds. The dog lost 2 and 10/16 pounds. How much did the dog weigh then?
• Work the problem.
(Observe students and give feedback.)
e. Check your work.
• How much did the dog weigh? (Signal.) *16 and 1/16 pounds.*
(Display:) [82:4B]

$$\begin{array}{r} \text{b.} \quad 18\frac{11}{16} \\ - \ 2\frac{10}{16} \\ \hline \boxed{16\frac{1}{16}} \text{ pounds} \end{array}$$

Here's what you should have.

EXERCISE 5: AVERAGE
TEMPERATURE TABLES

a. You're going to learn about averages.
(Display:) [82:5A]

The numbers show the highest temperature for 4 days.

Mon	Tues	Wed	Thurs
26°	40°	38°	32°

Here's a table. The numbers show the highest temperature for 4 days.
The temperatures are: 26 degrees, 40 degrees, 38 degrees, and 32 degrees.
(Add to show:) [82:5B]

The numbers show the highest temperature for 4 days.
What is the average temperature for these 4 days?

Mon	Tues	Wed	Thurs
26°	40°	38°	32°

Here's the question: What is the average temperature for these 4 days?
That means: If the temperature was the same on every day, what would that temperature be?
b. Here's how you find the average: You add the numbers and divide by 4. Once more: You add the numbers and divide by 4.
• What do you do first? (Signal.) *Add the numbers.*
• Then what do you do? (Signal.) *Divide by 4.*
(Add to show:) [82:5C]

The numbers show the highest temperature for 4 days.
What is the average temperature for these 4 days?

Mon	Tues	Wed	Thurs
26°	40°	38°	32°

$$\begin{array}{r} \overset{1}{2}\,6 \\ 4\,0 \\ 3\,8 \\ + \ 3\,2 \\ \hline 1\,3\,6 \end{array}$$

The total for the numbers is 136.
c. (Add to show:) [82:5D]

The numbers show the highest temperature for 4 days.
What is the average temperature for these 4 days?

Mon	Tues	Wed	Thurs
26°	40°	38°	32°

$$\begin{array}{r} \overset{1}{2}\,6 \\ 4\,0 \\ 3\,8 \\ + \ 3\,2 \\ \hline 1\,3\,6 \end{array} \qquad \begin{array}{r} \boxed{3\,4°} \\ 4\overline{)1\,3{,}6} \end{array}$$

We divide by 4. The answer is 34.
That's the average temperature: 34 degrees.

- So if the temperature was the same on all days, what would the temperature be? (Signal.) *34 degrees.*

d. (Add to show:) [82:5E]

The numbers show the highest temperature for 4 days. What is the average temperature for these 4 days?	Mon	Tues	Wed	Thurs	Fri
	26°	40°	38°	32°	24°

$$\begin{array}{r} \overset{1}{2}\,6 \\ 4\,0 \\ 3\,8 \\ +\ 3\,2 \\ \hline 1\,3\,6 \end{array} \qquad \boxed{3\,4°} \\ 4\overline{)1\,3{,}6}$$

- How many temperatures are shown now? (Signal.) *5.*
- What's the last temperature? (Signal.) *24 degrees.*
- Is that temperature **above** the average temperature of 34 degrees? (Signal.) *No.*
- Is that temperature **below** the average temperature of 34 degrees? (Signal.) *Yes.*
  So if we figure out the average for the 5 temperatures, it will be lower than 34.
e. Listen: How many temperatures do we have now? (Signal.) *5.*
  So we add the numbers and divide by 5.
- What do we do first? (Signal.) *Add the numbers.*
- What do we do next? (Signal.) *Divide by 5.*
- Why don't we divide by 4? (Call on a student. Idea: *We have 5 temperatures.*)
  Yes, we have 5 temperatures.
  (Change to show:) [82:5F]

The numbers show the highest temperature for 5 days. What is the average temperature for these 5 days?	Mon	Tues	Wed	Thurs	Fri
	26°	40°	38°	32°	24°

$$\begin{array}{r} \overset{2}{2}\,6 \\ 4\,0 \\ 3\,8 \\ 3\,2 \\ +\ 2\,4 \\ \hline 1\,6\,0 \end{array} \qquad \boxed{3\,2°} \\ 5\overline{)1\,6{,}0}$$

The total for the 5 days is 160.
160 divided by 5 is 32.

- What's the average temperature? (Signal.) *32 degrees.*

f. (Display:) [82:5G]

What is the average temperature for these 3 days?	Sat	Sun	Mon
	53°	57°	46°

Here are new temperatures.

- To find the average, what do we do first? (Signal.) *Add the numbers.*

- Then we divide. What do we divide by? (Signal.) *3.*
  Yes, there are 3 temperatures. So we divide by 3.
- Once more: What do we do first? (Signal.) *Add the numbers.*
- What do we do next? (Signal.) *Divide by 3.*
g. In part A, figure out the total for the temperatures. Divide by 3. Raise your hand when you know the average temperature for the 3 days. ✔
- What's the total for the 3 numbers? (Signal.) *156.*
- What's the average temperature? (Signal.) *52 degrees.*
  (Add to show:) [82:5H]

What is the average temperature for these 3 days?	Sat	Sun	Mon
	53°	57°	46°

$$\begin{array}{r} \overset{1}{5}\,3 \\ 5\,7 \\ +\ 4\,6 \\ \hline 1\,5\,6 \end{array} \qquad \boxed{5\,2°} \\ 3\overline{)1\,5\,6}$$

Here's what you should have.

h. (Change to show:) [82:5I]

What is the average temperature for these 4 days?	Sat	Sun	Mon	Tues
	53°	57°	46°	56°

New problem.

- What's the first thing you do to find the average? (Signal.) *Add the numbers.*
- Then what do you divide by? (Signal.) *4.*
  Yes, there are now 4 temperatures.
i. Listen: The average you figured for 3 temperatures was 52. The new temperature is 56. Is that above or below the average? (Signal.) *Above the average.*
- So will the new average temperature be higher than 52 or lower than 52? (Signal.) *Higher than 52.*
- To figure out the new average temperature, what do you do first? (Signal.) *Add the numbers.*
- Then what do you do? (Signal.) *Divide by 4.*
  Yes, divide by 4.
- Work the problem. ✔
- Everybody, what's the total of the 4 temperatures? (Signal.) *212.*
- What's the average temperature? (Signal.) *53 degrees.*

(Add to show:) [82:5J]

What is the average temperature for these 4 days?	Sat	Sun	Mon	Tues
	53°	57°	46°	56°

$$\begin{array}{r} \overset{2}{5}\,3 \\ 5\,7 \\ 4\,6 \\ +\ 5\,6 \\ \hline 2\,1\,2 \end{array} \qquad \boxed{5\,3°} \\ 4\overline{)2\,1\,2}$$

Here's what you should have.

EXERCISE 6: FRACTIONS
AS MIXED NUMBERS   REMEDY

a. Find part 2 in your textbook. ✔
    (Teacher reference:)

a. $9\frac{27}{20}$ b. $15\frac{17}{12}$ c. $2\frac{13}{10}$ d. $1\frac{19}{13}$ e. $19\frac{7}{5}$

These mixed numbers have a fraction that is more than 1. You're going to copy the mixed numbers and rewrite them so they have a fraction that is less than 1.

b. Copy mixed number A and rewrite it. ✔
•   Everybody, what does 9 and 27/20 equal? (Signal.) *10 and 7/20.*
c. Copy and rewrite the rest of the mixed numbers.
    (Observe students and give feedback.)
d. Check your work.
•   Mixed number B. What does 15 and 17/12 equal? (Signal.) *16 and 5/12.*
•   Mixed number C. What does 2 and 13/10 equal? (Signal.) *3 and 3/10.*
•   Mixed number D. What does 1 and 19/13 equal? (Signal.) *2 and 6/13.*
•   Mixed number E. What does 19 and 7/5 equal? (Signal.) *20 and 2/5.*

EXERCISE 7: SENTENCES
PARTS OF A GROUP

a. (Display:) [82:7A]

$\frac{2}{3}$ of the students are girls.

Here's a sentence: 2/3 of the students are girls.
•   My turn: If 2/3 of the students are girls, what is the fraction for boys? 1/3. 2/3 and 1/3 equals 3/3. That's the fraction for all the students.
•   Your turn: If 2/3 of the students are girls, what's the fraction for boys? (Signal.) *1/3.*
•   What's the fraction for all the students? (Signal.) *3/3.*
b. (Display:) [82:7B]

$\frac{3}{8}$ of the trees had leaves.

New sentence: 3/8 of the trees had leaves.
•   If 3/8 of the trees had leaves, what's the fraction for trees that did **not** have leaves? (Signal.) *5/8.*
•   What's the fraction for **all** the trees? (Signal.) *8/8.*
    (Repeat until firm.)
c. (Display:) [82:7C]

$\frac{1}{9}$ of the pencils were red.

New sentence: 1/9 of the pencils were red.
•   If 1/9 of the pencils were red, what's the fraction for the pencils that were **not** red? (Signal.) *8/9.*
•   What's the fraction for **all** of the pencils? (Signal.) *9/9.*
    (Repeat until firm.)
d. (Display:) [82:7D]

$\frac{8}{10}$ of the trees had leaves.

New sentence: 8/10 of the trees had leaves.
•   If 8/10 of the trees had leaves, what's the fraction for trees that did **not** have leaves? (Signal.) *2/10.*
•   What's the fraction for **all** the trees? (Signal.) *10/10.*
    (Repeat until firm.)

a. Find part 3 in your textbook. ✔
   (Teacher reference:)

a. $\frac{4}{5}$ of the glasses were dirty.
   • What's the fraction for all the glasses?
   • What's the fraction for clean glasses?

b. $\frac{11}{15}$ of the bricks were stacked.
   • What fraction of the bricks were not stacked?
   • What's the fraction for all the bricks?

c. $\frac{3}{8}$ of the horses were for sale.
   • What's the fraction for all the horses?
   • What's the fraction for horses that were not for sale?

b. Sentence A: 4/5 of the glasses were dirty.
   The questions are: What's the fraction for all the glasses? What's the fraction for clean glasses?
   • Write answers to the questions. ✔

c. Check your work.
   • First question: What's the fraction for all the glasses? (Signal.) *5/5.*
   • Next question: What's the fraction for clean glasses? (Signal.) *1/5.*

d. Write answers to the rest of the questions. (Observe students and give feedback.)

e. Check your work.
   • Sentence B: 11/15 of the bricks were stacked.
   • Everybody, what fraction of the bricks were not stacked? (Signal.) *4/15.*
   • What's the fraction for all the bricks? (Signal.) *15/15.*

f. Sentence C: 3/8 of the horses were for sale.
   • Everybody, what's the fraction for all the horses? (Signal.) *8/8.*
   • What's the fraction for horses that were not for sale? (Signal.) *5/8.*

> Assign Independent Work, Textbook parts 4–9 and Workbook parts 3–5.

> Optional extra math-fact practice worksheets are available on ConnectED.

Lesson 83

EXERCISE 1: MENTAL MATH

a. Time for some mental math.
- Listen: 60 divided by 2. (Pause.) (Signal.) *30.*
- 60 ÷ 3. (Pause.) (Signal.) *20.*
- 90 ÷ 9. (Pause.) (Signal.) *10.*
- 90 ÷ 3. (Pause.) (Signal.) *30.*
- (Repeat until firm.)

b. Listen: 270 divided by 3. (Pause.) (Signal.) *90.*
- 180 ÷ 2. (Pause.) (Signal.) *90.*
- 360 ÷ 4. (Pause.) (Signal.) *90.*
- 360 ÷ 2. (Pause.) (Signal.) *180.*
- (Repeat until firm.)

c. Listen: 6 × 30. (Pause.) (Signal.) *180.*
- 3 × 60. (Pause.) (Signal.) *180.*
- 6 × 60. (Pause.) (Signal.) *360.*
- 4 × 90. (Pause.) (Signal.) *360.*
- (Repeat until firm.)

d. Listen: 2 × 90. (Pause.) (Signal.) *180.*
- 2 × 80. (Pause.) (Signal.) *160.*
- 4 × 50. (Pause.) (Signal.) *200.*
- 6 × 50. (Pause.) (Signal.) *300.*
- 6 × 60. (Pause.) (Signal.) *360.*
- (Repeat until firm.)

e. Listen: 120 divided by 2. (Pause.) (Signal.) *60.*
- 120 ÷ 4. (Pause.) (Signal.) *30.*
- 140 ÷ 2. (Pause.) (Signal.) *70.*
- 80 times 4. (Pause.) (Signal.) *320.*
- 60 × 4. (Pause.) (Signal.) *240.*
- (Repeat until firm.)

EXERCISE 2: MIXED-NUMBER OPERATIONS
ADDITION/SUBTRACTION

a. (Display:)   [83:2A]

$$4 \frac{4}{12}$$
$$+ \ 5 \frac{10}{12}$$

Here's a new kind of problem.

- Read the problem. (Signal.)
  4 and 4/12 + 5 and 10/12.
- Say the problem for the fractions. (Signal.)
  4/12 + 10/12.
- What's the answer? (Signal.) *14/12.*
  (Add to show:)   [83:2B]

$$4 \frac{4}{12}$$
$$+ \ 5 \frac{10}{12}$$
$$\overline{ \frac{14}{12}}$$

- Say the problem for the whole numbers.
  (Signal.) *4 + 5.*
- What's the answer? (Signal.) *9.*
  (Add to show:)   [83:2C]

$$4 \frac{4}{12}$$
$$+ \ 5 \frac{10}{12}$$
$$\overline{9 \frac{14}{12}}$$

b. The fraction in the answer is more than 1. So we have to rewrite the fraction.
- Raise your hand when you know what 14/12 equals. ✔
- What does it equal? (Signal.) *1 and 2/12.*
  So we write 14/12 as 1 and 2/12.
  (Add to show:)   [83:2D]

$$4 \frac{4}{12}$$
$$+ \ 5 \frac{10}{12}$$
$$\overline{9 \ \overset{1}{} \frac{14}{12}}$$
$$\frac{2}{12}$$

Now we add the whole numbers: 1 + 9.

- What's the answer? (Signal.) *10.*
  (Add to show:) [83:2E]

$$4 \frac{4}{12}$$
$$+ 5 \frac{10}{12}$$
$$9 \overset{1}{} \frac{14}{12}$$
$$\boxed{10 \frac{2}{12}}$$

Remember, if the answer has a fraction that is more than 1, rewrite the answer.

c. (Display:) [83:2F]

$$3 \frac{4}{5}$$
$$+ 8 \frac{3}{5}$$

- Say the problem and the answer for the fractions. (Signal.) *4/5 + 3/5 = 7/5.*
  (Add to show:) [83:2G]

$$3 \frac{4}{5}$$
$$+ 8 \frac{3}{5}$$
$$\frac{7}{5}$$

- Say the problem and the answer for the whole numbers. (Signal.) *3 + 8 = 11.*
  (Add to show:) [83:2H]

$$3 \frac{4}{5}$$
$$+ 8 \frac{3}{5}$$
$$11 \frac{7}{5}$$

d. The mixed number in the answer has a fraction that is more than 1. Copy the answer in part A. Below, write the answer with a fraction that is less than 1. ✔
- Everybody, what's the new answer? (Signal.) *12 and 2/5.*

(Add to show:) [83:2I]

$$3 \frac{4}{5}$$
$$+ 8 \frac{3}{5}$$
$$11 \overset{1}{} \frac{7}{5}$$
$$\boxed{12 \frac{2}{5}}$$

Yes, 11 and 7/5 equals 12 and 2/5.

e. (Display:) [83:2J]

$$9 \frac{2}{7}$$
$$+ \frac{11}{7}$$

- Say the problem for the fractions. (Signal.) *2/7 + 11/7.*
- What's the answer? (Signal.) *13/7.*
  The whole number in the answer is 9.
  (Add to show:) [83:2K]

$$9 \frac{2}{7}$$
$$+ \frac{11}{7}$$
$$9 \frac{13}{7}$$

f. Copy the answer. Below, write the answer with a fraction that is less than 1. ✔
- Everybody, what's the new answer? (Signal.) *10 and 6/7.*
  (Add to show:) [83:2L]

$$9 \frac{2}{7}$$
$$+ \frac{11}{7}$$
$$9 \overset{1}{} \frac{13}{7}$$
$$\boxed{10 \frac{6}{7}}$$

Yes, 9 and 13/7 equals 10 and 6/7.

g. (Display:) [83:2M]

$$17 \tfrac{5}{11}$$
$$+ 10 \tfrac{8}{11}$$

- Say the problem for the fractions. (Signal.) *5/11 + 8/11.*
- What's the answer? (Signal.) *13/11.*
- Say the problem for the whole numbers. (Signal.) *17 + 10.*
- What's the answer? (Signal.) *27.* (Add to show:) [83:2N]

$$17 \tfrac{5}{11}$$
$$+ 10 \tfrac{8}{11}$$
$$\overline{27 \tfrac{13}{11}}$$

h. Copy the answer. Below, write the answer with a fraction that is less than 1. ✔
- Everybody, what's the new answer? (Signal.) *28 and 2/11.* (Add to show:) [83:2O]

$$17 \tfrac{5}{11}$$
$$+ 10 \tfrac{8}{11}$$
$$\overline{27 \overset{1}{} \tfrac{13}{11}}$$
$$\boxed{28 \tfrac{2}{11}}$$

Yes, 27 and 13/11 equals 28 and 2/11.

EXERCISE 3: EXPONENTS
DIVISION BY BASE 10 [REMEDY]

a. You're going to work problems that divide by 10, 100, or 1000.
(Display:) [83:3A]

$$\tfrac{8}{10}$$

Here's 8 divided by 10. When you divide by 10, you move the decimal point one place to the left to make the number smaller.

(Add to show:) [83:3B]

$$\tfrac{8}{10} = .8$$

So 8 divided by 10 equals 8 tenths.
- What does 8 ÷ 10 equal? (Signal.) *8 tenths.*

b. (Add to show:) [83:3C]

$$\tfrac{8}{10} = .8$$
$$\tfrac{8}{100}$$

Here's 8 divided by one hundred.
- How many places do you move the decimal point for 100? (Signal.) *2 places.* (Add to show:) [83:3D]

$$\tfrac{8}{10} = .8$$
$$\tfrac{8}{100} = .08$$

So 8 divided by 100 is 8 hundredths.

c. (Add to show:) [83:3E]

$$\tfrac{8}{10} = .8$$
$$\tfrac{8}{100} = .08$$
$$\tfrac{8}{1000}$$

Here's 8 divided by one thousand.
- How many places do you move the decimal point for 1000? (Signal.) *3 places.* (Add to show:) [83:3F]

$$\tfrac{8}{10} = .8$$
$$\tfrac{8}{100} = .08$$
$$\tfrac{8}{1000} = .008$$

So what's 8 divided by 1000? (Signal.) *8 thousandths.*

d. (Display:) [83:3G]

$$\text{a. } \frac{47}{10^3} \qquad \text{b. } \frac{2.81}{10^2}$$

$$\text{c. } \frac{6931}{10^2}$$

Problem A: 47 divided by 10^3.

- How many places do we move the decimal point? (Signal.) *3 places.*
  (Add to show:) [83:3H]

$$\text{a. } \frac{47}{10^3} = .047 \qquad \text{b. } \frac{2.81}{10^2}$$

$$\text{c. } \frac{6931}{10^2}$$

We move the decimal point three places, so we have to write a zero in front of the 4.

- What does 47 divided by 10^3 equal? (Signal.) *47 thousandths.*

e. Problem B: 2.81 divided by 10^2.

- How many places do we move the decimal point? (Signal.) *2 places.*
  (Add to show:) [83:3I]

$$\text{a. } \frac{47}{10^3} = .047 \qquad \text{b. } \frac{2.81}{10^2} = .02\,81$$

$$\text{c. } \frac{6931}{10^2}$$

We move the decimal point two places, so we have to write a zero in front of the 2.

f. Problem C: 6931 divided by 10^2.

- How many places do we move the decimal point? (Signal.) *2 places.*
  (Add to show:) [83:3J]

$$\text{a. } \frac{47}{10^3} = .047 \qquad \text{b. } \frac{2.81}{10^2} = .02\,81$$

$$\text{c. } \frac{6931}{10^2} = 69.31$$

- What does 6931 ÷ 10^2 equal? (Signal.) *69.31.*

a. Open your workbook to Lesson 83 and find part 1. ✔
   (Teacher reference:) [R] Part E

$$\text{a. } \frac{24}{10^3} = \qquad \text{d. } \frac{23.4}{10^3} =$$
$$\text{b. } \frac{3.08}{10^3} = \qquad \text{e. } \frac{89.6}{10^2} =$$
$$\text{c. } \frac{10.45}{10^4} =$$

b. Read problem A. (Signal.) *24 divided by 10^3.*
- How many places do you move the decimal point? (Signal.) *3 places.*
- Write the answer. ✔
- Everybody, what does 24 ÷ 10^3 equal? (Signal.) *24 thousandths.*
  (Display:) [83:3K]

$$\text{a. } \frac{24}{10^3} = .024$$

Here's what you should have.

c. Write answers to the rest of the problems. (Observe students and give feedback.)

d. Check your work.
   (Display:) [83:3L]

$$\text{b. } \frac{3.08}{10^3} = .003\,08 \qquad \text{d. } \frac{23.4}{10^3} = .023\,4$$

$$\text{c. } \frac{10.45}{10^4} = .0010\,45 \qquad \text{e. } \frac{89.6}{10^2} = .89\,6$$

Here's what you should have for items B through E.

EXERCISE 4: FRACTION ANALYSIS
1 MORE

a. (Display:) [83:4A]

$$\frac{4}{8} \qquad \frac{3}{5} \qquad \frac{2}{3} \qquad \frac{5}{9} \qquad \frac{2}{16}$$

You're going to add 1 to these fractions and tell me the fraction you end up with.

- (Point to **4/8**.) What fraction? (Signal.) *4/8.* The fraction is eighths. So if you add 1, you add 8 eighths.
- What fraction of 1 do you add? (Signal.) *8/8.*

b. (Point to **3/5.**) What fraction? (Signal.) *3/5.*
• The fraction is fifths. So what fraction of 1 do you add? (Signal.) *5/5.*
c. (Point to **2/3.**) What fraction? (Signal.) *2/3.*
• The fraction is thirds. So what fraction of 1 do you add? (Signal.) *3/3.*
d. (Point to **5/9.**) What fraction? (Signal.) *5/9.*
• The fraction is ninths. So what fraction of 1 do you add? (Signal.) *9/9.*
e. (Point to **2/16.**) What fraction? (Signal.) *2/16.*
• The fraction is 16ths. So what fraction of 1 do you add? (Signal.) *16/16.*
(Repeat until firm.)
f. This time, you'll tell me the fraction you end up with.
• (Point to **4/8.**) The fraction is eighths. So what fraction of 1 do you add? (Signal.) *8/8.*
• What fraction do you end up with? (Signal.) *12/8.*
g. (Point to **3/5.**) The fraction is fifths. So what fraction of 1 do you add? (Signal.) *5/5.*
• What fraction do you end up with? (Signal.) *8/5.*
h. (Point to **2/3.**) The fraction is thirds. So what fraction of 1 do you add? (Signal.) *3/3.*
• What fraction do you end up with? (Signal.) *5/3.*
i. (Point to **5/9.**) The fraction is ninths. So what fraction of 1 do you add? (Signal.) *9/9.*
• What fraction do you end up with? (Signal.) *14/9.*
j. (Point to **2/16.**) The fraction is 16ths. So what fraction of 1 do you add? (Signal.) *16/16.*
• What fraction do you end up with? (Signal.) *18/16.*

============ **WORKBOOK PRACTICE** ============

a. Find part 2 in your workbook. ✔
(Teacher reference:)

a. $\frac{5}{8}$ b. $\frac{2}{3}$ c. $\frac{6}{10}$ d. $\frac{4}{5}$ e. $\frac{7}{9}$

You'll write fractions that are 1 more than the fraction shown. Remember, if the fraction you start with is fourths, you'll add 4 fourths.
• If the fraction you start with is tenths, what will you add? (Signal.) *10/10.*
• If the fraction is halves, what will you add? (Signal.) *2/2.*
b. Item A. The fraction is 5/8. Next to it, write the fraction that is 1 more than 5/8. ✔
• Everybody, what fraction is 1 more than 5/8? (Signal.) *13/8.*
c. Item B. What's the fraction? (Signal.) *2/3.*
• Next to it, write the fraction that is 1 more than 2/3. ✔
• Everybody, what fraction is 1 more than 2/3? (Signal.) *5/3.*

d. Item C. What's the fraction? (Signal.) *6/10.*
• Next to it, write the fraction that is 1 more than 6/10. ✔
• Everybody, what fraction is 1 more than 6/10? (Signal.) *16/10.*
e. Item D. What's the fraction? (Signal.) *4/5.*
• Next to it, write the fraction that is 1 more than 4/5. ✔
• Everybody, what fraction is 1 more than 4/5? (Signal.) *9/5.*
f. Item E. What's the fraction? (Signal.) *7/9.*
• Next to it, write the fraction that is 1 more than 7/9. ✔
• Everybody, what fraction is 1 more than 7/9? (Signal.) *16/9.*

EXERCISE 5: EXPRESSIONS
EVALUATION

a. (Display:)   [83:5A]

$$3\,(14 - 6)$$

Here's a new kind of problem.
I'll read it: 3 times the quantity 14 minus 6.
Once more: 3 times the quantity 14 minus 6.
• Your turn: Read the problem. (Signal.) *3 times the quantity 14 – 6.*
b. (Add to show:)   [83:5B]

$$3\,(14 - 6)$$
$$3\,(40 + 60)$$
$$3\,(15 \div 5)$$
$$20\,(10 \times \tfrac{1}{2})$$
$$2\,(17 + 10 - 7)$$

• (Point to **3 (40 + 60).**) Read the problem. (Signal.) *3 times the quantity 40 + 60.*
• (Point to **3 (15 ÷ 5).**) Read the problem. (Signal.) *3 times the quantity 15 ÷ 5.*
• (Point to **20 (10 × 1/2.**) Read the problem. (Signal.) *20 times the quantity 10 × 1/2.*
• (Point to **2 (17 + 10 – 7.**) Read the problem. (Signal.) *2 times the quantity 17 + 10 – 7.*
(Repeat until firm.)

c. Here's how you work these problems. First, you figure out the quantity inside the parentheses. Then you multiply by the number outside the parentheses.

• What do you figure out first? (Signal.) *The quantity inside the parentheses.*
• Then what do you do? (Signal.) *Multiply by the number outside the parentheses.* (Repeat until firm.)

d. (Change to show:)   [83:5C]

$$3 (14 - 6)$$

• Read this problem again. (Signal.) *3 times the quantity 14 − 6.*
• What's the first thing we figure out? (Signal.) *The quantity inside the parentheses.*
• Read the quantity inside the parentheses. (Signal.) *14 − 6.*
• What's 14 − 6? (Signal.) *8.* (Add to show:)   [83:5D]

$$3 (14 - 6)$$
$$3 (8)$$

Now we multiply 3 times 8.
• Say the multiplication problem. (Signal.) *3 × 8.*
• What's the answer? (Signal.) *24.* (Add to show:)   [83:5E]

$$3 (14 - 6)$$
$$3 (8) = \boxed{24}$$

So 3 times the quantity 14 − 6 equals 24.

e. (Display:)   [83:5F]

$$3 (40 + 60)$$

• Read this problem. (Signal.) *3 times the quantity 40 + 60.*
• What's the first thing we figure out? (Signal.) *The quantity inside the parentheses.*

• Raise your hand when you know the quantity inside the parentheses. ✔
• What's the quantity inside the parentheses? (Signal.) *100.*
• What do we do now? (Signal.) *Multiply by 3.*
• Say the multiplication problem. (Signal.) *3 × 100.* (Add to show:)   [83:5G]

$$3 (40 + 60)$$
$$3 (100)$$

• What's the answer? (Signal.) *300.* (Add to show:)   [83:5H]

$$3 (40 + 60)$$
$$3 (100) = \boxed{300}$$

• So what does 3 times the quantity 40 + 60 equal? (Signal.) *300.*

f. (Display:)   [83:5I]

$$3 (15 \div 5)$$

• Read this problem. (Signal.) *3 times the quantity 15 ÷ 5.*
• What's the first thing we figure out? (Signal.) *The quantity inside the parentheses.*
• What's the quantity inside the parentheses? (Signal.) *3.*
• Say the multiplication problem we work next. (Signal.) *3 times 3.*
• What's the answer? (Signal.) *9.* (Repeat until firm.) (Add to show:)   [83:5J]

$$3 (15 \div 5)$$
$$3 (3) = \boxed{9}$$

• What does 3 times the quantity 15 ÷ 5 equal? (Signal.) *9.*

g. (Display:)   [83:5K]

$$20 (10 \times \tfrac{1}{2})$$

- Read this problem. (Signal.) *20 times the quantity 10 × 1/2.*
- Raise your hand when you know the quantity inside the parentheses. ✔
- What's the quantity inside the parentheses? (Signal.) *5.*
- Say the multiplication problem we work next. (Signal.) *20 × 5.*
- What's the answer? (Signal.) *100.*
  (Repeat until firm.)
  (Add to show:) [83:5L]

$$20 \left(10 \times \frac{1}{2}\right)$$
$$20 (5) = \boxed{100}$$

- What does 20 times the quantity 10 × 1/2 equal? (Signal.) *100.*

h. (Display:) [83:5M]

$$2 (17 + 10 - 7)$$

- Read this problem. (Signal.) *2 times the quantity 17 + 10 − 7.*
- Raise your hand when you know the quantity inside the parentheses. ✔
- What's the quantity inside the parentheses? (Signal.) *20.*
- Say the problem we work next. (Signal.) *2 × 20.*
- What's the answer? (Signal.) *40.*
  (Repeat until firm.)
  (Add to show:) [83:5N]

$$2 (17 + 10 - 7)$$
$$2 (20) = \boxed{40}$$

Remember how to read these problems and how to work them.

EXERCISE 6: FRACTION WORD PROBLEMS
DIVISION

a. (Display:) [83:6A]

A person starts out with $\frac{5}{3}$ quarts of milk. The person divides the milk into 8 equal servings. How much milk is in each serving?

Here's a word problem that tells about dividing a fraction: A person starts out with 5/3 quarts of milk. The person divides the milk into 8 equal servings. How much milk is in each serving?

- What does the person start out with? (Signal.) *5/3 quarts.*
- How many equal servings does the person divide 5/3 quarts into? (Signal.) *8.*
  So you work the problem 5/3 ÷ 8.
- Work the problem in part A and figure out how much milk is in each serving. The unit name in the answer is quart.
  (Observe students and give feedback.)
- Everybody, how much is each serving? (Signal.) *5/24 quart.*
  (Add to show:) [83:6B]

A person starts out with $\frac{5}{3}$ quarts of milk. The person divides the milk into 8 equal servings. How much milk is in each serving?

$$\frac{5}{3} \div 8 = \frac{5}{3} \times \frac{1}{8} = \boxed{\frac{5}{24}} \text{ quart}$$

Here's what you should have.

b. (Display:) [81:6C]

Millie has $\frac{7}{3}$ pounds of grapes. She divides them into 4 equal servings. How many pounds is each serving?

New problem: Millie has 7/3 pounds of grapes. She divides them into 4 equal servings. How many pounds is each serving?

- Write the problem and work it.
  (Observe students and give feedback.)

c. Check your work. You worked the problem 7/3 divided by 4. That's 7/3 times 1/4.

- How many pounds is each serving? (Signal.) *7/12 pound.*
  (Add to show:) [83:6D]

Millie has $\frac{7}{3}$ pounds of grapes. She divides them into 4 equal servings. How many pounds is each serving?

$$\frac{7}{3} \div 4 = \frac{7}{3} \times \frac{1}{4} = \boxed{\frac{7}{12}} \text{ pound}$$

Here's what you should have.

d. (Display:) [83:6E]

> A painter uses $\frac{5}{2}$ gallons of paint to cover 3 identical walls. How much paint is used on each wall?

New problem: A painter uses 5/2 gallons of paint to cover 3 identical walls. How much paint is used on each wall?

* Work the problem and write the answer. Remember the unit name.
  (Observe students and give feedback.)

e. Check your work. You worked the problem 5/2 divided by 3. That's 5/2 times 1/3.

* How much paint is used on each wall? (Signal.) *5/6 gallon.*
  (Add to show:) [83:6F]

> A painter uses $\frac{5}{2}$ gallons of paint to cover 3 identical walls. How much paint is used on each wall?
>
> $$\frac{5}{2} \div 3 = \frac{5}{2} \times \frac{1}{3} = \boxed{\frac{5}{6}} \text{ gallon}$$

Here's what you should have.

EXERCISE 7: AVERAGE

TABLES

a. Last time, you learned about averages. To find the average, you add and then divide.

* If you find the average of 4 things, what do you divide by? (Signal.) *4.*
* If you find the average of 11 things, what do you divide by? (Signal.) *11.*

b. (Display:) [83:7A]

What is the average distance Mia walked?		April	May	June
	Mia	18	24	27

Miles walked:

The table shows the number of miles Mia walked in 3 months—April, May, and June.

* To find the average, you first add numbers. Say the numbers. (Signal.) *18, 24, 27.*
* After you find the total, you divide. What do you divide by? (Signal.) *3.*

c. Work the problem in part A. Write the unit name in the answer. ✔

* Everybody, what's the number of total miles? (Signal.) *69.*
* What's the average distance for these 3 months? (Signal.) *23 miles.*

d. (Change to show:) [83:7B]

Miles walked:

What is the average distance Eli walked?		April	May	June	July
	Mia	18	24	27	
	Eli	25	31	15	21

You're going to figure out the average number of miles for Eli.

* How many numbers will you add? (Signal.) *4.*
* What will you divide the total by? (Signal.) *4.*
* Work the problem. Write the unit name in the answer. ✔
* Check your work. You added 4 numbers. What numbers? (Signal.) *25, 31, 15, 21.*
* What's the total? (Signal.) *92.*
* You divided 92 by a number. What number? (Signal.) *4.*
* What's the average number of miles for Eli? (Signal.) *23 miles.*

e. (Change to show:) [83:7C]

Miles walked:

What is the average distance Amy walked?		April	May	June	July
	Mia	18	24	27	
	Eli	25	31	15	21
	Amy			16	22

* How many numbers are there for Amy? (Signal.) *2.*
* Figure out her average for the 2 months.
  (Observe students and give feedback.)
* Everybody, what's the total number of miles for Amy? (Signal.) *38.*
* You divided 38 by a number. What number? (Signal.) *2.*
* What's the average number of miles she walks in a month? (Signal.) *19 miles.*

a. Open your textbook to Lesson 83 and find part 1. ✔
(Teacher reference:)

Number of Pounds Different Boxes Weigh

a.	22	45	12	50

Temperatures on Different Days

b.	Monday	Tuesday	Wednesday	Thursday	Friday
	38°	27°	22°	33°	40°

b. Problem A shows the weight of 4 boxes. The weights are 22 pounds, 45 pounds, 12 pounds, and 50 pounds.

• What's the first thing you do to find the average? (Signal.) *Add the numbers.*

• What's the next thing you do? (Signal.) *Divide by 4.*

• Figure out the average weight of the boxes. The answer will be a mixed number.
(Observe students and give feedback.)

• Check your work. You added the 4 weights. What's the total? (Signal.) *129.*

• What did you divide by? (Signal.) *4.*

• What's the average weight of a box? (Signal.) *32 and 1/4 pounds.*

c. Problem B. These are temperatures recorded on different days. Find the average temperature.
(Observe students and give feedback.)

• Check your work. You added 5 temperatures. What's the total? (Signal.) *160.*

• What did you divide by? (Signal.) *5.*

• What's the average temperature? (Signal.) *32 degrees.*
Make sure you have the degree symbol in your answer.

EXERCISE 8: SENTENCES
PARTS OF A GROUP

a. (Display:)   [83:8A]

$$\frac{3}{5}$$ of the papers were graded.

Here's a sentence: 3/5 of the papers were graded.

• Listen: What's the fraction for all the papers? (Signal.) *5/5.*

• What's the fraction for papers that were not graded? (Signal.) *2/5.*
(Repeat until firm.)

a. Find part 2 in your textbook. ✔
(Teacher reference:)

a. $\frac{1}{4}$ of the eggs were brown.
  • What's the fraction for all the eggs?
  • What's the fraction for eggs that were not brown?

b. $\frac{11}{12}$ of the windows were open.
  • What fraction of the windows were closed?
  • What's the fraction for all the windows?

c. $\frac{40}{100}$ of the eggs were large.
  • What fraction of the eggs were small?
  • What's the fraction for all the eggs?

b. Sentence A: 1/4 of the eggs were brown.

• Write the fractions that answer the questions. ✔

c. Check your work.

• Everybody, what's the fraction for all the eggs? (Signal.) *4/4.*

• What's the fraction for eggs that were not brown? (Signal.) *3/4.*

d. Write fractions for the rest of the items.
(Observe students and give feedback.)

e. Check your work.

• Sentence B: 11/12 of the windows were open.

• Everybody, what fraction of the windows were closed? (Signal.) *1/12.*

• What's the fraction for all the windows? (Signal.) *12/12.*

f. Sentence C: 40/100 of the eggs were large.

• Everybody, what fraction of the eggs were small? (Signal.) *60/100.*

• What's the fraction for all the eggs? (Signal.) *100/100.*

> Assign Independent Work, Textbook parts 3–7 and Workbook parts 3–7.

Optional extra math-fact practice worksheets are available on ConnectED.

Lesson 84

EXERCISE 1: MENTAL MATH

a. Time for some mental math.
- Listen: 180 divided by 3. (Pause.) (Signal.) *60.*
- 180 ÷ 2. (Pause.) (Signal.) *90.*
- 270 ÷ 9. (Pause.) (Signal.) *30.*
- 360 ÷ 2. (Pause.) (Signal.) *180.*
- (Repeat until firm.)

b. Listen: 360 divided by 6. (Pause.) (Signal.) *60.*
- 360 ÷ 4. (Pause.) (Signal.) *90.*
- 720 ÷ 9. (Pause.) (Signal.) *80.*
- 720 ÷ 8. (Pause.) (Signal.) *90.*
- (Repeat until firm.)

c. Listen: 7 × 30. (Pause.) (Signal.) *210.*
- 7 × 40. (Pause.) (Signal.) *280.*
- 9 × 40. (Pause.) (Signal.) *360.*
- 9 × 80. (Pause.) (Signal.) *720.*
- 9 × 20. (Pause.) (Signal.) *180.*
- (Repeat until firm.)

d. Listen: 160 divided by 4. (Pause.) (Signal.) *40.*
- 180 ÷ 2. (Pause.) (Signal.) *90.*
- 180 ÷ 6. (Pause.) (Signal.) *30.*
- 420 ÷ 6. (Pause.) (Signal.) *70.*
- 630 ÷ 9. (Pause.) (Signal.) *70.*
- 540 ÷ 9. (Pause.) (Signal.) *60.*
- (Repeat until firm.)

EXERCISE 2: MIXED-NUMBER OPERATIONS
ADDITION/SUBTRACTION   REMEDY

a. Open your workbook to Lesson 84 and find part 1. ✔
(Teacher reference:)   R Part A

| a. $10\frac{6}{9}$ | b. $8\frac{4}{5}$ | c. $7\frac{8}{10}$ | d. $23\frac{4}{7}$ |
| $+\ 3\frac{4}{9}$ | $+11\frac{3}{5}$ | $+4\frac{9}{10}$ | $+12\frac{5}{7}$ |

These problems add mixed numbers. If the answer has a fraction that is more than 1, you rewrite the answer.

b. Read problem A. (Signal.)
10 and 6/9 + 3 and 4/9.
- Do the addition for the fractions and the whole numbers. Then stop. ✔
- Everybody, what does 10 and 6/9 plus 3 and 4/9 equal? (Signal.) *13 and 10/9.*
- The fraction in the mixed number is more than 1. Rewrite the mixed number so the fraction is less than 1. ✔

(Display:)   [84:2A]

$$a.\quad 10\frac{6}{9}$$
$$+\ 3\frac{4}{9}$$
$$\overline{13\overset{1}{}\frac{10}{9}}$$
$$\boxed{14\frac{1}{9}}$$

Here's what you should have.
You rewrote 13 and 10/9 as 14 and 1/9.
c. Work the rest of the problems in part 1. (Observe students and give feedback.)
d. Check your work.
- Problem B: 8 and 4/5 + 11 and 3/5. The answer is 19 and 7/5.
- You rewrote the answer. What answer did you write? (Signal.) *20 and 2/5.*
e. C: 7 and 8/10 + 4 and 9/10. The answer is 11 and 17/10.
- You rewrote the answer. What answer did you write? (Signal.) *12 and 7/10.*
f. D: 23 and 4/7 + 12 and 5/7. The answer is 35 and 9/7.
- You rewrote the answer. What answer did you write? (Signal.) *36 and 2/7.*

EXERCISE 3: EXPONENTS
DIVISION BY BASE 10

a. (Display:)   [84:3A]

$$2\ 4.19$$

- Read this number. (Signal.) *24 and 19 hundredths.*
- If we divide, does the number get smaller or larger? (Signal.) *Smaller.*
If we divide by 10, we move the decimal point one place.

(Add to show:) [84:3B]

$$2.4\underset{\curvearrowright}{\,}19$$

- What's 24.19 divided by 10? (Signal.) *2 and 419 thousandths.*
b. (Display:) [84:3C]

$$13.2$$

- Read this number. (Signal.) *13 and 2 tenths.*
- If we divide this number by 10^2, how many places do we move the decimal point? (Signal.) *2 places.*
(Add to show:) [84:3D]

$$\underset{\curvearrowright}{.13}2$$

- What's 13.2 divided by 10^2? (Signal.) *132.*

══════════ **WORKBOOK PRACTICE** ══════════

a. Find part 2 in your workbook. ✔
(Teacher reference:)

a. $\frac{11.08}{10^2} =$ d. $\frac{.23}{10^2} =$

b. $\frac{100.5}{10^1} =$ e. $\frac{9652}{10^3} =$

c. $\frac{6.25}{10^3} =$

b. Complete each equation.
(Observe students and give feedback.)
c. Check your work.
(Display:) [84:3E]

a. $\frac{11.08}{10^2} = \underset{\curvearrowright}{.11}08$ d. $\frac{.23}{10^2} = \underset{\curvearrowright}{.00}23$

b. $\frac{100.5}{10^1} = 10.\underset{\curvearrowright}{0}5$ e. $\frac{9652}{10^3} = 9.\underset{\curvearrowright}{652}$

c. $\frac{6.25}{10^3} = .\underset{\curvearrowright}{006}25$

Here's what you should have for each item.

EXERCISE 4: FRACTION ANALYSIS
1 MORE

a. (Display:) [84:4A]

$$\frac{5}{6}$$

I'm going to rewrite this fraction so it is 1 more than 5/6. The fraction is sixths, so the fraction of 1 is 6/6.
- What's the fraction of 1? (Signal.) *6/6.*
- What fraction is 1 more than 5/6? (Signal.) *11/6.*
(Add to show:) [84:4B]

$$\frac{11}{\cancel{5}}\\\frac{}{6}$$

So I cross out the 5 and write 11.
b. (Display:) [84:4C]

$$\frac{4}{9}$$

We're going to rewrite this fraction so it is 1 more than 4/9.
- What's the fraction of 1 you add? (Signal.) *9/9.*
- What fraction is 1 more than 4/9? (Signal.) *13/9.*
(Add to show:) [84:4D]

$$\frac{13}{\cancel{4}}\\\frac{}{9}$$

So I cross out the 4 and write 13.
c. (Display:) [84:4E]

$$\frac{2}{10}$$

We're going to rewrite this fraction so it is 1 more than 2/10.
- What's the fraction of 1 you add? (Signal.) *10/10.*
- What fraction is 1 more than 2/10? (Signal.) *12/10.*
(Add to show:) [84:4F]

$$\frac{12}{\cancel{2}}\\\frac{}{10}$$

So I cross out the 2 and write 12.

a. Find part 3 in your workbook. ✔
(Teacher reference:)

a. $\frac{5}{7}$ b. $\frac{2}{9}$ c. $\frac{1}{4}$ d. $\frac{5}{8}$ e. $\frac{3}{5}$

- Read fraction A. (Signal.) *5/7.*
  You're going to rewrite the fraction so it is 1 more than 5/7.
- What's the fraction of 1 you add? (Signal.) *7/7.*
- Cross out the numerator and complete the fraction that is 1 more than 5/7. ✔
- Everybody, what fraction is 1 more than 5/7? (Signal.) *12/7.*
  (Display:)   [84:4G]

$$a. \quad \frac{\cancel{5}^{12}}{7}$$

Here's what you should have.
b. Read fraction B. (Signal.) *2/9.*
- What's the fraction of 1 you add? (Signal.) *9/9.*
- Cross out the numerator and complete the fraction that is 1 more than 2/9. ✔
- Everybody, what fraction is 1 more than 2/9? (Signal.) *11/9.*
c. Read fraction C. (Signal.) *1/4.*
- Cross out the numerator and complete the fraction that is 1 more than 1/4. ✔
- Everybody, what fraction is 1 more than 1/4? (Signal.) *5/4.*
d. Read fraction D. (Signal.) *5/8.*
- Cross out the numerator and complete the fraction that is 1 more than 5/8. ✔
- Everybody, what fraction is 1 more than 5/8? (Signal.) *13/8.*
e. Read fraction E. (Signal.) *3/5.*
- Cross out the numerator and complete the fraction that is 1 more than 3/5. ✔
- Everybody, what fraction is 1 more than 3/5? (Signal.) *8/5.*

EXERCISE 5: EXPRESSIONS
EVALUATION

a. (Display:)   [84:5A]

$$10 \, (46 + 4)$$

This problem shows 10 times a quantity.
- Read the problem. (Signal.) *10 times the quantity 46 + 4.*
- What's the first thing we figure out? (Signal.) *The quantity inside the parentheses.*
- What's the next thing we do? (Signal.) *Multiply by 10.*
  (Repeat until firm.)
  Yes, first we figure out the quantity inside the parentheses. Then we multiply by 10.
- Raise your hand when you know the quantity inside the parentheses. ✔
- What's the quantity inside the parentheses? (Signal.) *50.*
- Say the multiplication problem. (Signal.) *10 × 50.*
- What's the answer? (Signal.) *500.*
  (Repeat until firm.)
  (Add to show:)   [84:5B]

$$10 \, (46 + 4)$$
$$10 \, (50) = \boxed{500}$$

- What's 10 times the quantity 46 + 4? (Signal.) *500.*
b. (Display:)   [84:5C]

$$\frac{1}{3} \, (15 - 5 + 2)$$

- Read this problem. (Signal.) *1/3 times the quantity 15 – 5 + 2.*
- First you work the problem 15 minus 5. What's the answer? (Signal.) *10.*
- Say the next problem you work. (Signal.) *10 + 2.*
- What's the answer? (Signal.) *12.*
  (Add to show:)   [84:5D]

$$\frac{1}{3} \, (15 - 5 + 2)$$
$$\frac{1}{3} \, (12) =$$

- Say the next problem you work. (Signal.) *1/3 × 12.*
- What's the answer? (Signal.) *4.*
  (Add to show:) [84:5E]

$$\tfrac{1}{3}(15 - 5 + 2)$$
$$\tfrac{1}{3}(12) = \boxed{4}$$

Remember, first you work the problem inside the parentheses. Then you multiply by the value outside the parentheses.

━━━ **TEXTBOOK PRACTICE** ━━━

a. Open your textbook to Lesson 84 and find part 1. ✔
   (Teacher reference:)

a. 8 (12 ÷ 2)   b. 10 (22 + 3 − 5)
c. $3\left(\tfrac{8}{9} - \tfrac{7}{9}\right)$ d. 12 (5 × 2)

b. Read problem A. (Signal.) *8 times the quantity 12 ÷ 2.*
- Copy the problem and work it.
  (Observe students and give feedback.)
- Check your work. What does 8 times the quantity 12 ÷ 2 equal? (Signal.) *48.*
  (Display:) [84:5F]

a. 8 (12 ÷ 2)
8 (6) = $\boxed{48}$

Here's what you should have.

c. Read problem B. (Signal.) *10 times the quantity 22 + 3 − 5.*
- Copy the problem and work it. ✔
  Check your work. First you worked the problem 22 plus 3 minus 5.
- What's the quantity inside the parentheses? (Signal.) *20.*
- Everybody, what does 10 times 20 equal? (Signal.) *200.*
  (Display:) [84:5G]

b. 10 (22 + 3 − 5)
10 (20) = $\boxed{200}$

Here's what you should have.

d. Read problem C. (Signal.) *3 times the quantity 8/9 − 7/9.*
- Copy the problem and work it. Simplify your answer. ✔
- Everybody, what does 3 times the quantity 8/9 − 7/9 equal? (Signal.) *1/3.*
  (Display:) [84:5H]

c. $3\left(\tfrac{8}{9} - \tfrac{7}{9}\right)$
$\overset{1}{\cancel{3}}\left(\tfrac{1}{\underset{3}{\cancel{9}}}\right) = \boxed{\tfrac{1}{3}}$

Here's what you should have.

e. Read problem D. (Signal.) *12 times the quantity 5 × 2.*
- Copy the problem and work it. ✔
- Everybody, what does 12 times the quantity 5 × 2 equal? (Signal.) *120.*
  (Display:) [84:5I]

d. 12 (5 × 2)
12 (10) = $\boxed{120}$

Here's what you should have.

EXERCISE 6: FRACTION WORD PROBLEMS
DIVISION

a. Find part 2 in your textbook. ✔
   (Teacher reference:)

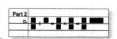

a. A doctor divides $\tfrac{7}{8}$ ounce of medicine into 5 equal doses. How many ounces is each dose?

b. A builder has $\tfrac{18}{5}$ tons of gravel. She divides the gravel into 4 equal piles. How many tons are in each pile?

These word problems tell about dividing a fraction.

b. Problem A: A doctor divides 7/8 ounce of medicine into 5 equal doses. How many ounces is each dose?
- Work the problem and write the answer with a unit name.
  (Observe students and give feedback.)

c. Check your work.
- Read the problem you started with. (Signal.)
  7/8 ÷ 5.
- How many ounces is each dose? (Signal.)
  7/40 ounce.
  (Display:) [84:6A]

$$\text{a. } \frac{7}{8} \div 5 = \frac{7}{8} \times \frac{1}{5} = \boxed{\frac{7}{40}} \text{ ounce}$$

Here's what you should have.

d. Problem B: A builder has 18/5 tons of gravel. She divides the gravel into 4 equal piles. How many tons are in each pile?
- Work the problem.
  (Observe students and give feedback.)

e. Check your work.
- Read the problem you started with. (Signal.)
  18/5 ÷ 4.
- How many tons are in each pile? (Signal.)
  9/10 ton.
  (Display:) [84:6B]

$$\text{b. } \frac{18}{5} \div 4 = \frac{\overset{9}{\cancel{18}}}{5} \times \frac{1}{\underset{2}{\cancel{4}}} = \boxed{\frac{9}{10}} \text{ ton}$$

Here's what you should have.

EXERCISE 7: AVERAGE
BAR GRAPHS

a. Find part 3 in your textbook. ✔
  (Teacher reference:)

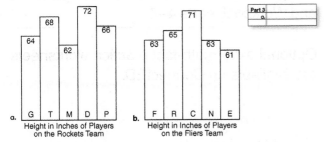

You'll work problems to find the average. Remember the steps. First add the numbers that you will average. Then divide by the number of things there are.

b. Problem A shows the height of players on the Rockets team.
- Touch the bar for player G. ✔
- What's the height of that player? (Signal.)
  64 inches.
- What's the height of player T? (Signal.)
  68 inches.
  You can see the heights of the other players. You'll figure out the average height of the players.

c. Work the problem. Add the heights. Then divide by the number of players. Write the unit name in the answer.
  (Observe students and give feedback.)

d. Check your work.
- Read the numbers you added. (Signal.) *64, 68, 62, 72, 66.*
- What's the total? (Signal.) *332.*
- What number did you divide by? (Signal.) *5.*
- What's the average height of a player?
  (Signal.) *66 and 2/5 inches.*

e. Problem B shows the heights of players on another team—the Fliers.
- Figure out the average height of a Flier.
  (Observe students and give feedback.)
- Everybody, read the numbers you added.
  (Signal.) *63, 65, 71, 63, 61.*
- What's the total? (Signal.) *323.*
- Read the division problem you worked.
  (Signal.) *323 ÷ 5.*
- What's the average height of a Flier? (Signal.)
  64 and 3/5 inches.
- Which team has the taller average player—the Rockets or the Fliers? (Signal.) *The Rockets.*

EXERCISE 8: SENTENCES
PARTS OF A GROUP

a. (Display:) [84:8A]

$$\frac{3}{5} \text{ of the windows were open.}$$

- Read the sentence. (Signal.) *3/5 of the windows were open.*
  You're going to say the sentence for the windows that were closed.
- What fraction of the windows were closed?
  (Signal.) *2/5.*
  Yes, 2/5 of the windows were closed.
- Say the sentence. (Signal.) *2/5 of the windows were closed.*

b. (Display:) [84:8B]

$$\frac{7}{8} \text{ of the students were sleeping.}$$

- Read the sentence. (Signal.) *7/8 of the students were sleeping.*
- What fraction of the students were awake? (Signal.) *1/8.*
- Say the sentence for the students that were awake. (Signal.) *1/8 of the students were awake.*

c. (Display:) [84:8C]

$$\frac{2}{9} \text{ of the cars were black.}$$

- Read the sentence. (Signal.) *2/9 of the cars were black.*
- What fraction of the cars were not black? (Signal.) *7/9.*
- Say the sentence for the cars that were not black. (Signal.) *7/9 of the cars were not black.*

d. (Display:) [84:8D]

$$\frac{6}{10} \text{ of the doors were locked.}$$

- Read the sentence. (Signal.) *6/10 of the doors were locked.*
- What fraction of the doors were unlocked? (Signal.) *4/10.*
- Say the sentence for the doors that were unlocked. (Signal.) *4/10 of the doors were unlocked.*

e. (Display:) [84:8E]

$$\frac{9}{11} \text{ of the trees were yellow.}$$

- Read the sentence. (Signal.) *9/11 of the trees were yellow.*
- What fraction of the trees were not yellow? (Signal.) *2/11.*
- Say the sentence for the trees that were not yellow. (Signal.) *2/11 of the trees were not yellow.*

f. (Display:) [84:8F]

$$\frac{80}{100} \text{ of the muffins were baked.}$$

- Read the sentence. (Signal.) *80/100 of the muffins were baked.*
- What fraction of the muffins were not baked? (Signal.) *20/100.*
- Say the sentence for the muffins that were not baked. (Signal.) *20/100 of the muffins were not baked.*

EXERCISE 9: INDEPENDENT WORK
VOLUME

a. Find part 5 in your textbook. ✔
(Teacher reference:)

Find the volume of each figure. Start with the letter equation.

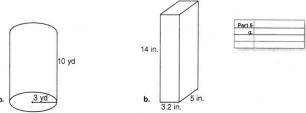

I'll read the directions: **Find the volume of each figure. Start with the letter equation.**

b. Remember, Volume = Area of the base times height.
- Say the equation. (Signal.) *Volume = Area of the base times height.*
Start with the letter equation when you work the problems.

Assign Independent Work, Textbook parts 4–9 and Workbook parts 4–7.

Optional extra math-fact practice worksheets are available on ConnectED.

Lesson

EXERCISE 1: MENTAL MATH

a. Time for mental math. I'll say a problem. You'll say the whole fact.
- • Listen: 90 times 3. Say the fact. (Signal.) *90 × 3 = 270.*
- • 90 × 2. Say the fact. (Signal.) *90 × 2 = 180.*
- • 90 × 4. Say the fact. (Signal.) *90 × 4 = 360.*
  (Repeat until firm.)
- b. Listen: 60 × 2. Say the fact. (Signal.) *60 × 2 = 120.*
- • 60 × 4. Say the fact. (Signal.) *60 × 4 = 240.*
- • 60 × 5. Say the fact. (Signal.) *60 × 5 = 300.*
  (Repeat until firm.)
- c. Listen: 80 × 2. Say the fact. (Signal.) *80 × 2 = 160.*
- • 80 × 3. Say the fact. (Signal.) *80 × 3 = 240.*
- • 80 × 5. Say the fact. (Signal.) *80 × 5 = 400.*
  (Repeat until firm.)
- d. Listen: 400 divided by 5. Say the fact. (Signal.) *400 ÷ 5 = 80.*
- • 350 ÷ 5. Say the fact. (Signal.) *350 ÷ 5 = 70.*
- • 200 ÷ 5. Say the fact. (Signal.) *200 ÷ 5 = 40.*
- • 500 ÷ 5. Say the fact. (Signal.) *500 ÷ 5 = 100.*
  (Repeat until firm.)
- e. Listen: 180 ÷ 2. Say the fact. (Signal.) *180 ÷ 2 = 90.*
- • 360 ÷ 4. Say the fact. (Signal.) *360 ÷ 4 = 90.*
- • 360 ÷ 90. Say the fact. (Signal.) *360 ÷ 90 = 4.*
- • 180 ÷ 90. Say the fact. (Signal.) *180 ÷ 90 = 2.*
- • 270 ÷ 90. Say the fact. (Signal.) *270 ÷ 90 = 3.*
  (Repeat until firm.)

EXERCISE 2: MIXED-NUMBER OPERATIONS
ADDITION/SUBTRACTION   REMEDY

a. (Display:)   [85:2A]

$$3\frac{4}{5}$$
$$+\;5\frac{1}{5}$$

- • Read this problem. (Signal.) *3 and 4/5 + 5 and 1/5.*
- • Say the problem and the answer for the fractions. (Signal.) *4/5 + 1/5 = 5/5.*
  (Add to show:)   [85:2B]

$$3\frac{4}{5}$$
$$+\;5\frac{1}{5}$$
$$\frac{5}{5}$$

- • Say the problem and the answer for the whole numbers. (Signal.) *3 + 5 = 8.*
  (Add to show:)   [85:2C]

$$3\frac{4}{5}$$
$$+\;5\frac{1}{5}$$
$$8\frac{5}{5}$$

5/5 is 1. So we add 1 to 8.
(Add to show:)   [85:2D]

$$3\frac{4}{5}$$
$$+\;5\frac{1}{5}$$
$$8\overset{1}{}\frac{5}{5}$$
$$\boxed{9}$$

We don't show a fraction.
- • What does 3 and 4/5 + 5 and 1/5 equal? (Signal.) *9.*

===== **WORKBOOK PRACTICE** =====

a. Open your workbook to Lesson 85 and find part 1. ✔
(Teacher reference:) $\boxed{\text{R} | \textit{Part B}}$

a. $5\frac{3}{12}$ b. $6\frac{6}{10}$ c. $22\frac{7}{8}$ d. $18\frac{5}{6}$
    $+2\frac{9}{12}$ $+13\frac{5}{10}$ $+1\frac{6}{8}$ $+10\frac{1}{6}$

b. Read problem A. (Signal.) *5 and 3/12 + 2 and 9/12.*
• Work the problem.
(Observe students and give feedback.)
• Everybody, what does 5 and 3/12 + 2 and 9/12 equal? (Signal.) *8.*
Yes, 7 and 12/12 equals 8.
(Display:) [85:2E]

$$\begin{array}{r} \mathbf{a.} \qquad 5\frac{3}{12} \\ +\ 2\frac{9}{12} \\ \hline 7\frac{\overset{1}{12}}{12} \\ \boxed{8} \end{array}$$

Here's what you should have.
c. Work the rest of the problems in part 1.
(Observe students and give feedback.)
d. Check your work.
• Problem B: 6 and 6/10 + 13 and 5/10.
What's the answer? (Signal.) *20 and 1/10.*
• C: 22 and 7/8 + 1 and 6/8.
What's the answer? (Signal.) *24 and 5/8.*
• D: 18 and 5/6 + 10 and 1/6.
What's the answer? (Signal.) *29.*

EXERCISE 3: EXPONENTS
DIVISION BY BASE 10

a. Open your textbook to Lesson 85 and find part 1. ✔
(Teacher reference:)

a. $\frac{119.08}{10^2}$ b. $\frac{382}{10^3}$ c. $\frac{12.06}{10^3}$ d. $\frac{.5}{10^2}$ e. $\frac{10.03}{10^1}$ 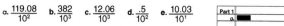

These problems divide by 10 with an exponent. You're going to write the answers.

b. Read problem A. (Signal.) *119.08 ÷ 10².*
• Write the answer. Make loops if you need to. ✔
(Display:) [85:3A]

> **a.** 1.1908

Here's what you should have.
c. Read problem B. (Signal.) *382 ÷ 10³.*
• Write the answer. ✔
(Display:) [85:3B]

> **b.** .382

Here's what you should have.
d. Work the rest of the problems in part 1.
(Observe students and give feedback.)
e. Check your work.
(Display:) [85:3C]

> **c.** .01206
> **d.** .005
> **e.** 1.003

Here's what you should have for problems C through E.

EXERCISE 4: EXPRESSIONS
EVALUATION $\boxed{\text{REMEDY}}$

a. (Display:) [85:4A]

$$4\ (5 + B + 6) \qquad \boxed{B = 9}$$

Here's a problem with a letter.
• Read the problem. (Signal.) *4 times the quantity 5 + B + 6.*
The box shows the number for B.
So we replace B with 9 below.
(Add to show:) [85:4B]

$$4\ (5 + B + 6) \qquad \boxed{B = 9}$$
$$4\ (5 + 9 + 6)$$

• Read the new problem. (Signal.) *4 times the quantity 5 + 9 + 6.*
• Raise your hand when you know the quantity inside the parentheses. ✔
• What's the quantity? (Signal.) *20.*

(Add to show:) [85:4C]

$$4 (5 + B + 6) \quad \boxed{B = 9}$$
$$4 (5 + 9 + 6)$$
$$4 (20)$$

- Say the next problem you work. (Signal.) *4 × 20.*
- What's the answer? (Signal.) *80.*
(Add to show:) [85:4D]

$$4 (5 + B + 6) \quad \boxed{B = 9}$$
$$4 (5 + 9 + 6)$$
$$4 (20) = \boxed{80}$$

Remember, first replace the letter with a number. Then work the problem.

━━━━ **TEXTBOOK PRACTICE** ━━━━

a. Find part 2 in your textbook. ✔
(Teacher reference:)

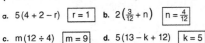

a. $5(4 + 2 - r)$ $\boxed{r = 1}$ b. $2\left(\frac{3}{12} + n\right)$ $\boxed{n = \frac{4}{12}}$

c. $m(12 ÷ 4)$ $\boxed{m = 9}$ d. $5(13 - k + 12)$ $\boxed{k = 5}$

All these problems have a letter.
b. Read problem A. (Signal.) *5 times the quantity 4 + 2 − R.*
- Copy the problem. Rewrite the quantity with a number for R. Then work the problem.
(Observe students and give feedback.)
- Everybody, what does 5 times the quantity 4 + 2 − 1 equal? (Signal.) *25.*
(Display:) [85:4E]

$$\text{a. } 5 (4 + 2 - r) \quad \boxed{r = 1}$$
$$5 (4 + 2 - 1)$$
$$5 (5) = \boxed{25}$$

Here's what you should have.
c. Read problem B. (Signal.) *2 times the quantity 3/12 + N.*
- The problem gives a value for N. What value? (Signal.) *4/12.*
- Rewrite the problem with 4/12 in place of N. Stop when you've done that much. ✔
- Everybody, read the new problem. (Signal.) *2 times the quantity 3/12 + 4/12.*
- Work the problem. First figure out the value inside the parentheses. Then multiply by 2. Pencils down when you're finished.

(Observe students and give feedback.)
- You worked the problem 3/12 + 4/12. What's the quantity? (Signal.) *7/12.*
Then you multiplied 2 × 7/12. The answer is 7/6.
- What's the mixed-number answer? (Signal.) *1 and 1/6.*
(Display:) [85:4F]

b. $2\left(\frac{3}{12} + n\right) \quad \boxed{n = \frac{4}{12}}$

$2\left(\frac{3}{12} + \frac{4}{12}\right)$

$\overset{1}{2}\left(\frac{7}{\underset{6}{\cancel{12}}}\right) = \frac{7}{6} = \boxed{1\frac{1}{6}}$

Here's what you should have.
d. Work the rest of the problems in part 2. Copy each problem. Rewrite the problem with a number in place of the letter. Then work the problem.
(Observe students and give feedback.)
e. Check your work.
- Problem C. You rewrote the problem as 9 times the quantity 12 ÷ 4. Then you multiplied 9 × 3.
- What's the answer? (Signal.) *27.*
f. Problem D. You rewrote the problem as 5 times the quantity 13 − 5 + 12. Then you multiplied 5 × 20.
- What's the answer? (Signal.) *100.*

EXERCISE 5: FRACTION WORD PROBLEMS
DIVISION

a. (Display:) [85:5A]

> Nancy has 3 pounds of peanuts. She divides the peanuts into servings that are $\frac{1}{2}$ pound each. How many servings does she make?

Here's a word problem that divides by a fraction: Nancy has 3 pounds of peanuts. She divides the peanuts into servings that are 1/2 pound each. How many servings does she make?
The problem tells how many pounds she starts with and the fraction you divide by.
- How many pounds does she start with? (Signal.) *3.*
- What do we divide 3 by? (Signal.) *1/2.*

(Add to show:)   [85:5B]

> Nancy has 3 pounds of peanuts. She divides the peanuts into servings that are $\frac{1}{2}$ pound each. How many servings does she make?
>
> $$3 \div \frac{1}{2}$$

Here's the problem.

- Copy the problem and work it in part A. Write the number and unit name in the answer. Pencils down when you know how many servings she makes.
  (Observe students and give feedback.)
- Everybody, how many servings does she make? (Signal.) *6 servings.*
  Yes, the answer is 6 servings.

(Add to show:)   [85:5C]

> Nancy has 3 pounds of peanuts. She divides the peanuts into servings that are $\frac{1}{2}$ pound each. How many servings does she make?
>
> $$3 \div \frac{1}{2} = 3 \times \frac{2}{1} = \boxed{6 \text{ servings}}$$

Here's what you should have.

b. (Display:)   [85:5D]

> A pile of gravel weighs $\frac{3}{4}$ ton. A builder wants to divide the gravel into piles that are $\frac{1}{8}$ ton each. How many of these piles will he be able to make?

New problem: A pile of gravel weighs 3/4 ton. The builder wants to divide the gravel into piles that are 1/8 ton each. How many of these piles will he be able to make?

- How much gravel does the builder start with? (Signal.) *3/4 ton.*
- What do we divide by? (Signal.) *1/8.*
- Work the problem and figure out how many piles he'll be able to make. Remember to simplify.
  (Observe students and give feedback.)

c. Check your work. You worked the problem 3/4 divided by 1/8. That's 3/4 times 8.
- What's the whole-number answer? (Signal.) *6.*
- What's the unit name in the answer? (Signal.) *Piles.*
  Yes, the builder will be able to make 6 piles that are 1/8 of a ton each.

d. (Display:)   [85:5E]

> A baker has 6 pounds of flour. She wants to make cakes. Each cake uses $\frac{3}{4}$ pound of flour. How many cakes will she be able to make with 6 pounds of flour?

New problem: A baker has 6 pounds of flour. She wants to make cakes. Each cake uses 3/4 pound of flour. How many cakes will she be able to make with 6 pounds of flour?

- Work the problem. Write the unit name in the answer.
  (Observe students and give feedback.)

e. Check your work. You worked the problem 6 divided by 3/4. That's 6 times 4/3.
- What's the answer? (Signal.) *8.*
  Yes, she will be able to make 8 cakes.

EXERCISE 6: AVERAGE
BAR GRAPHS

a. (Display:)   [85:6A]

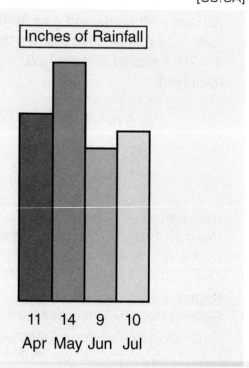

The bars show how much rain fell in 4 months. In April, there were 11 inches of rain.

- How many inches of rain fell in May? (Signal.) *14.*
- June? (Signal.) *9.*
- July? (Signal.) *10.*
- In part A, figure out the average number of inches for a month.
  (Observe students and give feedback.)
b. Check your work.
- You added the 4 numbers. What's the total? (Signal.) *44.*
- What did you divide by? (Signal.) *4.*
- What's the average number? (Signal.) *11.*
  Yes, if the same amount of rain fell each month, there would be 11 inches each month.
c. Look at the bars.
- One month has the average number of 11. Which month is that? (Signal.) *April.*
- Which month is above average? (Signal.) *May.*
- Which months are below average? (Signal.) *June and July.*
d. (Add to show:) [85:6B]

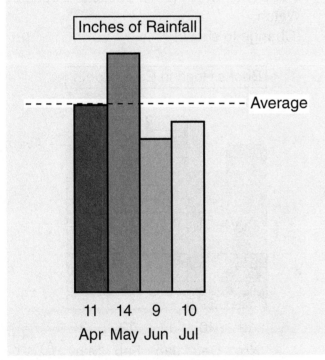

11 14 9 10
Apr May Jun Jul

This line shows the average. We can show that all months would have 11 inches just by moving the units that are above the line to below the line.

e. (Change to show:) [85:6C]

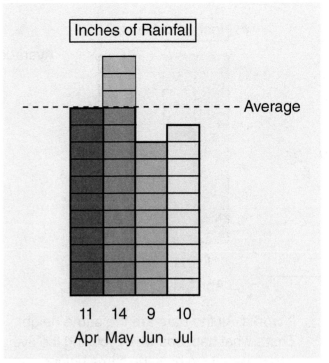

11 14 9 10
Apr May Jun Jul

Watch. We move 2 units to June.
(Change to show:) [85:6D]

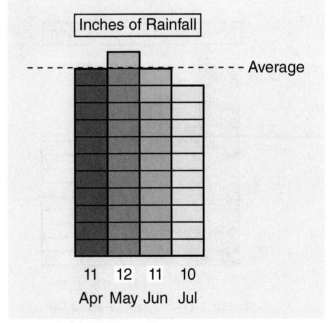

11 12 11 10
Apr May Jun Jul

We move one unit to July.

(Change to show:) [85:6E]

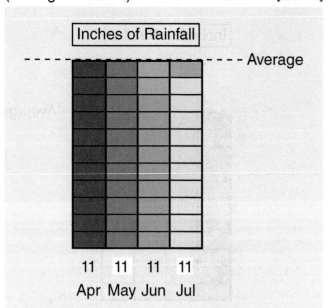

Inches of Rainfall

- Average

| 11 | 11 | 11 | 11 |
| Apr | May | Jun | Jul |

It works. All the bars are the same height. That's what happens when you find the average. You make all the bars the same height.

f. Let's do the same thing with another problem. (Display:) [85:6F]

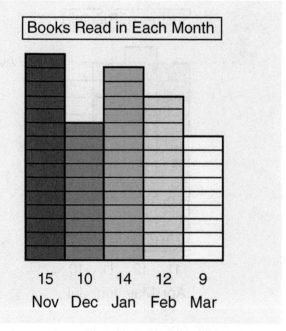

Books Read in Each Month

| 15 | 10 | 14 | 12 | 9 |
| Nov | Dec | Jan | Feb | Mar |

These bars show how many books a group of students read in 5 months—November through March.

• Figure out the number of books that were read in an average month.
  (Observe students and give feedback.)

g. Check your work.
• You added 5 numbers. What's the total? (Signal.) *60.*
• What did you divide by? (Signal.) *5.*
• What's the average number? (Signal.) *12.*

h. Let's draw a line to show the average. (Change to show:) [85:6G]

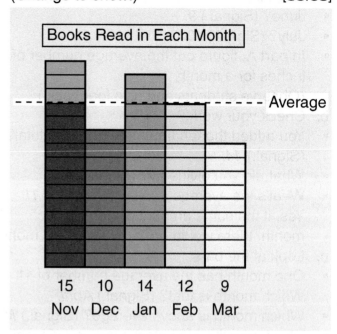

Books Read in Each Month

- Average

| 15 | 10 | 14 | 12 | 9 |
| Nov | Dec | Jan | Feb | Mar |

You can see the units that are above the line. We'll move them below the line. We'll move 3 books to March and 2 books to December. Watch.

(Change to show:) [85:6H]

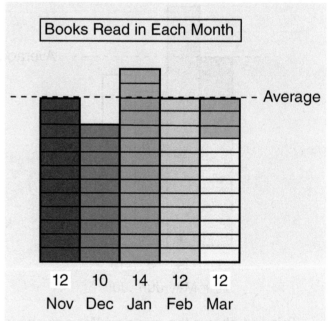

Books Read in Each Month

- Average

| 12 | 10 | 14 | 12 | 12 |
| Nov | Dec | Jan | Feb | Mar |

(Change to show:) [85:6I]

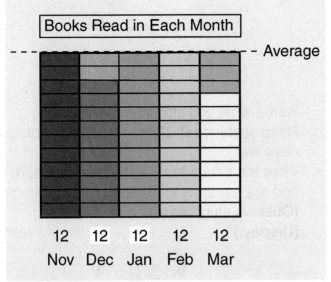

Books Read in Each Month

- - - - - - - - - - - Average

| 12 | 12 | 12 | 12 | 12 |
| Nov | Dec | Jan | Feb | Mar |

It works. That's what happens when you find the average. You make all the bars the same height.

EXERCISE 7: SENTENCES
PARTS OF A GROUP

a. (Display:) [85:7A]

$$\frac{2}{13} \text{ of the children were sick.}$$

- Read the sentence. (Signal.) *2/13 of the children were sick.*
  2/13 of the children were sick.
- Raise your hand when you know the fraction of the children that were well. ✔
- What fraction of the children were well? (Signal.) *11/13.*
- Say the sentence for the children who were well. (Signal.) *11/13 of the children were well.*

b. (Display:) [85:7B]

$$\frac{10}{100} \text{ of the shirts were wet.}$$

- Read the sentence. (Signal.) *10/100 of the shirts were wet.*
- Raise your hand when you know the fraction of the shirts that were dry. ✔
- What fraction of the shirts were dry? (Signal.) *90/100.*
- Say the sentence for the shirts that were dry. (Signal.) *90/100 of the shirts were dry.*

c. (Display:) [85:7C]

$$\frac{3}{11} \text{ of the cars were new.}$$

- Read the sentence. (Signal.) *3/11 of the cars were new.*
- Raise your hand when you know the fraction of the cars that were used. ✔
- What fraction of the cars were used? (Signal.) *8/11.*
- Say the sentence for the cars that were used. (Signal.) *8/11 of the cars were used.*

d. This time, we'll write equations for each sentence.
- 3/11 of the cars were new. Say the equation for that sentence. (Signal.) *3/11 C = N.*
  (Add to show:) [85:7D]

$$\frac{3}{11} \text{ of the cars were new.}$$

$$\frac{3}{11} c = n$$

- Say the equation for the cars that were new. (Signal.) *3/11 C = N.*
- Say the equation for the cars that were used. (Signal.) *8/11 C = U.*
  (Repeat until firm.)
  (Add to show:) [85:7E]

$$\frac{3}{11} \text{ of the cars were new.}$$

$$\frac{3}{11} c = n$$

$$\frac{8}{11} c = u$$

e. (Display:) [85:7F]

$$\frac{3}{4} \text{ of the boards were painted.}$$

- Read the sentence. (Signal.) *3/4 of the boards were painted.*
- Say the equation for the boards that were painted. (Signal.) *3/4 B = P.*
  (Add to show:) [85:7G]

$$\frac{3}{4} \text{ of the boards were painted.}$$

$$\frac{3}{4} b = p$$

f. Raise your hand when you can say the equation for the boards that were not painted. ✔
- Say the equation for the boards that were not painted. (Signal.) *1/4 B = N.*
  (Repeat until firm.)

(Add to show:)   [85:7H]

$$\frac{3}{4} \text{ of the boards were painted.}$$

$$\frac{3}{4} b = p$$

$$\frac{1}{4} b = n$$

g. (Display:)   [85:7I]

$$\frac{1}{2} \text{ of the bottles were empty.}$$

- Read the sentence. (Signal.) *1/2 of the bottles were empty.*
- Say the equation for the bottles that were empty. (Signal.) *1/2 B = E.*
- Raise your hand when you can say the equation for the bottles that were not empty. ✔
- Say the equation for the bottles that were not empty. (Signal.) *1/2 B = N.*
  (Repeat until firm.)
  (Add to show:)   [85:7J]

$$\frac{1}{2} \text{ of the bottles were empty.}$$

$$\frac{1}{2} b = e$$

$$\frac{1}{2} b = n$$

=== **TEXTBOOK PRACTICE** ===

a. Find part 3 in your textbook. ✔
   (Teacher reference:)

   a. $\frac{2}{5}$ of the cats were black.

   b. $\frac{4}{7}$ of the plates were wet.

   c. $\frac{1}{9}$ of the bolts were rusty.

   For each sentence, you'll write two equations.
b. Read sentence A. (Signal.) *2/5 of the cats were black.*
- Write the equation for the cats that were black. Then write the equation for the cats that were not black. Remember, write N for not.
  (Observe students and give feedback.)

(Display:)   [85:7K]

$$\text{a. } \frac{2}{5} c = b$$

$$\frac{3}{5} c = n$$

Here's what you should have.
c. Read sentence B. (Signal.) *4/7 of the plates were wet.*
- Write the equation for the plates that were wet and the equation for the plates that were dry.
  (Observe students and give feedback.)
  (Display:)   [85:7L]

$$\text{b. } \frac{4}{7} p = w$$

$$\frac{3}{7} p = d$$

Here's what you should have.
d. Read sentence C. (Signal.) *1/9 of the bolts were rusty.*
- Write the equation for the bolts that were rusty and the equation for the bolts that are were not rusty.
  (Observe students and give feedback.)
  (Display:)   [85:7M]

$$\text{c. } \frac{1}{9} b = r$$

$$\frac{8}{9} b = n$$

Here's what you should have.

EXERCISE 8: MIXED NUMBERS
REWRITING

a. (Display:)   [85:8A]

$$\text{a. } \frac{3}{4} \quad \text{b. } \frac{5}{9} \quad \text{c. } \frac{2}{10}$$

We're going to rewrite each fraction so it is 1 more.
- Read fraction A. (Signal.) *3/4.*
- Raise your hand when you know the fraction that is 1 more than 3/4. ✔
- What fraction is 1 more than 3/4? (Signal.) *7/4.*

(Add to show:)   [85:8B]

$$\text{a. } \frac{\cancel{3}^{\,7}}{4} \qquad \text{b. } \frac{5}{9} \qquad \text{c. } \frac{2}{10}$$

So I cross out the 3 and write 7.

b. Read fraction B. (Signal.) *5/9.*
- Raise your hand when you know the fraction that is 1 more than 5/9. ✔
- What fraction is 1 more than 5/9? (Signal.) *14/9.*
(Add to show:)   [85:8C]

$$\text{a. } \frac{\cancel{3}^{\,7}}{4} \qquad \text{b. } \frac{\cancel{5}^{\,14}}{9} \qquad \text{c. } \frac{2}{10}$$

So I cross out the 5 and write 14.

c. Read fraction C. (Signal.) *2/10.*
- Raise your hand when you know the fraction that is 1 more than 2/10. ✔
- What fraction is 1 more than 2/10? (Signal.) *12/10.*
(Add to show:)   [85:8D]

$$\text{a. } \frac{\cancel{3}^{\,7}}{4} \qquad \text{b. } \frac{\cancel{5}^{\,14}}{9} \qquad \text{c. } \frac{\cancel{2}^{\,12}}{10}$$

So I cross out the 2 and write 12.

d. (Display:)   [85:8E]

$$5\frac{2}{7}$$

Here's a new kind of problem.
- Read the mixed number. (Signal.) *5 and 2/7.*
We're going to rewrite the mixed number so the fraction is 1 more than 2/7.
- We take the 1 from the 5. What's 5 − 1? (Signal.) *4.*
So I cross out 5 and write 4 above it.
(Add to show:)   [85:8F]

$$\cancel{5}^{\,4}\frac{2}{7}$$

We took 1 from 5. We add that 1 to the fraction 2/7.
- Raise your hand when you know the fraction that is 1 more than 2/7. ✔
- What fraction is 1 more than 2/7? (Signal.) *9/7.*

(Add to show:)   [85:8G]

$$\cancel{5}^{\,4}\frac{\cancel{2}^{\,9}}{7}$$

- Read the new mixed number. (Signal.) *4 and 9/7.* Remember, that equals the mixed number you started with.
e. (Display:)   [85:8H]

$$5\frac{1}{6}$$

- Read the mixed number. (Signal.) *5 and 1/6.* We're going to rewrite the mixed number so the fraction is 1 more than 1/6.
- We take 1 from 5. What's the new whole number? (Signal.) *4.*
(Add to show:)   [85:8I]

$$\cancel{5}^{\,4}\frac{1}{6}$$

We add 1 to the fraction 1/6.
- Raise your hand when you know the new fraction. ✔
- What's the new fraction? (Signal.) *7/6.*
(Add to show:)   [85:8J]

$$\cancel{5}^{\,4}\frac{\cancel{1}^{\,7}}{6}$$

- Read the new mixed number. (Signal.) *4 and 7/6.*
f. (Display:)   [85:8K]

$$5\frac{4}{10}$$

- Read the mixed number. (Signal.) *5 and 4/10.* We're going to rewrite the mixed number so the fraction is 1 more than 4/10.
- We take 1 from 5. What's the new whole number? (Signal.) *4.*
(Add to show:)   [85:8L]

$$\cancel{5}^{\,4}\frac{4}{10}$$

We add the 1 to the fraction 4/10.
- Raise your hand when you know the new fraction. ✔

- What's the new fraction? (Signal.) *14/10.*
  (Add to show:) [85:8M]

$$4 \frac{\cancel{14}}{5 \, 10}$$

- Read the new mixed number. (Signal.) *4 and 14/10.*

g. (Display:) [85:8N]

$$3 \frac{1}{4}$$

- Read the mixed number. (Signal.) *3 and 1/4.*
  We're going to rewrite the mixed number so
  the fraction is 1 more than 1/4.
- We take one from 3. What's the new whole
  number? (Signal.) *2.*
  (Add to show:) [85:8O]

$$\frac{2}{\cancel{3}} \frac{1}{4}$$

We add the 1 to the fraction 1/4.
- Raise your hand when you know the new
  fraction. ✔
- What's the new fraction? (Signal.) *5/4.*
  (Add to show:) [85:8P]

$$\frac{2}{\cancel{3}} \frac{5}{\cancel{4}}$$

- Read the new mixed number. (Signal.) *2 and 5/4.*

h. (Display:) [85:8Q]

$$10 \frac{2}{5}$$

- Read the mixed number. (Signal.) *10 and 2/5.*
  We're going to rewrite the mixed number so
  the fraction is 1 more than 2/5.
- We take 1 from 10. What's the new whole
  number? (Signal.) *9.*
  (Add to show:) [85:8R]

$$\frac{9}{\cancel{10}} \frac{2}{5}$$

We add the 1 to the fraction 2/5.
- Raise your hand when you know the new
  fraction. ✔
- What's the new fraction? (Signal.) *7/5.*
  (Add to show:) [85:8S]

$$\frac{9}{\cancel{10}} \frac{7}{\cancel{5}}$$

- Read the new mixed number. (Signal.) *9 and 7/5.*

Assign Independent Work, Textbook parts 4–10
and Workbook parts 2–4.

Optional extra math-fact practice worksheets
are available on ConnectED.

Lessons 86-90 Planning Page

	Lesson 86	Lesson 87	Lesson 88	Lesson 89	Lesson 90
Student Learning Objectives	**Exercises** 1. Add, subtract, multiply, and divide using mental math 2. Solve multiplication and division problems that include powers of 10 3. Solve fraction word problems with division 4. Evaluate expressions that have parentheses and a letter 5. Write equations for parts of a group sentences 6. Find the average of values in a bar graph 7. **Write equations for parts-of-a-group sentences with two and three names** 8. **Rewrite mixed numbers so the fraction is 1 more** 9. Complete work independently	**Exercises** 1. Subtract and divide using mental math 2. **Compare products and factors in fraction multiplication problems** 3. Write equations for parts-of-a-group sentences with two and three names 4. Rewrite mixed numbers so the fraction is 1 more 5. **Find the average of values in a set with zero** 6. Solve fraction word problems with division 7. **Write expressions with parentheses from written directions** 8. Write equations for parts of a group sentences 9. Complete work independently	**Exercises** 1. Subtract, multiply, divide, and find whole numbers for fractions using mental math 2. Compare products and factors in fraction multiplication problems 3. **Subtract mixed numbers** 4. Write expressions with parentheses from written directions 5. Solve word problems that have fractions 6. Find the average distance 7. Write equations for parts-of-a-group sentences with three names Complete work independently	**Exercises** 1. Multiply, divide, and find whole numbers for fractions using mental math 2. Compare products and factors in fraction multiplication problems 3. Subtract mixed numbers 4. Write expressions with parentheses from written directions 5. Solve fraction word problems with division 6. Find averages 7. Write equations for parts-of-a-group sentences with three names Complete work independently	**Exercises** 1. Multiply, divide, and find whole numbers for fractions using mental math 2. Compare products and factors in fraction multiplication problems 3. **Understand balance beams** 4. Subtract mixed numbers 5. **Use multiplication to convert units of measurement** 6. Solve fraction word problems with division 7. **Solve parts-of-a-group word problems** 8. Complete work independently
Common Core State Standards for Mathematics					
5.OA 1	✔	✔	✔	✔	✔
5.OA 2		✔	✔	✔	✔
5.NBT 1–2	✔	✔	✔	✔	✔
5.NBT 5	✔	✔	✔	✔	✔
5.NBT 6	✔	✔	✔		✔
5.NF 2	✔	✔			✔
5.NF 3			✔	✔	
5.NF 4	✔	✔	✔	✔	✔
5.NF 7	✔	✔	✔	✔	✔
5.MD 1					✔
5.G 2	✔	✔	✔		
Teacher Materials	Presentation Book 2, Board Displays CD or chalk board				
Student Materials	Textbook, Workbook, pencil, lined paper, ruler				
Additional Practice	• Student Practice Software: Block 4: Activity 1 (5.NF 5), Activity 2, Activity 3 (5.NBT 4), Activity 4 (5.NBT 3), Activity 5 (5.NBT 7), Activity 6 (5.MD 3 and 5.MD 4) • Provide needed fact practice with Level D or E Math Fact Worksheets.				
Mastery Test					Student Assessment Book (Present Mastery Test 9 following Lesson 90.)

Connecting Math Concepts   *Lessons 86–90 Planning Page*   **189**

Lesson  86

EXERCISE 1: MENTAL MATH

a. Time for mental math. You'll say the whole fact.
- Listen: 156 minus 10. Say the fact. (Signal.)
  156 – 10 = 146.
- 156 – 20. Say the fact. (Signal.) *156 – 20 = 136.*
- 156 – 30. Say the fact. (Signal.) *156 – 30 = 126.*
  (Repeat until firm.)

b. Listen: 124 plus 10. Say the fact. (Signal.)
  124 + 10 = 134.
- 124 + 20. Say the fact. (Signal.) *124 + 20 = 144.*
- 124 + 60. Say the fact. (Signal.) *124 + 60 = 184.*
- 124 + 100. Say the fact. (Signal.) *124 + 100 = 224.*
  (Repeat until firm.)

c. Listen: 56 minus 10. Say the fact. (Signal.)
  56 – 10 = 46.
- 56 – 30. Say the fact. (Signal.) *56 – 30 = 26.*
- 56 – 40. Say the fact. (Signal.) *56 – 40 = 16.*
  (Repeat until firm.)

d. Listen: 70 times 2. Say the fact. (Signal.)
  70 × 2 = 140.
- 70 × 3. Say the fact. (Signal.) *70 × 3 = 210.*
- 70 × 5. Say the fact. (Signal.) *70 × 5 = 350.*
  (Repeat until firm.)

e. Listen: 60 divided by 10. Say the fact. (Signal.)
  60 ÷ 10 = 6.
- 60 ÷ 20. Say the fact. (Signal.) *60 ÷ 20 = 3.*
- 60 ÷ 30. Say the fact. (Signal.) *60 ÷ 30 = 2.*
- 180 ÷ 90. Say the fact. (Signal.) *180 ÷ 90 = 2.*
- 360 ÷ 90. Say the fact. (Signal.) *360 ÷ 90 = 4.*
  (Repeat until firm.)

f. Listen: 40 times 2. Say the fact. (Signal.)
  40 × 2 = 80.
- 40 × 4. Say the fact. (Signal.) *40 × 4 = 160.*
- 40 × 8. Say the fact. (Signal.) *40 × 8 = 320.*
  (Repeat until firm.)

EXERCISE 2: EXPONENTS
MULTIPLICATION/DIVISION BY BASE 10 REMEDY

a. Open your textbook to Lesson 86 and find part 1. ✔
  (Teacher reference:)

a. 4.6×10^3 b. $\dfrac{4.6}{10^3}$ c. $\dfrac{110.8}{10^2}$

d. 6.02×10^4 e. $\dfrac{285}{10^4}$

Some of these problems divide by 10 with an exponent. Some problems multiply by 10 with an exponent.

b. Read problem A. (Signal.) 4.6×10^3.
- Is the answer greater than or less than 4.6? (Signal.) *Greater than 4.6.*
- Write the answer. ✔
  (Display:)   [86:2A]

> **a.** 4600

Here's what you should have.

c. Read problem B. (Signal.) $4.6 \div 10^3$.
- Is the answer greater than or less than 4.6? (Signal.) *Less than 4.6.*
- Write the answer. ✔
  (Display:)   [86:2B]

> **b.** .0046

Here's what you should have.

d. Write answers to the rest of the problems.
  (Observe students and give feedback.)

e. Check your work.
  (Display:)   [86:2C]

> **c.** 1.108
>
> **d.** 60,200
>
> **e.** .0285

Here's what you should have for problems C through E.

EXERCISE 3: FRACTION WORD PROBLEMS
DIVISION

a. Find part 2 in your textbook. ✔
(Teacher reference:)

a. A farmer has $\frac{8}{3}$ quarts of honey. He wants to put that honey in jars that each hold $\frac{2}{3}$ quart. How many jars will he be able to fill?

b. Another farmer has $\frac{9}{4}$ quarts of honey. He wants to fill jars that each hold $\frac{2}{3}$ quart. How many jars will he fill?

c. Another farmer has 8 quarts of honey. She wants to fill jars that each hold $\frac{2}{3}$ quart. How many jars will she fill?

To work these problems, you'll divide by a fraction.

b. Problem A: A farmer has 8/3 quarts of honey. He wants to put that honey in jars that each hold 2/3 quart. How many jars will he be able to fill?
- Write the problem you'll work. Then stop. ✔
- Everybody, read the problem. (Signal.) *8/3 ÷ 2/3.*
Yes, 8/3 divided by 2/3.
- Work the problem. Write the answer as a number and a unit name.
(Observe students and give feedback.)

c. Check your work.
- What's the whole number in the answer? (Signal.) *4.*
- What's the unit name? (Signal.) *Jars.*
Yes, he could fill 4 jars.

d. Problem B: Another farmer has 9/4 quarts of honey. He wants to fill jars that each hold 2/3 quart. How many jars will he fill?
The answer to this problem is a mixed number. The last jar will not be filled to the top. Only a fraction of that jar will be filled.
- Work the problem. Remember the unit name in the answer.
(Observe students and give feedback.)

e. Check your work.
(Display:) [86:3A]

$$\text{b.} \quad \frac{9}{4} \div \frac{2}{3} = \frac{9}{4} \times \frac{3}{2} = \frac{27}{8} = \boxed{3\frac{3}{8} \text{ jars}}$$

Here's what you should have: 3 and 3/8 jars.
- How many jars will be completely filled? (Signal.) *3.*
- Raise your hand when you know if the fourth jar will be less than half filled or more than half filled. ✔
- Everybody, more or less than half filled? (Signal.) *Less than half filled.*
Yes, 1/2 is 4/8. So 3/8 is less than 1/2.

f. Problem C: Another farmer has 8 quarts of honey. She wants to fill jars that each hold 2/3 quart. How many jars will she fill?
- Write the problem you'll work. Then stop. ✔
- Everybody, read the problem. (Signal.) *8 ÷ 2/3.*
- Work the problem. Remember the unit name in the answer.
(Observe students and give feedback.)

g. Check your work.
- Everybody, how many jars will she fill? (Signal.) *12 jars.*

EXERCISE 4: EXPRESSIONS
EVALUATION   REMEDY

a. Find part 3 in your textbook. ✔
(Teacher reference:)

a. $t(15 \div 3)$ $\boxed{t = 10}$ b. $8\left(r + \frac{3}{5}\right)$ $\boxed{r = \frac{1}{5}}$

c. $k(12 + 5 - 7)$ $\boxed{k = \frac{1}{2}}$ d. $m(15 - 5 - 4)$ $\boxed{m = 20}$

b. Read problem A. (Signal.) *T times the quantity 15 ÷ 3.*
- Copy the problem and work it. Remember, first show the problem with a number in place of the letter. Then work the problem. ✔
- Check your work. You rewrote the problem as 10 times the quantity 15 ÷ 3. Then you multiplied 10 × 5.
- What's the answer? (Signal.) *50.*
(Display:) [86:4A]

$$\text{a.} \quad t(15 \div 3) \qquad \boxed{t = 10}$$
$$10(15 \div 3)$$
$$10(5) = \boxed{50}$$

Here's what you should have.
c. Work the rest of the problems in part 3.
(Observe students and give feedback.)
d. Check your work.
- Problem B. What's the quantity inside the parentheses? (Signal.) *4/5.*
- What did you multiply 4/5 by? (Signal.) *8.*
- What's the fraction answer? (Signal.) *32/5.*
- What's the mixed number? (Signal.) *6 and 2/5.*

e. Problem C. What's the quantity inside the parentheses? (Signal.) *10.*
• What did you multiply 10 by? (Signal.) *1/2.*
• What's the answer? (Signal.) *5.*
f. Problem D. What's the quantity inside the parentheses? (Signal.) *6.*
• What did you multiply 6 by? (Signal.) *20.*
• What's the answer? (Signal.) *120.*

EXERCISE 5: SENTENCES
PARTS OF A GROUP

a. Find part 4 in your textbook. ✔
(Teacher reference:)

a. $\frac{2}{9}$ of the men were sleeping.

b. $\frac{15}{19}$ of the girls were eating.

c. $\frac{3}{16}$ of the computers were used.

d. $\frac{10}{11}$ of the papers were blank.

You're going to write two equations for each sentence.
b. Read sentence A. (Signal.) *2/9 of the men were sleeping.*
• Write the equation for the men who were sleeping and the equation for men who were awake.
(Observe students and give feedback.)
c. Check your work.
• Read the equation for men who were sleeping. (Signal.) *2/9 M = S.*
• Read the equation for men who were awake. (Signal.) *7/9 M = A.*
(Display:) [86:5A]

$$a. \ \frac{2}{9} m = s$$
$$\frac{7}{9} m = a$$

Here's what you should have.

d. Read sentence B. (Signal.) *15/19 of the girls were eating.*
• Write equations for the girls who were eating and the girls who were not eating. ✔
• Everybody, read the equation for girls who were eating. (Signal.) *15/19 G = E.*
• Read the equation for girls who were not eating. (Signal.) *4/19 G = N.*
(Display:) [86:5B]

$$b. \ \frac{15}{19} g = e$$
$$\frac{4}{19} g = n$$

Here's what you should have.
e. Read sentence C. (Signal.) *3/16 of the computers were used.*
• Write the equations for the computers that were used and the computers that were new. ✔
(Display:) [86:5C]

$$c. \ \frac{3}{16} c = u$$
$$\frac{13}{16} c = n$$

Here's what you should have. 3/16 C = U; 13/16 C = N.
f. Read sentence D. (Signal.) *10/11 of the papers were blank.*
• Write the equations for the papers that were blank and the papers that were not blank. ✔
(Display:) [86:5D]

$$d. \ \frac{10}{11} p = b$$
$$\frac{1}{11} p = n$$

Here's what you should have. 10/11 P = B; 1/11 P = N.

EXERCISE 6: AVERAGE
BAR GRAPHS

a. Open your workbook to Lesson 86 and find part 1. ✔
(Teacher reference:)

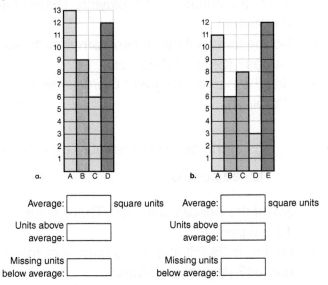

You're going to find the average number of square units in each column.

- What's the number of square units in column A of problem A? (Signal.) *13.*
- Column B? (Signal.) *9.*
- Column C? (Signal.) *6.*
- Column D? (Signal.) *12.*

b. Figure out the average for the columns.
(Observe students and give feedback.)

- You added the numbers for each column. What's the total number? (Signal.) *40.*
- What did you divide by? (Signal.) *4.*
- What's the average number of square units in a column? (Signal.) *10.*
- Draw a line to show the average is 10 and write 10 in the box. ✔

c. Count the number of units that are above the average of 10 and write the number. ✔
- How many units are above the average? (Signal.) *5.*

d. If there are 5 units above the average, there have to be 5 missing units below the average.
- Count the missing units below and write the number. ✔

- How many missing units are below the average? (Signal.) *5.*
There are 5 units above and 5 missing units below. So if you moved all the units that are above the average to empty spaces below the average, all the columns would be the same height. There would be 10 in every column.

e. Problem B. Figure out the average number of square units in each column. Draw a line to show the average for each column and write it in the box. Then write the number of units that are above the average and the number of missing units below the average.
(Observe students and give feedback.)

f. Check your work.
- What's the average number of square units in a column? (Signal.) *8 square units.*
- How many units are above the average? (Signal.) *7.*
- How many units are below the average? (Signal.) *7.*
So if you moved all the units above the average to fill in the spaces below the average, you would fill in every space, and you would have a rectangle that's 8 units high and 5 units wide.
(Display:)   [86:6A]

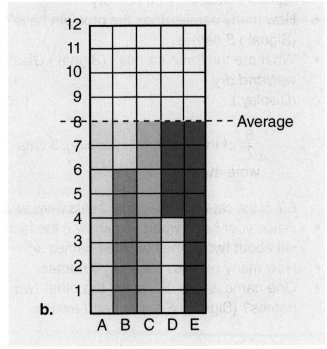

Here's what it would look like.

EXERCISE 7: SENTENCES
PARTS OF A GROUP

a. We'll work with facts. Some facts tell about two names. Some facts tell about three names.
(Display:) [86:7A]

> $\frac{3}{5}$ of the glasses were wet. There were 60 glasses.

3/5 of the glasses were wet. There were 60 glasses.
My turn: The problem has **two** names—glasses and wet.

- How many names does the problem have? (Signal.) *2 names.*
- What are the two names? (Signal.) *Glasses and wet.*

b. (Display:) [86:7B]

> $\frac{3}{5}$ of the glasses were wet. There were 60 dry glasses.

3/5 of the glasses were wet. There were 60 dry glasses.
My turn: The problem has **three** names—glasses, wet, and dry.

- How many names does the problem have? (Signal.) *3 names.*
- What are the three names? (Signal.) *Glasses, wet, and dry.*

c. (Display:) [86:7C]

> $\frac{6}{7}$ of the cats were sleeping. 3 cats were awake.

6/7 of the cats were sleeping. 3 cats were awake.

- Raise your hand when you know if the facts tell about two names or three names. ✔
- How many names? (Signal.) *3 names.*
- One name is cats. What are the other two names? (Signal.) *Sleeping and awake.*

d. (Display:) [86:7D]

> $\frac{4}{5}$ of the children were playing. There were 20 children.

4/5 of the children were playing. There were 20 children.

- Do the facts tell about two names or three names? (Signal.) *2 names.*
- Say the names. (Signal.) *Children and playing.*

e. (Display:) [86:7E]

> $\frac{3}{4}$ of the pencils were red. There were 12 red pencils.

3/4 of the pencils were red. There were 12 red pencils.

- Do the facts tell about two names or three names? (Signal.) *2 names.*
- Say the names. (Signal.) *Pencils and red.*

f. (Display:) [86:7F]

> $\frac{1}{3}$ of the buildings were old. There were 30 new buildings.

1/3 of the buildings were old. There were 30 new buildings.

- Do the facts tell about two names or three names? (Signal.) *3 names.*
- Say the names. (Signal.) *Buildings, old, and new.*
(Repeat until firm.)

WORKBOOK PRACTICE

a. Find part 2 in your workbook. ✔
   (Teacher reference:)

   a. $\frac{3}{5}$ of the glasses were wet. There were 60 glasses. _____

   b. $\frac{3}{5}$ of the glasses were wet. There were 60 dry glasses. _____

   c. $\frac{6}{7}$ of the cats were sleeping. 3 cats were awake. _____

   d. $\frac{4}{5}$ of the children were playing. There were 20 children. _____

   e. $\frac{3}{4}$ of the pencils were red. There were 12 red pencils. _____

   f. $\frac{1}{3}$ of the buildings were old. There were 30 new buildings. _____

   Here are the same facts we just worked with.

b. You're going to underline the different names in each item. Then you'll write the number of different names—2 or 3.

c. Work item A. ✔
   • Everybody, how many different names are there? (Signal.) *2 names.*
   • What are the names? (Signal.) *Glasses and wet.*

d. Work item B. ✔
   • Everybody, how many different names are there? (Signal.) *3 names.*
   • What are the names? (Signal.) *Glasses, wet, and dry.*

e. Work item C. ✔
   • Everybody, how many different names are there? (Signal.) *3 names.*
   • What are the names? (Signal.) *Cats, sleeping, and awake.*

f. Work item D. ✔
   • Everybody, how many different names are there? (Signal.) *2 names.*
   • What are the names? (Signal.) *Children and playing.*

g. Work item E. ✔
   • Everybody, how many different names are there? (Signal.) *2 names.*
   • What are the names? (Signal.) *Pencils and red.*

h. Work item F. ✔
   • Everybody, how many different names are there? (Signal.) *3 names.*
   • What are the names? (Signal.) *Buildings, old, and new.*

EXERCISE 8: MIXED NUMBERS
REWRITING

a. Find part 3 in your workbook. ✔
   (Teacher reference:)

   a. $6\frac{3}{7}$ b. $4\frac{2}{5}$ c. $9\frac{1}{3}$ d. $7\frac{4}{11}$

   You're going to rewrite each mixed number so the fraction is 1 more.
   • Read mixed number A. (Signal.) *6 and 3/7.*
   • How many do you take from 6? (Signal.) *1.*
   • What's the new whole number? (Signal.) *5.*
     And you add 1 to 3/7.

b. Read mixed number B. (Signal.) *4 and 2/5.*
   • How many do you take from 4? (Signal.) *1.*
   • What's the new whole number? (Signal.) *3.*
     And you add 1 to 2/5.

c. Read mixed number C. (Signal.) *9 and 1/3.*
   • How many do you take from 9? (Signal.) *1.*
   • What's the new whole number? (Signal.) *8.*
     And you add 1 to 1/3.

d. Read mixed number D. (Signal.) *7 and 4/11.*
   • How many do you take from 7? (Signal.) *1.*
   • What's the new whole number? (Signal.) *6.*
     And you add 1 to 4/11.

e. Rewrite mixed number A so the fraction is 1 more. ✔
   • Everybody, read the new mixed number. (Signal.) *5 and 10/7.*
   (Display:)   [86:8A]

   a. $6\frac{3}{7}$ → $5\frac{10}{7}$

   Here's what you should have.

f. Rewrite mixed number B so the fraction is 1 more. ✔
   • Everybody, read the new mixed number. (Signal.) *3 and 7/5.*
   (Display:)   [86:8B]

   b. $4\frac{2}{5}$ → $3\frac{7}{5}$

   Here's what you should have.

g. Rewrite mixed number C so the fraction is 1 more. ✔

- Everybody, read the new mixed number. (Signal.) *8 and 4/3.*
(Display:)   [86:8C]

$$\text{c.} \quad \overset{8}{\cancel{9}}\dfrac{\overset{4}{\cancel{4}}}{3}$$

Here's what you should have.

h. Rewrite mixed number D so the fraction is 1 more. ✔

- Everybody, read the new mixed number. (Signal.) *6 and 15/11.*
(Display:)   [86:8D]

$$\text{d.} \quad \overset{6}{\cancel{7}}\dfrac{\overset{15}{\cancel{4}}}{11}$$

Here's what you should have.

EXERCISE 9: INDEPENDENT WORK
SIMPLIFICATION

a. From now on, you won't see an S in a box to remind you to simplify answers. You'll just check all your fraction answers. If they are not simplified, you'll simplify them. Remember, **all** answers are to be simplified.

Assign Independent Work, Workbook part 4 and Textbook parts 5–11.

Optional extra math-fact practice worksheets are available on ConnectED.

Lesson 87

EXERCISE 1: MENTAL MATH

a. Time for mental math.
- Listen: 60 divided by 2. Say the fact. (Signal.)
  60 ÷ 2 = 30.
- 60 ÷ 3. Say the fact. (Signal.) *60 ÷ 3 = 20.*
- 60 ÷ 10. Say the fact. (Signal.) *60 ÷ 10 = 6.*
- 80 ÷ 10. Say the fact. (Signal.) *80 ÷ 10 = 8.*
- 80 ÷ 40. Say the fact. (Signal.) *80 ÷ 40 = 2.*
- 80 ÷ 20. Say the fact. (Signal.) *80 ÷ 20 = 4.*
- (Repeat until firm.)

b. 90 ÷ 3. Say the fact. (Signal.) *90 ÷ 3 = 30.*
- 90 ÷ 10. Say the fact. (Signal.) *90 ÷ 10 = 9.*
- 90 ÷ 30. Say the fact. (Signal.) *90 ÷ 30 = 3.*
- (Repeat until firm.)

c. Listen: 167 minus 10. Say the fact. (Signal.)
  167 − 10 = 157.
- 167 − 30. Say the fact. (Signal.) *167 − 30 = 137.*
- 167 − 40. Say the fact. (Signal.) *167 − 40 = 127.*
- (Repeat until firm.)

d. 283 − 10. Say the fact. (Signal.) *283 − 10 = 273.*
- 283 − 20. Say the fact. (Signal.) *283 − 20 = 263.*
- 283 − 50. Say the fact. (Signal.) *283 − 50 = 233.*
- 283 − 80. Say the fact. (Signal.) *283 − 80 = 203.*
- (Repeat until firm.)

e. Listen: 360 divided by 90. Say the fact.
  (Signal.) *360 ÷ 90 = 4.*
- 270 ÷ 90. Say the fact. (Signal.) *270 ÷ 90 = 3.*
- 180 ÷ 90. Say the fact. (Signal.) *180 ÷ 90 = 2.*
- (Repeat until firm.)

EXERCISE 2: FRACTION MULTIPLICATION
PRODUCT/FACTOR COMPARISON

a. (Display:)   [87:2A]

$$\frac{1}{2} \times \frac{4}{4} = \frac{4}{8}$$

- (Point to $\frac{1}{2}$.) Read the equation. (Signal.)
  1/2 × 4/4 = 4/8.
b. The problem multiplies 1/2 by a fraction that
  equals 1. What fraction? (Signal.) *4/4.*
  You know that when you multiply by a fraction
  that equals **1**, the value you end with **equals**
  the value you start with.

- What fraction do you end with? (Signal.) *4/8.*
  So 4/8 equals 1/2.
  (Add to show:)   [87:2B]

$$\frac{1}{2} \times \frac{4}{4} = \frac{4}{8}$$
$$\frac{1}{2} \times \frac{5}{4} =$$
$$\frac{1}{2} \times \frac{3}{4} =$$

c. Here's another rule about multiplying: If the
  fraction we multiply by is **more** than 1, we'll
  end up with **more** than we start with.
  If the fraction we multiply by is **less** than 1,
  we'll end up with **less** than we start with.
d. (Point to $\frac{1}{2} \times \frac{5}{4}$.) Read the problem. (Signal.)
  1/2 × 5/4.
- What fraction do we multiply 1/2 by?
  (Signal.) *5/4.*
- Is 5/4 more than 1 or less than 1? (Signal.)
  More than 1.
- So do we end up with more than 1/2 or less
  than 1/2? (Signal.) *More than 1/2.*
e. (Point to $\frac{1}{2} \times \frac{3}{4}$.) Read the problem. (Signal.)
  1/2 × 3/4.
- What fraction do we multiply 1/2 by?
  (Signal.) *3/4.*
- Is 3/4 more than 1 or less than 1? (Signal.)
  Less than 1.
- So do we end up with more than 1/2 or less
  than 1/2? (Signal.) *Less than 1/2.*
f. (Display:)   [87:2C]

a. $\frac{2}{3} \times \frac{9}{10}$

b. $\frac{2}{3} \times \frac{10}{9}$

c. $\frac{2}{3} \times 6$

d. $\frac{2}{3} \times \frac{4}{5}$

Here are some problems that have 2/3 as the
first fraction.

g. (Point to **A.**) Read the problem. (Signal.)
   2/3 × 9/10.
- What do you multiply 2/3 by? (Signal.) *9/10.*
- Is 9/10 more than 1 or less than 1? (Signal.)
  Less than 1.
- So do you end up with more than 2/3 or less
  than 2/3? (Signal.) *Less than 2/3.*
h. (Point to **B.**) Read the problem. (Signal.)
   2/3 × 10/9.
- What do you multiply 2/3 by? (Signal.) *10/9.*
- Is 10/9 more than 1 or less than 1? (Signal.)
  More than 1.
- So do you end up with more than 2/3 or less
  than 2/3? (Signal.) *More than 2/3.*
i. (Point to **C.**) Read the problem. (Signal.)
   2/3 × 6.
- What do you multiply 2/3 by? (Signal.) *6.*
- Is 6 more than 1 or less than 1? (Signal.)
  More than 1.
- So do you end up with more than 2/3 or less
  than 2/3? (Signal.) *More than 2/3.*
j. (Point to **D.**) Read the problem. (Signal.)
   2/3 × 4/5.
- What do you multiply 2/3 by? (Signal.) *4/5.*
- Is 4/5 more than 1 or less than 1? (Signal.)
  Less than 1.
- So do you end up with more than 2/3 or less
  than 2/3? (Signal.) *Less than 2/3.*
k. Remember, if you multiply by **more than 1,**
   you end up with **more** than you start with.
- If you multiply by **less than 1,** what do you end
  up with? (Signal.) *Less than you start with.*

EXERCISE 3: SENTENCES
PARTS OF A GROUP

a. We'll work with facts for different problems.
   Some facts tell about two names. Some facts
   tell about three names.
   (Display:) [87:3A]

   $\frac{2}{7}$ of the workers were eating. 42 workers
   were not eating.

   2/7 of the workers were eating. 42 workers
   were not eating.

- Raise your hand when you know if the facts
  tell about two names or three names. ✔
- How many names do the facts tell about?
  (Signal.) *3 names.*
- What are the three names? (Signal.) *Workers,
  eating, and not eating.*
b. (Display:) [87:3B]

   $\frac{2}{3}$ of the glasses were empty. There were
   60 full glasses.

   2/3 of the glasses were empty. There were
   60 full glasses.

- Raise your hand when you know if the facts
  tell about two names or three names. ✔
- How many names do the facts tell about?
  (Signal.) *3 names.*
- What are the three names? (Signal.) *Glasses,
  empty, and full.*
c. (Display:) [87:3C]

   $\frac{1}{3}$ of the books were open. 12 books
   were open.

   1/3 of the books were open. 12 books were open.

- Raise your hand when you know if the facts
  tell about two names or three names. ✔
- How many names do the facts tell about?
  (Signal.) *2 names.*
- What are the two names? (Signal.) *Books
  and open.*
d. (Display:) [87:3D]

   $\frac{3}{5}$ of the girls were playing. There were
   21 girls.

   3/5 of the girls were playing. There were
   21 girls.

- Raise your hand when you know if the facts
  tell about two names or three names. ✔
- How many names do the facts tell about?
  (Signal.) *2 names.*
- What are the two names? (Signal.) *Girls
  and playing.*

a. Open your workbook to Lesson 87 and find part 1. ✔
(Teacher reference:)

a. $\frac{2}{7}$ of the workers were eating. 42 workers were not eating. _____

b. $\frac{2}{3}$ of the glasses were empty. There were 60 full glasses. _____

c. $\frac{1}{3}$ of the books were open. 12 books were open. _____

d. $\frac{3}{5}$ of the girls were playing. There were 21 girls. _____

e. $\frac{3}{10}$ of the pencils were short. There were 14 long pencils. _____

f. $\frac{8}{11}$ of the buildings were old. There were 30 old buildings. _____

Here are the same facts we just worked with. You're going to underline the different names in each item. Then you'll write the number of names—2 or 3.

b. Work item A. ✔
• Everybody, how many names are there? (Signal.) *3 (names).*
• What are the names? (Signal.) *Workers, eating, and not eating.*

c. Work item B. ✔
• Everybody, how many names are there? (Signal.) *3.*
• What are the names? (Signal.) *Glasses, empty, and full.*

d. Work item C. ✔
• Everybody, how many names are there? (Signal.) *2.*
• What are the names? (Signal.) *Books and open.*

e. Work item D. ✔
• Everybody, how many names are there? (Signal.) *2.*
• What are the names? (Signal.) *Girls and playing.*

f. Work item E. ✔
• Everybody, how many names are there? (Signal.) *3.*
• What are the names? (Signal.) *Pencils, short, and long.*

g. Work item F. ✔
• Everybody, how many names are there? (Signal.) *2.*
• What are the names? (Signal.) *Buildings and old.*

EXERCISE 4: MIXED NUMBERS
REWRITING

a. Find part 2 in your workbook. ✔
(Teacher reference:)

a. $8\frac{2}{4}$ b. $2\frac{3}{7}$ c. $6\frac{1}{8}$ d. $4\frac{5}{9}$

You're going to rewrite each mixed number so the fraction is 1 more.
• Read mixed number A. (Signal.) *8 and 2/4.*
• Take one from the 8 and add it to 2/4. Raise your hand when you've written the new mixed number. ✔
• Everybody, read the new mixed number. (Signal.) *7 and 6/4.*
(Display:)   [87:4A]

$$\text{a. } 8\!\!\!\!\diagdown\,^{7}\frac{\diagup\!\!\!\!2\,^{6}}{4}$$

Here's what you should have.

b. Rewrite mixed number B. ✔
• Everybody, read the new mixed number. (Signal.) *1 and 10/7.*
(Display:)   [87:4B]

$$\text{b. } 2\!\!\!\!\diagdown\,^{1}\frac{\diagup\!\!\!\!3\,^{10}}{7}$$

Here's what you should have.

c. Rewrite mixed number C. ✔
• Everybody, read the new mixed number. (Signal.) *5 and 9/8.*
(Display:)   [87:4C]

$$\text{c. } 6\!\!\!\!\diagdown\,^{5}\frac{\diagup\!\!\!\!1\,^{9}}{8}$$

Here's what you should have.

d. Rewrite mixed number D. ✔
• Everybody, read the new mixed number. (Signal.) *3 and 14/9.*
(Display:)   [87:4D]

$$\text{d. } 4\!\!\!\!\diagdown\,^{3}\frac{\diagup\!\!\!\!5\,^{14}}{9}$$

Here's what you should have.

e. Remember, the new mixed number equals the mixed number you started with.

EXERCISE 5: AVERAGE
WITH ZEROS

a. (Display:)   [87:5A]

Jim	Fran	Alex	Liz	Ann
2	3	0	6	4

Trips to a Lake or Ocean Last Year

This table shows the number of trips people took to a lake or ocean last year.
* How many people are there? (Signal.) 5.
* So what number will you divide by? (Signal.) 5. Yes, 5. It doesn't matter that one of the numbers is zero.
* Raise your hand when you know the average for this group. ✔
* Everybody, what's the average? (Signal.) 3.

TEXTBOOK PRACTICE

a. Open your textbook to Lesson 87 and find part 1. ✔
(Teacher reference:)

Number of Chin-ups		
	Last Month	This Month
Jim	4	4
Fran	0	1
Alex	0	0
Liz	8	7
Paul	18	13

The chart shows the number of chin-ups each person could do last month and this month.
* Read the numbers for last month. (Signal.) *4, zero, zero, 8, 18.*
* Figure out the average for last month. (Observe students and give feedback.)

b. Check your work.
* What's the total for the group? (Signal.) *30.*
* What number did you divide by? (Signal.) *5.*
* What's the average number for last month? (Signal.) *6.*

c. Figure out the average number of chin-ups for this month. (Observe students and give feedback.)

d. Check your work.
* What's the total for the group? (Signal.) *25.*
* What number did you divide by? (Signal.) *5.*
* What's the average number for this month? (Signal.) *5.*

e. Raise your hand when you know which person improved the most from last month. ✔
* Which person? (Signal.) *Fran.*
* Raise your hand when you know which person dropped the most from last month. ✔
* Which person? (Signal.) *Paul.*

EXERCISE 6: FRACTION WORD PROBLEMS
DIVISION

a. Find part 2 in your textbook. ✔
(Teacher reference:)

a. A store owner has $\frac{9}{4}$ pounds of syrup. She pours the syrup into jars that each hold $\frac{3}{8}$ pound. How many jars can she fill?

b. Fran has 26 pounds of walnuts. She wants to put them in bags that hold $\frac{3}{4}$ pound. How many bags can she fill?

c. A tank held 35 gallons of gasoline. The gasoline was poured into containers that each held $\frac{2}{3}$ gallon. How many containers were filled?

To work these problems, you'll divide by a fraction.

b. Problem A: A store owner has 9/4 pounds of syrup. She pours the syrup into jars that each hold 3/8 pound. How many jars can she fill?
* Write the problem you'll work. Then stop. ✔
* Everybody, read the problem. (Signal.) *9/4 ÷ 3/8.*
Yes, 9/4 divided by 3/8.
* Work the problem. Write the answer as a number and a unit name. (Observe students and give feedback.)

c. Check your work.
* What's the whole number in the answer? (Signal.) *6.*
* What's the unit name? (Signal.) *Jars.*
Yes, she can fill 6 jars with syrup.

d. Problem B: Fran has 26 pounds of walnuts. She wants to put them in bags that each hold 3/4 pound. How many bags can she fill?
* Work the problem. Show your answer as a whole number or mixed number with a unit name. (Observe students and give feedback.)

e. Check your work. You worked the problem 26 divided by 3/4. That's 26 times 4/3.
- How many bags can she fill? (Signal.) *34 and 2/3 bags.*
f. Problem C: A tank held 35 gallons of gasoline. The gasoline was poured into containers that each held 2/3 gallon. How many containers were filled?
- Work the problem.
  (Observe students and give feedback.)
g. Check your work. You worked the problem 35 divided by 2/3. That's 35 times 3/2.
- Everybody, what's the answer? (Signal.) *52 and 1/2 containers.*

EXERCISE 7: EXPRESSIONS
FROM WRITTEN DIRECTIONS

a. I'm going to show you directions for writing problems. All the problems will have values inside the parentheses and a number you multiply by outside the parentheses.
  (Display:)   [87:7A]

> Show the quantity 4 + 5 + 1.
> Then multiply by 9.

Listen: We'll show the quantity 4 + 5 + 1. Then we'll multiply by 9.
- I'll show the quantity with parentheses. What goes inside the parentheses? (Signal.) *4 + 5 + 1.*
  (Add to show:)   [87:7B]

> Show the quantity 4 + 5 + 1.
> Then multiply by 9.
> $$(4 + 5 + 1)$$

- What do I show next? (Signal.) *9.*
- Where do I write 9? (Signal.) *Outside the parentheses.*
  (Add to show:)   [87:7C]

> Show the quantity 4 + 5 + 1.
> Then multiply by 9.
> $$9(4 + 5 + 1)$$

- Read the problem we wrote for the directions. (Signal.) *9 times the quantity 4 + 5 + 1.*

b. (Display:)   [87:7D]

> Show the quantity 70 ÷ 7.
> Then multiply by 4.

Show the quantity 70 ÷ 7. Then multiply by 4.
- What do I write inside the parentheses? (Signal.) *70 ÷ 7.*
- What do I multiply by? (Signal.) *4.*
  (Add to show:)   [87:7E]

> Show the quantity 70 ÷ 7.
> Then multiply by 4.
> $$4(70 ÷ 7)$$

- Read the problem we wrote for the directions. (Signal.) *4 times the quantity 70 ÷ 7.*
c. (Display:)   [87:7F]

> Show the quantity 16 – 2 + 4.
> Then multiply by 3.

Show the quantity 16 – 2 + 4. Then multiply by 3.
- What do I write inside the parentheses? (Signal.) *16 – 2 + 4.*
- What do I multiply by? (Signal.) *3.*
  (Add to show:)   [87:7G]

> Show the quantity 16 – 2 + 4.
> Then multiply by 3.
> $$3(16 – 2 + 4)$$

- Read the problem we wrote for the directions. (Signal.) *3 times the quantity 16 – 2 + 4.*
d. You're going to write problems in part A.
  (Display:)   [87:7H]

> a. Show the quantity 36 – 15.
>   Then multiply by $\frac{1}{3}$.

Problem A: Show the quantity 36 – 15. Then multiply by 1/3.
- Write the problem in part A. ✔
- Everybody, read the problem you wrote. (Signal.) *1/3 times the quantity 36 – 15.*
  (Add to show:)   [87:7I]

> a. Show the quantity 36 – 15.
>   Then multiply by $\frac{1}{3}$.
> $$\frac{1}{3}(36 – 15)$$

Here's what you should have.

e. (Display:) [87:7J]

> **b.** Show the quantity 15 ÷ 3.
> Then multiply by $\frac{1}{3}$.

Problem B: Show the quantity 15 ÷ 3. Then multiply by 1/3.

• Write the problem. ✔
• Everybody, read the problem you wrote. (Signal.) *1/3 times the quantity 15 ÷ 3.* (Add to show:) [87:7K]

> **b.** Show the quantity 15 ÷ 3.
> Then multiply by $\frac{1}{3}$.
> $$\frac{1}{3}(15 \div 3)$$

Here's what you should have.

f. (Display:) [87:7L]

> **c.** Show the quantity 30 + 50 + 10.
> Then multiply by $\frac{4}{5}$.

Problem C: Show the quantity 30 + 50 + 10. Then multiply by 4/5.

• Write the problem. ✔
• Everybody, read the problem you wrote. (Signal.) *4/5 times the quantity 30 + 50 + 10.* (Add to show:) [87:7M]

> **c.** Show the quantity 30 + 50 + 10.
> Then multiply by $\frac{4}{5}$.
> $$\frac{4}{5}(30 + 50 + 10)$$

Here's what you should have.

g. (Display:) [87:7N]

> **d.** Show the quantity 11 + 89.
> Then multiply by 10.

Problem D: Show the quantity 11 + 89. Then multiply by 10.

• Write the problem. ✔
• Everybody, read the problem you wrote. (Signal.) *10 times the quantity 11 + 89.* (Add to show:) [87:7O]

> **d.** Show the quantity 11 + 89.
> Then multiply by 10.
> $$10(11 + 89)$$

Here's what you should have.

EXERCISE 8: SENTENCES
PARTS OF A GROUP

> REMEDY

a. Find part 3 in your textbook. ✔
(Teacher reference:)

> a. $\frac{3}{4}$ of the fish were alive.
> • Write the equation for fish that were alive and the equation for fish that were dead.
> b. $\frac{7}{8}$ of the boxes are upstairs.
> • Write the equation for the boxes upstairs and the equation for the boxes downstairs.
> c. $\frac{2}{3}$ of the papers are white.
> • Write the equation for the papers that are white and the equation for the papers that are not white.
> d. $\frac{5}{9}$ of the animals were hungry.
> • Write the equation for the animals that were hungry and the equation for the animals that were not hungry.

For each item, you'll write two equations.

b. Read sentence A. (Signal.) *3/4 of the fish were alive.*
• Write the equation for the fish that were alive and the equation for the fish that were dead. (Observe students and give feedback.) (Display:) [87:8A]

> a. $\frac{3}{4} f = a$
> $\frac{1}{4} f = d$

Here's what you should have: 3/4 F = A; 1/4 F = D.

c. Write two equations for the rest of the items in part 3.
(Observe students and give feedback.)
d. Check your work.
• Sentence B: 7/8 of the boxes are upstairs.
• Read the equation for the boxes upstairs. (Signal.) *7/8 B = U.*
• Read the equation for the boxes downstairs. (Signal.) *1/8 B = D.*
e. Sentence C: 2/3 of the papers are white.
• Read the equation for the papers that are white. (Signal.) *2/3 P = W.*
• Read the equation for the papers that are not white. (Signal.) *1/3 P = N.*
f. Sentence D: 5/9 of the animals were hungry.
• Read the equation for the animals that were hungry. (Signal.) *5/9 A = H.*
• Read the equation for the animals that were not hungry. (Signal.) *4/9 A = N.*

a. Find part 4 in your textbook. ✔
   (Teacher reference:)

a. 4006 − 1247   b. 4818 + 3919   c. 8.02 × .7
d. 3.99 × .05   e. 35 − 24.99

Some of these problems multiply. Remember, when you multiply, you add the decimal places in the numbers you multiply.

b. Touch problem C. ✔
• Raise your hand when you know how many decimal places are in the numbers you multiply. ✔
• How many decimal places? (Signal.) *3 places.*
• So how many decimal places will be in the answer? (Signal.) *3 places.*

c. Touch problem D. ✔
• Raise your hand when you know how many decimal places are in the numbers you multiply. ✔
• How many decimal places? (Signal.) *4 places.*
• So how many decimal places will be in the answer? (Signal.) *4 places.*
   (Repeat until firm.)

d. Find part 5 in your textbook. ✔
   (Teacher reference:)

| Number of People Visiting a Campground Each Day | | | | | Part 5 a. |
Sunday	Monday	Tuesday	Wednesday	Thursday	Friday
30	44	70	11	99	100

a. What is the average number of people that visited the campground each day?
b. How many days are above the average?
c. How many days are below the average?

This table shows the number of **people** visiting a campground each day from Sunday through Friday.

As part of your independent work, you'll figure out the average number of people who visited the campground each day.

You'll also write the number of days that are above the average and the number below the average.

Assign Independent Work, Textbook parts 4–12 and Workbook part 3.

Optional extra math-fact practice worksheets are available on ConnectED.

Lesson

EXERCISE 1: MENTAL MATH

a. Time for mental math. You'll say the whole fact.
- Listen: 294 minus 10. Say the fact. (Signal.) *294 − 10 = 284.*
- 294 − 30. Say the fact. (Signal.) *294 − 30 = 264.*
- 294 − 60. Say the fact. (Signal.) *294 − 60 = 234.* (Repeat until firm.)

b. Listen: 450 divided by 5. Say the fact. (Signal.) *450 ÷ 5 = 90.*
- 350 ÷ 5. Say the fact. (Signal.) *350 ÷ 5 = 70.*
- 200 ÷ 5. Say the fact. (Signal.) *200 ÷ 5 = 40.*
- 100 ÷ 5. Say the fact. (Signal.) *100 ÷ 5 = 20.* (Repeat until firm.)

c. Listen: 30 times 2. Say the fact. (Signal.) *30 × 2 = 60.*
- 30 × 4. Say the fact. (Signal.) *30 × 4 = 120.*
- 30 × 7. Say the fact. (Signal.) *30 × 7 = 210.*
- 30 × 8. Say the fact. (Signal.) *30 × 8 = 240.*
- 30 × 10. Say the fact. (Signal.) *30 × 10 = 300.* (Repeat until firm.)

d. Tell me the whole number each fraction equals.
- Listen: 20 fourths. What whole number? (Signal.) *5.*
- 20 fifths. What whole number? (Signal.) *4.*
- 16 halves. What whole number? (Signal.) *8.*
- 16 fourths. What whole number? (Signal.) *4.*
- 12 fourths. What whole number? (Signal.) *3.*
- 54 ninths. What whole number? (Signal.) *6.*
- 63 ninths. What whole number? (Signal.) *7.*
- 90 ninths. What whole number? (Signal.) *10.* (Repeat until firm.)

EXERCISE 2: FRACTION MULTIPLICATION
PRODUCT/FACTOR COMPARISON

a. Last time, you learned a rule about multiplying by fractions. If the fraction we multiply by is less than 1, we end up with less than we start with.
- If we multiply by more than 1, what do we end up with? (Signal.) *More than we start with.*

b. (Display:) [88:2A]

$$\text{a. } \frac{5}{2} \times \frac{2}{3}$$
$$\text{b. } \frac{5}{2} \times \frac{2}{1}$$
$$\text{c. } \frac{5}{2} \times \frac{4}{5}$$

All these problems start with 5/2.

c. (Point to **A**.) Read the problem. (Signal.) *5/2 × 2/3.*
- Do you multiply by a fraction that is more than 1 or less than 1? (Signal.) *Less than 1.*
- So do you end up with more than 5/2 or less than 5/2? (Signal.) *Less than 5/2.*

d. (Point to **B**.) Read the problem. (Signal.) *5/2 × 2 over 1.*
- Do you multiply by a fraction that is more than 1 or less than 1? (Signal.) *More than 1.*
- So do you end up with more than 5/2 or less than 5/2? (Signal.) *More than 5/2.*

e. (Point to **C**.) Read the problem. (Signal.) *5/2 × 4/5.*
- Do you multiply by a fraction that is more than 1 or less than 1? (Signal.) *Less than 1.*
- So do you end up with more than 5/2 or less than 5/2? (Signal.) *Less than 5/2.*

f. In part A, copy problem A and work it. Don't simplify. Just multiply. ✔
- Read the equation for problem A. (Signal.) *5/2 × 2/3 = 10/6.* (Add to show:) [88:2B]

$$\text{a. } \frac{5}{2} \times \frac{2}{3} = \boxed{\frac{10}{6}}$$
$$\text{b. } \frac{5}{2} \times \frac{2}{1}$$
$$\text{c. } \frac{5}{2} \times \frac{4}{5}$$

- The problem starts with 5/2. Do you multiply by a fraction that is more than 1 or less than 1? (Signal.) *Less than 1.*
- So is 10/6 more than 5/2 or less than 5/2? (Signal.) *Less than 5/2.*

g. Copy problem B and do the multiplication. ✔
• Read the equation for problem B. (Signal.)
$5/2 \times 2/1 = 10/2$.
(Add to show:) [88:2C]

$$\textbf{a. } \frac{5}{2} \times \frac{2}{3} = \boxed{\frac{10}{6}}$$

$$\textbf{b. } \frac{5}{2} \times \frac{2}{1} = \boxed{\frac{10}{2}}$$

$$\textbf{c. } \frac{5}{2} \times \frac{4}{5}$$

• The problem starts with 5/2. Do you multiply by a fraction that is more than 1 or less than 1? (Signal.) *More than 1.*
• So is 10/2 more than 5/2 or less than 5/2? (Signal.) *More than 5/2.*
h. Copy problem C and do the multiplication. ✔
• Read the equation for problem C. (Signal.)
$5/2 \times 4/5 = 20/10$.
(Add to show:) [88:2D]

$$\textbf{a. } \frac{5}{2} \times \frac{2}{3} = \boxed{\frac{10}{6}}$$

$$\textbf{b. } \frac{5}{2} \times \frac{2}{1} = \boxed{\frac{10}{2}}$$

$$\textbf{c. } \frac{5}{2} \times \frac{4}{5} = \boxed{\frac{20}{10}}$$

• The problem starts with 5/2. Do you multiply by a fraction that is more than 1 or less than 1? (Signal.) *Less than 1.*
• So is 20/10 more than 5/2 or less than 5/2? (Signal.) *Less than 5/2.*

EXERCISE 3: MIXED-NUMBER OPERATIONS
SUBTRACTION
[REMEDY]

a. We're going to work subtraction problems by rewriting a mixed number.
(Display:) [88:3A]

$$\textbf{a. } \begin{array}{r} 5\frac{3}{8} \\ -\ 2\frac{6}{8} \\ \hline \end{array}$$

• Read the problem. (Signal.) *5 and 3/8 – 2 and 6/8.*

• Read the problem for the fractions. (Signal.)
3/8 – 6/8.
• Can we work that problem? (Signal.) *No.*
So we rewrite 5 and 3/8. We take one from the 5 and add it to 3/8. The whole number we start with is 5.
• What's the new whole number? (Signal.) *4.*
(Add to show:) [88:3B]

$$\textbf{a. } \begin{array}{r} \overset{4}{\cancel{5}}\frac{3}{8} \\ -\ 2\frac{6}{8} \\ \hline \end{array}$$

• Raise your hand when you know the new fraction. ✔
• What's the new fraction? (Signal.) *11/8.*
(Add to show:) [88:3C]

$$\textbf{a. } \begin{array}{r} \overset{4}{\cancel{5}}\overset{11}{\cancel{\frac{3}{8}}} \\ -\ 2\frac{6}{8} \\ \hline \end{array}$$

The new problem for the fractions is 11/8 – 6/8.
• Say the new problem for the fractions. (Signal.) *11/8 – 6/8.*
• Can we work that problem? (Signal.) *Yes.*
b. (Add to show:) [88:3D]

$$\textbf{a. } \begin{array}{r} \overset{4}{\cancel{5}}\overset{11}{\cancel{\frac{3}{8}}} \\ -\ 2\frac{6}{8} \\ \hline \end{array} \qquad \textbf{b. } \begin{array}{r} 7\frac{1}{4} \\ -\ 5\frac{2}{4} \\ \hline \end{array}$$

• New problem. Read it. (Signal.) *7 and 1/4 – 5 and 2/4.*
• Read the problem for the fractions. (Signal.) *1/4 – 2/4.*
• Can we work that problem? (Signal.) *No.*
So we rewrite 7 and 1/4.
• What's the new whole number? (Signal.) *6.*
• Raise your hand when you know the new fraction. ✔
• What's the new fraction? (Signal.) *5/4.*
(Repeat until firm.)

(Add to show:)   [88:3E]

a. $5\overset{4}{\cancel{5}}\dfrac{\overset{11}{\cancel{3}}}{8}$ b. $\cancel{7}\dfrac{\overset{5}{\cancel{1}}}{4}$

   $-2\dfrac{6}{8}$ $-5\dfrac{2}{4}$

- Say the new problem for the fractions. (Signal.) *5/4 – 2/4.*
- Can we work that problem? (Signal.) *Yes.*

c. (Add to show:)   [88:3F]

a. $5\overset{4}{\cancel{5}}\dfrac{\overset{11}{\cancel{3}}}{8}$ b. $\cancel{7}\dfrac{\overset{5}{\cancel{1}}}{4}$ c. $9\dfrac{2}{5}$

   $-2\dfrac{6}{8}$ $-5\dfrac{2}{4}$ $-6\dfrac{3}{5}$

- New problem. Read it. (Signal.) *9 and 2/5 – 6 and 3/5.*
- Read the problem for the fractions. (Signal.) *2/5 – 3/5.*
- Can we work that problem? (Signal.) *No.* So we rewrite 9 and 2/5.
- What's the new whole number? (Signal.) *8.*
- Raise your hand when you can say the new fraction. ✔
- What's the new fraction? (Signal.) *7/5.* (Repeat until firm.) (Add to show:)   [88:3G]

a. $5\overset{4}{\cancel{5}}\dfrac{\overset{11}{\cancel{3}}}{8}$ b. $\cancel{7}\dfrac{\overset{5}{\cancel{1}}}{4}$ c. $\overset{8}{\cancel{9}}\dfrac{\overset{7}{\cancel{2}}}{5}$

   $-2\dfrac{6}{8}$ $-5\dfrac{2}{4}$ $-6\dfrac{3}{5}$

- Say the new problem for the fractions. (Signal.) *7/5 – 3/5.*
- Can we work that problem? (Signal.) *Yes.*

d. Now we'll work the problems.
- (Point to **A.**) Say the new problem for the fractions. (Signal.) *11/8 – 6/8.*
- What's the answer? (Signal.) *5/8.*
- Say the new problem for the whole numbers. (Signal.) *4 – 2.*
- What's the answer? (Signal.) *2.*

(Add to show:)   [88:3H]

a. $5\overset{4}{\cancel{5}}\dfrac{\overset{11}{\cancel{3}}}{8}$ b. $\cancel{7}\dfrac{\overset{5}{\cancel{1}}}{4}$ c. $\overset{8}{\cancel{9}}\dfrac{\overset{7}{\cancel{2}}}{5}$

   $-2\dfrac{6}{8}$ $-5\dfrac{2}{4}$ $-6\dfrac{3}{5}$

   $\boxed{2\dfrac{5}{8}}$

- Say the problem we started with and the answer. (Signal.) *5 and 3/8 – 2 and 6/8 = 2 and 5/8.*
- e. (Point to **B.**) Say the new problem for the fractions. (Signal.) *5/4 – 2/4.*
- What's the answer? (Signal.) *3/4.*
- Say the new problem for the whole numbers. (Signal.) *6 – 5.*
- What's the answer? (Signal.) *1.* (Add to show:)   [88:3I]

a. $5\overset{4}{\cancel{5}}\dfrac{\overset{11}{\cancel{3}}}{8}$ b. $\cancel{7}\dfrac{\overset{5}{\cancel{1}}}{4}$ c. $\overset{8}{\cancel{9}}\dfrac{\overset{7}{\cancel{2}}}{5}$

   $-2\dfrac{6}{8}$ $-5\dfrac{2}{4}$ $-6\dfrac{3}{5}$

   $\boxed{2\dfrac{5}{8}}$ $\boxed{1\dfrac{3}{4}}$

- Say the problem we started with and the answer. (Signal.) *7 and 1/4 – 5 and 2/4 = 1 and 3/4.*
- f. (Point to **C.**) Say the new problem for the fractions. (Signal.) *7/5 – 3/5.*
- What's the answer? (Signal.) *4/5.*
- Say the new problem for the whole numbers. (Signal.) *8 – 6.*
- What's the answer? (Signal.) *2.* (Add to show:)   [88:3J]

a. $5\overset{4}{\cancel{5}}\dfrac{\overset{11}{\cancel{3}}}{8}$ b. $\cancel{7}\dfrac{\overset{5}{\cancel{1}}}{4}$ c. $\overset{8}{\cancel{9}}\dfrac{\overset{7}{\cancel{2}}}{5}$

   $-2\dfrac{6}{8}$ $-5\dfrac{2}{4}$ $-6\dfrac{3}{5}$

   $\boxed{2\dfrac{5}{8}}$ $\boxed{1\dfrac{3}{4}}$ $\boxed{2\dfrac{4}{5}}$

- Say the problem we started with and the answer. (Signal.) *9 and 2/5 – 6 and 3/5 = 2 and 4/5.*

EXERCISE 4: EXPRESSIONS
FROM WRITTEN DIRECTIONS   `REMEDY`

a. Open your textbook to Lesson 88 and find
   part 1. ✔
   (Teacher reference:)

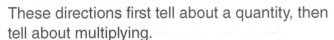

   a. Show the quantity 300 ÷ 6. Then multiply by 7.
   b. Show the quantity 27 − 7 − 10. Then multiply by 8.
   c. Show the quantity 65 − 40. Then multiply by 4.
   d. Show the quantity 80 ÷ 2. Then multiply by $\frac{1}{2}$.
   e. Show the quantity 7 + 3 − 5. Then multiply by 30.

   These directions first tell about a quantity, then
   tell about multiplying.

b. Problem A: Show the quantity 300 ÷ 6. Then
   show the number you multiply by in front.
   That's 7.
- Write the problem. Then stop. ✔
   (Display:)   [88:4A]

   $$\text{a. } 7(300 \div 6)$$

   You showed 7 times the quantity 300 ÷ 6.
- Work the problem and write the answer. ✔
- Everybody, what's 7 × 50? (Signal.) *350.*

c. Write problem B. Then stop. ✔
   (Display:)   [88:4B]

   $$\text{b. } 8(27 - 7 - 10)$$

   You showed 8 times the quantity 27 − 7 − 10.
- Work the problem and write the answer. ✔
   You multiplied 8 times the quantity 27 − 7 − 10.
- Everybody, what's the answer? (Signal.) *80.*

d. Work the rest of the problems in part 1.
   (Observe students and give feedback.)

e. Check your work.
- Problem C. You multiplied 4 times the quantity
   65 − 40.
- What's the answer? (Signal.) *100.*

f. D. You multiplied 1/2 times the quantity 80 ÷ 2.
- What's the answer? (Signal.) *20.*

g. E. You multiplied 30 times the quantity 7 + 3 − 5.
- What's the answer? (Signal.) *150.*

EXERCISE 5: FRACTION WORD PROBLEMS
DIVISION

a. Find part 2 in your textbook. ✔
   (Teacher reference:)

   a. There are $\frac{28}{3}$ gallons of paint. The paint is divided
   equally into 4 containers. How many gallons are in each
   container?

   b. There are $\frac{28}{3}$ gallons of paint. The paint is poured
   into containers that each hold $\frac{4}{3}$ gallons. How many
   containers will be filled?

   c. $\frac{7}{4}$ pounds of shrimp make 4 servings. How much does
   each serving weigh?

   d. $\frac{8}{3}$ pounds of shrimp is divided into servings that each
   weigh $\frac{2}{5}$ pound. How many servings can be made?

   All these problems divide. For one type of
   problem, you divide by a whole number. For
   the other type, you divide by a fraction.

b. Problem A: There are 28/3 gallons of paint.
   The paint is divided equally into 4 containers.
   How many gallons are in each container?
- Raise your hand when you know the problem
   you'll write. ✔
- What problem will you write? (Signal.) *28/3 ÷ 4.*
   Yes, you'll divide by a whole number.

c. Problem B: There are 28/3 gallons of paint.
   The paint is poured into containers that each
   hold 4/3 gallons. How many containers will be
   filled?
- Raise your hand when you know the problem
   you'll write. ✔
- What problem will you write? (Signal.) *28/3 ÷ 4/3.*

d. Work problem A. Remember to simplify and
   write your answer as a mixed number and a
   unit name.
   (Observe students and give feedback.)

e. Check your work.
- Problem A: 28/3 ÷ 4.
- You worked the problem 28/3 × 1/4. How
   many gallons are in each container? (Signal.)
   2 and 1/3 gallons.
   Yes, 2 and 1/3 gallons.
   (Display:)   [88:5A]

   $$\text{a. } \frac{28}{3} \div 4 = \frac{\overset{7}{\cancel{28}}}{3} \times \frac{1}{\cancel{4}} = \frac{7}{3} = \boxed{2\frac{1}{3} \text{ gallons}}$$

   Here's what you should have.

f. Work problem B. Remember to simplify.
(Observe students and give feedback.)

g. Check your work.

• Problem B: 28/3 ÷ 4/3.

• You worked the problem 28/3 × 3/4. How many containers will be filled? (Signal.) *7 containers.*
Yes, 7 containers.
(Display:) [88:5B]

b. $\dfrac{28}{3} \div \dfrac{4}{3} = \dfrac{\overset{7}{\cancel{28}}}{\cancel{3}} \times \dfrac{\cancel{3}^{\,1}}{\cancel{4}} = \boxed{7 \text{ containers}}$

Here's what you should have.

h. Work problem C.
(Observe students and give feedback.)

i. Check your work.

• Problem C: 7/4 ÷ 4. You don't have to simplify.

• You worked the problem 7/4 × 1/4. How much does each serving weigh? (Signal.) *7/16 pound.*
Yes, 7/16 pound.
(Display:) [88:5C]

c. $\dfrac{7}{4} \div 4 = \dfrac{7}{4} \times \dfrac{1}{4} = \boxed{\dfrac{7}{16} \text{ pound}}$

Here's what you should have.

j. Work problem D. Remember to simplify.
(Observe students and give feedback.)

k. Check your work.

• Problem D: 8/3 ÷ 2/5.

• You worked the problem 8/3 × 5/2. How many servings can be made? (Signal.) *6 and 2/3 servings.*

(Display:) [88:5D]

d. $\dfrac{8}{3} \div \dfrac{2}{5} = \dfrac{\overset{4}{\cancel{8}}}{3} \times \dfrac{5}{\cancel{2}} = \dfrac{20}{3} = \boxed{6\dfrac{2}{3} \text{ servings}}$

Here's what you should have.

EXERCISE 6: AVERAGE

DISTANCE

a. (Display:) [88:6A]

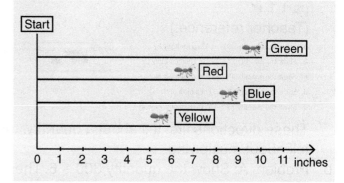

This diagram shows how far different bugs are from where they started.

• What's the color of the bug that is farthest from the starting line? (Signal.) *Green.*

• What's the color of the bug that is closest to the starting line? (Signal.) *Yellow.*

b. We're going to figure out the average distance of bugs from the starting line. So we add the number of inches for each bug and then divide by the number of bugs.

• Say the numbers we'll add: (Call on a student. Idea: *10, 7, 9, and 6.*)

• Everybody, in part B, figure out the average distance.
(Observe students and give feedback.)

• Everybody, what's the total of the distances? (Signal.) *32.*

• What did you divide 32 by? (Signal.) *4.*

• What's the average distance? (Signal.) *8 inches.*
(Add to show:) [88:6B]

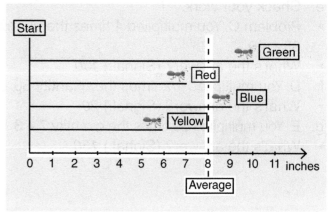

Yes, the average is 8.
(Point to average line.) You can see that there are 3 inches above the average. And the number of empty spaces below the average is 3. The amount above the average and the amount below the average is the same.

a. Find part 3 in your textbook. ✔
(Teacher reference:)

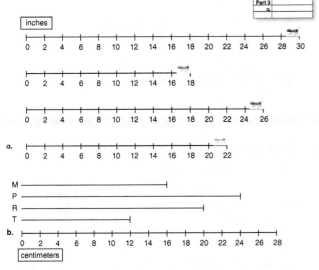

These are like the problem we just worked.

b. Problem A. Find the average distance for the bugs.
(Observe students and give feedback.)

c. Check your work.
• You added 30, 18, 26, and 22. What's the total? (Signal.) *96.*
• What number did you divide by? (Signal.) *4.*
• What's the average distance? (Signal.) *24 inches.*

d. Problem B. These are just lines. You can see the length of each line.
• Touch line M. ✔
What's its length? (Signal.) *16 centimeters.*
• Touch line P. ✔
What's its length? (Signal.) *24 centimeters*
• Figure out the average of all the lines.
(Observe students and give feedback.)

e. Check your work.
• Read the numbers you added. (Signal.) *16, 24, 20, and 12.*
• What's the total? (Signal.) *72.*
• What's the average length of the lines? (Signal.) *18 centimeters.*

EXERCISE 7: SENTENCES
PARTS OF A GROUP

a. We'll work with facts from word problems. We're going to write a letter equation for the facts.
(Display:)   [88:7A]

> $\frac{4}{7}$ of the buildings were tall. There were 21 buildings that were not tall.

4/7 of the buildings were tall. There were 21 buildings that were not tall.
• Do these facts give two names or three names? (Signal.) *3 names.*
• What are the names? (Signal.) *Buildings, tall, and not tall.*

b. Here's the rule: If the facts have three different names, you write the equation for the name that has a number.
• Which name has a number? (Signal.) *Not tall.*
• Raise your hand when you can say the letter equation for not tall. ✔
• Everybody, say the equation. (Signal.) *3/7 B = N.*
(Add to show:)   [88:7B]

> $\frac{4}{7}$ of the buildings were tall. There were 21 buildings that were not tall.
>
> $$\frac{3}{7} b = n$$

Yes, 4/7 of the buildings were tall. So 3/7 of the buildings were **not** tall. The equation is 3/7 B = N.

c. (Display:)   [88:7C]

> $\frac{2}{7}$ of the cars were old. There were 15 new cars.

2/7 of the cars were old. There were 15 new cars.
• What are the three names? (Signal.) *Cars, old, and new.*
We write the equation for the name that has a number.
• Which name has a number? (Signal.) *New.*
• Raise your hand when you can say the letter equation for new. ✔
• Say the letter equation. (Signal.) *5/7 C = N.*

(Add to show:)   [88:7D]

> $\frac{2}{7}$ of the cars were old. There were 15 new cars.
>
> $$\frac{5}{7}\,c = n$$

Yes, 5/7 of the cars were new. So the equation is 5/7 C = N.

d. (Display:)   [88:7E]

> $\frac{9}{10}$ of the flowers are yellow. 6 flowers are not yellow.

9/10 of the flowers are yellow. 6 flowers are not yellow.
- What are the three names? (Signal.) *Flowers, yellow, and not yellow.*
  We write the equation for the name that has a number.
- Which name has a number? (Signal.) *Not yellow.*
- Raise your hand when you can say the letter equation for not yellow. ✔
- Say the letter equation. (Signal.) *1/10 F = N.*
  (Repeat until firm.)

(Add to show:)   [88:7F]

> $\frac{9}{10}$ of the flowers are yellow. 6 flowers are not yellow.
>
> $$\frac{1}{10}\,f = n$$

Yes, 1/10 of the flowers were **not** yellow. Here's the letter equation.

e. (Display:)   [88:7G]

> $\frac{3}{5}$ of the doors are open. 12 doors are closed.

3/5 of the doors are open. 12 doors are closed.
- What are the three names? (Signal.) *Doors, open, and closed.*
  We write the equation for the name that has a number.
- Which name has a number? (Signal.) *Closed.*
- Raise your hand when you can say the letter equation for closed. ✔
- Say the letter equation. (Signal.) *2/5 D = C.*
  (Repeat until firm.)

(Add to show:)   [88:7H]

> $\frac{3}{5}$ of the doors are open. 12 doors are closed.
>
> $$\frac{2}{5}\,d = c$$

Here's the letter equation.

──────── **TEXTBOOK PRACTICE** ────────

a. Find part 4 in your textbook. ✔
(Teacher reference:)

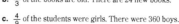

a. $\frac{3}{8}$ of the children were eating. 40 children were not eating.
b. $\frac{2}{3}$ of the books are old. There are 24 new books.
c. $\frac{4}{9}$ of the students were girls. There were 360 boys.
d. $\frac{2}{5}$ of the animals were not eating. 15 animals were eating.
e. $\frac{1}{10}$ of the eggs were broken. There were 27 unbroken eggs.

For each item, you'll write the equation with two letters. Below, you'll write the equation with the number the problem gives.

b. Item A: 3/8 of the children were eating. 40 children were not eating.
- What are the three names? (Signal.) *Children, eating, and not eating.*
- Which name has a number? (Signal.) *Not eating.*
- Write the letter equation for not eating. Below, write the equation with a number for not eating.
  (Observe students and give feedback.)
- Everybody, read the equation with two letters. (Signal.) *5/8 C = N.*
- Read the equation with a number. (Signal.) *5/8 C = 40.*

c. Item B: 2/3 of the books are old. There are 24 new books.
- What are the three names? (Signal.) *Books, old, and new.*
- Which name has a number? (Signal.) *New.*
- Write the letter equation for new. Below, write the equation with a number for new.
  (Observe students and give feedback.)
- Everybody, read the equation with two letters. (Signal.) *1/3 B = N.*
- Read the equation with a number. (Signal.) *1/3 B = 24.*

d. C: 4/9 of the students were girls. There were 360 boys.
- What are the three names? (Signal.) *Students, girls, and boys.*
- Which name has a number? (Signal.) *Boys.*
- Write the letter equation for boys. Below, write the equation with a number for boys. (Observe students and give feedback.)
- Everybody, read the equation with two letters. (Signal.) *5/9 S = B.*
- Read the equation with a number. (Signal.) *5/9 S = 360.*

e. D: 2/5 of the animals were not eating. 15 animals were eating.
- What are the three names? (Signal.) *Animals, not eating, and eating.*
- Which name has a number? (Signal.) *Eating.*
- Write the letter equation for eating. Below, write the equation with a number for eating. (Observe students and give feedback.)
- Everybody, read the equation with two letters. (Signal.) *3/5 A = E.*
- Read the equation with a number. (Signal.) *3/5 A = 15.*

f. E: 1/10 of the eggs were broken. There were 27 unbroken eggs.
- What are the three names? (Signal.) *Eggs, broken, and unbroken.*
- Which name has a number? (Signal.) *Unbroken.*
- Write the letter equation for unbroken. Below, write the equation with a number for unbroken. (Observe students and give feedback.)
- Everybody, read the equation with two letters. (Signal.) *9/10 E = U.*
- Read the equation with a number. (Signal.) *9/10 E = 27.*

Assign Independent Work, Textbook parts 5–11 and Workbook parts 1–3.

Optional extra math-fact practice worksheets are available on ConnectED.

Lesson 89

EXERCISE 1: MENTAL MATH

a. Time for some mental math.
- You'll tell me the whole number each fraction equals.
- Listen: 10 fifths. What whole number? (Signal.) *2.*
- 30 fifths. What whole number? (Signal.) *6.*
- 45 fifths. What whole number? (Signal.) *9.*
- 50 fifths. What whole number? (Signal.) *10.*
- 3 thirds. What whole number? (Signal.) *1.*
- 9 thirds. What whole number? (Signal.) *3.*
- (Repeat until firm.)

b. 18 thirds. What whole number? (Signal.) *6.*
- 27 thirds. What whole number? (Signal.) *9.*
- 8 fourths. What whole number? (Signal.) *2.*
- 12 fourths. What whole number? (Signal.) *3.*
- 20 fourths. What whole number? (Signal.) *5.*
- 28 fourths. What whole number? (Signal.) *7.*
- (Repeat until firm.)

c. You'll say the whole fact.
- Listen: 90 times 2. Say the fact. (Signal.) *90 × 2 = 180.*
- 90 × 3. Say the fact. (Signal.) *90 × 3 = 270.*
- 90 × 4. Say the fact. (Signal.) *90 × 4 = 360.*
- 60 × 4. Say the fact. (Signal.) *60 × 4 = 240.*
- 40 × 4. Say the fact. (Signal.) *40 × 4 = 160.*
  20 × 4. Say the fact. (Signal.) *20 × 4 = 80.*
- (Repeat until firm.)

d. Listen: 80 divided by 40. Say the fact. (Signal.) *80 ÷ 40 = 2.*
- 80 ÷ 20. Say the fact. (Signal.) *80 ÷ 20 = 4.*
- 80 ÷ 10. Say the fact. (Signal.) *80 ÷ 10 = 8.*
- 70 ÷ 10. Say the fact. (Signal.) *70 ÷ 10 = 7.*
- 60 ÷ 10. Say the fact. (Signal.) *60 ÷ 10 = 6.*
- 60 ÷ 30. Say the fact. (Signal.) *60 ÷ 30 = 2.*
- 60 ÷ 6. Say the fact. (Signal.) *60 ÷ 6 = 10.*
- (Repeat until firm.)

EXERCISE 2: FRACTION MULTIPLICATION
PRODUCT/FACTOR COMPARISON

a. (Display:) [89:2A]

$$a.\ \frac{3}{7} \times \frac{1}{4} \qquad b.\ \frac{8}{3} \times \frac{5}{4}$$

You're going to figure out if you end up with more than you start with or less than you start with. Remember, if you multiply by more than 1, you end up with more than you start with.

b. (Point to **A**.) Read the problem. (Signal.) *3/7 × 1/4.*
- What do you multiply 3/7 by? (Signal.) *1/4.*
- Is 1/4 more than 1 or less than 1? (Signal.) *Less than 1.*
- So do you end up with more than 3/7 or less than 3/7? (Signal.) *Less than 3/7.*
- Copy 3/7 × 1/4 in part A and work it. Don't simplify. Just multiply. ✔
- Read the equation for problem A. (Signal.) *3/7 × 1/4 = 3/28.*
(Add to show:) [89:2B]

$$a.\ \frac{3}{7} \times \frac{1}{4} = \frac{3}{28} \qquad b.\ \frac{8}{3} \times \frac{5}{4}$$

I'm going to write the sign to show if we end up with more or less than we start with.
(Add to show:) [89:2C]

$$a.\ \frac{3}{7} \times \frac{1}{4} = \frac{3}{28} \qquad b.\ \frac{8}{3} \times \frac{5}{4}$$

$$\frac{3}{7} \qquad\qquad \frac{3}{28}$$

- The problem starts with 3/7 and ends with 3/28. Do you multiply 3/7 by a fraction that is more than 1 or less than 1? (Signal.) *Less than 1.*

- So is 3/28 more than 3/7 or less than 3/7? (Signal.) *Less than 3/7.*
  (Add to show:) [89:2D]

$$\text{a. } \frac{3}{7} \times \frac{1}{4} = \frac{3}{28} \quad \text{b. } \frac{8}{3} \times \frac{5}{4}$$
$$\frac{3}{7} > \frac{3}{28}$$

Here's the sign. It shows that 3/28 is less.
c. (Point to **B**.) Read the problem. (Signal.) *8/3 × 5/4.*
- What do you multiply 8/3 by? (Signal.) *5/4.*
- Is 5/4 more than 1 or less than 1? (Signal.) *More than 1.*
- So do you end up with more than 8/3 or less than 8/3? (Signal.) *More than 8/3.*
- Copy 8/3 × 5/4 and work it. Don't simplify. Just multiply. ✔
- Read the equation for problem B. (Signal.) *8/3 × 5/4 = 40/12.*
  (Add to show:) [89:2E]

$$\text{a. } \frac{3}{7} \times \frac{1}{4} = \frac{3}{28} \quad \text{b. } \frac{8}{3} \times \frac{5}{4} = \frac{40}{12}$$
$$\frac{3}{7} > \frac{3}{28}$$

I'm going to write the statement to show if we end up with more or less than we start with.
(Add to show:) [89:2F]

$$\text{a. } \frac{3}{7} \times \frac{1}{4} = \frac{3}{28} \quad \text{b. } \frac{8}{3} \times \frac{5}{4} = \frac{40}{12}$$
$$\frac{3}{7} > \frac{3}{28} \qquad \frac{8}{3} \qquad \frac{40}{12}$$

The problem starts with 8/3. You multiply by a fraction that is more than 1. So you end up with more than you start with.
- Listen: Which is more—8/3 or 40/12? (Signal.) *40/12.*

(Add to show:) [89:2G]

$$\text{a. } \frac{3}{7} \times \frac{1}{4} = \frac{3}{28} \quad \text{b. } \frac{8}{3} \times \frac{5}{4} = \frac{40}{12}$$
$$\frac{3}{7} > \frac{3}{28} \qquad \frac{8}{3} < \frac{40}{12}$$

Here's the sign. It shows that 40/12 is more.
d. (Display:) [89:2H]

$$\text{c. } \frac{3}{2} \times \frac{5}{7}$$

(Point to **C**.) Read the problem. (Signal.) *3/2 × 5/7.*
- What do you multiply 3/2 by? (Signal.) *5/7.*
- Is 5/7 more than 1 or less than 1? (Signal.) *Less than 1.*
- So do you end up with more than 3/2 or less than 3/2? (Signal.) *Less than 3/2.*
- Copy 3/2 × 5/7 and work it. ✔
- Read the equation for problem C. (Signal.) *3/2 × 5/7 = 15/14.*
  (Add to show:) [89:2I]

$$\text{c. } \frac{3}{2} \times \frac{5}{7} = \frac{15}{14}$$

I'm going to write the statement to show if we end up with more or less than we started with.
(Add to show:) [89:2J]

$$\text{c. } \frac{3}{2} \times \frac{5}{7} = \frac{15}{14}$$
$$\frac{3}{2} \qquad \frac{15}{14}$$

- Did you multiply 3/2 by a fraction that is more than 1 or less than 1? (Signal.) *Less than 1.*
- So is 15/14 more than 3/2 or less than 3/2? (Signal.) *Less than 3/2.*

(Add to show:)   [89:2K]

$$\mathbf{c.}\ \frac{3}{2} \times \frac{5}{7} = \frac{15}{14}$$

$$\frac{3}{2} \quad > \quad \frac{15}{14}$$

Here's the sign. It shows that 3/2 is more.

━━━━━━━ **WORKBOOK PRACTICE** ━━━━━━━

a. Open your workbook to Lesson 89 and find part 1. ✔
(Teacher reference:)

a. $\frac{5}{3} \times \frac{4}{2} = $ ▢   b. $\frac{2}{7} \times \frac{6}{9} = $ ▢   c. $\frac{10}{9} \times \frac{1}{2} = $ ▢

$\frac{5}{3}$ ▢ $\frac{2}{7}$ ▢ $\frac{10}{9}$ ▢

You're going to work each problem. Below, you'll complete the statement to show if you end up with more or less than you start with.

b. Read problem A. (Signal.) *5/3 × 4/2.*
• Write the answer to the problem. ✔
• Everybody, what does 5/3 × 4/2 equal? (Signal.) *20/6.*
• Now you'll complete the statement below. Write 20/6 in the box below. ✔
(Display:)   [89:2L]

$$\mathbf{a.}\ \frac{5}{3} \times \frac{4}{2} = \frac{20}{6}$$

$$\frac{5}{3} \qquad \frac{20}{6}$$

Here's what you should have.
• Now write the sign to show if you end up with more than 5/3 or less than 5/3. ✔
(Add to show:)   [89:2M]

$$\mathbf{a.}\ \frac{5}{3} \times \frac{4}{2} = \frac{20}{6}$$

$$\frac{5}{3} \quad < \quad \frac{20}{6}$$

Here's what you should have. You multiplied by more than 1, so 20/6 is more than 5/3.

c. Read problem B. (Signal.) *2/7 × 6/9.*
• Write the answer to the problem. ✔
• Everybody, what does 2/7 × 6/9 equal? (Signal.) *12/63.*
• Copy 12/63 below and write the sign to show if you end up with more than 2/7 or less than 2/7. ✔
Check your work. You multiplied by less than 1, so 12/63 is less than 2/7.
(Display:)   [89:2N]

$$\mathbf{b.}\ \frac{2}{7} \times \frac{6}{9} = \frac{12}{63}$$

$$\frac{2}{7} \quad > \quad \frac{12}{63}$$

Here's what you should have.

d. Read problem C. (Signal.) *10/9 × 1/2.*
• Write the answer to the problem. Then complete the statement below.
(Observe students and give feedback.)
• Everybody, what does 10/9 × 1/2 equal? (Signal.) *10/18.*
• Is 10/18 more than 10/9 or less than 10/9? (Signal.) *Less than 10/9.*
Yes, you multiplied by less than 1, so 10/18 is less than 10/9.
(Display):   [89:2O]

$$\mathbf{c.}\ \frac{10}{9} \times \frac{1}{2} = \frac{10}{18}$$

$$\frac{10}{9} \quad > \quad \frac{10}{18}$$

Here's what you should have.

EXERCISE 3: MIXED-NUMBER OPERATIONS
SUBTRACTION   [REMEDY]

a. (Display:)   [89:3A]

$$8\frac{3}{10}$$
$$-4\frac{6}{10}$$

This mixed-number problem subtracts.
• Read the problem. (Signal.) *8 and 3/10 – 4 and 6/10.*
• Read the problem for the fractions. (Signal.) *3/10 – 6/10.*
• Can you work that problem? (Signal.) *No.*
So we rewrite 8 and 3/10.

- What's the new whole number? (Signal.) *7.*
- Raise your hand when you know the new fraction. ✔
- What's the new fraction? (Signal.) *13/10.*
  (Add to show:)   [89:3B]

$$\begin{array}{r} \overset{7}{\cancel{8}}\,\overset{13}{\cancel{\tfrac{3}{10}}} \\[2pt] -\,4\,\tfrac{6}{10} \\ \hline \end{array}$$

- Say the new problem for the fractions. (Signal.) *13/10 – 6/10.*
- What's the answer? (Signal.) *7/10.*
- Say the new problem for the whole numbers. (Signal.) *7 – 4.*
- What's the answer? (Signal.) *3.*
  (Add to show:)   [89:3C]

$$\begin{array}{r} \overset{7}{\cancel{8}}\,\overset{13}{\cancel{\tfrac{3}{10}}} \\[2pt] -\,4\,\tfrac{6}{10} \\ \hline 3\,\tfrac{7}{10} \end{array}$$

- Read the problem we started with and the answer. (Signal.) *8 and 3/10 – 4 and 6/10 = 3 and 7/10.*

=== **WORKBOOK PRACTICE** ===

a. Find part 2 in your workbook. ✔
  (Teacher reference:)   R Part C

a. $6\tfrac{2}{9}$	b. $8\tfrac{2}{10}$	c. $4\tfrac{1}{5}$
$-1\tfrac{4}{9}$	$-5\tfrac{9}{10}$	$-2\tfrac{4}{5}$

  These problems are like the one we just worked.
b. Read problem A. (Signal.) *6 and 2/9 – 1 and 4/9.*
- Can you work the problem the way it is written? (Signal.) *No.*
- Rewrite 6 and 2/9 so the fraction is one more. Stop when you've done that much. ✔
- Everybody, what's the new mixed number? (Signal.) *5 and 11/9.*

(Display:)   [89:3D]

$$\text{a.}\quad \overset{5}{\cancel{6}}\,\overset{11}{\cancel{\tfrac{2}{9}}} \\[2pt] -\,1\,\tfrac{4}{9}$$

Here's what you should have.
- Work the new problem for the fractions and the new problem for the whole numbers. (Observe students and give feedback.)
c. Check your work.
- You worked the problem 11/9 – 4/9. What's the answer? (Signal.) *7/9.*
- What's 6 and 2/9 minus 1 and 4/9? (Signal.) *4 and 7/9.*
  (Add to show:)   [89:3E]

$$\text{a.}\quad \overset{5}{\cancel{6}}\,\overset{11}{\cancel{\tfrac{2}{9}}} \\[2pt] -\,1\,\tfrac{4}{9} \\ \hline 4\,\tfrac{7}{9}$$

Here's what you should have.
d. Read problem B. (Signal.) *8 and 2/10 – 5 and 9/10.*
- Can you work the problem the way it is written? (Signal.) *No.*
- Rewrite the top number and work the problem. (Observe students and give feedback.)
e. Check your work.
- You worked the problem 12/10 – 9/10. What's the answer? (Signal.) *3/10.*
- What's 8 and 2/10 – 5 and 9/10? (Signal.) *2 and 3/10.*
  (Display:)   [89:3F]

$$\text{b.}\quad \overset{7}{\cancel{8}}\,\overset{12}{\cancel{\tfrac{2}{10}}} \\[2pt] -\,5\,\tfrac{9}{10} \\ \hline 2\,\tfrac{3}{10}$$

Here's what you should have.
f. Read problem C. (Signal.) *4 and 1/5 – 2 and 4/5.*
- Can you work the problem the way it is written? (Signal.) *No.*
- Rewrite the top number and work the problem. (Observe students and give feedback.)

g. Check your work.
- You worked the problem 6/5 – 4/5. What's the answer? (Signal.) *2/5.*
- What's 4 and 1/5 – 2 and 4/5? (Signal.) *1 and 2/5.*
(Display:) [89:3G]

$$c. \quad 4 \frac{\cancel{3}^{\cancel{6}}}{5} \\ -2 \frac{4}{5} \\ \overline{1 \frac{2}{5}}$$

Here's what you should have.

EXERCISE 4: EXPRESSIONS
FROM WRITTEN DESCRIPTIONS
[REMEDY]

a. (Display:) [89:4A]

Add 15, 3, and 7. Multiply by 2.

This problem tells the quantity we start with and what we multiply it by.
We add 15, 3, and 7. Write that quantity in parentheses in part A. Then show the quantity multiplied by 2. ✔
(Add to show:) [89:4B]

Add 15, 3, and 7. Multiply by 2.
$$2(15 + 3 + 7)$$

Here's what you should have.

b. (Display:) [89:4C]

Divide 35 by 5. Multiply by 2.

New problem: Divide 35 by 5. Multiply by 2.
- Write the problem. Show the quantity for the first part in parentheses. ✔
(Add to show:) [89:4D]

Divide 35 by 5. Multiply by 2.
$$2(35 ÷ 5)$$

Here's what you should have.

c. (Display:) [89:4E]

Add 12 and 3 and subtract 10.
Then multiply by 5.

The first sentence tells about the quantity inside the parentheses. The second sentence tells what you multiply by.
- Write the problem. ✔
(Add to show:) [89:4F]

Add 12 and 3 and subtract 10.
Then multiply by 5.
$$5(12 + 3 - 10)$$

Here's what you should have.

━━━━ TEXTBOOK PRACTICE ━━━━

a. Open your textbook to Lesson 89 and find part 1. ✔
(Teacher reference:)

> a. Add $\frac{3}{2}$ and $\frac{7}{2}$. Multiply by 9.
> b. Add 80 and 10. Subtract 60. Then multiply the quantity by 2.
> c. Add 15 and 11. Subtract 6. Then multiply by 5.
> d. Subtract $\frac{4}{6}$ from $\frac{40}{6}$. Then multiply by 3.
> e. Add 12 and 12. Subtract 10. Then multiply by 10.

You're going to write problems like the ones we just worked.
b. Problem A: Add 3/2 and 7/2. Multiply by 9.
- Write the problem. Stop when you've done that much.
(Observe students and give feedback.)
(Display:) [89:4G]

$$\text{a. } 9 \left(\frac{3}{2} + \frac{7}{2} \right)$$

Here's what you should have.
- Figure out the value inside the parentheses. Simplify if you can. Stop when you've done that much.
(Point to $\frac{3}{2}$.) Here's the problem you started with, 3/2 + 7/2 = 10/2.
- What does 10/2 equal? (Signal.) *5.*

(Add to show:) [89:4H]

$$a.\ 9\left(\frac{3}{2}+\frac{7}{2}\right)$$

$$9\left(\frac{10}{2}\right)$$

$$9\ (5)\ =$$

- Complete the problem. ✔
- Everybody, what's the answer? (Signal.) *45.*
c. Problem B: Add 80 and 10. Subtract 60. Then multiply the quantity by 2.
- Write the problem and figure out the answer. (Observe students and give feedback.)
Problem B: 2 times the quantity 80 + 10 − 60. That's 2 × 30.
- What's the answer? (Signal.) *60.*
(Display:) [89:4I]

$$b.\ 2\ (80+10-60)$$

$$2\ (30)\ =\boxed{60}$$

Here's what you should have.
d. Work the rest of the items in part 1. Write the problem and work it. Remember, when you figure out the value inside the parentheses, simplify if you can.
(Observe students and give feedback.)
e. Check your work.
- Problem C: Add 15 and 11. Subtract 6. Then multiply by 5.
- What's the answer? (Signal.) *100.*
(Display:) [89:4J]

$$c.\ 5\ (15+11-6)$$

$$5\ (20)\ =\boxed{100}$$

Here's what you should have.
f. Problem D: Subtract 4/6 from 40/6. Then multiply by 3.
- What's the answer? (Signal.) *18.*
(Display:) [89:4K]

$$d.\ 3\left(\frac{40}{6}-\frac{4}{6}\right)$$

$$3\left(\frac{36}{6}\right)$$

$$3\ (6)\ =\boxed{18}$$

Here's what you should have.

g. Problem E: Add 12 and 12. Subtract 10. Then multiply by 10.
- What's the answer? (Signal.) *140.*
(Display:) [89:4L]

$$e.\ 10\ (12+12-10)$$

$$10\ (14)\ =\boxed{140}$$

Here's what you should have.

EXERCISE 5: FRACTION WORD PROBLEMS
DIVISION

a. Find part 2 in your textbook. ✔
(Teacher reference:)

> a. Robert wants to put raisins into bags that each hold $\frac{7}{8}$ pound. He has 42 pounds of raisins. How many bags can be filled?
>
> b. Each dose of medicine weighs $\frac{1}{18}$ ounce. How many doses can be made from $\frac{2}{3}$ ounce of medicine?
>
> c. $\frac{35}{2}$ cups of flour make 7 cake recipes. How many cups of flour does each recipe call for?
>
> d. A pie is divided into 9 slices. The pie weighs $\frac{81}{2}$ ounces. How many ounces does each slice weigh?

You have to read these problems carefully to figure out the problem you work.
b. Problem A: Robert wants to put raisins into bags that each hold 7/8 pound. He has 42 pounds of raisins. How many bags can be filled?
- Raise your hand when you know the amount Robert has before he divides. ✔
- What's the amount? (Signal.) *42 pounds.*
- Start with 42 and say the problem you'll work. (Signal.) *42 ÷ 7/8.*
c. Problem B: Each dose of medicine weighs 1/18 ounce. How many doses can be made from 2/3 ounce of medicine?
- Raise your hand when you know the amount there is before you divide. ✔
- What's the amount? (Signal.) *2/3 ounce.*
- Say the division problem we'll work. (Signal.) *2/3 ÷ 1/18.*
(Repeat until firm.)

d. Problem C: 35/2 cups of flour make 7 cake recipes. How many cups of flour does each recipe call for?

- Raise your hand when you know the amount there is before you divide. ✔
- What's the amount? (Signal.) *35/2 cups.*
- Say the division problem. (Signal.) *35/2 ÷ 7.*

e. Problem D: A pie is divided into 9 slices. The pie weighs 81/2 ounces. How many ounces does each slice weigh?

- Raise your hand when you know the amount there is before you divide. ✔
- What's the amount? (Signal.) *81/2 ounces.*
- Say the division problem. (Signal.) *81/2 ÷ 9.* (Repeat until firm.)

f. Go back to problem A. Write the division problem. Then stop. ✔

- Check your work. Read the division problem. (Signal.) *42 ÷ 7/8.*
- Work the problem and write the answer with a unit name. (Observe students and give feedback.)
- Everybody, how many bags can be filled? (Signal.) *48 bags.*

g. Problem B. Write the division problem. Then stop. ✔

- Check your work. Read the division problem. (Signal.) *2/3 ÷ 1/18.*
- Work the problem and write the answer with a unit name. (Observe students and give feedback.)
- Everybody, how many doses can be made? (Signal.) *12 doses.*

h. Problem C. Write the division problem. Then stop. ✔

- Check your work. Read the division problem. (Signal.) *35/2 ÷ 7.*
- Work the problem. (Observe students and give feedback.)
- Everybody, how many cups of flour are in each recipe? (Signal.) *2 and 1/2 cups.*

i. Problem D. Write the division problem. Then stop. ✔

- Check your work. Read the division problem. (Signal.) *81/2 ÷ 9.*
- Work the problem. (Observe students and give feedback.)
- Everybody, how many ounces does each slice weigh? (Signal.) *4 and 1/2 ounces.*

EXERCISE 6: AVERAGE REMEDY

a. Find part 3 in your textbook. ✔ (Teacher reference:)

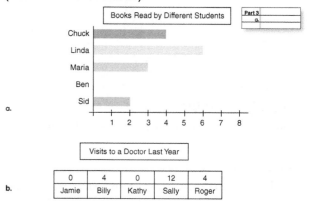

Graph A shows the number of books read last school year by 5 students.

- There is no bar for Ben. So how many books did he read? (Signal.) *Zero.*
- Figure out the average for these 5 students. (Observe students and give feedback.)

b. Check your work.

- What's the total number of books? (Signal.) *15.*
- What did you divide by? (Signal.) *5.*
- What's the average? (Signal.) *3.*

c. Table B shows how many times different people went to a doctor last year.

- Two people went to see a doctor 4 times last year. What are their names? (Signal.) *Billy and Roger.* Yes, Billy and Roger.
- Two people did not go see a doctor last year. What are their names? (Signal.) *Jamie and Kathy.* Yes, Jamie and Kathy.
- Figure out the average number for the members of this group. (Observe students and give feedback.)

d. Everybody, what's the total number of visits? (Signal.) *20.*

- What did you divide by? (Signal.) *5.*
- What's the average number of visits? (Signal.) *4.*

e. There's only one person above the average. Who is that? (Signal.) *Sally.*

- Raise your hand when you know how many visits above the average she had. ✔
- How many visits above the average? (Signal.) *8 visits.*

EXERCISE 7: SENTENCES
PARTS OF A GROUP

REMEDY

a. Find part 4 in your textbook. ✔
(Teacher reference:)

R Test 10

a. $\frac{3}{4}$ of the meals are cooked. There are 48 uncooked meals.

b. $\frac{7}{10}$ of the assignments are not corrected. 30 assignments are corrected.

c. $\frac{1}{9}$ of the horses are male. 72 of the horses are female.

d. $\frac{8}{15}$ of the apples are eaten. 56 apples are not eaten.

e. $\frac{13}{20}$ of the hats are clean. There are 7 dirty hats.

- Item A: 3/4 of the meals are cooked. There are 48 uncooked meals.
- What are the three names? (Signal.) *Meals, cooked, and uncooked.*
- Which name has a number? (Signal.) *Uncooked.*

b. Write the letter equation for uncooked. Below, write the equation with a number for uncooked.
(Observe students and give feedback.)
- Everybody, read the equation with two letters. (Signal.) *1/4 M = U.*
- Read the equation with a number. (Signal.) *1/4 M = 48.*

c. Write two equations for the rest of the problems. Remember, first the letter equation, then the equation with a number for one of the letters.
(Observe students and give feedback.)

d. Check your work.
- Item B: 7/10 of the assignments are not corrected. 30 assignments are corrected.
- Read the equation with two letters. (Signal.) *3/10 A = C.*
- Read the equation with a number. (Signal.) *3/10 A = 30.*

e. Item C: 1/9 of the horses are male. 72 of the horses are female.
- Read the equation with two letters. (Signal.) *8/9 H = F.*
- Read the equation with a number. (Signal.) *8/9 H = 72.*

f. Item D: 8/15 of the apples are eaten. 56 apples are not eaten.
- Read the equation with two letters. (Signal.) *7/15 A = N.*
- Read the equation with a number. (Signal.) *7/15 A = 56.*

g. Item E: 13/20 of the hats are clean. There are 7 dirty hats.
- Read the equation with two letters. (Signal.) *7/20 H = D.*
- Read the equation with a number. (Signal.) *7/20 H = 7.*

Assign Independent Work, Textbook parts 5–12 and Workbook part 3.

Optional extra math-fact practice worksheets are available on ConnectED.

Lesson 90

EXERCISE 1: MENTAL MATH

a. Time for some mental math.
   You'll tell me the whole number each fraction equals.
   - Listen: 63 ninths. What whole number? (Signal.) *7.*
   - 24 eighths. What whole number? (Signal.) *3.*
   - 24 sixths. What whole number? (Signal.) *4.*
   - 35 fifths. What whole number? (Signal.) *7.*
   (Repeat until firm.)

b. Listen: 18 ninths. What whole number? (Signal.) *2.*
   - 28 fourths. What whole number? (Signal.) *7.*
   - 40 eighths. What whole number? (Signal.) *5.*
   (Repeat until firm.)

c. Listen: 160 divided by 20. Say the fact. (Signal.) *160 ÷ 20 = 8.*
   - 180 ÷ 20. Say the fact. (Signal.) *180 ÷ 20 = 9.*
   - 360 ÷ 90. Say the fact. (Signal.) *360 ÷ 90 = 4.*
   - 180 ÷ 90. Say the fact. (Signal.) *180 ÷ 90 = 2.*
   (Repeat until firm.)

d. Listen: 120 divided by 60. Say the fact. (Signal.) *120 ÷ 60 = 2.*
   - 240 ÷ 60. Say the fact. (Signal.) *240 ÷ 60 = 4.*
   - 480 ÷ 60. Say the fact. (Signal.) *480 ÷ 60 = 8.*
   - 540 ÷ 60. Say the fact. (Signal.) *540 ÷ 60 = 9.*
   (Repeat until firm.)

e. Listen: 90 times 6. Say the fact. (Signal.) *90 × 6 = 540.*
   - 90 × 8. Say the fact. (Signal.) *90 × 8 = 720.*
   - 50 × 9. Say the fact. (Signal.) *50 × 9 = 450.*
   - 30 × 9. Say the fact. (Signal.) *30 × 9 = 270.*
   - 20 × 9. Say the fact. (Signal.) *20 × 9 = 180.*
   - 20 × 6. Say the fact. (Signal.) *20 × 6 = 120.*
   (Repeat until firm.)

EXERCISE 2: FRACTION MULTIPLICATION
PRODUCT/FACTOR COMPARISON

a. Open your workbook to Lesson 90 and find part 1. ✔
   (Teacher reference:)

   a. $\frac{4}{1}\left(\frac{5}{6}\right)=$ ▢ $\frac{4}{1}$ ▢ ▢

   b. $\frac{7}{5}\left(\frac{10}{9}\right)=$ ▢ $\frac{7}{5}$ ▢ ▢

   c. $\frac{3}{8}\left(\frac{5}{4}\right)=$ ▢ $\frac{3}{8}$ ▢ ▢

   d. $\frac{2}{7}\left(\frac{7}{9}\right)=$ ▢ $\frac{2}{7}$ ▢ ▢

   You're going to write answers to each problem and complete the statement to show if the fraction you end with is more or less than the fraction you start with.

b. Read problem A. (Signal.) *4 over 1 × 5/6.*
   - What fraction are you multiplying by? (Signal.) *5/6.*
   - Is 5/6 more than 1 or less than 1? (Signal.) *Less than 1.*
   - So will you end with a fraction that's more or less than 4 over 1? (Signal.) *Less than 4 over 1.*

c. Read problem B. (Signal.) *7/5 × 10/9.*
   - What fraction are you multiplying by? (Signal.) *10/9.*
   - Is 10/9 more than 1 or less than 1? (Signal.) *More than 1.*
   - So will you end with a fraction that's more than 7/5 or less than 7/5? (Signal.) *More than 7/5.*
   (Repeat until firm.)

d. Read problem C. (Signal.) *3/8 × 5/4.*
- What fraction are you multiplying by? (Signal.) *5/4.*
- Is 5/4 more than 1 or less than 1? (Signal.) *More than 1.*
- So will you end with a fraction that's more than 3/8 or less than 3/8? (Signal.) *More than 3/8.*

e. Read problem D. (Signal.) *2/7 × 7/9.*
- What fraction are you multiplying by? (Signal.) *7/9.*
- Is 7/9 more than 1 or less than 1? (Signal.) *Less than 1.*
- So will you end with a fraction that's more than 2/7 or less than 2/7? (Signal.) *Less than 2/7.*
(Repeat until firm.)

f. Go back to problem A and write the answer. Don't simplify. Then complete the statement to show if you end with more or less than you start with.
(Observe students and give feedback.)
- Everybody, what does 4 over 1 × 5/6 equal? (Signal.) *20/6.*
- Is 20/6 more than 4 over 1 or less than 4 over 1? (Signal.) *Less than 4 over 1.*
(Display:)   [90:2A]

a. $\dfrac{4}{1}\left(\dfrac{5}{6}\right)=\dfrac{20}{6}$

$\dfrac{4}{1}\ >\ \dfrac{20}{6}$

Here's what you should have for problem A.

g. Write the answer for problem B. Then complete the statement to show if you end with more or less than 7/5.
(Observe students and give feedback.)
- Everybody, what does 7/5 × 10/9 equal? (Signal.) *70/45.*
- Is 70/45 more than 7/5 or less than 7/5? (Signal.) *More than 7/5.*
(Display:)   [90:2B]

b. $\dfrac{7}{5}\left(\dfrac{10}{9}\right)=\dfrac{70}{45}$

$\dfrac{7}{5}\ <\ \dfrac{70}{45}$

Here's what you should have for problem B.

h. Work problem C. Then complete the statement to show if you end with more or less than 3/8.
(Observe students and give feedback.)
- Everybody, what does 3/8 × 5/4 equal? (Signal.) *15/32.*

- Is 15/32 more than 3/8 or less than 3/8? (Signal.) *More than 3/8.*
(Display:)   [90:2C]

c. $\dfrac{3}{8}\left(\dfrac{5}{4}\right)=\dfrac{15}{32}$

$\dfrac{3}{8}\ <\ \dfrac{15}{32}$

Here's what you should have for problem C.

i. Work problem D. Then complete the statement to show if you end with more or less than 2/7.
(Observe students and give feedback.)
- Everybody, what does 2/7 × 7/9 equal? (Signal.) *14/63.*
- Is 14/63 more than 2/7 or less than 2/7? (Signal.) *Less than 2/7.*
(Display:)   [90:2D]

d. $\dfrac{2}{7}\left(\dfrac{7}{9}\right)=\dfrac{14}{63}$

$\dfrac{2}{7}\ >\ \dfrac{14}{63}$

Here's what you should have for problem D.

EXERCISE 3: BALANCE BEAMS
TOTAL DISTANCE

a. (Display:)   [90:3A]

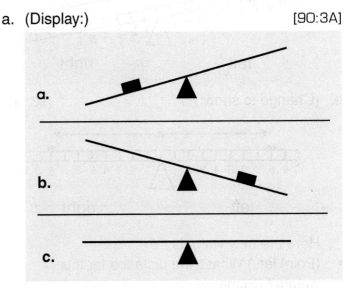

These diagrams show balance beams. If there is a weight on only one side, that side goes down.

b. (Point to **a.**) The left side has the weight on it. Which side is down? (Signal.) *The left side.*
- (Point to **b.**) Which side has a weight on it? (Signal.) *The right side.*
  Which side is down? (Signal.) *The right side.*
- (Point to **c.**) Neither side has a weight on it. So neither side goes down.

c. Here's the rule about balance beams. We figure out the distance of weights from zero on the balance beam. If the distance is the same on both sides, the beam will balance.

d. (Display:) [90:3B]

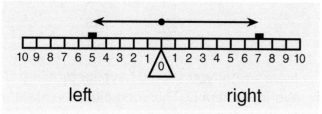

left   right

The arrows show the distance of the weights from zero.
- (Point left.) What's the distance for this weight? (Signal.) *5.*
- (Point right.) What's the distance for this weight? (Signal.) *7.*
- Are the distances the same? (Signal.) *No.*
- So will the beam balance? (Signal.) *No.*
- The weight that is farther from zero will go down. Is that the weight on the left side or on the right side? (Signal.) *The right side.*
- So which side will go down? (Signal.) *The right side.*
(Change to show:) [90:3C]

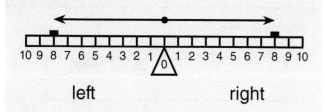

left   right

e. (Change to show:) [90:3D]

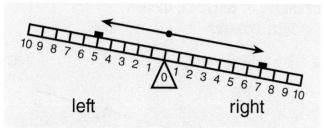

left   right

Here are new weights.
- (Point left.) What's the distance for this weight? (Signal.) *8.*
- (Point right.) What's the distance for this weight? (Signal.) *8.*
- Are the distances the same? (Signal.) *Yes.*
- So will the beam balance? (Signal.) *Yes.*

f. (Display:) [90:3E]

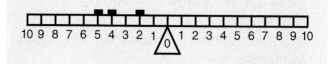

(Point left.) There is more than one weight on this side of the balance beam. So we have to figure out the total distance for the weights.
- (Point to weight at **5**.) What's the distance for this weight? (Signal.) *5.*
- (Point to weight at **4**.) What's the distance for this weight? (Signal.) *4.*
- (Point to weight at **2**.) What's the distance for this weight? (Signal.) *2.*
So the total distance on this side is 5 + 4 + 2.
- What do we add to find the total distance? (Signal.) *5 + 4 + 2.*
- Raise your hand when you know the total distance on the left. ✔
- Everybody, what's the total distance on the left? (Signal.) *11.*
- So what total distance do we need on the right side to make the beam balance? (Signal.) *11.*

g. (Add to show:) [90:3F]

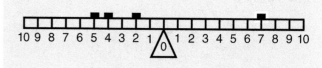

- What's the total on the right now? (Signal.) *7.*
- There is 11 on the left and 7 on the right. Will the beam balance? (Signal.) *No.*
- Which side has the greater distance? (Signal.) *The left side.*
- So which side will go down? (Signal.) *The left side.*
(Change to show:) [90:3G]

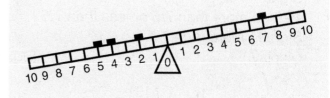

Here's what will happen.

h. I'm going to add a weight at 5.
(Add to show:) [90:3H]

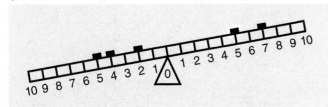

- Raise your hand when you know the total distance on the right now. ✔
- What's the total distance on the right? (Signal.) *12.*
- There is 11 on the left and 12 on the right. Which side has the greater total distance? (Signal.) *The right side.*
- So which side will go down? (Signal.) *The right side.*
(Change to show:) [90:3I]

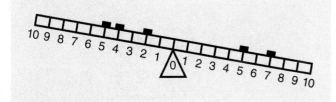

Here's what will happen.
i. (Add to show:) [90:3J]

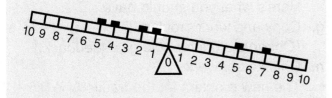

We changed the total distance on the left side.
- What's the total distance on the left? (Signal.) *12.*
- What's the total distance on the right? (Signal.) *12.*
- Are the totals the same? (Signal.) *Yes.*
- So what will the beam do? (Signal.) *Balance.*
(Change to show:) [90:3K]

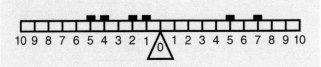

Here's what will happen.
j. Remember, if the total distance is the same on both sides, the beam will balance.

EXERCISE 4: MIXED-NUMBER OPERATIONS
SUBTRACTION

a. (Display:) [90:4A]

$$6$$
$$-\ 4\frac{1}{8}$$

- Read the problem. (Signal.) *6 – 4 and 1/8.* To work the problem, we have to rewrite the top number.
- The problem for the fraction is zero – 1/8. Can we work that problem? (Signal.) *No.* So we rewrite the top number. We take one from 6 and write it as a fraction.
- What's the new whole number? (Signal.) *5.*
- What's the fraction that equals one? (Signal.) *8/8.*
(Add to show:) [90:4B]

$$\overset{5}{\cancel{6}}\ \frac{8}{8}$$
$$-\ 4\frac{1}{8}$$

- The problem for the fractions is 8/8 – 1/8. What's the answer? (Signal.) *7/8.*
- The problem for the whole numbers is 5 – 4. What's the answer? (Signal.) *1.*
(Add to show:) [90:4C]

$$\overset{5}{\cancel{6}}\ \frac{8}{8}$$
$$-\ 4\frac{1}{8}$$
$$\boxed{1\frac{7}{8}}$$

- What's 6 – 4 and 1/8? (Signal.) *1 and 7/8.*
b. Remember, if the top number does not have a fraction, you rewrite the top number.

a. Open your textbook to Lesson 90 and find
   part 1. ✔
   (Teacher reference:)

a.	b.	c.	d.
9	$7\frac{4}{8}$	$14\frac{2}{7}$	10
$-2\frac{3}{4}$	$-3\frac{5}{8}$	$-2\frac{5}{7}$	$-3\frac{2}{5}$

- Read problem A. (Signal.) *9 – 2 and 3/4.*
- Copy the problem. Then stop. ✔
- Everybody, say the problem for the fractions.
  (Signal.) *Zero – 3/4.*
- Can you work that problem? (Signal.) *No.*
- Rewrite the top number and work the problem.
  (Observe students and give feedback.)

b. Check your work.
- You rewrote the top number. What's the new
  whole number? (Signal.) *8.*
- What's the fraction? (Signal.) *4/4.*
- What does 9 – 2 and 3/4 equal? (Signal.)
  6 and 1/4.
  (Display:)   [90:4D]

Here's what you should have.

c. Read problem B. (Signal.) *7 and 4/8 –
   3 and 5/8.*
- Can you work that problem? (Signal.) *No.*
- Copy the problem. Rewrite the top number
  and work the problem.
  (Observe students and give feedback.)

d. Check your work.
- The new problem for the fraction is 12/8 – 5/8.
  What's the answer? (Signal.) *7/8.*
- Read the new problem and the answer for the
  whole numbers. (Signal.) *6 – 3 = 3.*
- So what's 7 and 4/8 – 3 and 5/8? (Signal.)
  3 and 7/8.

(Display: [90:4E]

Here's what you should have.

e. Copy and work problem C.
   (Observe students and give feedback.)

f. Check your work.
- The new problem for the fractions is 9/7 – 5/7.
  What's the answer? (Signal.) *4/7.*
- What does 14 and 2/7 – 2 and 5/7 equal?
  (Signal.) *11 and 4/7.*
  (Display:)   [90:4F]

Here's what you should have.

g. Copy and work problem D.
   (Observe students and give feedback.)

h. Check your work.
- The new problem for the fractions is 5/5 – 2/5.
  What's the answer? (Signal.) *3/5.*
- What does 10 – 3 and 2/5 equal? (Signal.)
  6 and 3/5.
  (Display:)   [90:4G]

Here's what you should have.

EXERCISE 5: UNIT CONVERSION

MULTIPLICATION

a. (Display:) [90:5A]

> 1 yard is 36 inches.
>
> 1 hour is 60 minutes.
>
> 1 day is 24 hours.

Read these measurement facts.

- (Point to **1 yard.**) Get ready. (Signal.) *1 yard is 36 inches.*
- (Point to **1 hour.**) Get ready. (Signal.) *1 hour is 60 minutes.*
- (Point to **1 day.**) Get ready. (Signal.) *1 day is 24 hours.*

b. (Point to **1 yard.**) What's the number for yards? (Signal.) *1.*
- What's the number for inches? (Signal.) *36.*
- Are there more yards or inches? (Signal.) *Inches.*
- How many times more? (Signal.) *36.*
  So if you change yards to inches, you end up with more than 1.

c. (Change to show:) [90:5B]

> 1 yard is 36 inches.
>
> Change 5 yards into inches.

- Read the problem. (Signal.) *Change 5 yards into inches.*
  The related units are yards and inches.
- What are the related units? (Signal.) *Yards and inches.*
  If you change yards into inches, you'll end up with more than 5.
- So do you multiply or divide? (Signal.) *Multiply.*
  Yes, you multiply 5 × 36.
- Say the multiplication problem. (Signal.) *5 × 36.*

d. (Change to show:) [90:5C]

> 1 yard is 36 inches.
>
> Change 3 yards into inches.

- Read the problem. (Signal.) *Change 3 yards into inches.*
- What are the related units? (Signal.) *Yards and inches.*

- If you change 3 yards into inches, will you end up with more than 3? (Signal.) *Yes.*
- So do you multiply or divide? (Signal.) *Multiply.*
- Say the problem you'll work. (Signal.) *3 × 36.*

e. (Change to show:) [90:5D]

> 1 yard is 36 inches.
>
> Change 48 yards into inches.

- Read the problem. (Signal.) *Change 48 yards into inches.*
- What are the related units? (Signal.) *Yards and inches.*
- If you change 48 yards into inches, will you end up with more than 48? (Signal.) *Yes.*
- So do you multiply or divide? (Signal.) *Multiply.*
- Say the problem you'll work. (Signal.) *48 × 36.*

f. (Display:) [90:5E]

> 1 hour is 60 minutes.

- Read this fact. (Signal.) *1 hour is 60 minutes.*
- What's the number for hours? (Signal.) *1.*
- What's the number for minutes? (Signal.) *60.*
- How many times more minutes than hours? (Signal.) *60.*
- So if you change hours into minutes, what do you multiply by? (Signal.) *60.*

g. (Add to show:) [90:5F]

> 1 hour is 60 minutes.
>
> Change 7 hours into minutes.

- Read the problem. (Signal.) *Change 7 hours into minutes.*
- Will you end up with more than 7? (Signal.) *Yes.*
- Say the multiplication problem. (Signal.) *7 × 60.*

h. (Change to show:) [90:5G]

> 1 hour is 60 minutes.
>
> Change $\frac{1}{2}$ hour into minutes.

- Read the problem. (Signal.) *Change 1/2 hour into minutes.*
- Will you end up with more than 1/2? (Signal.) *Yes.*
- Say the problem. (Signal.) *1/2 × 60.*

i. (Display:) [90:5H]

> 1 day is 24 hours.

- Read this fact. (Signal.) *1 day is 24 hours.*
- Are there more days or hours? (Signal.) *Hours.*
- How many times more? (Signal.) *24.*

j. (Add to show:) [90:5I]

> 1 day is 24 hours.
>
> Change 13 days into hours.

- Read the problem. (Signal.) *Change 13 days into hours.*
- Will you end up with more than 13? (Signal.) *Yes.*
- Say the problem you work. (Signal.) *13 × 24.*

k. (Change to show:) [90:5J]

> 1 day is 24 hours.
>
> Change 3 days into hours.

- Read the problem. (Signal.) *Change 3 days into hours.*
- Will you end up with more than 3? (Signal.) *Yes.*
- Write the multiplication problem and the answer in part A. Put the unit name **hours** in the answer.
  (Observe students and give feedback.)
- You worked the problem 3 × 24. How many hours is 3 days? (Signal.) *72 hours.*
  (Add to show:) [90:5K]

> 1 day is 24 hours.
>
> Change 3 days into hours.
>
> $$\begin{array}{r} \overset{1}{2}\,4 \\ \times\quad 3 \\ \hline \boxed{7\,2 \text{ hours}} \end{array}$$

Here's what you should have.

l. (Display:) [90:5L]

> 1 hour is 60 minutes.
>
> Change 5 hours into minutes.

- New problem. Read the fact. (Signal.) *1 hour is 60 minutes.*
- Read the problem. (Signal.) *Change 5 hours into minutes.*
- Will you end up with more than 5? (Signal.) *Yes.*
- What will you multiply 5 by to change hours to minutes? (Signal.) *60.*

- Write the problem and the answer.
  (Observe students and give feedback.)
- Check your work. You worked the problem 5 × 60.
- How many minutes is 5 hours? (Signal.) *300 minutes.*

EXERCISE 6: FRACTION WORD PROBLEMS
DIVISION

REMEDY

a. Find part 2 in your textbook. ✔
- Pencils down. ✔
  (Teacher reference:)

R Test 10

> a. How many servings of $\frac{3}{4}$ pound each would somebody get from 6 pounds of salad?
>
> b. Bill has $\frac{45}{7}$ gallons of cream. He wants to fill 5 containers. How much cream will be in each container?
>
> c. You want to make 3 servings from $\frac{55}{2}$ ounces of salad. How many ounces would be in each serving?
>
> d. Each container holds $\frac{7}{8}$ pound of nuts. How many containers can be filled with 14 pounds of nuts?

Part 2
a.

To work these problems, you have to divide. But the problems are written so you have to read them carefully.

b. Problem A: How many servings of 3/4 pound each would somebody get from 6 pounds of salad?
- Raise your hand when you know the amount there is before you divide—3/4 pound or 6 pounds. ✔
- What's the amount? (Signal.) *6 pounds.*
- Raise your hand when you can say the division problem. ✔
- Say the problem. (Signal.) *6 ÷ 3/4.*

c. Problem B: Bill has 45/7 gallons of cream. He wants to fill 5 containers. How much cream will be in each container?
- Raise your hand when you know the amount there is before you divide. ✔
- What's the amount? (Signal.) *45/7 gallons.*
- Say the division problem. (Signal.) *45/7 ÷ 5.*

d. Problem C: You want to make 3 servings from 55/2 ounces of salad. How many ounces would be in each serving?
- Raise your hand when you know the amount there is before you divide. ✔
- What's the amount? (Signal.) *55/2 ounces.*
- Say the division problem. (Signal.) *55/2 ÷ 3.*
  (Repeat until firm.)

e. Go back to problem A. ✔
   How many servings of 3/4 pound each would somebody get from 6 pounds of salad?
- Write the problem you'll work. Stop when you've done that much. Remember, you find the amount you have before you divide. ✔
- Everybody, read the problem you'll work. (Signal.) *6 ÷ 3/4.*
- Work the problem and write the answer with a unit name.
   (Observe students and give feedback.)
- Check your work. You worked the problem 6 ÷ 3/4.
- What's the whole answer? (Signal.) *8 servings.*
   Yes, there are 8 servings of salad.
f. Problem B: Bill has 45/7 gallons of cream. He wants to fill 5 containers. How much cream will be in each container?
- Write the problem you'll work. ✔
- Everybody, read the problem. (Signal.) *45/7 ÷ 5.*
   Yes, Bill has 45/7 gallons. That's what he divides by 5.
- Work the problem and write the answer with a unit name.
   (Observe students and give feedback.)
- Check your work. You worked the problem 45/7 ÷ 5.
- What's the whole answer? (Signal.) *1 and 2/7 gallons.*
   Yes, each container will have 1 and 2/7 gallons.
g. Problem C: You want to make 3 servings from 55/2 ounces of salad. How many ounces would be in each serving?
- Write the problem you'll work. Stop when you've done that much. ✔
- Everybody, read the problem. (Signal.) *55/2 ÷ 3.*
- Work the problem and write the answer with a unit name.
   (Observe students and give feedback.)
- Check your work. You worked the problem 55/2 ÷ 3.
- What's the whole answer? (Signal.) *9 and 1/6 ounces.*
   Yes, each serving is 9 and 1/6 ounces.
h. Problem D: Each container holds 7/8 pound of nuts. How many containers can be filled with 14 pounds of nuts?
- Write the problem you'll work. ✔
- Everybody, read the problem. (Signal.) *14 ÷ 7/8.*

- Work the problem and write the answer with a unit name.
   (Observe students and give feedback.)
- Check your work. You worked the problem 14 ÷ 7/8.
- What's the whole answer? (Signal.)
   16 containers.
   Yes, 16 containers can be filled with nuts.

EXERCISE 7: WORD PROBLEMS
PARTS OF A GROUP

a. (Display:)   [90:7A]

> $\frac{3}{4}$ of the books were open.
> There were 20 closed books.
> How many books were there
> in all?

3/4 of the books were open. There were 20 closed books. How many books were there in all?
You're going to write an equation for the facts.
- Raise your hand when you know if the facts give two names or three names. ✔
- How many names do the facts give? (Signal.) *3 names.*
b. In part B, write the equation with letters. Below, write the equation with the number the problem gives. ✔
- Everybody, read the equation with letters. (Signal.) *1/4 B = C.*
- Read the equation with a number. (Signal.) *1/4 B = 20.*
c. Work the problem and write the answer. ✔
- Everybody, how many books were there in all? (Signal.) *80 books.*
d. (Display:)   [90:7B]

> $\frac{5}{8}$ of the flowers were blooming.
> 15 flowers were not blooming.
> How many flowers were there?

New problem: 5/8 of the flowers were blooming. 15 flowers were not blooming. How many flowers were there?
- Does the problem give two names or three names? (Signal.) *3 names.*

e. Write the equation with letters. Below, write the equation with the number the problem gives. ✔
• Everybody, read the equation with letters. (Signal.) *3/8 F = N.*
• Read the equation with a number. (Signal.) *3/8 F = 15.*
f. Work the problem and write the answer. ✔
• Everybody, how many flowers were there? (Signal.) *40 flowers.*
g. (Display:) [90:7C]

> $\frac{2}{7}$ of the dogs were running. 20 dogs were not running. How many dogs were running? How many dogs were there in all?

New problem: 2/7 of the dogs were running. 20 dogs were not running. How many dogs were running? How many dogs were there in all?
This problem asks two questions. You work it the same way you work the other problems with three names. You write the equation for the name that has a number and solve the equation. That answers one of the questions.
h. Write the equation and solve it. Write the answer with a unit name. Stop when you've done that much.
(Observe students and give feedback.)
i. I'll read the questions the problem asks. You'll tell me which question you can now answer:
How many dogs were running?
How many dogs were there in all?
• Which question can you answer? (Signal.) *How many dogs were there in all?*
• What's the answer? (Signal.) *28 dogs.*
j. (Add to show:) [90:7D]

> $\frac{2}{7}$ of the dogs were running. 20 dogs were not running. How many dogs were running? How many dogs were there in all?
>
> $$\frac{5}{7} d = n$$
> $$\left(\frac{7}{5}\right) \frac{5}{7} d = \overset{4}{\cancel{20}} \left(\frac{7}{\cancel{5}}\right)$$
> $$d = 28$$
>
> 28 dogs

Here's what you should have.

k. The other question is: How many dogs were running?
Now you can figure out how many dogs were running.
• Raise your hand when you can say the problem. ✔
• Say the problem. (Signal.) *28 – 20.*
l. Figure out the answer and write the unit name—running dogs. ✔
• Everybody, how many dogs were running? (Signal.) *8 running dogs.*
m. (Display:) [90:7E]

> 36 athletes were women. $\frac{3}{5}$ of the athletes were men. How many athletes were there in all? How many athletes were men?

New problem: 36 athletes were women. 3/5 of the athletes were men. How many athletes were there in all? How many athletes were men?
n. Write the letter equation for the facts. Stop when you've done that much. ✔
• Everybody, read the letter equation. (Signal.) *2/5 A = W.*
(Add to show:) [90:7F]

> 36 athletes were women. $\frac{3}{5}$ of the athletes were men. How many athletes were there in all? How many athletes were men?
>
> $$\frac{2}{5} a = w$$

o. Put in the number the problem gives and solve for the other letter. Raise your hand when you know which question you have answered. ✔
• Everybody, which question have you answered? (Signal.) *How many athletes were there in all?*
• What's the answer? (Signal.) *90 athletes.*
p. Now figure out the answer to the other question. Work the problem and write the answer with a unit name. ✔
• You worked the problem 90 minus 36. Everybody, how many athletes were men? (Signal.) *54 men.*

EXERCISE 8: INDEPENDENT WORK
AVERAGE

a. Find part 3 in your textbook. ✔
(Teacher reference:)

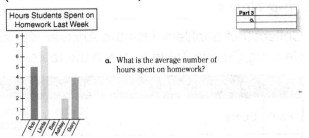

You're going to figure out the average for this problem. Remember, you divide by the number of people—even if one person has a score of zero.

- How many people does this problem show? (Signal.) *5.*
- So what are you going to divide by to find the average? (Signal.) *5.*
  You'll work the problem as part of your independent work. Write the answer as a mixed number with a unit name.

Assign Independent Work, Textbook parts 3–10.

Optional extra math-fact practice worksheets are available on ConnectED.

Mastery Test ⑨

Teacher Presentation

a. Find Test 9 in your test booklet. ✔
   You'll work the test by yourself. Read the directions carefully and do your best work. Put your pencil down when you've finished the test.
   (Observe but do not give feedback.)

Scoring Notes

a. Collect test booklets. Use the *Answer Key* and Passing Criteria Table to score the tests.

Passing Criteria Table—Mastery Test 9			
Part	Score	Possible Score	Passing Score
1	1 for each item	3	3
2	1 for each letter equation	6	5
3	3 for each item (substitution, simplification, answer)	9	8
4	2 for each item (addition answer, simplified answer)	6	5
5	3 for each item (addition answer, division problem, answer with unit name)	3	3
6	2 for each item (rewritten top number, answer)	6	5
7	1 for each item	4	3
8	3 for each item (expression, simplification, answer)	12	10
9	2 for each item (mixed-number problem, answer with unit name)	8	6
	Total	57	

b. Complete the Mastery Test 9 Remedy Summary Sheet to determine whether group remedies are needed. Reproducible Remedy Summary Sheets are at the back of the *Answer Key* and the back of the *Teacher's Guide.*

• If ¼ or more of the students did not pass a test part, present the remedy for that part before beginning Lesson 91. The Remedy Table follows and also appears at the end of the Mastery Test 9 *Answer Key.* Remedies worksheets follow Mastery Test 9 in the *Student Assessment Book.*

Part	Test Items	Remedy Lesson	Remedy Ex.	Remedies Worksheet	Textbook
1	Fractions (As Mixed Numbers)	81	6	—	Part 3
		82	6	—	Part 2
2	Sentences (Parts of a group)	86	5	—	Part 4
		87	8	—	Part 3
3	Expressions (Evaluation)	85	4	—	Part 2
		86	4	—	Part 3
4	Mixed-Number Operations (Addition)	84	2	Part A	—
		85	2	Part B	—
5	Average	88	6	—	Part 3
		89	6	—	Part 3
6	Mixed-Number Operations (Subtraction)	88	3	—	—
		89	3	Part C	—
7	Exponents (Multiplication/ Division by Base 10)	81	2	Part D	—
		83	3	Part E	—
		86	2	—	Part 1
8	Expressions (From Written Directions)	88	4	—	Part 1
		89	4	—	Part 1
9	Mixed-Number Operations (Word Problems)	81	4	—	Part 1
		82	4	—	Part 1

Table header: Remedy Table—Mastery Test 9

Retest
Retest individual students on any part failed.

	Lesson 91	Lesson 92	Lesson 93	Lesson 94	Lesson 95
Student Learning Objectives	**Exercises** 1. Multiply, divide, and find whole numbers for fractions using mental math 2. Compare products and factors in fraction multiplication problems 3. Understand balance beams **4. Add mixed numbers with unlike denominators** 5. Use multiplication to convert units of measurement 6. Solve fraction word problems with division 7. Solve parts-of-a-group word problems 8. Complete work independently	**Exercises** 1. Multiply, divide, and find whole numbers for fractions using mental math 2. Understand balance beams **3. Add and subtract mixed numbers with unlike denominators** 4. Solve fraction word problems with division 5. Use multiplication to convert units of measurement 6. Compare products and factors in fraction multiplication problems 7. Say equations for parts-of-a-group sentences Complete work independently	**Exercises** 1. Add using mental math **2. Solve long division problems with 2-digit quotients** **3. Find average distance on balance beams** 4. Compare products and factors in fraction multiplication problems **5. Use the associative property of multiplication to solve volume problems** **6. Use multiplication and division to convert units of measurement** 7. Solve parts-of-a-group word problems 8. Complete work independently	**Exercises** 1. Add using mental math 2. Solve long division problems with 2-digit quotients 3. Find average distance on balance beams 4. Compare products and factors in fraction multiplication problems 5. Use multiplication and division to convert units of measurement 6. Solve parts-of-a-group word problems 7. Use the associative property of multiplication to solve volume problems **8. Solve mixed-number word problems** Complete work independently	**Exercises** 1. Add and subtract using mental math 2. Solve long division problems with 2-digit quotients **3. Use fractions to show probability** 4. Find average distance on balance beams 5. Use multiplication and division to convert units of measurement **6. Solve complex volume problems** 7. Solve mixed-number word problems Complete work independently

Common Core State Standards for Mathematics					
5.OA 1	✔	✔	✔	✔	✔
5.NBT 1–2			✔		✔
5.NBT 3		✔	✔		✔
5.NBT 5	✔	✔	✔	✔	✔
5.NBT 6	✔		✔		✔
5.NF 1	✔	✔	✔	✔	✔
5.NF 2		✔			
5.NF 3				✔	
5.NF 4–5	✔	✔	✔	✔	✔
5.NF 6	✔		✔	✔	✔
5.NF 7	✔	✔	✔	✔	✔
5.MD 1	✔	✔	✔	✔	✔
5.MD 5			✔	✔	✔
5.G 2	✔			✔	
Teacher Materials	Presentation Book 2, Board Displays CD or chalk board				
Student Materials	Textbook, Workbook, pencil, lined paper, ruler				
Additional Practice	• Student Practice Software: Block 4: Activity 1 (5.NF 5), Activity 2, Activity 3 (5.NBT 4), Activity 4 (5.NBT 3), Activity 5 (5.NBT 7), Activity 6 (5.MD 3 and 5.MD 4) • Provide needed fact practice with Level D or E Math Fact Worksheets.				
Mastery Test					

Lesson

EXERCISE 1: MENTAL MATH

a. Time for some mental math.
    You'll tell me the whole number each fraction
    equals.
- Listen: 20 fifths. What whole number?
    (Signal.) *4.*
- 30 fifths. What whole number? (Signal.) *6.*
- 45 fifths. What whole number? (Signal.) *9.*
- 60 tenths. What whole number? (Signal.) *6.*
- 90 tenths. What whole number? (Signal.) *9.*
    (Repeat until firm.)
b. Listen: 28 sevenths. What whole number?
    (Signal.) *4.*
- 36 ninths. What whole number? (Signal.) *4.*
- 48 eighths. What whole number? (Signal.) *6.*
    (Repeat until firm.)
c. Listen: 90 divided by 3. Say the fact. (Signal.)
    90 ÷ 3 = 30.
- 360 ÷ 4. Say the fact. (Signal.) *360 ÷ 4 = 90.*
- 180 ÷ 2. Say the fact. (Signal.) *180 ÷ 2 = 90.*
    (Repeat until firm.)
d. Listen: 450 ÷ 9. Say the fact. (Signal.)
    450 ÷ 9 = 50.
- 480 ÷ 8. Say the fact. (Signal.) *480 ÷ 8 = 60.*
- 320 ÷ 8. Say the fact. (Signal.) *320 ÷ 8 = 40.*
    (Repeat until firm.)
e. Listen: 30 times 6. Say the fact. (Signal.)
    30 × 6 = 180.
- 20 × 8. Say the fact. (Signal.) *20 × 8 = 160.*
- 50 × 3. Say the fact. (Signal.) *50 × 3 = 150.*
- 70 × 3. Say the fact. (Signal.) *70 × 3 = 210.*
    (Repeat until firm.)
f. Listen: 40 × 6. Say the fact. (Signal.)
    40 × 6 = 240.
- 40 × 7. Say the fact. (Signal.) *40 × 7 = 280.*
- 60 × 7. Say the fact. (Signal.) *60 × 7 = 420.*
- 60 × 8. Say the fact. (Signal.) *60 × 8 = 480.*
    (Repeat until firm.)

EXERCISE 2: FRACTION MULTIPLICATION
PRODUCT/FACTOR COMPARISON

a. Open your workbook to Lesson 91 and find
    part 1. ✔
    (Teacher reference:)

$$a.\ \frac{6}{5}\left(\frac{9}{8}\right)= \qquad b.\ \frac{3}{7}\left(\frac{5}{4}\right)= \qquad c.\ \frac{9}{4}\times\frac{6}{7}=$$

$$\frac{6}{5} \qquad\qquad \frac{3}{7} \qquad\qquad \frac{9}{4}$$

You're going to write the answer to each
problem and complete the statement to show
if the fraction you end with is more than or
less than the fraction you start with.
b. Read problem A. (Signal.) *6/5 × 9/8.*
- What fraction are you multiplying 6/5 by?
    (Signal.) *9/8.*
- Is 9/8 more than 1 or less than 1? (Signal.)
    More than 1.
- So will you end with a fraction that's more than
    6/5 or less than 6/5? (Signal.) *More than 6/5.*
    (Repeat until firm.)
c. Read problem B. (Signal.) *3/7 × 5/4.*
- What fraction are you multiplying by?
    (Signal.) *5/4.*
- Is 5/4 more than 1 or less than 1? (Signal.)
    More than 1.
- So will you end with more than 3/7 or less
    than 3/7? (Signal.) *More than 3/7.*
    (Repeat until firm.)
d. Read problem C. (Signal.) *9/4 × 6/7.*
- What fraction are you multiplying by?
    (Signal.) *6/7.*
- Is 6/7 more than 1 or less than 1? (Signal.)
    Less than 1.
- So will you end with more than 9/4 or less
    than 9/4? (Signal.) *Less than 9/4.*
    (Repeat until firm.)

e. Go back and write the answer to each problem. Don't simplify. Just do the multiplication. Below, complete the statement to show if the fraction you end with is more or less than the fraction you start with. **(Observe students and give feedback.)**
f. Check your work.
- Problem A. Everybody, what does 6/5 × 9/8 equal? (Signal.) *54/40.*
- Do you multiply by more than 1 or less than 1? (Signal.) *More than 1.*
- So do you end with more or less than you start with? (Signal.) *More than you start with.* (Display:) [91:2A]

$$a. \quad \frac{6}{5}\left(\frac{9}{8}\right) = \frac{54}{40}$$

$$\frac{6}{5} < \frac{54}{40}$$

Here's what you should have for problem A.
g. Problem B. Everybody, what does 3/7 × 5/4 equal? (Signal.) *15/28.*
- Do you multiply by more than 1 or less than 1? (Signal.) *More than 1.*
- So do you end with more or less than you start with? (Signal.) *More than you start with.* (Display:) [91:2B]

$$b. \quad \frac{3}{7}\left(\frac{5}{4}\right) = \frac{15}{28}$$

$$\frac{3}{7} < \frac{15}{28}$$

Here's what you should have for problem B.
h. Problem C. Everybody, what does 9/4 × 6/7 equal? (Signal.) *54/28.*
- Do you multiply by more than 1 or less than 1? (Signal.) *Less than 1.*
- So do you end with more or less than you start with? (Signal.) *Less than you start with.* (Display:) [91:2C]

$$c. \quad \frac{9}{4} \times \frac{6}{7} = \frac{54}{28}$$

$$\frac{9}{4} > \frac{54}{28}$$

Here's what you should have for problem C.

EXERCISE 3: BALANCE BEAM
TOTAL DISTANCE

a. (Display:) [91:3A]

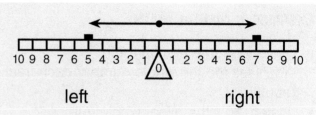

You learned that a balance beam will balance if the total distance on each side is the same.
- (Point left.) What's the distance of the weight on the left side? (Signal.) *5.*
- (Point right.) What's the distance of the weight on the right side? (Signal.) *7.*
- Is the distance the same on both sides? (Signal.) *No.*
- So will the beam balance? (Signal.) *No.*
- Which side has the greater total distance? (Signal.) *The right side.*
- So which side will go down? (Signal.) *The right side.*
(Change to show:) [91:3B]

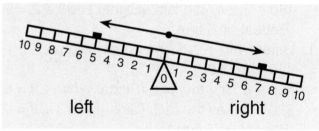

b. Let's change the total distance on the left side so the beam will balance.
- How much distance do I add on that side? (Signal.) *2.*
Yes, if we add 2 units of distance on the left, the beam will balance.
(Change to show:) [91:3C]

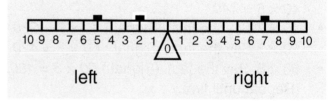

a. Find part 2 in your workbook. ✔
(Teacher reference:)

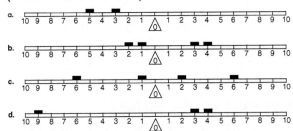

For each balance beam, you're going to add one weight to the side that has less distance. You'll make the total distance the same on both sides.

b. Problem A. Add one weight so the distances on both sides are the same. ✔
• Where did you place the weight on the right side? (Signal.) *8.*
Yes, that gives a total distance of 8 on both sides.

c. Problem B. Raise your hand when you know the total distance on each side. ✔
• What's the total distance on the left? (Signal.) *3.*
• What's the total distance on the right? (Signal.) *7.*
• Which side needs more distance? (Signal.) *The left side.*
• How much more distance? (Signal.) *4.*

d. Show the weight for B. Then fix up the rest of the beams so they balance. Remember, you can only add **one weight** to the side that has less distance.
(Observe students and give feedback.)

e. Check your work.
• Problem C. Which side needed more distance? (Signal.) *The left side.*
• Where did you place your weight on that side? (Signal.) *At 1.*

f. Problem D. Which side needed more distance? (Signal.) *The right side.*
• Where did you place your weight on that side? (Signal.) *At 2.*

EXERCISE 4: MIXED-NUMBER OPERATIONS
UNLIKE DENOMINATORS REMEDY

a. (Display:) [91:4A]

$$2\frac{1}{5}$$
$$+4\frac{2}{3}$$

Here's a new kind of mixed-number problem.
• Read the problem. (Signal.) *2 and 1/5 + 4 and 2/3.*
We can't add the fractions the way they are written. So we find the lowest common denominator for 1/5 and 2/3.
Remember, the lowest common denominator is the first common number for counting by 5 and 3.
• Raise your hand when you know the lowest common denominator for fifths and thirds. ✔
• What's the lowest common denominator? (Signal.) *15.*
(Add to show:) [91:4B]

$$2\frac{1}{5}\left(\frac{}{}\right) = \frac{}{15}$$
$$+4\frac{2}{3}\left(\frac{}{}\right) = +\frac{}{15}$$

b. You multiply 1/5 by a fraction of 1. What fraction? (Signal.) *3/3.*
(Add to show:) [91:4C]

$$2\frac{1}{5}\left(\frac{3}{3}\right) = \frac{}{15}$$
$$+4\frac{2}{3}\left(\frac{}{}\right) = +\frac{}{15}$$

• You multiply 2/3 by a fraction of 1. What fraction? (Signal.) *5/5.*
(Add to show:) [91:4D]

$$2\frac{1}{5}\left(\frac{3}{3}\right) = \frac{}{15}$$
$$+4\frac{2}{3}\left(\frac{5}{5}\right) = +\frac{}{15}$$

• (Point to **1/5**.) What's 1 × 3? (Signal.) *3.*
• (Point to **2/3**.) What's 2 × 5? (Signal.) *10.*

(Add to show:) [91:4E]

$$2\frac{1}{5}\left(\frac{3}{3}\right) = \frac{3}{15}$$
$$+4\frac{2}{3}\left(\frac{5}{5}\right) = +\frac{10}{15}$$

c. Read the new problem for the fractions. (Signal.) *3/15 + 10/15.*
- What's the answer? (Signal.) *13/15.*
(Add to show:) [91:4F]

$$2\frac{1}{5}\left(\frac{3}{3}\right) = \frac{3}{15}$$
$$+4\frac{2}{3}\left(\frac{5}{5}\right) = +\frac{10}{15}$$
$$\frac{13}{15}$$

- Read the problem for the whole numbers. (Signal.) *2 + 4.*
- What's the answer? (Signal.) *6.*
(Add to show:) [91:4G]

$$2\frac{1}{5}\left(\frac{3}{3}\right) = \frac{3}{15}$$
$$+4\frac{2}{3}\left(\frac{5}{5}\right) = +\frac{10}{15}$$
$$6\frac{13}{15}$$

So the whole answer is 6 and 13/15.

=== **WORKBOOK PRACTICE** ===

a. Find part 3 in your workbook. ✔
(Teacher reference:) [R] **Part A**

a. $10\frac{3}{8}$
  $+ 2\frac{1}{4}$

b. $9\frac{5}{6}$
  $- 8\frac{3}{4}$

c. $6\frac{2}{7}$
  $+ 11\frac{2}{3}$

These are like the problem we just worked.
b. Read problem A. (Signal.) *10 and 3/8 + 2 and 1/4.*
- Work the problem for the fractions. Stop when you've done that much.
(Observe students and give feedback.)

(Display:) [91:4H]

a. $$10\frac{3}{8}\left(\frac{1}{1}\right) = \frac{3}{8}$$
$$+2\frac{1}{4}\left(\frac{2}{2}\right) = +\frac{2}{8}$$
$$\frac{5}{8}$$

Here's what you should have.
You changed 1/4 into 2/8. You worked the problem 3/8 + 2/8.
- What's the answer? (Signal.) *5/8.*
- Write the answer for the whole numbers. ✔
- Everybody, what's the answer for the whole numbers? (Signal.) *12.*
(Add to show:) [91:4I]

a. $$10\frac{3}{8}\left(\frac{1}{1}\right) = \frac{3}{8}$$
$$+2\frac{1}{4}\left(\frac{2}{2}\right) = +\frac{2}{8}$$
$$12\frac{5}{8}$$

So the whole answer is 12 and 5/8.
c. Read problem B. (Signal.) *9 and 5/6 – 8 and 3/4.*
- Work the problem. Remember, first figure out the lowest common denominator for the fractions.
(Observe students and give feedback.)
d. Check your work.
(Display:) [91:4J]

b. $$9\frac{5}{6}\left(\frac{2}{2}\right) = \frac{10}{12}$$
$$-8\frac{3}{4}\left(\frac{3}{3}\right) = -\frac{9}{12}$$
$$1\frac{1}{12}$$

Here's what you should have.
You rewrote 5/6 as 10/12.
You rewrote 3/4 as 9/12.
So the answer is 1 and 1/12.
e. Read problem C. (Signal.) *6 and 2/7 + 11 and 2/3.*
- Work the problem. Remember, first figure out the lowest common denominator for the fractions.
(Observe students and give feedback.)

f. Check your work.
(Display:) [91:4K]

$$c. \quad 6\frac{2}{7}\left(\frac{3}{3}\right) = \frac{6}{21}$$
$$+ 11\frac{2}{3}\left(\frac{7}{7}\right) = +\frac{14}{21}$$
$$\overline{\qquad\qquad 17\frac{20}{21}}$$

Here's what you should have.
You rewrote 2/7 as 6/21.
You rewrote 2/3 as 14/21.
So the answer is 17 and 20/21.

EXERCISE 5: UNIT CONVERSION
MULTIPLICATION

a. (Display:) [91:5A]

> 1 day is 24 hours.

This fact shows related units.
- (Point to **1 day.**) Read this fact. (Signal.) *1 day is 24 hours.*
- Are there more days or hours? (Signal.) *Hours.*
- How many times more? (Signal.) *24.*
- So if you change days into hours, do you multiply or divide? (Signal.) *Multiply.*
- What do you multiply by? (Signal.) *24.*
b. (Add to show:) [91:5B]

> 1 day is 24 hours.
> Change 9 days into hours.

- Read the problem. (Signal.) *Change 9 days into hours.*
- Will you end up with more than 9? (Signal.) *Yes.*
- Say the problem you work. (Signal.) *9 × 24.*
- Work the problem in part A. Write it as a column problem: 24 × 9. Remember the unit name in the answer.
  (Observe students and give feedback.)
- Everybody, how many hours is 9 days? (Signal.) *216 hours.*
c. (Display:) [91:5C]

> 1 year is 12 months.

- (Point to **1 year.**) Read this fact. (Signal.) *1 year is 12 months.*
- Are there more years or months? (Signal.) *Months.*
- How many times more? (Signal.) *12.*
d. (Add to show:) [91:5D]

> 1 year is 12 months.
> Change 3 years into months.

- Read the problem. (Signal.) *Change 3 years into months.*
- Will you end up with more than 3? (Signal.) *Yes.*
- Work the problem. Remember the unit name. ✔
- Everybody, how many months is 3 years? (Signal.) *36 months.*
  (Display:) [91:5E]

> 1 dollar is 4 quarters.

e. (Point to **1 dollar.**) Read this fact. (Signal.) *1 dollar is 4 quarters.*
- Are there more dollars or quarters? (Signal.) *Quarters.*
- How many times more? (Signal.) *4.*
f. (Add to show:) [91:5F]

> 1 dollar is 4 quarters.
> Change 12 dollars into quarters.

- Read the problem. (Signal.) *Change 12 dollars into quarters.*
- Work the problem. Remember the unit name. ✔
- Everybody, how many quarters is 12 dollars? (Signal.) *48 quarters.*

TEXTBOOK PRACTICE

a. Open your textbook to Lesson 91 and find part 1. ✔
(Teacher reference:)

a. 1 week is 7 days. Change 11 weeks into days.
b. 1 year is 12 months. Change 10 years into months.
c. 1 hour is 60 minutes. Change 8 hours into minutes.

b. Problem A: 1 week is 7 days. Change 11 weeks into days.
- What do you multiply weeks by to get days? (Signal.) *7.*
- Change 11 weeks into days. Work the problem.
 (Observe students and give feedback.)
- Check your work. You worked the problem 11 × 7.
- How many days is 11 weeks? (Signal.) *77 days.*

c. Problem B: 1 year is 12 months. Change 10 years into months.
- Work the problem.
 (Observe students and give feedback.)
- Check your work. You worked the problem 10 × 12.
- How many months is 10 years? (Signal.) *120 months.*

d. Problem C: 1 hour is 60 minutes. Change 8 hours into minutes.
- Work the problem.
 (Observe students and give feedback.)
- Check your work. You worked the problem 8 × 60.
- How many minutes is 8 hours? (Signal.) *480 minutes.*

EXERCISE 6: FRACTION WORD PROBLEMS
DIVISION
REMEDY

a. Find part 2 in your textbook. ✔
- Pencils down. ✔
 (Teacher reference:)

a. Mary uses $\frac{18}{5}$ pounds of potatoes to make mashed potatoes. She needs to make 6 meals. How many pounds is each serving of mashed potatoes?

b. Each perfume bottle holds $\frac{2}{5}$ ounce. How many bottles can be filled if there are 4 ounces of perfume?

c. How many $\frac{5}{8}$-pound servings of shrimp can be made from 10 pounds?

d. A whole pizza weighs $\frac{7}{2}$ pounds. The pizza is cut into 8 equal slices. What does each slice weigh?

To work these problems, you have to divide. But the problems are written so you have to read them carefully.

b. Problem A: Mary uses 18/5 pounds of potatoes to make mashed potatoes. She needs to make 6 meals. How many pounds is each serving of mashed potatoes?
- Raise your hand when you know the amount there is before you divide. ✔
- What's the amount? (Signal.) *18/5 pounds.*
- Say the division problem. (Signal.) *18/5 ÷ 6.*

c. Problem B: Each perfume bottle holds 2/5 ounce. How many bottles can be filled if there are 4 ounces of perfume?
- Raise your hand when you know the amount there is before you divide. ✔
- What's the amount? (Signal.) *4 ounces.*
- Say the division problem. (Signal.) *4 ÷ 2/5.*

d. Problem C: How many 5/8-pound servings of shrimp can be made from 10 pounds?
- Raise your hand when you know the amount there is before you divide. ✔
- What's the amount? (Signal.) *10 pounds.*
- Say the division problem. (Signal.) *10 ÷ 5/8.*

e. Problem D: A whole pizza weighs 7/2 pounds. The pizza is cut into 8 equal slices. What does each slice weigh?
- Raise your hand when you know the amount there is before you divide. ✔
- What's the amount? (Signal.) *7/2 pounds.*
- Say the division problem. (Signal.) *7/2 ÷ 8.*
 (Repeat until firm.)

f. Go back and work problem A. Remember the unit name in the answer.
 (Observe students and give feedback.)
- Check your work. You worked the problem 18/5 ÷ 6.
- What's the whole answer? (Signal.) *3/5 pound.* Yes, each serving size is 3/5 of a pound.

g. Work problem B.
 (Observe students and give feedback.)
- Check your work. You worked the problem 4 ÷ 2/5.
- What's the whole answer? (Signal.) *10 bottles.* Yes, 10 bottles can be filled with perfume.

h. Work problem C.
   (Observe students and give feedback.)
- Check your work. You worked the problem
  10 ÷ 5/8.
- What's the whole answer? (Signal.) *16 servings.*
  Yes, 16 servings can be made.
i. Work problem D.
   (Observe students and give feedback.)
- Check your work. You worked the problem
  7/2 ÷ 8.
- What's the whole answer? (Signal.) *7/16 pound.*
  Yes, each slice weighs 7/16 of a pound. That's
  7 ounces.

EXERCISE 7: WORD PROBLEMS
PARTS OF A GROUP
REMEDY

a. Find part 3 in your textbook. ✔
   (Teacher reference:)

a. $\frac{3}{4}$ of the flowers were roses. 28 flowers were not roses. How many flowers were there? How many roses were there?

b. $\frac{3}{10}$ of the doors were open. There were 28 closed doors. How many doors were there? How many doors were open?

c. $\frac{4}{7}$ of the plates were dry. There were 21 wet plates. How many dry plates were there? How many plates were there in all?

All these problems have three names and two
questions.

b. Problem A: 3/4 of the flowers were roses.
   28 flowers were not roses. How many flowers
   were there? How many roses were there?
- Write the letter equation for the facts. Stop
  when you've done that much. ✔
- Everybody, read the letter equation. (Signal.)
  1/4 F = N.
  (Display:)   [91:7A]

$$a.\ \frac{1}{4}f = n$$

Yes, 1/4 F = N.
c. Put in the number the problem gives and
   work that problem. Pencils down when you've
   answered one of the questions.
   (Observe students and give feedback.)

d. Check your work. You put in a number for N
   and figured out F.
- How many flowers were there? (Signal.)
  112 flowers.
e. Now you can figure out the answer to the
   other question the problem asks.
- Work the problem. ✔
- You worked the problem 112 minus 28. What's
  the whole answer? (Signal.) *84 roses.*
  Yes, there were 84 roses.
f. Problem B: 3/10 of the doors were open. There
   were 28 closed doors. How many doors were
   there? How many doors were open?
- Write the letter equation. Work the problem
  and answer both questions.
  (Observe students and give feedback.)
- The first question asks: How many doors were
  there? What's the answer? (Signal.) *40 doors.*
- The second question asks: How many doors
  were open? What's the answer? (Signal.) *12
  open doors.*
  (Display:)   [91:7B]

b. $\frac{7}{10}d = c$ $\overset{3}{\cancel{4}}\overset{}{10}$
                                  -28
$\left(\frac{10}{7}\right)\frac{7}{10}d = \overset{4}{\cancel{28}}\left(\frac{10}{\cancel{7}}\right)$ $\boxed{12\text{ open doors}}$

$$d = 40$$
$\boxed{40\text{ doors}}$

Here's what you should have.
g. Problem C: 4/7 of the plates were dry. There
   were 21 wet plates. How many dry plates were
   there? How many plates were there in all?
h. Work the problem. Write answers to both
   questions. Remember to put the unit names in
   the answers.
   (Observe students and give feedback.)
i. Check your work.
   You started with the letter equation 3/7 P = W.
   You put in the number for W.
- How many dry plates were there? (Signal.)
  28 dry plates.
- How many plates were there in all? (Signal.)
  49 plates.

EXERCISE 8: INDEPENDENT WORK

MIXED-NUMBER SUBTRACTION

a. Find part 4 in your textbook. ✔
   (Teacher reference:)

a. $5\frac{2}{9}$ $-2\frac{7}{9}$ b. 7 $-3\frac{2}{5}$ c. $5\frac{2}{8}$ $-1\frac{7}{8}$ d. $12\frac{1}{5}$ $-3\frac{3}{5}$

These problems are part of your independent work. You have to rewrite a number before you work each problem.

b. Read problem A. (Signal.) *5 and 2/9 – 2 and 7/9.*
   • Can you work the problem for the fraction? (Signal.) *No.*
     So you rewrite the top number— 5 and 2/9.
   • What will the new whole number be? (Signal.) *4.*
   • Raise your hand when you know the new fraction. ✔
   • What's the new fraction? (Signal.) *11/9.*
     (Repeat until firm.)

c. Read problem B. (Signal.) *7 – 3 and 2/5.*
   • Can you work that problem? (Signal.) *No.*
     So you rewrite the top number.
   • What will the new whole number be? (Signal.) *6.*
   • Raise your hand when you know the new fraction. ✔
   • What's the new fraction? (Signal.) *5/5.*
     (Repeat until firm.)

Assign Independent Work, Textbook parts 4–12 and Workbook part 4.

Optional extra math-fact practice worksheets are available on ConnectED.

Lesson 92

EXERCISE 1: MENTAL MATH

a. Time for some mental math.
   You'll tell me the whole number each fraction equals.
   - Listen: 30 fifths. What whole number? (Signal.) *6.*
   - 40 fifths. What whole number? (Signal.) *8.*
   - 45 fifths. What whole number? (Signal.) *9.*
   - 70 tenths. What whole number? (Signal.) *7.*
   - 90 tenths. What whole number? (Signal.) *9.*
   (Repeat until firm.)

b. Listen: 21 sevenths. What whole number? (Signal.) *3.*
   - 36 sixths. What whole number? (Signal.) *6.*
   - 48 sixths. What whole number? (Signal.) *8.*
   (Repeat until firm.)

c. Listen: 90 divided by 3. Say the fact. (Signal.) $90 \div 3 = 30.$
   - $270 \div 3$. Say the fact. (Signal.) $270 \div 3 = 90.$
   - $180 \div 90$. Say the fact. (Signal.) $180 \div 90 = 2.$
   - $450 \div 5$. Say the fact. (Signal.) $450 \div 5 = 90.$
   - $320 \div 4$. Say the fact. (Signal.) $320 \div 4 = 80.$
   - $420 \div 6$. Say the fact. (Signal.) $420 \div 6 = 70.$
   (Repeat until firm.)

d. Listen: 40 times 6. Say the fact. (Signal.) $40 \times 6 = 240.$
   - 30×8. Say the fact. (Signal.) $30 \times 8 = 240.$
   - 30×3. Say the fact. (Signal.) $30 \times 3 = 90.$
   - 60×3. Say the fact. (Signal.) $60 \times 3 = 180.$
   - 60×6. Say the fact. (Signal.) $60 \times 6 = 360.$
   (Repeat until firm.)

e. Listen: 40×7. Say the fact. (Signal.) $40 \times 7 = 280.$
   - 60×8. Say the fact. (Signal.) $60 \times 8 = 480.$
   - 60×9. Say the fact. (Signal.) $60 \times 9 = 540.$
   (Repeat until firm.)

EXERCISE 2: BALANCE BEAM
TOTAL DISTANCE

a. Open your workbook to Lesson 92 and find part 1. ✔
   (Teacher reference:)

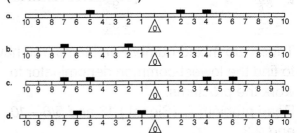

   You're going to make the sides balance by adding one weight to the side that has less total distance.

b. Fix up beam A so it balances. ✔
   - Everybody, which side had less total distance? (Signal.) *The left side.*
   - How much more distance did that side need? (Signal.) *1.*
   So you added a weight at 1.
   (Display:)   [92:2A]

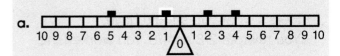

   Here's what you should have.

c. Fix up the rest of the beams so they balance.
   (Observe students and give feedback.)

d. Check your work.
   - Problem B. Which side needed more distance? (Signal.) *The right side.*
   - Where did you place the weight on that side? (Signal.) *At 9.*

e. C. Which side needed more distance? (Signal.) *The right side.*
   - Where did you place the weight on that side? (Signal.) *At 2.*

f. D. Which side needed more distance? (Signal.) *The left side.*
   - Where did you place the weight on that side? (Signal.) *At 3.*

EXERCISE 3: MIXED-NUMBER OPERATIONS

UNLIKE DENOMINATORS

REMEDY

a. Find part 2 in your workbook. ✔
(Teacher reference:) R Part B

a. $13\frac{5}{6}$
   $-10\frac{2}{3}$

b. $4\frac{2}{5}$
   $+15\frac{1}{4}$

c. $6\frac{5}{8}$
   $-2\frac{1}{6}$

d. $9\frac{2}{5}$
   $+3\frac{3}{10}$

These are mixed-number problems. You have to find the lowest common denominator to work the problems.

b. Read problem A. (Signal.) *13 and 5/6 – 10 and 2/3.*

• Work the problem for the fractions. Stop when you've done that much.
(Observe students and give feedback.)
(Display:) [92:3A]

$$a. \quad 13\frac{5}{6}\left(\frac{1}{1}\right) = \frac{5}{6}$$
$$-10\frac{2}{3}\left(\frac{2}{2}\right) = -\frac{4}{6}$$
$$\frac{1}{6}$$

Here's what you should have.
You rewrote 2/3 as 4/6. You worked the problem 5/6 – 4/6.

• What's the answer? (Signal.) *1/6.*
• Work the problem for the whole numbers. ✔
• Everybody, what's the answer for the whole numbers? (Signal.) *3.*
So the answer to the whole problem is 3 and 1/6.

c. Read problem B. (Signal.) *4 and 2/5 + 15 and 1/4.*

• Work the problem. Remember, first figure out the lowest common denominator for the fractions.
(Observe students and give feedback.)
(Display:) [92:3B]

$$b. \quad 4\frac{2}{5}\left(\frac{4}{4}\right) = \frac{8}{20}$$
$$+15\frac{1}{4}\left(\frac{5}{5}\right) = +\frac{5}{20}$$
$$19\frac{13}{20}$$

Here's what you should have.
You rewrote 2/5 as 8/20 and 1/4 as 5/20.

• What's 8/20 + 5/20? (Signal.) *13/20.*
• What's 4 + 15? (Signal.) *19.*
So the answer to the whole problem is 19 and 13/20.

d. Read problem C. (Signal.) *6 and 5/8 – 2 and 1/6.*

• Work the problem. Remember, first figure out the lowest common denominator for the fractions.
(Observe students and give feedback.)
(Display:) [92:3C]

$$c. \quad 6\frac{5}{8}\left(\frac{3}{3}\right) = \frac{15}{24}$$
$$-2\frac{1}{6}\left(\frac{4}{4}\right) = -\frac{4}{24}$$
$$4\frac{11}{24}$$

Here's what you should have.
You rewrote 5/8 as 15/24 and 1/6 as 4/24.

• What's 15/24 – 4/24? (Signal.) *11/24.*
• What's 6 – 2? (Signal.) *4.*
So the answer to the whole problem is 4 and 11/24.

e. Read problem D. (Signal.) *9 and 2/5 + 3 and 3/10.*

• Work the problem. Remember, first figure out the lowest common denominator for the fractions.
(Observe students and give feedback.)
(Display:) [92:3D]

$$d. \quad 9\frac{2}{5}\left(\frac{2}{2}\right) = \frac{4}{10}$$
$$+3\frac{3}{10}\left(\frac{1}{1}\right) = +\frac{3}{10}$$
$$12\frac{7}{10}$$

Here's what you should have.
You rewrote 2/5 as 4/10.

• What's 4/10 + 3/10? (Signal.) *7/10.*
• What's 9 + 3? (Signal.) *12.*
So the answer to the whole problem is 12 and 7/10.

EXERCISE 4: FRACTION WORD PROBLEMS
DIVISION

a. Open your textbook to Lesson 92 and find part 1. ✔
   (Teacher reference:)

   a. A bag has 14 pounds of sand in it. Sarah makes piles that are each $\frac{2}{7}$ pound. How many piles does she make?

   b. Each cup holds $\frac{1}{5}$ quart. How many cups can be filled if there are 3 quarts of water?

   c. A cook divided a salad into 4 equal servings. There were $\frac{7}{2}$ pounds of salad. How much did each serving weigh?

   d. A pie weighs $\frac{8}{3}$ pounds. It is cut into slices that each weigh $\frac{1}{9}$ pound. How many slices are there?

Part 1	
a.	

   You divide to work these problems.
   Remember, you find the amount you have before you divide.

- Problem A: A bag has 14 pounds of sand in it. Sarah makes piles that are each 2/7 pound. How many piles does she make?
- Raise your hand when you know the amount there is before you divide. ✔
- What amount? (Signal.) *14 pounds.*

b. Problem B: Each cup holds 1/5 quart. How many cups can be filled if there are 3 quarts of water?
- Raise your hand when you know the amount there is before you divide. ✔
- What amount? (Signal.) *3 quarts.*
- Say the division problem. (Signal.) *3 ÷ 1/5.*

c. Problem C: A cook divided a salad into 4 equal servings. There were 7/2 pounds of salad. How much did each serving weigh?
- Raise your hand when you know the amount there is before you divide. ✔
- What amount? (Signal.) *7/2 pounds.*
- Say the division problem. (Signal.) *7/2 ÷ 4.*

d. Problem D: A pie weighs 8/3 pounds. It is cut into slices that each weigh 1/9 pound. How many slices are there?
- Raise your hand when you know the amount there is before you divide. ✔
- What amount? (Signal.) *8/3 pounds.*
   (Repeat until firm.)

e. Go back to problem A. Read the problem to yourself and work it.
   (Observe students and give feedback.)

f. Check your work.
- Read the division problem you started with. (Signal.) *14 ÷ 2/7.*
- How many piles does she make? (Signal.) *49 piles.*

g. Read Problem B to yourself and work it.
   (Observe students and give feedback.)

h. Check your work.
- Read the division problem you started with. (Signal.) *3 ÷ 1/5.*
- How many cups are filled with 3 quarts of water? (Signal.) *15 cups.*

i. Read problem C to yourself and work it.
   (Observe students and give feedback.)

j. Check your work.
- Read the division problem you started with. (Signal.) *7/2 ÷ 4.*
- How much did each serving weigh? (Signal.) *7/8 pound.*

k. Read problem D to yourself and work it.
   (Observe students and give feedback.)

l. Check your work.
- Read the division problem you started with. (Signal.) *8/3 ÷ 1/9.*
- How many slices are there? (Signal.) *24 slices.*

EXERCISE 5: UNIT CONVERSION
MULTIPLICATION

a. Find part 2 in your textbook. ✔
   (Teacher reference:)

   a. Change 3 years into seasons.

   b. Change 7 minutes into seconds.

   c. Change 9 weeks into days.

   d. Change 6 feet into inches.

   Facts:
   - 1 foot is 12 inches.
   - 1 week is 7 days.
   - 1 year is 4 seasons.
   - 1 minute is 60 seconds.

   The problems tell about related units.
   Each fact shows the number you multiply by.

b. Problem A: Change 3 years into seasons.
- What are the related units? (Signal.) *Years and seasons.*
- Find the fact that tells about those units. ✔
- Read the fact. (Signal.) *1 year is 4 seasons.*
- So what do you multiply years by to get seasons? (Signal.) *4.*

c. Problem B: Change 7 minutes into seconds.
- What are the related units? (Signal.) *Minutes and seconds.*
- Find the fact that tells about those units. ✔
- Read the fact. (Signal.) *1 minute is 60 seconds.*
- So what do you multiply minutes by to get seconds? (Signal.) *60.*

d. Problem C: Change 9 weeks into days.
- What are the related units? (Signal.) *Weeks and days.*
- Find the fact that tells about those units. ✔
- Read the fact. (Signal.) *1 week is 7 days.*
- So what do you multiply weeks by to get days? (Signal.) *7.*

e. Go back and work problem A. Remember the unit name in the answer.
(Observe students and give feedback.)
• Everybody, read the problem you worked. (Signal.) *3 × 4.*
• How many seasons is 3 years? (Signal.) *12 seasons.*
f. Work problem B.
(Observe students and give feedback.)
• You worked the problem 7 times 60.
• How many seconds is 7 minutes? (Signal.) *420 seconds.*
g. Work problem C.
(Observe students and give feedback.)
• Everybody, read the problem you worked. (Signal.) *9 × 7.*
• How many days is 9 weeks? (Signal.) *63 days.*
h. Work problem D.
(Observe students and give feedback.)
• You worked the problem 6 times 12.
• How many inches is 6 feet? (Signal.) *72 inches.*

EXERCISE 6: FRACTION MULTIPLICATION
PRODUCT/FACTOR COMPARISON

REMEDY

a. Find part 3 in your textbook. ✔
(Teacher reference:)

a. $\frac{8}{2}\left(\frac{3}{3}\right)=$ ▨   b. $\frac{8}{3}\left(\frac{4}{3}\right)=$ ▨

c. $\frac{3}{5}\left(\frac{9}{10}\right)=$ ▨   d. $\frac{7}{4}\left(\frac{5}{8}\right)=$ ▨

b. Read problem A. (Signal.) *8/2 × 3/3.*
• What fraction are you multiplying 8/2 by? (Signal.) *3/3.*
• Is 3/3 more than 1, less than 1, or equal to 1? (Signal.) *Equal to 1.*
• So will you end with a fraction that's more than, less than, or equal to 8/2? (Signal.) *Equal to 8/2.*

c. Read problem B. (Signal.) *8/3 × 4/3.*
• What fraction are you multiplying by? (Signal.) *4/3.*
• Is 4/3 more than 1, less than 1, or equal to 1? (Signal.) *More than 1.*
• So will you end with a fraction that's more or less than 8/3? (Signal.) *More than 8/3.*
d. Read problem C. (Signal.) *3/5 × 9/10.*
• What fraction are you multiplying by? (Signal.) *9/10.*
• Is 9/10 more than 1, less than 1, or equal to 1? (Signal.) *Less than 1.*
• So will you end with a fraction that's more or less than 3/5? (Signal.) *Less than 3/5.*
e. Read problem D. (Signal.) *7/4 × 5/8.*
• What fraction are you multiplying by? (Signal.) *5/8.*
• Is 5/8 more than 1, less than 1, or equal to 1? (Signal.) *Less than 1.*
• So will you end with a fraction that's more or less than 7/4? (Signal.) *Less than 7/4.*
(Repeat steps b–e for problems that were not firm.)
f. Copy and work problem A. Complete the multiplication equation. Then write the statement below to show if the fraction you end with is more than, less than, or equal to the number you start with.
(Observe students and give feedback.)
g. Check your work.
• Read the multiplication equation for A. (Signal.) *8/2 × 3/3 = 24/6.*
• Read the statement for A. (Signal.) *8/2 = 24/6.*
(Display:)   [92:6A]

a. $\dfrac{8}{2}\left(\dfrac{3}{3}\right)=\dfrac{24}{6}$

$\dfrac{8}{2}=\dfrac{24}{6}$

Here's what you should have for problem A.
h. Problem B. Copy and work the problem. Then write the statement below to show if the fraction you end with is more than, less than, or equal to the fraction you start with.
(Observe students and give feedback.)

i. Check your work.
- Read the multiplication equation for B. (Signal.) *8/3 × 4/3 = 32/9.*
- Read the statement for B. (Signal.) *8/3 is less than 32/9.*
(Display:) [92:6B]

$$\textbf{b.}\ \ \frac{8}{3}\left(\frac{4}{3}\right)=\frac{32}{9}$$

$$\frac{8}{3}\ <\ \frac{32}{9}$$

Here's what you should have for problem B.
j. Problem C. Copy and work the problem.
Then write the statement below to show if the fraction you end with is more than, less than, or equal to the fraction you start with.
(Observe students and give feedback.)
k. Check your work.
- Read the multiplication equation for C. (Signal.) *3/5 × 9/10 = 27/50.*
- Read the statement for C. (Signal.) *3/5 is more than 27/50.*
(Display:) [92:6C]

$$\textbf{c.}\ \ \frac{3}{5}\left(\frac{9}{10}\right)=\frac{27}{50}$$

$$\frac{3}{5}\ >\ \frac{27}{50}$$

Here's what you should have for problem C.
l. Work problem D.
(Observe students and give feedback.)
m. Check your work.
- Read the multiplication equation for D. (Signal.) *7/4 × 5/8 = 35/32.*
- Read the statement for D. (Signal.) *7/4 is more than 35/32.*
(Display:) [92:6D]

$$\textbf{d.}\ \ \frac{7}{4}\left(\frac{5}{8}\right)=\frac{35}{32}$$

$$\frac{7}{4}\ >\ \frac{35}{32}$$

Here's what you should have for problem D.

EXERCISE 7: SENTENCES
PARTS OF A GROUP

a. You're going to say equations for the facts. Some of the facts give two names. Some give three names. If a fact gives three names, you write the equation for the name that has a number.
(Display:) [92:7A]

$$\frac{2}{7}\ \text{of the shirts were wet. There were 25 dry shirts.}$$

2/7 of the shirts were wet. There were 25 dry shirts.
- Raise your hand when you know if the facts tell about two names or three names. ✔
- How many names? (Signal.) *3 names.*
- Which name has a number? (Signal.) *Dry.*
- Raise your hand when you can say the letter equation for dry. ✔
- Say the letter equation. (Signal.) *5/7 S = D.*

b. (Display:) [92:7B]

$$\frac{2}{7}\ \text{of the shirts were wet. There were 28 shirts.}$$

2/7 of the shirts were wet. There were 28 shirts.
- Raise your hand when you know if the facts tell about two names or three names. ✔
- How many names? (Signal.) *2 names.*
- What are the two names? (Signal.) *Shirts and wet.*
- Which name has a number? (Signal.) *Shirts.*
- Raise your hand when you can say the letter equation for those two names. ✔
- Say the letter equation. (Signal.) *2/7 S = W.*

c. (Display:) [92:7C]

$$\frac{4}{5}\ \text{of the shirts were wet. There were 20 wet shirts.}$$

4/5 of the shirts were wet. There were 20 wet shirts.

- Raise your hand when you know if the facts tell about two names or three names. ✔
- How many names? (Signal.) *2 names.*
- What are the two names? (Signal.) *Shirts and wet.*
- Which name has a number? (Signal.) *Wet.*
- Raise your hand when you can say the letter equation for those two names. ✔
- Say the letter equation. (Signal.) *4/5 S = W.*

d. (Display:) [92:7D]

$\frac{4}{9}$ of the trees were maples. 50 trees were not maples.

4/9 of the trees were maples. 50 trees were not maples.

- Raise your hand when you know if the facts tell about two names or three names. ✔
- How many names? (Signal.) *3 names.*
- What are the three names? (Signal.) *Trees, maples, and not maples.*
- Which name has a number? (Signal.) *Not maples.*
- Raise your hand when you can say the letter equation for not maples. ✔
- Say the letter equation. (Signal.) *5/9 T = N.*

a. Find part 4 in your textbook. ✔
   (Teacher reference:)

   a. $\frac{3}{5}$ of the trees were firs. There were 20 trees.
   b. $\frac{7}{8}$ of the dogs were sleeping. 24 dogs were awake.
   c. $\frac{2}{9}$ of the apples were ripe. There were 18 apples.
   d. $\frac{5}{8}$ of the children were boys. There were 27 girls.
   e. $\frac{3}{5}$ of the tigers were sleeping. There were 30 sleeping tigers.

b. Problem A: 3/5 of the trees were firs. There were 20 trees.
- Write the letter equation for the facts. ✔
- Everybody, read the letter equation. (Signal.) *3/5 T = F.*

c. Problem B: 7/8 of the dogs were sleeping. 24 dogs were awake.
- Write the letter equation. ✔
- Everybody, read the letter equation. (Signal.) *1/8 D = A.*

d. Problem C: 2/9 of the apples were ripe. There were 18 apples.
- Write the letter equation. ✔
- Everybody, read the letter equation. (Signal.) *2/9 A = R.*

e. Problem D: 5/8 of the children were boys. There were 27 girls.
- Write the letter equation. ✔
- Everybody, read the letter equation. (Signal.) *3/8 C = G.*

f. Problem E: 3/5 of the tigers were sleeping. There were 30 sleeping tigers.
- Write the letter equation. ✔
- Everybody, read the letter equation. (Signal.) *3/5 T = S.*

Assign Independent Work, Textbook parts 5–11 and Workbook parts 3 and 4.

Optional extra math-fact practice worksheets are available on ConnectED.

Lesson  93

EXERCISE 1: MENTAL MATH

a. Time for some mental math.
- Listen: 46 plus 4. (Pause.) What's the answer? (Signal.) *50.*
- 46 + 6. (Pause.) What's the answer? (Signal.) *52.*
- 26 + 6. (Pause.) What's the answer? (Signal.) *32.*
- 86 + 6. (Pause.) What's the answer? (Signal.) *92.*
- 86 + 8. (Pause.) What's the answer? (Signal.) *94.*
- 86 + 9. (Pause.) What's the answer? (Signal.) *95.*
  (Repeat until firm.)
b. Listen: 24 plus 6. (Pause.) What's the answer? (Signal.) *30.*
- 24 + 7. (Pause.) What's the answer? (Signal.) *31.*
- 24 + 9. (Pause.) What's the answer? (Signal.) *33.*
- 84 + 9. (Pause.) What's the answer? (Signal.) *93.*
- 53 + 7. (Pause.) What's the answer? (Signal.) *60.*
- 53 + 9. (Pause.) What's the answer? (Signal.) *62.*
  (Repeat until firm.)
c. Listen: 50 plus 7. (Pause.) What's the answer? (Signal.) *57.*
- 50 + 17. (Pause.) What's the answer? (Signal.) *67.*
- 50 + 18. (Pause.) What's the answer? (Signal.) *68.*
- 30 + 18. (Pause.) What's the answer? (Signal.) *48.*
- 70 + 18. (Pause.) What's the answer? (Signal.) *88.*
- 70 + 14. (Pause.) What's the answer? (Signal.) *84.*
- 70 + 11. (Pause.) What's the answer? (Signal.) *81.*
  (Repeat until firm.)

d. Listen: 60 plus 12. (Pause.) What's the answer? (Signal.) *72.*
- 60 + 15. (Pause.) What's the answer? (Signal.) *75.*
- 60 + 19. (Pause.) What's the answer? (Signal.) *79.*
- 80 + 19. (Pause.) What's the answer? (Signal.) *99.*
  (Repeat until firm.)

EXERCISE 2: LONG DIVISION
2-DIGIT QUOTIENT

a. (Display:) [93:2A]

$$\begin{array}{r} 6 \\ 12\overline{)756} \\ -72 \\ \hline 3 \end{array}$$

Here's 756 divided by 12. The problem is partly worked. The first digit of the answer is written.
- What's that digit? (Signal.) *6.*
- What's the remainder? (Signal.) *3.*
b. We're going to work the problem to figure out the ones digit of the answer. Here's how we do that: We bring down the ones digit of 756.
- What do we bring down? (Signal.) *The ones digit of 756.*
- What's the ones digit? (Signal.) *6.*
  (Add to show:) [93:2B]

$$\begin{array}{r} 6 \\ 12\overline{)756} \\ -72 \\ \hline 36 \end{array}$$

- Now we work the new division problem for the ones digit of the answer. The new problem is 36 divided by 12.
- Say the new division problem. (Signal.) *36 ÷ 12.*
- Raise your hand when you know the answer. ✔

- What's the answer? (Signal.) *3.*
  (Add to show:)   [93:2C]

$$
\begin{array}{r}
6\,3 \\
12\overline{)7\,5\,6} \\
-7\,2 \\
\hline
3\,6
\end{array}
$$

The new multiplication problem is 12 times 3.
That's 36.
(Add to show:)   [93:2D]

$$
\begin{array}{r}
6\,3 \\
12\overline{)7\,5\,6} \\
-7\,2 \\
\hline
3\,6 \\
-3\,6
\end{array}
$$

c. Say the subtraction problem for figuring out the remainder. (Signal.) *36 – 36.*
- What's the remainder? (Signal.) *Zero.*
  (Add to show:)   [93:2E]

$$
\begin{array}{r}
\boxed{6\,3} \\
12\overline{)7\,5\,6} \\
-7\,2 \\
\hline
3\,6 \\
-3\,6 \\
\hline
0
\end{array}
$$

- Read the whole answer. (Signal.) *63.*
d. (Display:)   [93:2F]

$$
\begin{array}{r}
4 \\
71\overline{)2\,9\,8\,2} \\
-2\,8\,4 \\
\hline
1\,4
\end{array}
$$

- Read this problem. (Signal.) *2982 ÷ 71.*
  The first digit of the answer is shown. We're going to work the problem for the ones digit of the answer.
- The first thing we do is bring down the ones digit.
  What's the ones digit? (Signal.) *2.*
  Yes, we bring down the 2.
  (Add to show:)   [93:2G]

$$
\begin{array}{r}
4 \\
71\overline{)2\,9\,8\,2} \\
-2\,8\,4 \\
\hline
1\,4\,2
\end{array}
$$

e. Say the division problem for 142. (Signal.)
  142 ÷ 71.
  The estimation problem is 14 divided by 7.
- Say the estimation problem. (Signal.) *14 ÷ 7.*
- What's the answer? (Signal.) *2.*
  So we write 2 as the ones digit in the answer.
  (Add to show:)   [93:2H]

$$
\begin{array}{r}
4\,2 \\
71\overline{)2\,9\,8\,2} \\
-2\,8\,4 \\
\hline
1\,4\,2
\end{array}
$$

f. Now we multiply.
- Say the multiplication problem. (Signal.) *71 × 2.*
  The answer is 142.
  (Add to show:)   [93:2I]

$$
\begin{array}{r}
4\,2 \\
71\overline{)2\,9\,8\,2} \\
-2\,8\,4 \\
\hline
1\,4\,2 \\
-1\,4\,2
\end{array}
$$

g. Say the subtraction problem for the remainder.
  (Signal.) *142 – 142.*
- What's the remainder? (Signal.) *Zero.*
  (Add to show:)   [93:2J]

$$
\begin{array}{r}
\boxed{4\,2} \\
71\overline{)2\,9\,8\,2} \\
-2\,8\,4 \\
\hline
1\,4\,2 \\
-1\,4\,2 \\
\hline
0
\end{array}
$$

We worked the problem 2982 ÷ 71.
- Read the whole answer. (Signal.) *42.*
h. Remember the steps for the ones digit of the answer. First, you bring down the ones digit. You work the new division problem and subtract again for the remainder.

Exercise 3: Balance Beam

Average Distance

a. (Display:) [93:3A]

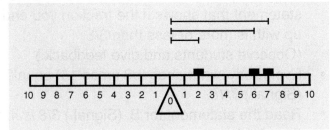

You can show averages on balance beams. (Point to lines.) The lines show the distance of each weight from zero.
To find the average, we figure the total distance. Then we divide by the number of weights.
We find the total distance by adding 2 plus 6 plus 7.

- Raise your hand when you know the total distance. ✔
- What's the total distance? (Signal.) *15.*
- What do you divide by to find the average? (Signal.) *3.*
  Yes, you work the problem 15 divided by 3.
- What's the average distance? (Signal.) *5.*
  (Add to show:) [93:3B]

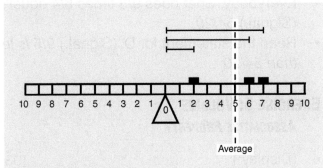

b. Listen: We can show that the average is 5 by putting three weights at 5 on the other side of the balance beam.
   (Add to show:) [93:3C]

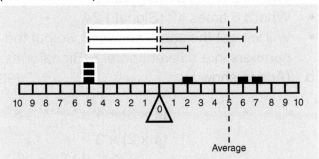

If the total distance on both sides is the same, the beam will balance.

- (Point left.) We have 5 plus 5 plus 5 on this side. What's the total distance? (Signal.) *15.*
  There is a distance of 15 on both sides, so the beam will balance.

c. (Display:) [93:3D]

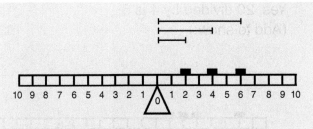

- New problem. (Point right.) Figure the average for this side in part A. ✔
- Everybody, what's the total distance for this side? (Signal.) *12.*
- What did you divide by? (Signal.) *3.*
- What's the average for this side? (Signal.) *4.*
  (Add to show:) [93:3E]

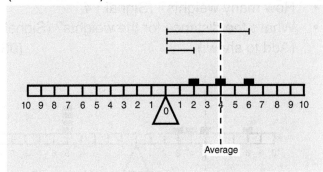

The average is 4, so the beam will balance if we place the same number of weights at 4 on the other side.

- How many weights? (Signal.) *3.*
  (Add to show:) [93:3F]

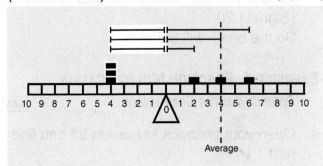

- (Point left.) What's the total distance for this side? (Signal.) *12.*
  Both sides have the same total distance, so the beam will balance.

d. (Display:) [93:3G]

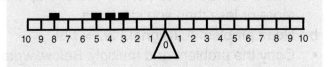

- Figure out the average distance on the left side. ✔
- Check your work. What's the total distance? (Signal.) *20.*
- What did you divide by? (Signal.) *4.*

- What's the average distance? (Signal.) *5.*
  Yes, 20 divided by 4 is 5.
  (Add to show:) [93:3H]

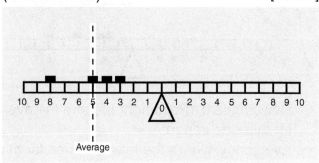

e. So we could balance the beam by putting the same number of weights at one place on the other side.
- How many weights? (Signal.) *4.*
- What's the distance for the weights? (Signal.) *5.*
  (Add to show:) [93:3I]

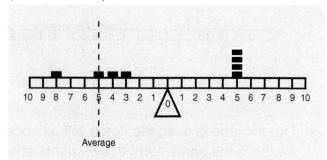

- Raise your hand when you know the total distance on each side. ✔
- What's the total distance on each side? (Signal.) *20.*
  So the beam will balance.

EXERCISE 4: FRACTION MULTIPLICATION
PRODUCT/FACTOR COMPARISON
REMEDY

a. Open your textbook to Lesson 93 and find part 1. ✔
  (Teacher reference:)

a. $\frac{3}{2}\left(\frac{5}{9}\right) =$ ☐   b. $\frac{3}{8}\left(\frac{1}{2}\right) =$ ☐

c. $\frac{2}{9}\left(\frac{3}{3}\right) =$ ☐   d. $\frac{9}{5}\left(\frac{6}{4}\right) =$ ☐

You're going to work these problems and write statements below to show if you end up with more or less than you start with.

b. Read problem A. (Signal.) *3/2 × 5/9.*
- Copy the problem and multiply. Below, write the statement that compares 3/2 to the fraction you end up with.
  (Observe students and give feedback.)
- Everybody, what does 3/2 times 5/9 equal? (Signal.) *15/18.*

- Read the statement for A. (Signal.) *3/2 is more than 15/18.*
c. Read problem B. (Signal.) *3/8 × 1/2.*
- Copy the problem and work it. Below, write the statement that shows if the fraction you end up with is more or less than 3/8.
  (Observe students and give feedback.)
- Everybody, what does 3/8 times 1/2 equal? (Signal.) *3/16.*
- Read the statement for B. (Signal.) *3/8 is more than 3/16.*
d. Read problem C. (Signal.) *2/9 × 3/3.*
- Copy the problem and work it. Below, write the statement that compares 2/9 to the fraction you end up with.
  (Observe students and give feedback.)
- Everybody, what does 2/9 times 3/3 equal? (Signal.) *6/27.*
- Read the statement for C. (Signal.) *2/9 equals 6/27.*
e. Read problem D. (Signal.) *9/5 × 6/4.*
- Copy the problem and work it. Below, write the statement that compares 9/5 to the fraction you end up with.
  (Observe students and give feedback.)
- Everybody, what does 9/5 times 6/4 equal? (Signal.) *54/20.*
- Read the statement for D. (Signal.) *9/5 is less than 54/20.*

EXERCISE 5: VOLUME
ASSOCIATIVE PROPERTY

a. (Display:) [93:5A]

$$(2 \times 3) \times 4$$

- Read this problem. (Signal.) *2 × 3 × 4.*
- What's 2 times 3? (Signal.) *6.*
- What's 6 times 4? (Signal.) *24.*
- Will we get the same answer if we put the numbers in a different order? (Signal.) *Yes.*
b. (Add to show:) [93:5B]

$$(2 \times 3) \times 4$$
$$(4 \times 2) \times 3$$

- Read this problem. (Signal.) *4 × 2 × 3.*
- What's 4 times 2? (Signal.) *8.*
- What's 8 times 3? (Signal.) *24.*

c. (Add to show:) [93:5C]

$$(2 \times 3) \times 4$$
$$(4 \times 2) \times 3$$
$$(4 \times 3) \times 2$$

- Read this problem. (Signal.) *4 × 3 × 2.*
- What's 4 times 3? (Signal.) *12.*
- What's 12 times 2? (Signal.) *24.*

d. We can do the same thing when we work with rectangular prisms.
   (Display:) [93:5D]

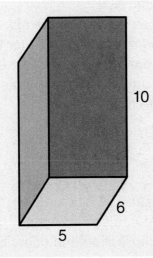

We can show the same multiplication patterns with rectangular prisms by rotating them, which changes the base and height.
- What are the numbers for the base now? (Signal.) *6 and 5.*
- Say the multiplication fact for the base. (Signal.) *6 × 5 = 30.*
- Say the multiplication fact for the base times the height. (Signal.) *30 × 10 = 300.*
- What's the volume? (Signal.) *300 cubic units.*
  (Repeat until firm.)

e. (Display:) [93:5E]

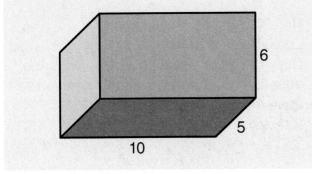

Here's the same prism rotated.

- What are the numbers for the base? (Signal.) *10 and 5.*
- Say the multiplication fact for the base. (Signal.) *10 × 5 = 50.*
- Say the multiplication fact for the base times the height. (Signal.) *50 × 6 = 300.*
- What's the volume? (Signal.) *300 cubic units.*
  (Repeat until firm.)

f. (Display:) [93:5F]

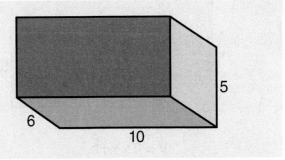

Here's the same prism rotated again.
- What are the numbers for the base? (Signal.) *6 and 10.*
- Say the multiplication fact for the base. (Signal.) *6 × 10 = 60.*
- Say the multiplication fact for the base times the height. (Signal.) *60 × 5 = 300.*
- What's the volume? (Signal.) *300 cubic units.*
  (Repeat until firm.)

g. (Add to show:) [93:5G]

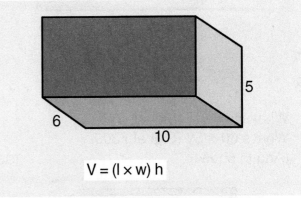

$$V = (l \times w)\, h$$

Here's a new equation for the volume of a rectangular prism. This equation does not work for finding the volume of a cylinder. It says: Volume equals length times width times height.

- Which measurements are in the parentheses? (Signal.) *Length times width.*
  They give the area of the base.
- What's the third measurement? (Signal.) *Height.*
  That's the value outside the parentheses.
- Say the multiplication for the area of the base. (Signal.) *6 × 10.*
  (Add to show:)   [93:5H]

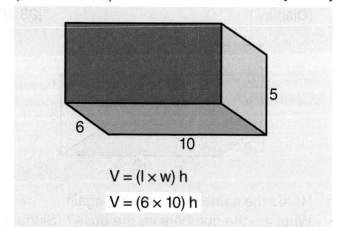

$$V = (l \times w)\, h$$
$$V = (6 \times 10)\, h$$

- What's the height? (Signal.) *5.*
  (Add to show:)   [93:5I]

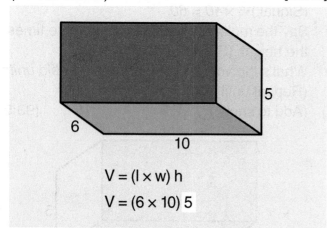

$$V = (l \times w)\, h$$
$$V = (6 \times 10)\, 5$$

- What's 6 × 10? (Signal.) *60.*
- What's 60 × 5? (Signal.) *300.*
  (Add to show:)   [93:5J]

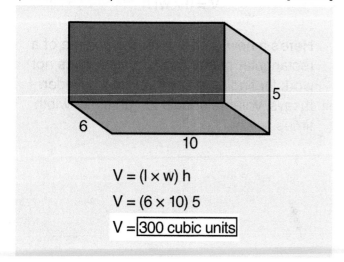

$$V = (l \times w)\, h$$
$$V = (6 \times 10)\, 5$$
$$V = \boxed{300 \text{ cubic units}}$$

That's the volume—300 cubic units.

TEXTBOOK PRACTICE

a. Find part 2 in your textbook. ✔
   (Teacher reference:)

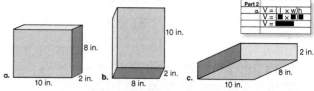

Here's a new rectangular prism in three different positions. You're going to show the numbers you multiply for the base and height for each position.

b. Look at prism A. ✔
- Copy the new letter equation for volume and write the numbers multiplied for the base inside the parentheses. Write the number for the height outside the parentheses. Stop when you've done that much. ✔
  (Display:)   [93:5K]

> **a.** $V = (l \times w)\, h$
> $$V = (10 \times 2)\, 8$$

Here's what you should have.
- Do the multiplication and complete the equation below. ✔
- Everybody, say the multiplication for the base. (Signal.) *10 × 2.*
- What's the answer? (Signal.) *20.*
- Say the multiplication for the base times the height. (Signal.) *20 × 8.*
- What's the volume? (Signal.) *160 cubic inches.*
  (Add to show:)   [93:5L]

> **a.** $V = (l \times w)\, h$
> $$V = (10 \times 2)\, 8$$
> $$V = \boxed{160 \text{ cu in.}}$$

Here's what you should have.
c. Look at prism B. ✔
- Copy the letter equation. Below, write the equation with numbers. Stop when you've done that much. ✔
  (Display:)   [93:5M]

> **b.** $V = (l \times w)\, h$
> $$V = (8 \times 2)\, 10$$

Here's what you should have.
- Work the problem.
  (Observe students and give feedback.)

d. Check your work.
- Say the multiplication for the base. (Signal.) *8 × 2.*
- What's the answer? (Signal.) *16.*
- Say the multiplication for the base times the height. (Signal.) *16 × 10.*
- What's the volume? (Signal.) *160 cubic inches.* (Add to show:) [93:5N]

$$V = (l \times w)\, h$$
$$V = (8 \times 2)\, 10$$
$$V = \boxed{160 \text{ cu in.}}$$

Here's what you should have.

e. Work the problem for prism C. ✔
- Everybody, say the multiplication fact for the base. (Signal.) *10 × 8 = 80.* You multiplied 80 times 2 for the volume.
- What's the volume? (Signal.) *160 cubic inches.* (Display:) [93:5O]

$$V = (l \times w)\, h$$
$$V = (10 \times 8)\, 2$$
$$V = \boxed{160 \text{ cu in.}}$$

Here's what you should have.
- Is the volume the same in all three positions? (Signal.) *Yes.* So we can use any two numbers for the base.

EXERCISE 6: UNIT CONVERSION
MULTIPLICATION/DIVISION

a. (Display:) [93:6A]

1 week is 7 days.

- Say the fact. (Signal.) *1 week is 7 days.*
- Are there more weeks or days? (Signal.) *Days.*
- How many times more? (Signal.) *7.* So to end up with days, you multiply by 7. To end up with **weeks,** you **divide** by 7.

b. Once more: Are there more weeks or days? (Signal.) *Days.*
- How many times more? (Signal.) *7.*
- So to end up with days, what do you do? (Signal.) *Multiply by 7.*
- To end up with weeks, what do you do? (Signal.) *Divide by 7.* (Repeat until firm.)

c. (Display:) [93:6B]

1 year is 12 months.

- Say the fact. (Signal.) *1 year is 12 months.*
- Are there more years or months? (Signal.) *Months.*
- How many times more? (Signal.) *12.*
- So to end up with months, what do you do? (Signal.) *Multiply by 12.*
- To end up with years, what do you do? (Signal.) *Divide by 12.* (Repeat until firm.)

d. (Display:) [93:6C]

1 year is 4 seasons.

- Say the fact. (Signal.) *1 year is 4 seasons.*
- Are there more years or seasons? (Signal.) *Seasons.*
- How many times more? (Signal.) *4.*
- So to end up with seasons, what do you do? (Signal.) *Multiply by 4.*
- To end up with years, what do you do? (Signal.) *Divide by 4.* (Repeat until firm.)

e. (Display:) [93:6D]

1 hour is 60 minutes.

- Say the fact. (Signal.) *1 hour is 60 minutes.*
- Are there more hours or minutes? (Signal.) *Minutes.*
- How many times more? (Signal.) *60.*
- So to end up with **hours,** what do you do? (Signal.) *Divide by 60.*
- To end up with **minutes,** what do you do? (Signal.) *Multiply by 60.* (Repeat until firm.)

f. (Display:) [93:6E]

> 1 year is 4 seasons.
>
> Change 9 seasons into years.

Here's a fact and a problem.
(Point to **1 year.**) Read the fact. (Signal.)
1 year is 4 seasons.

- Are there more seasons or years? (Signal.)
  Seasons.
- (Point to **change.**) Read the problem. (Signal.)
  Change 9 seasons into years.
- What are you going to end up with?
  (Signal.) *Years.*
- To end up with years, do you multiply by 4 or
  divide by 4? (Signal.) *Divide by 4.*
- To end up with years, what do you do?
  (Signal.) *Divide by 4.*
  So the problem you work is 9 ÷ 4.
- Say the problem you work. (Signal.) *9 ÷ 4.*

g. (Change to show:) [93:6F]

> 1 year is 4 seasons.
>
> Change 12 seasons into years.

- (Point to **1.**) Read the fact. (Signal.) *1 year is
  4 seasons.*
- Are there more seasons or years? (Signal.)
  Seasons.
- Read the problem. (Signal.) *Change
  12 seasons into years.*
- What are you going to end up with?
  (Signal.) *Years.*
- To end up with years, do you multiply by 4 or
  divide by 4? (Signal.) *Divide by 4.*
- Say the problem you work. (Signal.) *12 ÷ 4.*

h. (Change to show:) [93:6G]

> 1 year is 4 seasons.
>
> Change 12 years into seasons.

- Read the fact. (Signal.) *1 year is 4 seasons.*
- Read the problem. (Signal.) *Change 12 years
  into seasons.*
- To end up with seasons, do you multiply by 4
  or divide by 4? (Signal.) *Multiply by 4.*
- Work the problem and write the answer in part
  B. Remember the unit name. ✔
- Everybody, say the problem you worked.
  (Signal.) *12 × 4.*
- How many seasons is 12 years? (Signal.)
  48 seasons.

i. (Change to show:) [93:6H]

> 1 year is 4 seasons.
>
> Change 12 seasons into years.

- Read the fact. (Signal.) *1 year is 4 seasons.*
- Read the problem. (Signal.) *Change 12
  seasons into years.*
- To end up with years, what do you do?
  (Signal.) *Divide by 4.*
- Work the problem and write the answer. ✔
- Everybody, read the problem you worked.
  (Signal.) *12 ÷ 4.*
- How many years is 12 seasons? (Signal.)
  3 years.

j. (Display:) [93:6I]

> 1 week is 7 days.

- Read the fact. (Signal.) *1 week is 7 days.*
- Are there more days or weeks? (Signal.) *Days.*
- So what do you do to end up with days?
  (Signal.) *Multiply by 7.*
- What do you do to end up with weeks?
  (Signal.) *Divide by 7.*
  (Repeat until firm.)

k. (Add to show:) [93:6J]

> 1 week is 7 days.
>
> Change 28 days into weeks.

- Read the problem. (Signal.) *Change 28 days
  into weeks.*
- Do you end up with days or weeks?
  (Signal.) *Weeks.*
- What do you do to end up with weeks?
  (Signal.) *Divide by 7.*
  (Repeat until firm.)
- Work the problem. ✔
- How many weeks is 28 days? (Signal.) *4 weeks.*
  (Add to show:) [93:6K]

> 1 week is 7 days.   4 weeks
>   ⎯⎯⎯
> Change 28 days into weeks.   7⟌2 8

Here's the problem you worked.

l. (Change to show:)   [93:6L]

> 1 week is 7 days.
>
> Change 28 weeks into days.

- Read the problem. (Signal.) *Change 28 weeks into days.*
- What do you end up with? (Signal.) *Days.*
- Work the problem.
  (Observe students and give feedback.)
- Everybody, how many days is 28 weeks?
  (Signal.) *196 days.*
  (Add to show:)   [93:6M]

> 1 week is 7 days. $\overset{5}{2}\,8$
>
> Change 28 weeks into days. $\times\ \ \ 7$
>
> $\boxed{1\,9\,6\ \text{days}}$

Here's the problem you worked.

EXERCISE 7: WORD PROBLEMS
PARTS OF A GROUP   REMEDY

a. Find part 3 in your textbook. ✔
    (Teacher reference:)

> a. $\frac{2}{9}$ of the flowers were not white. 54 flowers were not white. How many flowers were there in all? How many flowers were white?
>
> b. $\frac{2}{9}$ of the windows were broken. 28 windows were not broken. How many windows were there? How many broken windows were there?
>
> c. $\frac{5}{8}$ of the fish were large. There were 40 fish. How many small fish were there? How many large fish were there?
>
> d. $\frac{3}{5}$ of the doors in a building were open. There were 60 closed doors. How many doors were in the building? How many doors were open?

Some of these problems have facts that tell about two names. The facts for the other problems tell about three names.

b. Problem A: 2/9 of the flowers were not white. 54 flowers were not white. How many flowers were there in all? How many flowers were white?
- Raise your hand when you know if the facts for this problem give two names or three names. ✔
- How many names? (Signal.) *2 names.*

c. Write the letter equation for those names. Below, write the equation with the number the problem gives. Stop when you've done that much.
    (Observe students and give feedback.)

- Everybody, read the letter equation. (Signal.)
  2/9 F = N.
- Read the equation with a number for not white. (Signal.) *2/9 F = 54.*
  (Display:)   [93:7A]

> a. $\frac{2}{9}$ f = n
>
> $\frac{2}{9}$ f = 54

Here's what you should have.
d. Work the problem. Answer both questions.
    (Observe students and give feedback.)
e. Check your work.
- How many flowers were there in all? (Signal.)
  243 flowers.
- How many flowers were white? (Signal.)
  189 white flowers.

f. Problem B: 2/9 of the windows were broken. 28 windows were not broken. How many windows were there? How many broken windows were there?
- Raise your hand when you know if the facts for this problem give two names or three names. ✔
- How many names? (Signal.) *3 names.*

g. Write the letter equation for the name that has a number. Below, write the equation with the number the problem gives. Stop when you've done that much.
    (Observe students and give feedback.)
- Everybody, read the letter equation. (Signal.)
  7/9 W = N.
- Read the equation with a number for not broken. (Signal.) *7/9 W = 28.*
  (Display:)   [93:7B]

> b. $\frac{7}{9}$ w = n
>
> $\frac{7}{9}$ w = 28

Here's what you should have.
h. Work the problem. Answer both questions.
    (Observe students and give feedback.)
i. Check your work.
- Everybody, how many windows were there?
  (Signal.) *36 windows.*
- How many broken windows were there?
  (Signal.) *8 broken windows.*

j. Problem C: 5/8 of the fish were large. There were 40 fish. How many small fish were there? How many large fish were there?
- Raise your hand when you know if the facts for this problem give two names or three names. ✔
- How many names? (Signal.) *2 names.*

k. Write the letter equation for those names. Below, write the equation with the number the problem gives. Stop when you've done that much.
(Observe students and give feedback.)
- Everybody, read the letter equation. (Signal.) *5/8 F = L.*
- Read the equation with a number for fish. (Signal.) *5/8 × 40 = L.*
(Display:)   [93:7C]

$$\text{c. } \frac{5}{8} f = l$$
$$\frac{5}{8} (40) = l$$

Here's what you should have.

l. Work the problem. Answer both questions.
(Observe students and give feedback.)

m. Check your work.
- How many small fish were there? (Signal.) *15 small fish.*
- How many large fish were there? (Signal.) *25 large fish.*

n. Problem D: 3/5 of the doors in a building were open. There were 60 doors that were not open. How many doors were in the building? How many doors were open?
- Raise your hand when you know if the facts for this problem give two names or three names. ✔
- How many names? (Signal.) *3 names.*

o. Write the letter equation for the name that has a number. Below, write the equation with the number the problem gives. Stop when you've done that much.
(Observe students and give feedback.)

- Everybody, read the letter equation. (Signal.) *2/5 D = N.*
- Read the equation with a number for not open. (Signal.) *2/5 D = 60.*
(Display:)   [93:7D]

$$\text{d. } \frac{2}{5} d = n$$
$$\frac{2}{5} d = 60$$

Here's what you should have.

p. Work the problem. Answer both questions.
(Observe students and give feedback.)

q. Check your work.
- How many doors were in the building? (Signal.) *150 doors.*
- How many doors were open? (Signal.) *90 open doors.*

EXERCISE 8: INDEPENDENT WORK
MIXED-NUMBER OPERATIONS/FRACTION WORD PROBLEMS

a. Find part 4 in your textbook. ✔
(Teacher reference:)

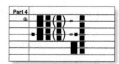

a. $12\frac{1}{4}$ b. $27\frac{5}{8}$ c. $38\frac{1}{6}$
$+18\frac{1}{5}$ $-11\frac{1}{4}$ $+ 2\frac{4}{9}$

You'll work these problems as part of your independent work. You first find the lowest common denominator for the fractions.

b. Read problem A. (Signal.) *12 and 1/4 + 18 and 1/5.*
The diagram shows how to set up the problem. First find the lowest common denominator for 1/4 and 1/5. Then work the problem.

c. Find part 5 in your textbook. ✔
(Teacher reference:)

a. A factory uses $\frac{22}{3}$ pounds of iron to make 13 spikes. How much iron is used to make each spike?

b. 5 equal servings are made from a roast that weighs $\frac{13}{4}$ pounds. How much does each serving weigh?

You have to read each problem carefully before you work it.

d. Problem A: A factory uses 22/3 pounds of iron to make 13 spikes. How much iron is used to make each spike?
- Raise your hand when you know the amount that will be divided. ✔
- What amount will be divided? (Signal.) *22/3 pounds.*
- Say the division problem. (Signal.) *22/3 ÷ 13.*

e. Problem B: 5 equal servings are made from a roast that weighs 13/4 pounds. How much does each serving weigh?
- Raise your hand when you know the amount that is divided. ✔
- What amount is divided? (Signal.) *13/4 pounds.*
- Say the division problem. (Signal.) *13/4 ÷ 5.*
  (Repeat until firm.)

f. You'll work these problems as part of your independent work. Remember to read them carefully before you write the division problem.

Assign Independent Work, Textbook parts 4–11 and Workbook parts 1 and 2.

Optional extra math-fact practice worksheets are available on ConnectED.

Lesson

EXERCISE 1: MENTAL MATH

a. Time for some mental math.
- Listen: 76 plus 4. (Pause.) What's the answer? (Signal.) *80.*
- 76 + 6. (Pause.) What's the answer? (Signal.) *82.*
- 36 + 6. (Pause.) What's the answer? (Signal.) *42.*
- 96 + 6. (Pause.) What's the answer? (Signal.) *102.*
- 96 + 8. (Pause.) What's the answer? (Signal.) *104.*
- 86 + 9. (Pause.) What's the answer? (Signal.) *95.*
(Repeat until firm.)

b. Listen: 34 plus 6. (Pause.) What's the answer? (Signal.) *40.*
- 54 + 6. (Pause.) What's the answer? (Signal.) *60.*
- 54 + 9. (Pause.) What's the answer? (Signal.) *63.*
- 84 + 9. (Pause.) What's the answer? (Signal.) *93.*
- 33 + 7. (Pause.) What's the answer? (Signal.) *40.*
- 33 + 9. (Pause.) What's the answer? (Signal.) *42.*
(Repeat until firm.)

c. Listen: 60 plus 7. (Pause.) What's the answer? (Signal.) *67.*
- 60 + 17. (Pause.) What's the answer? (Signal.) *77.*
- 60 + 27. (Pause.) What's the answer? (Signal.) *87.*
- 60 + 47. (Pause.) What's the answer? (Signal.) *107.*
- 70 + 18. (Pause.) What's the answer? (Signal.) *88.*
- 50 + 14. (Pause.) What's the answer? (Signal.) *64.*
(Repeat until firm.)

d. Listen: 80 plus 11. (Pause.) What's the answer? (Signal.) *91.*
- 40 + 12. (Pause.) What's the answer? (Signal.) *52.*
- 80 + 15. (Pause.) What's the answer? (Signal.) *95.*
- 60 + 19. (Pause.) What's the answer? (Signal.) *79.*
- 20 + 17. (Pause.) What's the answer? (Signal.) *37.*
(Repeat until firm.)

EXERCISE 2: LONG DIVISION
2-DIGIT QUOTIENT

a. (Display:) [94:2A]

$$
\begin{array}{r}
2 \\
59\overline{)1367} \\
-118 \\
\hline
18
\end{array}
$$

- Read the problem. (Signal.) *1367 ÷ 59.* The first digit of the answer is shown. Remember the steps for the ones digit of the answer.
- What's the first thing you do? (Signal.) *Bring down the ones digit.*
(Add to show:) [94:2B]

$$
\begin{array}{r}
2 \\
59\overline{)1367} \\
-118 \\
\hline
187
\end{array}
$$

b. You work the new division problem. Then you subtract to find the remainder.
- Say the new division problem. (Signal.) *187 ÷ 59.*
- Say the estimation problem for 187 ÷ 59. (Signal.) *19 ÷ 6.*

(Add to show:)   [94:2C]

```
        2
  5 9 ⟌1 3 6 7      6 ⟌1 9
    − 1 1 8
      1 8 7
```

- What's the answer? (Signal.) *3.*
  (Add to show:)   [94:2D]

```
        2 3            3
  5 9 ⟌1 3 6 7      6 ⟌1 9
    − 1 1 8
      1 8 7
```

c. Say the multiplication problem. (Signal.) *59 × 3.*
  Here's the problem with the answer.
  (Add to show:)   [94:2E]

```
                                  2
        2 3            3          5 9
  5 9 ⟌1 3 6 7      6 ⟌1 9      ×   3
    − 1 1 8                      1 7 7
      1 8 7
```

d. There's a remainder. Say the subtraction
  problem for the remainder. (Signal.) *187 − 177.*
  (Add to show:)   [94:2F]

```
                                  2
        2 3            3          5 9
  5 9 ⟌1 3 6 7      6 ⟌1 9      ×   3
    − 1 1 8                      1 7 7
      1 8 7
    − 1 7 7
```

- Raise your hand when you know the answer. ✔
- What's the remainder? (Signal.) *10.*
  (Add to show:)   [94:2G]

```
        2 3 10/59      3          2
  5 9 ⟌1 3 6 7      6 ⟌1 9        5 9
    − 1 1 8                     ×   3
      1 8 7                      1 7 7
    − 1 7 7
      1 0
```

We worked the problem 1367 divided by 59.
- What's the whole answer? (Signal.) *23 and 10/59.*

---

a. Open your workbook to Lesson 94 and find
   part 1. ✔
   (Teacher reference:)

```
        2                          3
a. 4 3 ⟌9 9 6            b. 8 7 ⟌2 7 2 7
     − 8 6                     − 2 6 1
       1 3                        1 1
```

The problem for the first digit of the answer is
already worked. You'll work the problem for the
ones digit. Both problems have a remainder.

b. Problem A: 996 ÷ 43.
- What's the first thing you do? (Signal.) *Bring
  down the ones digit.*
  Yes, bring down the ones digit.
- Do it. ✔
- Everybody, say the new division problem.
  (Signal.) *136 ÷ 43.*
- Say the estimation problem. (Signal.) *14 ÷ 4.*
  (Repeat until firm.)
- Work the estimation problem and complete
  problem A.
  (Observe students and give feedback.)

c. Check your work.
  You worked the estimation problem 14 ÷ 4.
- What's the answer? (Signal.) *3.*
- Say the multiplication for the ones digit.
  (Signal.) *43 × 3.*
- What's the answer? (Signal.) *129.*
  So you worked the subtraction problem
  136 − 129.
- What's the remainder? (Signal.) *7.*
  (Display:)   [94:2H]

```
         2 3 7/43          3          4 3
 a. 4 3 ⟌9 9 6       4 ⟌1 4        ×   3
      − 8 6                         1 2 9
        1 3 6
      − 1 2 9
          7
```

Here's what you should have.
- What's the answer to the whole problem?
  (Signal.) *23 and 7/43.*

---

d. Problem B: 2727 ÷ 87.
- What's the first thing you do? (Signal.) *Bring down the ones digit.*
  Yes, bring down the ones digit.
- Do it. ✔
- Everybody, say the new division problem. (Signal.) *117 ÷ 87.*
- Say the new estimation problem. (Signal.) *12 ÷ 9.*
  (Repeat until firm.)
- Work the estimation problem and complete problem B.
  (Observe students and give feedback.)

e. Check your work.
  You worked the estimation problem 12 ÷ 9.
- What's the answer? (Signal.) *1.*
- Say the multiplication for the ones digit. (Signal.) *87 × 1.*
- What's the answer? (Signal.) *87.*
  So you worked the subtraction problem 117 − 87.
- What's the remainder? (Signal.) *30.*
  (Display:)   [94:2I]

$$\begin{array}{r} 31\frac{30}{87} \\ 87\overline{\smash)2727} \\ -261 \\ \hline 117 \\ -87 \\ \hline 30 \end{array} \qquad 9\overline{\smash)12}^{\;1}$$

b.

Here's what you should have.
- What's the answer to the whole problem? (Signal.) *31 and 30/87.*

EXERCISE 3: BALANCE BEAM
AVERAGE DISTANCE

REMEDY

a. Find part 2 in your workbook. ✔
  (Teacher reference:)   R Part C

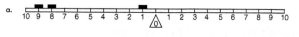

a.

b.

For each problem, you'll figure out the average for the left side. Then you'll show the same total distance on the other side by showing all the weights at the average distance.

b. Problem A. Figure out the average distance for the weights on the left side. Stop when you've done that much. ✔
- Everybody, what's the total distance on the left? (Signal.) *18.*
- What's the average distance on the left? (Signal.) *6.*
- On the right side, show all the weights at the average distance.
  (Observe students and give feedback.)
  (Display:)   [94:3A]

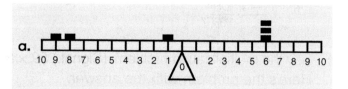

a.

Here's what you should have.
- Do you have the same total distance on both sides? (Signal.) *Yes.*
- So will the beam balance? (Signal.) *Yes.*

c. Work problem B. Figure out the average for the left side. Show the weights at the average distance on the right side.
  (Observe students and give feedback.)

d. Check your work.
- What's the total distance for the left side? (Signal.) *28.*
- What's the average distance? (Signal.) *7.*
- Where did you show all the weights on the right side? (Signal.) *7.*
- How many weights? (Signal.) *4.*
  (Display:)   [94:3B]

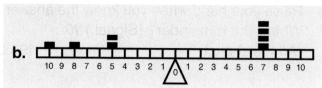

b.

Here's what you should have.

EXERCISE 4: FRACTION MULTIPLICATION
PRODUCT/FACTOR COMPARISON

a. You've learned that if you multiply a fraction by less than 1, you end up with less than you start with.
- If you multiply the fraction by more than 1, what do you end up with? (Signal.) *More than you start with.*
- If you multiply the fraction by **1,** what do you end up with? (Call on a student. Idea: *The same value you start with.*)

b. (Display:) [94:4A]

$$\frac{1}{2} \times \frac{2}{3} =$$

We're going to make diagrams of the fractions and see if the rules work.

- Read the problem. (Signal.) *1/2 × 2/3.*
- Is 2/3 more than 1, less than 1, or equal to 1? (Signal.) *Less than 1.*
- So will we end up with more than 1/2 or less than 1/2? (Signal.) *Less than 1/2.*
- Raise your hand when you know the fraction we'll end up with. ✔
- What fraction? (Signal.) *2/6.*
  (Add to show:) [94:4B]

Here are diagrams that show 1/2 and 2/6.

- Is 2/6 more than 1/2 or less than 1/2? (Signal.) *Less than 1/2.*

c. (Add to show:) [94:4C]

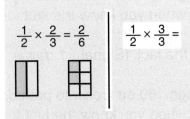

New problem: 1/2 × 3/3.

- Is 3/3 more than 1, less than 1, or equal to 1? (Signal.) *Equal to 1.*
- So will we end up with a value that is more than 1/2, less than 1/2, or equal to 1/2? (Signal.) *Equal to 1/2.*
- Raise your hand when you know the fraction we'll end up with. ✔
- What fraction? (Signal.) *3/6.*
  (Add to show:) [94:4D]

Here are diagrams that show 1/2 and 3/6.

- Is 3/6 more than 1/2, less than 1/2, or equal to 1/2? (Signal.) *Equal to 1/2.*

d. (Add to show:) [94:4E]

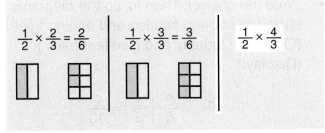

New problem: 1/2 × 4/3.

- Is 4/3 more than 1, less than 1, or equal to 1? (Signal.) *More than 1.*
- So will we end up with more than 1/2 or less than 1/2? (Signal.) *More than 1/2.*
- Raise your hand when you know the fraction we'll end up with. ✔
- What fraction? (Signal.) *4/6.*
  (Add to show:) [94:4F]

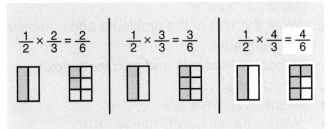

Here are diagrams that show 1/2 and 4/6.

- Is 4/6 more than 1/2 or less than 1/2? (Signal.) *More than 1/2.*
  So the rule works.

=== **WORKBOOK PRACTICE** ===

a. Find part 3 in your workbook. ✔
   (Teacher reference:)

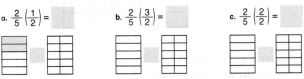

a. $\frac{2}{5}\left(\frac{1}{2}\right) =$ b. $\frac{2}{5}\left(\frac{3}{2}\right) =$ c. $\frac{2}{5}\left(\frac{2}{2}\right) =$

You're going to complete the equations. Then you'll shade the diagrams to show the starting fraction and the ending fraction.

b. Read problem A. (Signal.) *2/5 × 1/2.*
•   Write the answer. Then fix up the diagrams to show the starting fraction and ending fraction. **(Observe students and give feedback.)** (Display:)   [94:4G]

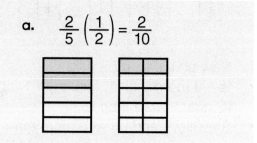

a. $\dfrac{2}{5}\left(\dfrac{1}{2}\right) = \dfrac{2}{10}$

Here's what you should have.
You multiplied by less than 1 and ended up with less than 2/5.
•   Write the sign between the diagrams to show which is more. ✔
c. Work the rest of the problems and complete the diagrams.
    (Observe students and give feedback.)
d. Check your work.
•   Problem B: 2/5 × 3/2.
•   What's the answer? (Signal.) *6/10.*
•   Is that more or less than you started with? (Signal.) *More.*
    (Display:)   [94:4H]

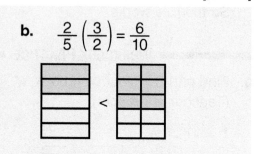

b. $\dfrac{2}{5}\left(\dfrac{3}{2}\right) = \dfrac{6}{10}$ <

Here's what you should have.
e. Problem C: 2/5 × 2/2.
•   What's the answer? (Signal.) *4/10.*
    That equals 2/5.
    (Display:)   [94:4I]

c. $\dfrac{2}{5}\left(\dfrac{2}{2}\right) = \dfrac{4}{10}$ =

Here's what you should have.

---

a. (Display:)   [94:5A]

> 1 gallon is 4 quarts.
>
> 1 day is 24 hours.
>
> 1 year is 12 months.
>
> 1 pound is 16 ounces.

a. Change 64 ounces into pounds.

b. Change 16 days into hours.

c. Change 160 ounces into pounds.

d. Change 24 years into months.

These are facts you'll use to work problems that change one unit into a related unit.
b. Problem A: Change 64 ounces into pounds.
•   Raise your hand when you know the fact for this problem. ✔
•   Everybody, read the fact. (Signal.) *1 pound is 16 ounces.*
c. Problem B: Change 16 days into hours.
•   Raise your hand when you know the fact for this problem. ✔
•   Everybody, read the fact. (Signal.) *1 day is 24 hours.*
d. Problem C: Change 160 ounces into pounds.
•   Raise your hand when you know the fact for this problem. ✔
•   Everybody, read the fact. (Signal.) *1 pound is 16 ounces.*
e. Problem D: Change 24 years into months.
•   Raise your hand when you know the fact for this problem. ✔
•   Everybody, read the fact. (Signal.) *1 year is 12 months.*
f. Go back to problem A: Change 64 ounces into pounds.
•   Look at the fact. Are there more pounds or ounces? (Signal.) *Ounces.*
•   What do you do to end up with ounces? (Signal.) *Multiply by 16.*
•   What do you do to end up with pounds? (Signal.) *Divide by 16.*
    (Repeat until firm.)
•   Look at the problem. What do you end up with? (Signal.) *Pounds.*
•   Work the problem in part A.
    (Observe students and give feedback.)

- Everybody, how many pounds is 64 ounces? (Signal.) *4 pounds.*
  (Display:) [94:5B]

```
                  4 pounds
      a. 16⟌64
           −64
             0
```

Here's what you should have.

g. (Display:) [94:5C]

> 1 gallon is 4 quarts.
>
> 1 day is 24 hours.
>
> 1 year is 12 months.
>
> 1 pound is 16 ounces.

   a. Change 64 ounces into pounds.

   b. Change 16 days into hours.

   c. Change 160 ounces into pounds.

   d. Change 24 years into months.

- Problem B: Change 16 days into hours.
- Look at the fact. Are there more days or hours? (Signal.) *Hours.*
- What do you do to end up with days? (Signal.) *Divide by 24.*
- What do you do to end up with hours? (Signal.) *Multiply by 24.*
  (Repeat until firm.)
- Look at the problem. What do you end up with? (Signal.) *Hours.*
- Work the problem.
  (Observe students and give feedback.)
- Everybody, read the problem you worked. (Signal.) *16 × 24.*
- How many hours is 16 days? (Signal.) *384 hours.*
  (Display:) [94:5D]

```
                 1
                 2
      b.        1 6
             × 2 4
             ─────
               6 4
           + 3 2 0
           ─────────
           3 8 4 hours
```

Here's what you should have.

h. (Display:) [94:5E]

> 1 gallon is 4 quarts.
>
> 1 day is 24 hours.
>
> 1 year is 12 months.
>
> 1 pound is 16 ounces.

   a. Change 64 ounces into pounds.

   b. Change 16 days into hours.

   c. Change 160 ounces into pounds.

   d. Change 24 years into months.

- Work problem C. Remember, find the fact and figure out if you multiply or divide.
  (Observe students and give feedback.)
- Everybody, read the problem you worked. (Signal.) *160 ÷ 16.*
- How many pounds is 160 ounces? (Signal.) *10 pounds.*
  (Display:) [94:5F]

```
                1 0 pounds
      c. 16⟌1 6 0
```

Here's what you should have.

i. (Display:) [94:5G]

> 1 gallon is 4 quarts.
>
> 1 day is 24 hours.
>
> 1 year is 12 months.
>
> 1 pound is 16 ounces.

   a. Change 64 ounces into pounds.

   b. Change 16 days into hours.

   c. Change 160 ounces into pounds.

   d. Change 24 years into months.

- Work problem D.
  (Observe students and give feedback.)
- Everybody, read the problem you worked. (Signal.) *24 × 12.*
- How many months is 24 years? (Signal.) *288 months.*

d.
$$\begin{array}{r} 24 \\ \times\ 12 \\ \hline 48 \\ +240 \\ \hline \boxed{288\ \text{months}} \end{array}$$

Here's what you should have.

EXERCISE 6: WORD PROBLEMS
PARTS OF A GROUP

a. Open your textbook to Lesson 94 and find part 1. ✔
(Teacher reference:)

a. $\frac{2}{7}$ of the students were in the gym. 25 students were not in the gym. How many students were there? How many students were in the gym?

b. $\frac{1}{3}$ of the cookies were burned. There were 24 cookies in all. How many cookies were burned? How many cookies were not burned?

c. After the storm, $\frac{3}{11}$ of the windows were not broken. There were 18 windows that were not broken. How many windows were there in all? How many windows were broken?

d. $\frac{3}{8}$ of the horses were sleeping. 45 horses were awake. How many horses were sleeping? How many horses were there in all?

Some of these problems have facts that tell about two names. The facts for the other problems tell about three names.

b. Problem A: 2/7 of the students were in the gym. 25 students were **not** in the gym. How many students were there? How many students were in the gym?
- Write the letter equation. Stop when you've done that much. ✔
- Everybody, read the letter equation. (Signal.) *5/7 S = N.*
- Work the problem. Answer both questions. (Observe students and give feedback.)
c. Check your work.
- How many students were there? (Signal.) *35 students.*
- How many students were in the gym? (Signal.) *10 students.*

d. Problem B: 1/3 of the cookies were burned. There were 24 cookies in all. How many cookies were burned? How many were not burned?
- Write the letter equation. Stop when you've done that much. ✔
- Everybody, read the letter equation. (Signal.) *1/3 C = B.*
- Work the problem. Answer both questions. (Observe students and give feedback.)
e. Check your work.
- How many cookies were burned? (Signal.) *8 burned cookies.*
- How many cookies were not burned? (Signal.) *16 not-burned cookies.*
f. Problem C: After the storm, 3/11 of the windows were not broken. There were 18 windows that were not broken. How many windows were there in all? How many windows were broken?
- Work the problem. Answer both questions. (Observe students and give feedback.)
g. Check your work.
- Everybody, read the equation with two letters. (Signal.) *3/11 W = N.*
- How many windows were there? (Signal.) *66 windows.*
- How many windows were broken? (Signal.) *48 broken windows.*
h. Problem D: 3/8 of the horses were sleeping. 45 horses were awake. How many horses were sleeping? How many horses were there in all?
- Work the problem. Answer both questions. (Observe students and give feedback.)
i. Check your work.
- Everybody, read the equation with two letters. (Signal.) *5/8 H = A.*
- How many horses were sleeping? (Signal.) *27 sleeping horses.*
- How many horses were there in all? (Signal.) *72 horses.*

Exercise 7: Volume

Associative Property

a. Find part 2 in your textbook. ✔
(Teacher reference:)

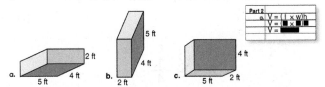

You're going to write equations for the base and height of the same rectangular prism when it is in different positions.
The equation shows the area of the base in parentheses times the height.
The area of the base is length times width.
Remember, show the multiplication for the base inside the parentheses and the height outside.

b. Prism A. Say the multiplication for the base. (Signal.) *5 × 4.*
• What's the number for the height? (Signal.) *2.*
• Copy the letter equation. Below, write the equation with numbers. Stop when you've done that much. ✔
• Everybody, read the multiplication for the area of the base. (Signal.) *5 × 4.*
• What's the height? (Signal.) *2.*
(Display:) [94:7A]

> **a.** V = (1 × w) h
> V = (5 × 4) 2

Here's what you should have.
• Figure out the volume.
(Observe students and give feedback.)
Check your work. You worked the problem 20 × 2.
• What's the volume? (Signal.) *40 cubic feet.*
(Add to show:) [94:7B]

> **a.** V = (1 × w) h
> V = (5 × 4) 2
> V = 40 cu ft

Here's what you should have.

c. Work the problems for prism B.
(Observe students and give feedback.)
d. Check your work.
• Prism B. Say the multiplication for the base. (Signal.) *2 × 4.*
• Say the multiplication for the base times the height. (Signal.) *8 × 5.*
• What's the volume? (Signal.) *40 cubic feet.*
(Display:) [94:7C]

> **b.** V = (1 × w) h
> V = (2 × 4) 5
> V = 40 cu ft

Here's what you should have.
e. Work the problems for prism C.
(Observe students and give feedback.)
f. Check your work.
• Prism C. Say the multiplication for the base. (Signal.) *5 × 2.*
• Say the multiplication for the base times the height. (Signal.) *10 × 4.*
• What's the volume? (Signal.) *40 cubic feet.*
(Display:) [94:7D]

> **c.** V = (1 × w) h
> V = (5 × 2) 4
> V = 40 cu ft

Here's what you should have.

Exercise 8: Mixed-Number Operations

Word Problems

REMEDY

a. Find part 3 in your textbook. ✔
(Teacher reference:)

a. Last week, Tom used up $3\frac{2}{5}$ gallons of gas. This week he used up $7\frac{4}{5}$ gallons. How many gallons did he use up during the two-week period?

b. Juanita had $9\frac{1}{4}$ cups of flour. She used $4\frac{3}{4}$ cups making bread. How many cups of flour does she have left?

These are word problems that tell about mixed numbers. To work one of them, you have to rewrite one of the mixed numbers.

b. Problem A: Last week, Tom used up 3 and 2/5 gallons of gas. This week he used up 7 and 4/5 gallons. How many gallons did he use up during the two-week period?

- Work the problem. Remember, if the answer is a mixed number with a fraction that is more than 1, rewrite the mixed number. Then write the unit name.
(Observe students and give feedback.)
- Everybody, how many gallons did Tom use during the two-week period? (Signal.) *11 and 1/5 gallons.*
(Display:) [94:8A]

$$a. \quad 3\frac{2}{5}$$
$$+ 7\frac{4}{5}$$
$$10\frac{\overset{1}{6}}{5}$$
$$\boxed{11\frac{1}{5}} \text{ gallons}$$

Here's what you should have.

c. Problem B: Juanita had 9 and 1/4 cups of flour. She used 4 and 3/4 cups making bread. How many cups of flour does she have left?

- Write the problem you'll work. Stop when you've done that much. ✔
- Read the problem. (Signal.) *9 and 1/4 – 4 and 3/4.*
- Can you work the problem for the fractions? (Signal.) *No.*
- So tell me the number you'll rewrite. (Signal.) *9 and 1/4.*
- What will the new whole number be? (Signal.) *8.*
- Raise your hand when you know the new fraction. ✔
- What's the new fraction? (Signal.) *5/4.*
(Repeat until firm.)

- Work the problem. Write the answer as a simplified mixed number and a unit name.
(Observe students and give feedback.)
- Everybody, how many cups of flour does she have left? (Signal.) *4 and 1/2 cups.*
(Display:) [94:8B]

$$b. \quad \overset{8}{\cancel{9}}\overset{5}{\cancel{\frac{1}{4}}}$$
$$- 4\frac{3}{4}$$
$$4\frac{2}{4} = \boxed{4\frac{1}{2}} \text{ cups}$$

Here's what you should have.
You rewrote 9 and 1/4 as 8 and 5/4. Then you subtracted. You simplified the answer to 4 and 1/2.

Assign Independent Work, Textbook parts 4–10 and Workbook parts 4 and 5.

Optional extra math-fact practice worksheets are available on ConnectED.

Lesson 95

EXERCISE 1: MENTAL MATH

a. Time for some mental math.
- Listen: 40 plus 17. (Pause.) What's the answer? (Signal.) *57.*
- 60 + 14. (Pause.) What's the answer? (Signal.) *74.*
- 80 + 25. (Pause.) What's the answer? (Signal.) *105.*
- 30 + 25. (Pause.) What's the answer? (Signal.) *55.*
  (Repeat until firm.)

b. Listen: 70 plus 27. (Pause.) What's the answer? (Signal.) *97.*
- 90 + 27. (Pause.) What's the answer? (Signal.) *117.*
- 90 + 37. (Pause.) What's the answer? (Signal.) *127.*
- 80 + 39. (Pause.) What's the answer? (Signal.) *119.*
- 50 + 41. (Pause.) What's the answer? (Signal.) *91.*
  (Repeat until firm.)

c. Listen: 30 minus 10. (Pause.) What's the answer? (Signal.) *20.*
- 30 − 11. (Pause.) What's the answer? (Signal.) *19.*
- 30 − 15. (Pause.) What's the answer? (Signal.) *15.*
- 70 − 15. (Pause.) What's the answer? (Signal.) *55.*
- 90 − 15. (Pause.) What's the answer? (Signal.) *75.*
- 90 − 16. (Pause.) What's the answer? (Signal.) *74.*
  (Repeat until firm.)

d. Listen: 34 plus 8. (Pause.) What's the answer? (Signal.) *42.*
- 36 + 8. (Pause.) What's the answer? (Signal.) *44.*
- 86 + 8. (Pause.) What's the answer? (Signal.) *94.*
- 97 + 8. (Pause.) What's the answer? (Signal.) *105.*
  (Repeat until firm.)

EXERCISE 2: LONG DIVISION
2-DIGIT QUOTIENT

a. Open your workbook to Lesson 95 and find part 1. ✔
(Teacher reference:)

$$\begin{array}{c} \ \ \ \ \ 4 \\ a.\ 53\overline{)2491} \\ \ \ \ \ -212 \\ \hline \ \ \ \ \ 37 \end{array} \qquad \begin{array}{c} \ \ \ 2 \\ b.\ 39\overline{)967} \\ \ \ -78 \\ \hline \ \ \ 18 \end{array} \qquad \begin{array}{c} \ \ \ \ 3 \\ c.\ 81\overline{)2780} \\ \ \ \ -243 \\ \hline \ \ \ \ 35 \end{array}$$

These are division problems that are partially worked. You'll work the problem for the ones digit of the answer and complete each problem.

b. Read problem A. (Signal.) *2491 ÷ 53.*
- What's the first thing you do to work the problem? (Signal.) *Bring down the ones digit.*
- Bring down the ones digit. Then work the estimation problem and complete the problem. (Observe students and give feedback.)

c. Check your work.
- Say the division problem for the ones digit. (Signal.) *371 ÷ 53.*
- Say the estimation problem. (Signal.) *37 ÷ 5.*
- What's the answer? (Signal.) *7.*
- Say the multiplication for the ones digit. (Signal.) *53 × 7.*
- What's the answer? (Signal.) *371.*
  So you worked the subtraction problem 371 − 371 for the remainder.
- What's the remainder? (Signal.) *Zero.*
  Yes, there is no remainder.
  (Display:)   [95:2A]

$$\begin{array}{c} \ \ \ \ \ \ 47 \\ a.\ 53\overline{)2491} \\ \ \ \ \ -212 \\ \hline \ \ \ \ \ 371 \\ \ \ \ -371 \\ \hline \ \ \ \ \ \ \ 0 \end{array} \qquad \begin{array}{c} \ \ \ 7 \\ 5\overline{)37} \end{array} \qquad \begin{array}{c} \ \ 2 \\ \ 53 \\ \times\ \ 7 \\ \hline 371 \end{array}$$

Here's what you should have.

d. Read problem B. (Signal.) *967 ÷ 39.*
- What's the first thing you do to work the problem? (Signal.) *Bring down the ones digit.*
- Bring down the ones digit. Then work the estimation problem and complete the problem. (Observe students and give feedback.)
e. Check your work.
- Say the division problem for the ones digit. (Signal.) *187 ÷ 39.*
- Say the estimation problem. (Signal.) *19 ÷ 4.*
- What's the answer? (Signal.) *4.*
- Say the multiplication for the ones digit. (Signal.) *39 × 4.*
- What's the answer? (Signal.) *156.*
  So you worked the subtraction problem 187 − 156 for the remainder.
- What's the remainder? (Signal.) *31.*
  (Display:) [95:2B]

$$\begin{array}{r} 24\frac{31}{39} \\ \textbf{b. } 39\overline{)967} \\ -78 \\ \hline 187 \\ -156 \\ \hline 31 \end{array} \qquad \begin{array}{r} 4 \\ 4\overline{)19} \end{array} \qquad \begin{array}{r} 3 \\ 39 \\ \times\ 4 \\ \hline 156 \end{array}$$

Here's what you should have.
f. Read problem C. (Signal.) *2780 ÷ 81.*
- What's the first thing you do to work the problem? (Signal.) *Bring down the ones digit.*
- Bring down the ones digit. Then work the estimation problem and complete the problem. (Observe students and give feedback.)
g. Check your work.
- Say the division problem for the ones digit. (Signal.) *350 ÷ 81.*
- Say the estimation problem. (Signal.) *35 ÷ 8.*
- What's the answer? (Signal.) *4.*
- Say the multiplication for the ones digit. (Signal.) *81 × 4.*
- What's the answer? (Signal.) *324.*
  So you worked the subtraction problem 350 − 324 for the remainder.
- What's the remainder? (Signal.) *26.*
  (Display:) [95:2C]

$$\begin{array}{r} 34\frac{26}{81} \\ \textbf{c. } 81\overline{)2780} \\ -243 \\ \hline 350 \\ -324 \\ \hline 26 \end{array} \qquad \begin{array}{r} 4 \\ 8\overline{)35} \end{array} \qquad \begin{array}{r} 81 \\ \times\ 4 \\ \hline 324 \end{array}$$

Here's what you should have.

EXERCISE 3: PROBABILITY
FRACTIONS

a. You're going to learn about probability.
- Say **probability.** (Signal.) *Probability.*
  The probability of something happening doesn't tell that it **will** happen. The probability tells **how likely** it is to happen.
- Is it very likely that a brick will fly through the window in a few seconds? (Signal.) *No.*
- Is it likely that we'll have a lunch period tomorrow? (Signal.) *Yes.*
- Is it likely that it will get dark tonight? (Signal.) *Yes.*
- Is it likely that every student will bring a cold lunch to school tomorrow? (Signal.) *No.*
- Is it likely that you'll have a math test before the end of the year? (Signal.) *Yes.*
b. (Display:) [95:3A]

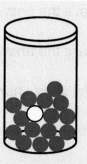

Here's a jar with blue marbles and one white marble. If you reached in the jar without looking at the marbles and pulled out one marble, would it probably be a blue marble or a white marble? (Signal.) *A blue marble.*
- Is it possible that you could pull out the white marble? (Signal.) *Yes.*
  But it's more probable that you'll pull out a blue marble.
c. We can use fractions to show the probability of pulling a blue marble or a white marble from the jar.
  (Add to show:) [95:3B]

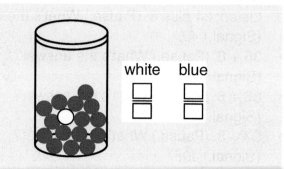

white   blue

We'll show the fraction for pulling out a white marble. The denominator of the fraction tells the number for all the marbles. The numerator tells the number for marbles you're interested in.

- Which part of the fraction tells about all the marbles? (Signal.) *The denominator.*
- Which part tells about the marbles you're interested in? (Signal.) *The numerator.*

d. There are 16 marbles in the jar. So the denominator of the fraction is 16.
   (Add to show:) [95:3C]

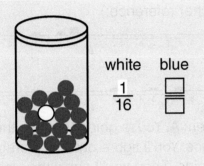

- How many white marbles are in the jar? (Signal.) *1.*
   If we're interested in the white marble, the numerator of the fraction is 1.
   (Add to show:) [95:3D]

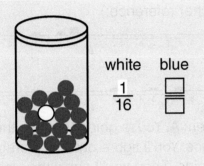

e. What fraction shows the probability of pulling out a white marble? (Signal.) *1/16.*
- Raise your hand when you know the fraction that shows the probability of pulling out a blue marble. ✔
- What fraction shows the probability of pulling out a blue marble? (Signal.) *15/16.*
   (Add to show:) [95:3E]

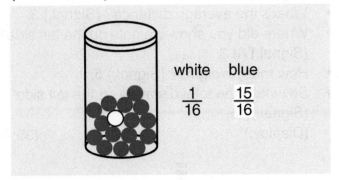

Yes, 15/16.
f. The fraction that is closer to 1 tells which event is more likely to happen.
- Which fraction is closer to 1? (Signal.) *15/16.*

- So is pulling out a blue marble or pulling out a white marble more probable? (Signal.) *Pulling out a blue marble.*
   Yes, pulling out a blue marble is much more likely to happen.

g. (Display:) [95:3F]

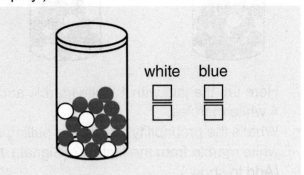

Here's a jar with 16 marbles. 4 are white. We're going to write fractions for pulling out a white marble and pulling out a blue marble.
- Tell me the number for the denominator. (Signal.) *16.*
   (Add to show:) [95:3G]

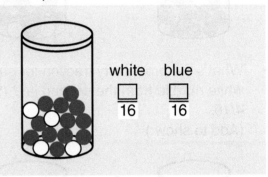

- Tell me the number for the numerator of white marbles. (Signal.) *4.*
- What's the fraction for pulling out a white marble? (Signal.) *4/16.*
- Raise your hand when you know the numerator for pulling out a blue marble. ✔
- What's the fraction for pulling out a blue marble? (Signal.) *12/16.*
   (Add to show:) [95:3H]

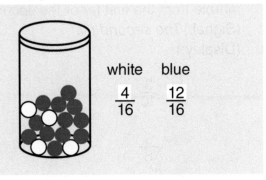

- Which fraction is closer to 1? (Signal.) *12/16.*
- So which event is more likely—pulling out a blue marble or pulling out a white marble? (Signal.) *Pulling out a blue marble.*

h. (Display:) [95:3I]

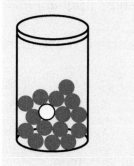

Here are the jars with 1 white marble and 4 white marbles.

• What's the probability fraction for pulling a white marble from the first jar? (Signal.) *1/16.*
(Add to show:) [95:3J]

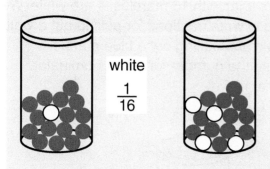

• What's the probability fraction for pulling a white marble from the second jar? (Signal.) *4/16.*
(Add to show:) [95:3K]

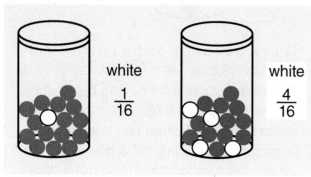

Which fraction is closer to 1? (Signal.) *4/16.*

• So which event is more likely—pulling a white marble from the first jar or the second jar? (Signal.) *The second jar.*

i. (Display:) [95:3L]

$$\frac{3}{10}$$

$$\frac{5}{8}$$

Here are two fractions. One is less than one half and one is more than one half.

• Raise your hand when you know which fraction is more than one half. ✔

• Which fraction is more than one half? (Signal.) *5/8.*

• So which fraction is closer to 1? (Signal.) *5/8.*
(Add to show:) [95:3M]

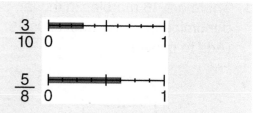

Here are the fractions. You can see that 5/8 is closer to 1 than 3/10.

• So which fraction shows a probability that is more likely? (Signal.) *5/8.*

j. Remember, the fraction that is closer to 1 tells about the event that is more probable.

EXERCISE 4: BALANCE BEAMS
AVERAGE DISTANCE   REMEDY

a. Find part 2 in your workbook. ✔
(Teacher reference:)   R Part D

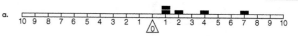

b. Problem A. You're going to make the beam balance. You'll figure out the average for the right side. Then you'll show the weights at the average distance on the left side. Remember, you have to use the same number of weights on both sides.

• Work problem A.
(Observe students and give feedback.)

c. Check your work.

• What's the total distance for the right side? (Signal.) *15.*

• What's the average distance? (Signal.) *3.*

• Where did you show weights on the left side? (Signal.) *At 3.*

• How many weights? (Signal.) *5.*

• So what's the total distance on the left side? (Signal.) *15.*
(Display:) [95:4A]

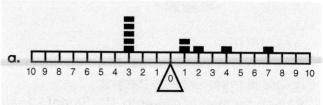

So the beam will balance.

d. Work problem B.
(Observe students and give feedback.)
e. Check your work.
• What's the total distance for the left side? (Signal.) *24.*
• What's the average distance? (Signal.) *6.*
• Where did you show weights on the right side? (Signal.) *At 6.*
• How many weights? (Signal.) *4.*
• So what's the total distance on the left side? (Signal.) *24.*
(Display:)   [95:4B]

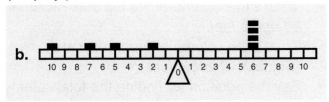

So the beam will balance.

EXERCISE 5: UNIT CONVERSION
MULTIPLICATION/DIVISION

a. (Display:)   [95:5A]

> • 1 year is 12 months.
> • 1 gallon is 4 quarts.
> • 1 day is 24 hours.
> • 1 hour is 60 minutes.

Read the facts.
• (Point to **1 year.**) *1 year is 12 months.*
• (Point to **1 gallon.**) *1 gallon is 4 quarts.*
• (Point to **1 day.**) *1 day is 24 hours.*
• (Point to **1 hour.**) *1 hour is 60 minutes.*

━━━━━━ TEXTBOOK PRACTICE ━━━━━━

a. Open your textbook to Lesson 95 and find part 1. ✔
(Teacher reference:)

a. Change 300 minutes into hours.
b. Change 27 quarts into gallons.
c. Change 48 years into months.
d. Change 48 hours into days.

Facts:
• 1 year is 12 months.
• 1 gallon is 4 quarts.
• 1 day is 24 hours.
• 1 hour is 60 minutes.

Part 1
a.

You will use the facts to work the problems.
b. Problem A: Change 300 minutes into hours.
• What units does the problem name? (Signal.) *Minutes and hours.*
• Find the fact and work the problem.
(Observe students and give feedback.)

• Everybody, read the problem you worked. (Signal.) *300 ÷ 60.*
• How many hours is 300 minutes? (Signal.) *5 hours.*
c. Work problem B. Write the answer as a mixed number and a unit name. ✔
• Everybody, read the problem you worked. (Signal.) *27 ÷ 4.*
• How many gallons is 27 quarts? (Signal.) *6 and 3/4 gallons.*
d. Work problem C. ✔
• Everybody, read the problem you worked. (Signal.) *48 × 12.*
• How many months is 48 years? (Signal.) *576 months.*
e. Work problem D. ✔
• Everybody, read the problem you worked. (Signal.) *48 ÷ 24.*
• How many days is 48 hours? (Signa.) *2 days.*

EXERCISE 6: VOLUME
COMPLEX FIGURES

a. (Display:)   [95:6A]

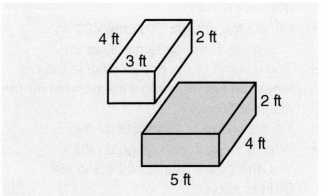

Here are stairs that are made out of concrete. Each stair is a rectangular prism. The bottom stair is shaded. The two stairs will be put together.

(Change to show:) [95:6B]

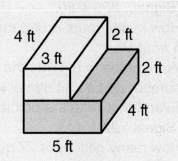

The top stair rests on the bottom stair. We're going to figure out the volume of the concrete. We do that by working two problems. We figure out the volume of the bottom stair. Then we add that volume to the volume of the top stair. So you figure out the total volume of the stairs by adding the volume of the bottom rectangular prism and the volume of the top rectangular prism.

b. The numbers are shown for the base of the bottom stair.
  • What are the numbers for the base of the bottom stair? (Signal.) *5 and 4.*
  • Say the multiplication for the area of the base. (Signal.) *5 × 4.*
  • What's the answer? (Signal.) *20.*
    The base of the bottom stair is 20.
  • The volume of the bottom stair is the base times the height. What's the number for the height? (Signal.) *2.*
    So the volume of the bottom stair is 20 × 2.
  • What's the answer? (Signal.) *40.*
    Yes, the bottom stair is 40 cubic feet.
(Add to show:) [95:6C]

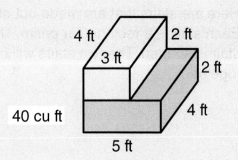

c. The top stair is not shaded. The base of that stair is 3 × 4.
  • Say the multiplication for the area of the base. (Signal.) *3 × 4.*
  • What's the answer? (Signal.) *12.*
  • What's the number for the height? (Signal.) *2.*
    So we find the volume by multiplying 12 × 2.
  • What's the volume of the top stair? (Signal.) *24 cubic feet.*

(Add to show:) [95:6D]

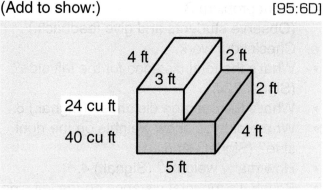

d. To find the volume of both stairs, we add the two volumes.
  • What's the volume of the top stair? (Signal.) *24 cubic feet.*
  • What's the volume of the bottom stair? (Signal.) *40 cubic feet.*
  • Say the addition for finding the total volume. (Signal.) *24 + 40.*
  • What's the total volume of the stairs? (Signal.) *64 cubic feet.*
(Add to show:) [95:6E]

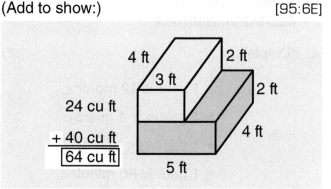

Remember, figure out the volume of each stair. Then add the volumes to get the total volume.

e. (Change to show:) [95:6F]

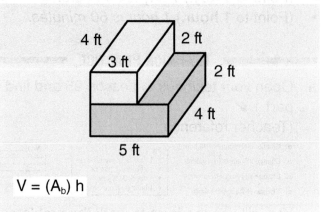

$$V = (A_b) h$$

  • Your turn to work the problem. In part A, figure out the volume of the bottom stair. Start with the equation for volume. Pencils down when you've done that much. ✔

(Add to show:)   [95:6G]

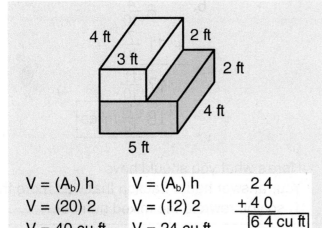

V = (A_b) h
V = (20) 2
V = 40 cu ft

Here's what you should have.
• What's the volume of the bottom stair?
  (Signal.) *40 cubic feet.*
f. Figure out the volume of the top stair. Start
    with the letter equation.
    (Observe students and give feedback.)
• Everybody, what's the volume of the top stair?
  (Signal.) *24 cubic feet.*
g. Find the total volume by adding the volumes
    of the two stairs. ✔
• Everybody, what's the volume of the whole
  stairway? (Signal.) *64 cubic feet.*
(Add to show:)   [95:6H]

V = (A_b) h   V = (A_b) h   2 4
V = (20) 2   V = (12) 2   + 4 0
V = 40 cu ft   V = 24 cu ft   6 4 cu ft

Here's what you should have.

================ **TEXTBOOK PRACTICE** ================

a. Find part 2 in your textbook. ✔
    (Teacher reference:)

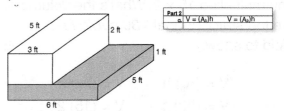

You'll figure out the total volume of these two
stairs. First, you'll figure out the volume of the
bottom stair. That stair is shaded.
• What are the numbers for the base of that
  stair? (Signal.) *6 and 5.*
• Start with the letter equation. Find the area of
  the base of the bottom stair. Multiply the area of
  the base by the height of the bottom stair. Stop
  when you have the volume of the bottom stair.
  (Observe students and give feedback.)
b. Check your work.
    (Display:)   [95:6I]

V = (A_b) h
V = (30) 1
V = 30 cu ft

Here's what you should have.
• You multiplied 6 × 5 for the area of the base.
  What's the area of the base? (Signal.) *30.*
• What's the height? (Signal.) *1.*
• What's the volume? (Signal.) *30 cubic feet.*
c. Now you'll figure out the volume of the top stair.
• Raise your hand when you know the numbers
  for the area of the base. ✔
• What are the numbers for the area of the
  base? (Signal.) *5 and 3.*
• What do you multiply the area of the base by?
  (Signal.) *2.*
  (Repeat until firm.)
• Figure out the volume of the top stair.
  (Observe students and give feedback.)

d. Check your work.
- You multiplied 5 × 3. What's the area of the base? (Signal.) *15.*
- You multiplied 15 × 2. What's the volume of the top stair? (Signal.) *30 cubic feet.*
  (Add to show:) [95:6J]

$$V = (A_b)\ h \qquad V = (A_b)\ h$$
$$V = (30)\ 1 \qquad V = (15)\ 2$$
$$V = 30\ \text{cu ft} \qquad V = 30\ \text{cu ft}$$

Here's what you should have.
e. Now add the two volumes to figure out the total volume of the stairs. ✔
- You worked the problem 30 + 30. What's the total volume? (Signal.) *60 cubic feet.*

EXERCISE 7: MIXED-NUMBER OPERATIONS
WORD PROBLEMS
[REMEDY]

a. Find part 3 in your textbook. ✔
   (Teacher reference:)

a. The school had $24\frac{1}{8}$ gallons of milk in the kitchen. The students drank $21\frac{6}{8}$ gallons of milk at lunch. How many gallons of milk were left?

b. On Wednesday, Shen ran $6\frac{4}{10}$ miles. On Saturday, Shen ran $11\frac{7}{10}$ miles. How many miles did he run in all?

These are word problems that tell about mixed numbers. To work one of them, you have to rewrite one of the mixed numbers.
b. Problem A: The school had 24 and 1/8 gallons of milk in the kitchen. The students drank 21 and 6/8 gallons of milk at lunch. How many gallons of milk were left?
- Write the problem you'll work. Stop when you've done that much. ✔
- Read the problem. (Signal.) *24 and 1/8 – 21 and 6/8.*
- Can you work the problem for the fractions? (Signal.) *No.*
- So tell me the number you'll rewrite. (Signal.) *24 and 1/8.*
- What will the new whole number be? (Signal.) *23.*
- Raise your hand when you know the new fraction. ✔
- What's the new fraction? (Signal.) *9/8.*
  (Repeat until firm.)

- Work the problem. Write the answer as a mixed number and a unit name.
  (Observe students and give feedback.)
- Everybody, how many gallons of milk were left? (Signal.) *2 and 3/8 gallons.*
  (Display:) [95:7A]

a. $24\frac{\cancel{1}^{\,9}}{8}^{\,3}$
   $-\ 21\frac{6}{8}$
   $\boxed{2\frac{3}{8}}$ gallons

Here's what you should have.
You rewrote 24 and 1/8 as 23 and 9/8. Then you subtracted.
c. Problem B: On Wednesday, Shen ran 6 and 4/10 miles. On Saturday, Shen ran 11 and 7/10 miles. How many miles did he run in all?
- Work the problem. Then write the unit name in the answer.
  (Observe students and give feedback.)
- Everybody, how many miles did he run in all? (Signal.) *18 and 1/10 miles.*
  (Display:) [95:7B]

b. $6\frac{4}{10}$
   $+\ 11\frac{7}{10}$
   $17\frac{{}^{1}11}{10}$
   $\boxed{18\frac{1}{10}}$ miles

Here's what you should have.
Your answer had a fraction that was more than 1, so you rewrote the mixed number.

Assign Independent Work, Textbook parts 4–10 and Workbook parts 3–5.

Optional extra math-fact practice worksheets are available on ConnectED.

Lessons 96–100 Planning Page

	Lesson 96	Lesson 97	Lesson 98	Lesson 99	Lesson 100
Student Learning Objectives	**Exercises** 1. Add and subtract using mental math 2. Use fractions to show probability 3. Solve long division problems with 2-digit quotients 4. Use multiplication and division to convert units of measurement 5. Add mixed numbers with unlike denominators 6. Solve complex volume problems 7. Find average distance on balance beams 8. Complete work independently	**Exercises** 1. Add and subtract using mental math 2. **Convert mixed numbers to fractions** 3. Find average distance on balance beams 4. **Complete equations to show probability** 5. Use multiplication and division to convert units of measurement 6. Add mixed numbers with unlike denominators 7. Solve complex volume problems 8. Complete work independently	**Exercises** 1. Add and subtract using mental math 2. Simplify fractions 3. Find average distance on balance beams 4. Convert mixed numbers to fractions 5. Use multiplication and division to convert units of measurement 6. Complete equations to show probability 7. Solve complex volume problems Complete work independently	**Exercises** 1. Solve for missing numbers and add using mental math 2. Simplify fractions 3. Find average distance on balance beams 4. Convert mixed numbers to fractions 5. Use multiplication and division to convert units of measurement 6. Complete equations to show probability 7. Solve long division problems with 2-digit quotients Complete work independently	**Exercises** 1. Solve for missing numbers and add using mental math 2. Find average distance on balance beams 3. Simplify fractions 4. Use multiplication and division to convert units of measurement 5. Complete equations to show probability 6. Solve long division problems with 2-digit quotients 7. Solve word problems for parts of a group 8. Complete work independently
Common Core State Standards for Mathematics					
5.OA 1	✔		✔	✔	✔
5.NBT 1–2	✔	✔	✔		
5.NBT 3		✔	✔		✔
5.NBT 5–6	✔	✔	✔	✔	✔
5.NBT 7	✔	✔	✔	✔	
5.NF 1	✔	✔	✔	✔	
5.NF 2	✔	✔	✔		✔
5.NF 4	✔	✔	✔	✔	✔
5.NF 5	✔	✔	✔		✔
5.NF 6	✔	✔	✔	✔	✔
5.NF 7	✔	✔	✔		
5.MD 1	✔	✔	✔	✔	✔
5.MD 5	✔	✔	✔	✔	
5.G 2	✔			✔	
Teacher Materials	Presentation Book 2, Board Displays CD or chalk board				
Student Materials	Textbook, Workbook, pencil, lined paper, ruler				
Additional Practice	• Student Practice Software: Block 4: Activity 1 (5.NF 5), Activity 2, Activity 3 (5.NBT 4), Activity 4 (5.NBT 3), Activity 5 (5.NBT 7), Activity 6 (5.MD 3 and 5.MD 4) • Provide needed fact practice with Level D or E Math Fact Worksheets.				
Mastery Test					Student Assessment Book (Present Mastery Test 10 following Lesson 100.)

Lesson 96

EXERCISE 1: MENTAL MATH

a. Time for some mental math.
- Listen: 40 plus 11. (Pause.) What's the answer? (Signal.) *51.*
- 140 + 11. (Pause.) What's the answer? (Signal.) *151.*
- 140 + 15. (Pause.) What's the answer? (Signal.) *155.*
- 160 + 15. (Pause.) What's the answer? (Signal.) *175.*
- 160 + 25. (Pause.) What's the answer? (Signal.) *185.*
(Repeat until firm.)

b. Listen: 80 plus 25. (Pause.) What's the answer? (Signal.) *105.*
- 30 + 25. (Pause.) What's the answer? (Signal.) *55.*
- 70 + 27. (Pause.) What's the answer? (Signal.) *97.*
- 70 + 37. (Pause.) What's the answer? (Signal.) *107.*
- 170 + 37. (Pause.) What's the answer? (Signal.) *207*
- 170 + 39. (Pause.) What's the answer? (Signal.) *209.*
(Repeat until firm.)

c. Listen: 60 minus 11. (Pause.) What's the answer? (Signal.) *49.*
- 40 – 11. (Pause.) What's the answer? (Signal.) *29.*
- 40 – 15. (Pause.) What's the answer? (Signal.) *25.*
- 80 – 15. (Pause.) What's the answer? (Signal.) *65.*
- 20 – 15. (Pause.) What's the answer? (Signal.) *5.*
- 90 – 15. (Pause.) What's the answer? (Signal.) *75.*
(Repeat until firm.)

d. Listen: 34 plus 8. (Pause.) What's the answer? (Signal.) *42.*
- 36 + 8. (Pause.) What's the answer? (Signal.) *44.*
- 86 + 8. (Pause.) What's the answer? (Signal.) *94.*
- 96 + 8. (Pause.) What's the answer? (Signal.) *104.*
(Repeat until firm.)

EXERCISE 2: PROBABILITY
FRACTIONS

a. (Display:) [96:2A]

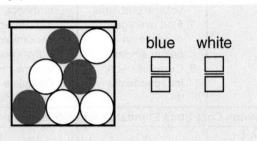

- Last time, you learned about probability. What word? (Signal.) *Probability.*
The probability of something happening is how likely it is to happen. You learned that you can make fractions to show probabilities. We're going to make fractions for the blue marbles and the white marbles.

b. Raise your hand when you know the denominator of both fractions. ✔
- What's the denominator? (Signal.) *7.*
Yes, there are 7 marbles, so the denominators are both 7.
(Change to show:) [96:2B]

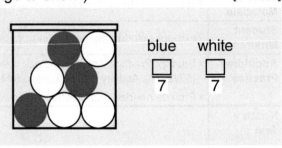

c. Raise your hand when you know the numerator for the blue marbles. ✔
- What's the numerator? (Signal.) *3.*
- What's the numerator for the white marbles? (Signal.) *4.*
  (Change to show:)   [96:2C]

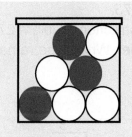

blue   white
$\frac{3}{7}$ $\frac{4}{7}$

- Which fraction is closer to 1? (Signal.) *4/7.*
- So which event is more probable—pulling a white marble from the box or pulling a blue marble from the box? (Signal.) *Pulling a white marble from the box.*

════════ **WORKBOOK PRACTICE** ════════

a. Open your workbook to Lesson 96 and find part 1. ✔
   (Teacher reference:)

gray white   black white   white black gray   black
a. __ __   b. __ __   c. __ __ __   d. $\frac{2}{7}$

b. Problem A: There are 9 marbles in the box. Some are gray and some are white. You're going to write the probability fraction for pulling a marble from the box without looking.
- Write the probability fraction for pulling a gray marble from box A. ✔
- What's the fraction for pulling a gray marble from box A? (Signal.) *8/9.*
- Write the probability fraction for pulling a white marble from box A. ✔
- What's the fraction for pulling a white marble from box A? (Signal.) *1/9.*
- Circle the fraction that has the higher probability. Remember, that's the fraction closer to 1. ✔
- Did you circle 1/9 or 8/9? (Signal.) *8/9.*
- So is pulling a white marble or pulling a gray marble more probable? (Signal.) *Pulling a gray marble.*

c. Problem B: There are 9 marbles in box B. Some of them are black.
- Write the probability fraction for pulling a black marble from box B, and write the probability for pulling a white marble from box B. Circle the fraction that shows the higher probability. (Observe students and give feedback.)
- Everybody, what's the fraction for pulling a black marble from box B? (Signal.) *4/9.*
- What's the fraction for pulling a white marble from box B? (Signal.) *5/9.*
- Which fraction did you circle? (Signal.) *5/9.*
- So which event is more likely—pulling a white marble or pulling a black marble? (Signal.) *Pulling a white marble.*

d. Problem C. There are white marbles, black marbles, and gray marbles in the box. You're going to write the fraction for each color.
- Write the fractions for the three colors. (Observe students and give feedback.)

e. Check your work.
- Read the fraction for white marbles. (Signal.) *3/10.*
- Read the fraction for black marbles. (Signal.) *2/10.*
- Read the fraction for gray marbles. (Signal.) *5/10.*

f. Circle the fraction that shows the most likely color to pull from the box. ✔
- Which fraction did you circle? (Signal.) *5/10.*
- So what color marble has the greatest probability? (Signal.) *Gray.*

g. Problem D. The probability fraction is shown for black marbles in the box.
- What's the probability fraction for black marbles? (Signal.) *2/7.*
  The rest of the marbles are white.
- Draw the black marbles and white marbles in the box. Remember, the fraction shows how many marbles there are and how many are black. ✔

h. Check your work.
- How many marbles did you draw in the box? (Signal.) *7.*
- How many did you color black? (Signal.) *2.*
  Yes, 7 marbles and 2 of them are black.

EXERCISE 3: LONG DIVISION
2-DIGIT QUOTIENT

a. Find part 2 in your workbook. ✔
    (Teacher reference:)

$$a.\ 38\overline{)925} \qquad b.\ 91\overline{)4823} \qquad c.\ 29\overline{)1601}$$
$$\begin{array}{c}2\\-76\\\hline16\end{array} \qquad \begin{array}{c}5\\-455\\\hline27\end{array} \qquad \begin{array}{c}5\\-145\\\hline15\end{array}$$

These are division problems that are partially worked. You'll work the problem for the ones digit of the answer and complete each problem.

b. Read problem A. (Signal.) *925 ÷ 38.*
•   Work the problem.
    (Observe students and give feedback.)
c. Check your work. You brought down the 5. You worked the problem 165 divided by 38.
•   Say the estimation problem for 165 divided by 38. (Signal.) *17 ÷ 4.*
•   What's the answer? (Signal.) *4.*
•   Then you multiplied 38 times 4. What's the answer? (Signal.) *152.*
•   You worked the subtraction problem 165 – 152 for the remainder. What's the remainder? (Signal.) *13.*
    (Display:)   [96:3A]

$$a.\ 38\overline{)925}\ \ 24\tfrac{13}{38} \qquad 4\overline{)17} \qquad \begin{array}{r}38\\\times\ \ 4\\\hline152\end{array}$$
$$\begin{array}{c}-76\\\hline165\\-152\\\hline13\end{array} \qquad 4 \qquad \overset{1}{}$$

Here's what you should have.
•   What's the whole answer? (Signal.) *24 and 13/38.*
d. Read problem B. (Signal.) *4823 ÷ 91.*
•   Work the problem.
    (Observe students and give feedback.)

e. Check your work. You brought down the 3. You worked the problem 273 ÷ 91.
•   Say the estimation problem for 273 ÷ 91. (Signal.) *27 ÷ 9.*
•   What's the answer? (Signal.) *3.*
•   Then you multiplied 91 times 3. What's the answer? (Signal.) *273.*
•   You worked the subtraction problem 273 – 273 for the remainder. What's the remainder? (Signal.) *Zero.*
    (Display:)   [96:3B]

$$b.\ 91\overline{)4823}\ \ 53 \qquad 9\overline{)27}\ \ 3 \qquad \begin{array}{r}91\\\times\ \ 3\\\hline273\end{array}$$
$$\begin{array}{c}-455\\\hline273\\-273\\\hline0\end{array}$$

Here's what you should have.
•   What's the whole answer? (Signal.) *53.*
f. Read problem C. (Signal.) *1601 ÷ 29.*
•   Work the problem.
    (Observe students and give feedback.)
g. Check your work. You brought down the 1. You worked the estimation problem 15 ÷ 3.
•   What's the answer? (Signal.) *5.*
•   Then you multiplied 29 times 5. What's the answer? (Signal.) *145.*
•   You worked the subtraction problem 151 – 145 for the remainder. What's the remainder? (Signal.) *6.*
    (Display:)   [96:3C]

$$c.\ 29\overline{)1601}\ \ 55\tfrac{6}{29} \qquad 3\overline{)15}\ \ 5 \qquad \begin{array}{r}\overset{4}{2}9\\\times\ \ 5\\\hline145\end{array}$$
$$\begin{array}{c}-145\\\hline151\\-145\\\hline6\end{array}$$

Here's what you should have.
•   What's the whole answer? (Signal.) *55 and 6/29.*

Exercise 4: Unit Conversion
Multiplication/Division

a. Open your textbook to Lesson 96 and find part 1. ✔
(Teacher reference:)

a. Change 5 years into months.
b. Change 12 days into hours.
c. Change 16 seasons into years.
d. Change 36 inches into feet.
e. Change 15 hours into minutes.

Facts:
• 1 hour is 60 minutes.
• 1 foot is 12 inches.
• 1 year is 12 months.
• 1 dollar is 20 nickels.
• 1 year is 4 seasons.
• 1 day is 24 hours.

Part 1
a.

You will use the measurement facts to work the problems.

b. Problem A: Change 5 years into months.
• Work the problem. ✔
• Everybody, read the problem you worked. (Signal.) 12×5.
• How many months is 5 years? (Signal.) *60 months.*
c. Work problem B.
(Observe students and give feedback.)
d. Check your work. You changed 12 days into hours.
• Read the problem you worked. (Signal.) 12×24.
• How many hours is 12 days? (Signal.) *288 hours.*
e. Work problem C. ✔
You changed 16 seasons into years.
• Read the problem you worked. (Signal.) $16 \div 4$.
• How many years is 16 seasons? (Signal.) *4 years.*
f. Work problem D. ✔
You changed 36 inches into feet.
• Read the problem you worked. (Signal.) $36 \div 12$.
• How many feet is 36 inches? (Signal.) *3 feet.*
g. Work problem E. ✔
You changed 15 hours into minutes.
• Read the problem you worked. (Signal.) 15×60.
• How many minutes is 15 hours? (Signal.) *900 minutes.*

Exercise 5: Mixed-Number Operations
Unlike Denominators   **REMEDY**

a. (Display:)   [96:5A]

$$5\frac{2}{3}$$
$$+\,8\frac{3}{4}$$

• Read the problem. (Signal.) *5 and 2/3 + 8 and 3/4.*
The denominators are not the same, so you have to find the lowest common denominator.
• Raise your hand when you know the lowest common denominator. ✔
• What's the lowest common denominator? (Signal.) *12.*
(Add to show:)   [96:5B]

$$5\frac{2}{3}\left(\frac{4}{4}\right) = \frac{8}{12}$$
$$+\,8\frac{3}{4}\left(\frac{3}{3}\right) = +\frac{9}{12}$$

• Say the problem for the fractions. (Signal.) *8/12 + 9/12.*
• What's the answer? (Signal.) *17/12.*
(Add to show:)   [96:5C]

$$5\frac{2}{3}\left(\frac{4}{4}\right) = \frac{8}{12}$$
$$+\,8\frac{3}{4}\left(\frac{3}{3}\right) = +\frac{9}{12}$$
$$\frac{17}{12}$$

b. Say the problem for the whole numbers. (Signal.) *5 + 8.*
• What's the answer? (Signal.) *13.*
(Add to show:)   [96:5D]

$$5\frac{2}{3}\left(\frac{4}{4}\right) = \frac{8}{12}$$
$$+\,8\frac{3}{4}\left(\frac{3}{3}\right) = +\frac{9}{12}$$
$$13\frac{17}{12}$$

c. The fraction in this mixed number is more than 1. So we have to rewrite the mixed number.
- Raise your hand when you know the mixed number for 17/12. ✔
- What's 17/12? (Signal.) *1 and 5/12.*
- What's the new whole number? (Signal.) *14.*
(Add to show:) [96:5E]

$$5 \frac{2}{3} \left(\frac{4}{4}\right) = \frac{8}{12}$$
$$+ 8 \frac{3}{4} \left(\frac{3}{3}\right) = + \frac{9}{12}$$
$$13 \overset{1}{} \frac{17}{12}$$
$$\boxed{14 \frac{5}{12}}$$

So the answer is 14 and 5/12.
d. Remember, if the fraction in the answer is more than 1, rewrite the mixed number.

━━━━━━ **TEXTBOOK PRACTICE** ━━━━━━

a. Find part 2 in your textbook. ✔
(Teacher reference:)

a. $5\frac{4}{5}$ b. $12\frac{5}{7}$ c. $10\frac{3}{4}$
  $+7\frac{3}{10}$ $+6\frac{1}{2}$ $+15\frac{5}{6}$

b. Read problem A. (Signal.) *5 and 4/5 + 7 and 3/10.*
- Copy and work the first part of the problem. Stop after you've added the fractions. ✔
- Check your work.
- You rewrote the 4/5 as 8/10 and worked the problem 8/10 + 3/10. What's the answer? (Signal.) *11/10.*
That's more than 1.
(Display:) [96:5F]

$$a. \quad 5 \frac{4}{5} \left(\frac{2}{2}\right) = \frac{8}{10}$$
$$+ 7 \frac{3}{10} \qquad = + \frac{3}{10}$$
$$\frac{11}{10}$$

Here's what you should have so far. If you showed 3/10 times 1 over 1, that's also correct.

c. Now work the rest of the problem. Add the whole numbers. Then rewrite the mixed number.
(Observe students and give feedback.)
(Add to show:) [96:5G]

$$a. \quad 5 \frac{4}{5} \left(\frac{2}{2}\right) = \frac{8}{10}$$
$$+ 7 \frac{3}{10} \qquad = + \frac{3}{10}$$
$$12 \overset{1}{} \frac{11}{10}$$
$$\boxed{13 \frac{1}{10}}$$

Here's what you should have. The fraction was 11/10. That's 1 and 1/10. The answer is 13 and 1/10.
d. Work problem B.
(Observe students and give feedback.)
Check your work.
(Display:) [96:5H]

$$b. \quad 12 \frac{5}{7} \left(\frac{2}{2}\right) = \frac{10}{14}$$
$$+ 6 \frac{1}{2} \left(\frac{7}{7}\right) = + \frac{7}{14}$$
$$18 \overset{1}{} \frac{17}{14}$$
$$\boxed{19 \frac{3}{14}}$$

Here's what you should have. The fraction in the answer was 17/14. So you rewrote the fraction as 1 and 3/14.
- What's the answer to the whole problem? (Signal.) *19 and 3/14.*
e. Work problem C.
(Observe students and give feedback.)
Check your work.
(Display:) [96:5I]

$$c. \quad 10 \frac{3}{4} \left(\frac{3}{3}\right) = \frac{9}{12}$$
$$+ 15 \frac{5}{6} \left(\frac{2}{2}\right) = + \frac{10}{12}$$
$$25 \overset{1}{} \frac{19}{12}$$
$$\boxed{26 \frac{7}{12}}$$

Here's what you should have. The fraction in the answer was 19/12. So you rewrote the fraction as 1 and 7/12.
- What's the answer to the whole problem? (Signal.) *26 and 7/12.*

EXERCISE 6: VOLUME
COMPLEX FIGURES

a. Find part 3 in your textbook. ✔
   (Teacher reference:)

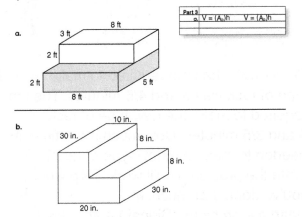

You'll figure out the total volume of the stairs.
b. Problem A shows the bottom stair shaded.
- Say the equation for finding the volume of the bottom stair. (Signal.) *Volume = Area of the base × height.*
- Figure out the volume of the bottom stair and top stair. Add the volumes to find the total volume of the stairs.
   (Observe students and give feedback.)
c. Check your work. You figured out the volume of the bottom stair.
- What's the number for the area of the base? (Signal.) *40.*
- What's the number for the height? (Signal.) *2.*
- What's the volume of the bottom stair? (Signal.) *80 cubic feet.*
   (Display:)   [96:6A]

> $V = (A_b)\ h$
> $V = (40)\ 2$
> $V = 80$ cu ft

Here's what you should have.
d. You figured out the volume of the top stair.
- What's the number for the area of the base? (Signal.) *24.*
- What's the number for the height? (Signal.) *2.*
- What's the volume of the top stair? (Signal.) *48 cubic feet.*

e. You added the two stairs. What's 80 + 48? (Signal.) *128.*
- So what's the total volume of the stairs? (Signal.) *128 cubic feet.*
   (Add to show:)   [96:6B]

> | $V = (A_b)\ h$ | $V = (A_b)\ h$ | $\begin{array}{r} 80 \\ +\ 48 \\ \hline \boxed{128\ \text{cu ft}} \end{array}$ |
> |---|---|---|
> | $V = (40)\ 2$ | $V = (24)\ 2$ | |
> | $V = 80$ cu ft | $V = 48$ cu ft | |

Here's what you should have.
f. Problem B doesn't show the bottom stair shaded.
- Figure out the volume of each stair. Then add to get the total volume.
   (Observe students and give feedback.)
g. Check your work. You figured out the volume of the bottom stair.
- The number for the area of the base is 600. What's the number for the height? (Signal.) *8.*
- What's the volume of the bottom stair? (Signal.) *4800 cubic inches.*
   (Display:)   [96:6C]

> $V = (A_b)\ h$
> $V = (600)\ 8$
> $V = 4800$ cu in.

Here's what you should have.
h. You figured out the volume of the top stair.
- What's the number for the area of the base? (Signal.) *300.*
- What's the height? (Signal.) *8.*
- What's the volume of the top stair? (Signal.) *2400 cubic inches.*
   You added the stairs—4800 + 2400.
- What's 4800 + 2400? (Signal.) *7200.*
- So what's the total volume of the stairs? (Signal.) *7200 cubic inches.*
   (Add to show:)   [96:6D]

> | $V = (A_b)\ h$ | $V = (A_b)\ h$ | $\begin{array}{r} \overset{1}{4}800 \\ +\,2400 \\ \hline \boxed{7200\ \text{cu in.}} \end{array}$ |
> |---|---|---|
> | $V = (600)\ 8$ | $V = (300)\ 8$ | |
> | $V = 4800$ cu in. | $V = 2400$ cu in. | |

Here's what you should have.

EXERCISE 7: BALANCE BEAMS
AVERAGE DISTANCES

a. Find part 4 in your textbook. ✔
   (Teacher reference:)

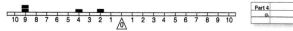

   a. At what distance would you show the weights on the right side?
   b. How many weights would you show on the right side?
   c. What would be the total distance on the right side?

   You'll figure out the average distance and then answer the questions.

b. Question A: At what distance would you show the weights on the right side?

• Figure out the average distance and write the answer.
   (Observe students and give feedback.)

c. Check your work.

• You figured out the total distance on the left side. What's the total distance? (Signal.) *24.*

• You divided by 4 to find the average distance. What's the average distance? (Signal.) *6.*
   You could make the beam balance by putting weights at the same distance on the right.

• Question A: At what distance would you show the weights on the right side? (Signal.) *6.*

d. Answer the other questions.
   (Observe students and give feedback.)

e. Check your work.

• Question B: How many weights would you show on the right side? (Signal.) *4.*

• C: What would be the total distance on the right side? (Signal.) *24.*

f. Find part 5 in your textbook. ✔
   (Teacher reference:)

   • You want to balance the beam by putting weights at one distance on the left side.

   a. At what distance would you show the weights on the left side?
   b. How many weights would you show on the left side?
   c. What would be the total distance on the left side?

   You'll work this problem as part of your Independent Work.

EXERCISE 8: INDEPENDENT WORK
MIXED-NUMBER WORD PROBLEMS

a. Find part 6 in your textbook. ✔

   a. The time required to make the old type of rack was $3\frac{4}{5}$ minutes. The time required to make the new type of rack is $4\frac{3}{5}$ minutes. How much more time is needed to make the new type of rack?

   b. Each cinder block weighed $3\frac{5}{8}$ pounds more than the concrete block. The concrete block weighed $17\frac{3}{8}$ pounds. How much did the cinder block weigh?

b. Problem A: The time required to make the old type of rack was 3 and 4/5 minutes. The time required to make the new type of rack was 4 and 3/5 minutes. How much more time is needed to make the new type of rack?

• Write the problem you'll work. Stop when you've done that much. ✔

c. Read the problem. (Signal.) *4 and 3/5 – 3 and 4/5.*

• Can you work the problem for the fractions? (Signal.) *No.*

• Tell me the number you'll rewrite. (Signal.) *4 and 3/5.*

• What will the new whole number be? (Signal.) *3.*

• Raise your hand when you know the new fraction. ✔

• What's the new fraction? (Signal.) *8/5.*
   (Repeat until firm.)
   You'll work the problems in part 6 as part of your independent work.

Assign Independent Work, Textbook parts 5–12 and Workbook parts 3–5.

Optional extra math-fact practice worksheets are available on ConnectED.

Lesson 97

EXERCISE 1: MENTAL MATH

a. Time for some mental math.
- Listen: 50 plus 12. (Pause.) (Signal.) *62.*
- 150 + 12. (Pause.) (Signal.) *162.*
- 160 + 15. (Pause.) (Signal.) *175.*
- 130 + 15. (Pause.) (Signal.) *145.*
- 130 + 25. (Pause.) (Signal.) *155.*
- 90 + 25. (Pause.) (Signal.) *115.*
- (Repeat until firm.)

b. Listen: 60 plus 25. (Pause.) (Signal.) *85.*
- 70 + 29. (Pause.) (Signal.) *99.*
- 170 + 27. (Pause.) (Signal.) *197.*
- 170 + 34. (Pause.) (Signal.) *204.*
- 170 + 38. (Pause.) (Signal.) *208.*
- (Repeat until firm.)

c. Listen: 90 minus 11. (Pause.) (Signal.) *79.*
- 90 − 12. (Pause.) (Signal.) *78.*
- 90 − 15. (Pause.) (Signal.) *75.*
- 190 − 15. (Pause.) (Signal.) *175.*
- 120 − 15. (Pause.) (Signal.) *105.*
- (Repeat until firm.)

d. Listen: 30 minus 15. (Pause.) (Signal.) *15.*
- 34 − 10. (Pause.) (Signal.) *24.*
- 34 − 8. (Pause.) (Signal.) *26.*
- (Repeat until firm.)

e. Listen: 87 plus 8. (Pause.) (Signal.) *95.*
- 88 + 8. (Pause.) (Signal.) *96.*
- 85 + 8. (Pause.) (Signal.) *93.*
- (Repeat until firm.)

EXERCISE 2: FRACTIONS
FOR MIXED NUMBERS

a. You know how to change fractions into mixed numbers. You're going to learn how to change mixed numbers into fractions.
(Display:)   [97:2A]

$$3\frac{2}{5} =$$

- Read this number. (Signal.) *3 and 2/5.*
  We're going to complete the equation to show the fraction that equals 3 and 2/5.
  The first thing we do is change the whole number into a fraction that has the denominator of the fraction.

- What's the denominator of the fraction? (Signal.) *5.*
  So we write 3 as a fraction with a denominator of 5.
  (Add to show:)   [97:2B]

$$3\frac{2}{5} =$$
$$\frac{}{5}$$

- The numerator is 3 times the denominator. What's 3 times 5? (Signal.) *15.*
  (Add to show:)   [97:2C]

$$3\frac{2}{5} =$$
$$\frac{15}{5}$$

Now we add 15/5 and 2/5.
(Add to show:)   [97:2D]

$$3\frac{2}{5} =$$
$$\frac{15}{5} + \frac{2}{5}$$

- What's the answer? (Signal.) *17/5.*
  (Add to show:)   [97:2E]

$$3\frac{2}{5} = \frac{17}{5}$$
$$\frac{15}{5} + \frac{2}{5}$$

- What fraction equals 3 and 2/5? (Signal.) *17/5.*

b. (Display:)   [97:2F]

$$7\frac{1}{3} =$$

We're going to complete the equation to show the fraction that equals 7 and 1/3.

- What will the fraction equal? (Signal.) *7 and 1/3.*
  We first change the whole number into a fraction that has a denominator of 3.

- What's the denominator? (Signal.) *3.*

- The numerator is 7 times 3. What's the answer? (Signal.) *21.*
  (Add to show:) [97:2G]

$$7\frac{1}{3} =$$

$$\frac{21}{3} + \frac{1}{3}$$

So 7 and 1/3 equals 21/3 plus 1/3.
- What's 21/3 + 1/3? (Signal.) *22/3.*
  (Add to show:) [97:2H]

$$7\frac{1}{3} = \frac{22}{3}$$

$$\frac{21}{3} + \frac{1}{3}$$

- Read the equation. (Signal.) *7 and 1/3 = 22/3.*
c. (Display:) [97:2I]

$$5\frac{2}{9} =$$

- Read the mixed number. (Signal.) *5 and 2/9.*
  We'll change 5 and 2/9 into a fraction.
- What's the denominator of that fraction?
  (Signal.) *9.*
  The fraction that equals 5 has a numerator that is 5 times the denominator.
- What's the numerator? (Signal.) *45.*
  (Add to show:) [97:2J]

$$5\frac{2}{9} =$$

$$\frac{45}{9} + \frac{2}{9}$$

- Raise your hand when you know the fraction that equals 5 and 2/9. ✔
- Everybody, what fraction equals 5 and 2/9? (Signal.) *47/9.*
  (Add to show:) [97:2K]

$$5\frac{2}{9} = \frac{47}{9}$$

$$\frac{45}{9} + \frac{2}{9}$$

━━━━━ **WORKBOOK PRACTICE** ━━━━━

a. Open your workbook to Lesson 97 and find part 1. ✔
  (Teacher reference:)

a. $3\frac{7}{10} =$ ☐   b. $2\frac{5}{8} =$ ☐

☐ + ☐   ☐ + ☐

b. Read mixed number A. (Signal.) *3 and 7/10.*
  Below the mixed number, you'll write the fraction addition and then complete the equation.
- First you write the fraction that equals 3. What's the denominator of that fraction? (Signal.) *10.*
- Write the fraction that equals 3. Then add 7/10 and complete the equation.
  (Observe students and give feedback.)
- Everybody, what fraction does 3 and 7/10 equal? (Signal.) *37/10.*
  (Display:) [97:2L]

a. $$3\frac{7}{10} = \frac{37}{10}$$

$$\frac{30}{10} + \frac{7}{10}$$

Here's what you should have.
The fraction for 3 is 30/10. You added 7/10.
c. Read mixed number B. (Signal.) *2 and 5/8.*
  Below the mixed number, you'll write the fraction addition and then complete the equation.
- First you write the fraction that equals 2. What's the denominator of that fraction? (Signal.) *8.*
- Write the fraction that equals 2. Then add 5/8 and complete the equation.
  (Observe students and give feedback.)
- Everybody, what fraction does 2 and 5/8 equal? (Signal.) *21/8.*
  (Display:) [97:2M]

b. $$2\frac{5}{8} = \frac{21}{8}$$

$$\frac{16}{8} + \frac{5}{8}$$

Here's what you should have.
The fraction for 2 is 16/8. You added 5/8.

EXERCISE 3: BALANCE BEAMS
AVERAGE DISTANCE

a. (Display:) [97:3A]

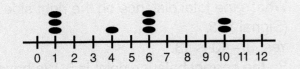

You're going to work a new kind of balance beam problem. This number line starts at zero and goes to 12. We want to find out where the balance point would be for these weights. The balance point is the average distance.
So we find the total distance from zero. Then we divide by the number of weights to find the average distance.
Remember, first we find the total distance from zero. Then we divide by the number of weights to find the average distance.

b. If we put the balance point at 4, it won't balance.
(Display:) [97:3B]

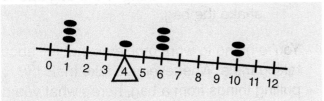

- Which side is heavier—the left or the right? (Signal.) *The right.*

c. If we put the balance point at 9, it won't balance.
(Display:) [97:3C]

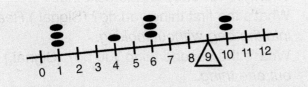

- Which side is heavier? (Signal.) *The left.*

d. (Display:) [97:3D]

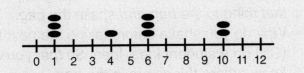

We'll find the average distance. At the average distance, we have the same distance on both sides.
- (Point to **1.**) What's the total distance at 1? (Signal.) *3.*
- (Point to **4.**) What's the total distance at 4? (Signal.) *4.*

- (Point to **6.**) Raise your hand when you know the total distance at 6. ✔
What's the total distance? (Signal.) *18.*
- (Point to **10.**) What's the total distance at 10? (Signal.) *20.*

e. (Add to show:) [97:3E]

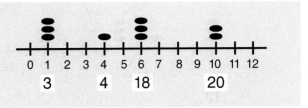

- We add the totals to find the total distance for all of the weights. Find the total distance in part A.
(Observe students and give feedback.)
- Everybody, what's the total distance? (Signal.) *45.*

f. Now you divide by the number of weights to find the average.
- Raise your hand when you know the number you divide by. ✔
- What do you divide by? (Signal.) *9.*
- Work the problem and write the average. (Observe students and give feedback.)
- You worked the problem 45 divided by 9. What's the answer? (Signal.) *5.*
- So where is the balance point for these weights? (Signal.) *At 5.*
(Change to show:) [97:3F]

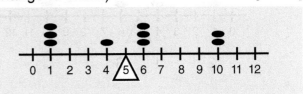

g. Let's check to see if the total distance is the same on each side of the balance point.
(Point to **1.**) These weights are 4 units from the balance point.
- (Point to **4.**) How many units is this weight from the balance point? (Signal.) *1.*
- (Point to **6.**) How many units are these weights from the balance point? (Signal.) *1.*
- (Point to **10.**) Raise your hand when you know how many units these weights are from the balance point. ✔
- How many units? (Signal.) *5.*
(Repeat until firm.)

h. (Point to **1**.) How many units are these weights from the balance point? (Signal.) *4.*
- How many weights are there? (Signal.) *3.*
- So what's the total distance for these weights? (Signal.) *12.*
  (Add to show:) [97:3G]

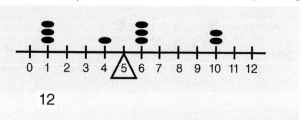

i. (Point to **4**.) How many units is this weight from the balance point? (Signal.) *1.*
(Add to show:) [97:3H]

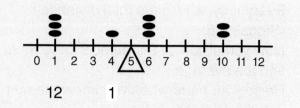

j. (Point to **6**.) How many units are these weights from the balance point? (Signal.) *1.*
- How many weights are there? (Signal.) *3.*
- So what's the total distance for these weights? (Signal.) *3.*
  (Add to show:) [97:3I]

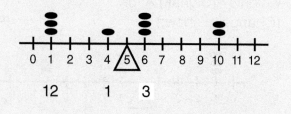

k. (Point to **10**.) How many units are these weights from the balance point? (Signal.) *5.*
- How many weights are there? (Signal.) *2.*
- So what's the total distance for the weights? (Signal.) *10.*
  (Add to show:) [97:3J]

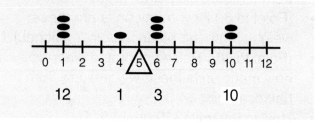

l. Raise your hand when you know the total distance on the left side. ✔
- What's the total distance? (Signal.) *13.*
  Yes, 12 + 1 is 13.
- What's the total distance on the right side? (Signal.) *13.*
  Yes, 3 + 10 is 13.
  So the correct balance point is 5. We have the same total distance on each side.

EXERCISE 4: PROBABILITY
EQUATIONS

a. (Display:) [97:4A]

> **Taking Trials**
>
> 1. You reach into the bag without looking.
> 2. You pull out one thing.
> 3. You record what that thing is.
> 4. You return that thing to the bag and shake the bag.

You're going to work problems that tell about taking trials. Whenever you take trials by pulling things from a bag, here's what you do:
1. You reach into the bag without looking.
2. You pull out one thing.
3. You record what that thing is.
4. You return that thing to the bag and shake the bag.
Then you're ready for the next trial.
- What's the first thing you do? (Signal.) *Reach into the bag without looking.*
- What's the second thing you do? (Signal.) *Pull out one thing.*
- What's the third thing you do? (Signal.) *Record what that thing is.*
- What's the fourth thing you do? (Signal.) *Return that thing to the bag and shake the bag.*
- Why do you shake before taking the next trial? (Call on a student idea. Idea: *So that you don't know where things are in the bag.*)

b. You can use probability fractions to estimate how often you'll pull out objects of a particular type. If the probability fraction is 2/3, you'd expect to pull out that type of object on about 2/3 of the trials. If the probability fraction is 1/2 you'd expect to pull out that type of object on about 1/2 of the trials.
(Display:) [97:4B]

$$\frac{3}{8}$$

The fraction shows the probability of pulling a white marble from a bag. On 3/8 of the trials you'd expect to pull a white marble from the bag.
- If you took 8 trials, you would expect to pull out a white marble on 3 trials.
- If you took 16 trials, you would expect to pull out a white marble on 6 trials.
- How many white marbles would you expect to pull from the bag after 8 trials? (Signal.) *3.*
- How many white marbles would you expect to pull from the bag after 16 trials? (Signal.) *6.*
- If you took 24 trials, how many white marbles would you expect to pull from the bag? (Signal.) *9.*

c. We can write equations that show probabilities. The equation for this problem is 3/8 T = W.
(Add to show:) [97:4C]

$$\frac{3}{8} t = w$$

T stands for trials and W stands for white.
- Read the equation. (Signal.) *3/8 T = W.*
- What does T stand for? (Signal.) *Trials.*
- What does W stand for? (Signal.) *White.*
The equation tells you that for 3/8 of the trials, you expect to pull out a white marble.
Once more: For 3/8 of the trials, you'd expect to pull out a white marble.

d. (Display:) [97:4D]

$$\frac{1}{9} t = w$$

- Read this equation. (Signal.) *1/9 T = W.*
This equation tells about the probability of pulling white marbles from a different bag. For 1/9 of the trials, you'd expect to pull out a white marble.
- What letter tells about trials? (Signal.) *T.*
- What letter tells about white marbles? (Signal.) *W.*
- For what fraction of trials would you expect to pull a white marble? (Signal.) *1/9.*

e. (Display:) [97:4E]

a. $\frac{3}{10}$

b. $\frac{3}{17}$

c. $\frac{2}{5}$

d. $\frac{10}{23}$

These are probability fractions for pulling out white marbles.
- Read probability fraction A. (Signal.) *3/10.*
So for 3/10 of the trials, you'd expect to pull out a white marble.
So the equation for trials is 3/10 T = W.
- Say the equation for trials. (Signal.) *3/10 T = W.*
(Add to show:) [97:4F]

a. $\frac{3}{10} t = w$

b. $\frac{3}{17}$

c. $\frac{2}{5}$

d. $\frac{10}{23}$

f. Fraction B. What's the probability fraction for pulling out a white marble? (Signal.) *3/17.*
- Say the equation for trials. (Signal.) *3/17 T = W.*
(Repeat until firm.)
(Add to show:)   [97:4G]

$$a. \ \frac{3}{10} t = w$$

$$b. \ \frac{3}{17} t = w$$

$$c. \ \frac{2}{5}$$

$$d. \ \frac{10}{23}$$

g. Fraction C. What's the probability fraction for pulling out a white marble? (Signal.) *2/5.*
- Say the equation for trials. (Signal.) *2/5 T = W.*
(Repeat until firm.)
(Add to show:)   [97:4H]

$$a. \ \frac{3}{10} t = w$$

$$b. \ \frac{3}{17} t = w$$

$$c. \ \frac{2}{5} t = w$$

$$d. \ \frac{10}{23}$$

h. Fraction D. What's the probability fraction for pulling out a white marble? (Signal.) *10/23.*
- Say the equation for trials. (Signal.)
10/23 T = W.
(Repeat until firm.)
(Add to show:)   [97:4I]

$$a. \ \frac{3}{10} t = w$$

$$b. \ \frac{3}{17} t = w$$

$$c. \ \frac{2}{5} t = w$$

$$d. \ \frac{10}{23} t = w$$

a. Open your textbook to Lesson 97 and find part 1. ✔
(Teacher reference:)

   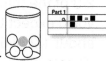

You're going to write the equation for trials. For all of these pictures, the probability fraction will tell about white marbles.

b. Picture A. Write the probability **fraction** for picture A. ✔
- What's the probability fraction for pulling out a white marble? (Signal.) *3/7.*
- Tell me the equation for trials. (Signal.) *3/7 T = W.*
(Repeat until firm.)
Yes, 3/7 T = W.
- Write the equation for A. ✔
- Everybody, read equation A. (Signal.) *3/7 T = W.*

c. Write the equation for B. ✔
- Read equation B. (Signal.) *4/9 T = W.*

d. Write equations for the rest of the items. (Observe students and give feedback.)

e. Check your work.
- Read equation C. (Signal.) *2/7 T = W.*
- Read equation D. (Signal.) *4/5 T = W.*

EXERCISE 5: UNIT CONVERSION
MULTIPLICATION/DIVISION   REMEDY

a. Find part 2 in your textbook. ✔
(Teacher reference:)

a. Change 200 nickels into dollars.
b. Change 8 years into months.
c. Change 30 hours into days.
d. Change 55 minutes into hours.
e. Change 14 feet into inches.

Facts:
- 1 hour is 60 minutes.
- 1 foot is 12 inches.
- 1 year is 12 months.
- 1 dollar is 20 nickels.
- 1 year is 4 seasons.
- 1 day is 24 hours.

You will use the measurement facts to work the problems.

b. Problem A: Change 200 nickels into dollars. Work the problem. ✔
- Read the problem you worked. (Signal.) *200 ÷ 20.*
- How many dollars is 200 nickels? (Signal.) *10 dollars.*

c. Problem B: Change 8 years into months. Work the problem. ✔
- Read the problem you worked. (Signal.) *12 × 8.*
- How many months is 8 years? (Signal.) *96 months.*
d. Problem C: Change 30 hours into days. The fact says that one day is 24 hours.
- Raise your hand when you can say the problem you'll work. ✔
- Say the problem. (Signal.) *30 ÷ 24.*
- Write it as a fraction and simplify the answer. ✔
- Everybody, how many days is 30 hours? (Signal.) *1 and 1/4 days.*
(Display:)   [97:5A]

$$c. \quad \frac{30}{24} = \frac{5}{4} = \boxed{1\frac{1}{4} \text{ days}}$$

Here's what you should have.
e. Problem D: Change 55 minutes into hours.
- Raise your hand when you can say the problem you'll work. ✔
- Say the problem you'll work. (Signal.) *55 ÷ 60.*
- Write it as a fraction and simplify the answer. ✔
- Everybody, how many hours is 55 minutes? (Signal.) *11/12 hour.*
f. Problem E: Change 14 feet into inches.
- Work the problem. ✔
- Everybody, read the problem you worked. (Signal.) *14 × 12.*
- How many inches is 14 feet? (Signal.) *168 inches.*

EXERCISE 6: MIXED-NUMBER OPERATIONS
UNLIKE DENOMINATORS   REMEDY

a. Find part 3 in your textbook. ✔
(Teacher reference:)

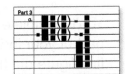

To work these problems, you have to rewrite fractions so they have the same denominator. If the mixed number in the answer has a fraction that is more than 1, you rewrite the answer.
b. Read problem A. (Signal.) *9 and 5/8 + 7 and 5/6.*
- Copy and work the problem.
(Observe students and give feedback.)
- Everybody, what does 9 and 5/8 plus 7 and 5/6 equal? (Signal.) *17 and 11/24.*

(Display:)   [97:6A]

$$a. \quad 9\frac{5}{8}\left(\frac{3}{3}\right) = \frac{15}{24}$$
$$+ 7\frac{5}{6}\left(\frac{4}{4}\right) = +\frac{20}{24}$$
$$16\frac{1}{}\frac{35}{24}$$
$$\boxed{17\frac{11}{24}}$$

Here's what you should have.
You rewrote 5/8 as 15/24 and 5/6 as 20/24. The fraction in the answer was 35/24. So you rewrote the answer as 17 and 11/24.
c. Read problem B. (Signal.) *10 and 4/5 + 4 and 7/20.*
- Work the problem.
(Observe students and give feedback.)
- Everybody, what does 10 and 4/5 plus 4 and 7/20 equal? (Signal.) *15 and 3/20.*

(Display:)   [97:6B]

$$b. \quad 10\frac{4}{5}\left(\frac{4}{4}\right) = \frac{16}{20}$$
$$+ 4\frac{7}{20} = +\frac{7}{20}$$
$$14\frac{1}{}\frac{23}{20}$$
$$\boxed{15\frac{3}{20}}$$

Here's what you should have.
You rewrote the problem so both fractions had denominators of 20.
d. Read problem C. (Signal.) *8 and 3/9 + 6 and 2/6.*
- Work the problem. Remember to simplify the fraction in the answer.
(Observe students and give feedback.)
- Everybody, what does 8 and 3/9 + 6 and 2/6 equal? (Signal.) *14 and 2/3.*

(Display:)   [97:6C]

$$c. \quad 8\frac{3}{9}\left(\frac{2}{2}\right) = \frac{6}{18}$$
$$+ 6\frac{2}{6}\left(\frac{3}{3}\right) = +\frac{6}{18}$$
$$14\frac{12}{18} = \boxed{14\frac{2}{3}}$$

Here's what you should have.
You rewrote the problem so both fractions had denominators of 18. The simplified answer is 14 and 2/3.

EXERCISE 7: VOLUME

COMPLEX FIGURES

REMEDY

a. Find part 4 in your textbook. ✔
   (Teacher reference:)

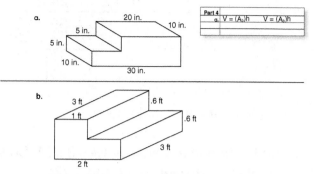

Item A. This is a stair problem. You'll figure out how much concrete is needed to make both stairs. Remember how to work the problem. You find the volume of one stair and add it to the volume of the other stair.

• Which stair has the greater volume—the top stair or the bottom stair? (Signal.) *The bottom stair.*

b. Figure out the volume of the bottom stair. Start with the equation Volume = Area of the base × height.
   (Observe students and give feedback.)
   (Display:)   [97:7A]

   a. $V = (A_b) h$
      $V = (300) 5$
      $V = 1500$ cu in.

Here's what you should have for the bottom stair. The volume is 1500 cubic inches.

c. Now figure out the volume for the top stair.
   (Observe students and give feedback.)
   (Display:)   [97:7B]

   a. $V = (A_b) h$ $V = (A_b) h$
      $V = (300) 5$ $V = (200) 5$
      $V = 1500$ cu in. $V = 1000$ cu in.

Here's what you should have for the top stair. The volume is 1000 cubic inches.

• Now add the two volumes. Raise your hand when you know the total volume for the stairs. ✔
   (Observe students and give feedback.)
   (Display:)   [97:7C]

   a. $V = (A_b) h$ $V = (A_b) h$ 1500
      $V = (300) 5$ $V = (200) 5$ +1000
      $V = 1500$ cu in. $V = 1000$ cu in.   2500 cu in.

Here's what you should have. You added 1500 and 1000. The volume is 2500 cubic inches.

d. Item B. Figure out the volume of the bottom stair. Then stop. The height is 6 tenths of a foot. ✔
   (Observe students and give feedback.)
   Check your work. The base of the stair is 2 feet times 3 feet. The height is 6 tenths of a foot. What's the volume? (Signal.) *3 and 6 tenths cubic feet.*

e. Now you'll figure out the volume of the top stair and add the volumes.

• Say the multiplication for the base. (Signal.) *3 × 1.*

• What's the height? (Signal.) *6 tenths of a foot.*

• Figure out the volume of the top stair. Raise your hand when you know the volume of the top stair. ✔

• Everybody, what's the volume? (Signal.) *1 and 8 tenths cubic feet.*

f. Add these two volumes. Write the unit name in the answer. ✔
   Check your work. You added 3 and 6 tenths and 1 and 8 tenths.

• What's the volume of the stairs? (Signal.) *5 and 4 tenths cubic feet.*
   (Display:)   [97:7D]

   $V = (A_b) h$ $V = (A_b) h$ 3.6
   $V = (6).6$ $V = (3).6$ +1.8
   $V = 3.6$ cu ft $V = 1.8$ cu ft   5.4 cu ft

Here's what you should have.

EXERCISE 8: INDEPENDENT WORK

LONG DIVISION—2-DIGIT QUOTIENT

a. Find part 2 in your workbook. ✔
   (Teacher reference:)

   $$a.\ 72\overline{)3041}\quad \begin{array}{r}4\\-288\\\hline 16\end{array}\qquad b.\ 89\overline{)5430}\quad \begin{array}{r}6\\-534\\\hline 9\end{array}$$

You'll complete these division problems as part of your independent work. Remember, first figure out the ones digit of the answer. Then subtract to find the remainder.

Assign Independent Work, Workbook parts 2–4 and Textbook parts 5–12.

Optional extra math-fact practice worksheets are available on ConnectED.

Lesson 98

EXERCISE 1: MENTAL MATH

a. Time for some mental math.
- Listen: 70 plus 12. (Pause.) (Signal.) *82.*
- 130 + 12. (Pause.) (Signal.) *142.*
- 120 + 15. (Pause.) (Signal.) *135.*
- 130 + 25. (Pause.) (Signal.) *155.*
- 180 + 25. (Pause.) (Signal.) *205.*
- 90 + 25. (Pause.) (Signal.) *115.*
- (Repeat until firm.)

b. Listen: 60 plus 27. (Pause.) (Signal.) *87.*
- 70 + 29. (Pause.) (Signal.) *99.*
- 170 + 21. (Pause.) (Signal.) *191.*
- 180 + 34. (Pause.) (Signal.) *214.*
- 180 + 38. (Pause.) (Signal.) *218.*
- (Repeat until firm.)

c. Listen: 50 minus 11. (Pause.) (Signal.) *39.*
- 50 − 13. (Pause.) (Signal.) *37.*
- 50 − 15. (Pause.) (Signal.) *35.*
- (Repeat until firm.)

d. Listen: 150 minus 11. (Pause.) (Signal.) *139.*
- 120 − 15. (Pause.) (Signal.) *105.*
- 130 − 15. (Pause.) (Signal.) *115.*
- 34 − 10. (Pause.) (Signal.) *24.*
- (Repeat until firm.)

e. Listen: 97 plus 8. (Pause.) (Signal.) *105.*
- 98 + 8. (Pause.) (Signal.) *106.*
- 95 + 8. (Pause.) (Signal.) *103.*
- (Repeat until firm.)

EXERCISE 2: FRACTION OPERATIONS
SIMPLIFY MULTIPLICATION

a. (Display:)   [98:2A]

$$\frac{20}{100}\,(25)$$

You're going to simplify this problem as much as possible. The first thing you do is cross out zeros.
- How many zeros can we cross out in each number? (Signal.) *1.*

(Add to show:)   [98:2B]

$$\frac{2\cancel{0}}{10\cancel{0}}\,(25)$$

Now we can simplify 2/10.
- What does 2/10 simplify to? (Signal.) *1/5.*
Yes, 1/5.
(Add to show:)   [98:2C]

$$\frac{\overset{1}{\cancel{2\cancel{0}}}}{\underset{5}{\cancel{10\cancel{0}}}}\,(25)$$

Now we can simplify the fraction 25/5.
- What does 25/5 simplify to? (Signal.) *5.*
Yes, 5.
(Add to show:)   [98:2D]

$$\frac{\overset{1}{\cancel{2\cancel{0}}}}{\underset{5}{\cancel{10\cancel{0}}}}\,(\overset{5}{\cancel{25}})$$

1 times 5 equals 5. That's the answer to the problem.
(Add to show:)   [98:2E]

$$\frac{\overset{1}{\cancel{2\cancel{0}}}}{\underset{5}{\cancel{10\cancel{0}}}}\,(\overset{5}{\cancel{25}}) = \boxed{5}$$

Remember, first cross out zeros. Then do the other simplifications.

b. (Display:)   [98:2F]

$$\frac{40}{100}\,(6) =$$

New problem.
- Can we cross out zeros? (Signal.) *Yes.*
(Add to show:)   [98:2G]

$$\frac{4\cancel{0}}{10\cancel{0}}\,(6) =$$

Now we have 4/10 times 6.

- The first fraction is 4/10. Can we simplify that fraction? (Signal.) *Yes.*
- 4 and 10 have a common factor. What factor is that? (Signal.) *2.*
- Raise your hand when you know what 4/10 simplifies to. ✔
- What does 4/10 simplify to? (Signal.) *2/5.* Yes, 2/5.
  (Add to show:) [98:2H]

$$\frac{\overset{2}{\cancel{4\cancel{0}}}}{\underset{5}{\cancel{10\cancel{0}}}} (6) =$$

- Can you do any other simplification? (Signal.) *No.*
  So you multiply 2/5 times 6.
- What's 2 times 6? (Signal.) *12.*
- What's the denominator? (Signal.) *5.*
  (Add to show:) [98:2I]

$$\frac{\overset{2}{\cancel{4\cancel{0}}}}{\underset{5}{\cancel{10\cancel{0}}}} (6) = \frac{12}{5}$$

Now we write the answer as a mixed number.
- Raise your hand when you know the mixed number. ✔
- What's the mixed number? (Signal.) *2 and 2/5.*
  (Add to show:) [98:2J]

$$\frac{\overset{2}{\cancel{4\cancel{0}}}}{\underset{5}{\cancel{10\cancel{0}}}} (6) = \frac{12}{5} = \boxed{2\frac{2}{5}}$$

Yes, 2 and 2/5.

c. (Display:) [98:2K]

$$\frac{30}{60} (8) =$$

- Copy this problem in part A and simplify it as much as possible. Remember, start with the zeros. Then look for the common factors of the first fraction.
  (Observe students and give feedback.)
d. Check your work.
  You crossed out zeros in the first fraction. You got the fraction 3/6.
- Did the numbers in that fraction have a common factor? (Signal.) *Yes.*
- What's the simplified fraction? (Signal.) *1/2.*

(Add to show:) [98:2L]

$$\frac{\overset{1}{\cancel{3\cancel{0}}}}{\underset{2}{\cancel{6\cancel{0}}}} (8) =$$

- Could you simply again? (Signal.) *Yes.* Yes, 8/2 can be simplified.
- What's the simplified value? (Signal.) *4.*
  (Add to show:) [98:2M]

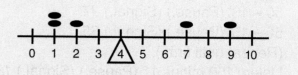

Yes, on top we have 1 x 4 so the answer is 4.

EXERCISE 3: BALANCE BEAMS
AVERAGE DISTANCE

a. (Display:) [98:3A]

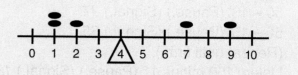

This balance beam starts at zero. The balance point is at 4.
We are going to figure out the distance on each side of the balance point. To do that, we count units from 4.
b. (Point to 1.) Raise your hand when you know how many units these weights are from the balance point. ✔
- What's the distance from the balance point? (Signal.) *3.*
  To find the total distance of these weights from the balance point, we work the problem 3 + 3.
- What's the total distance from the balance point? (Signal.) *6.*
  (Add to show:) [98:3B]

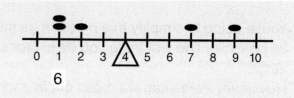

c. (Point to **2**.) Raise your hand when you know the distance of this weight from the balance point. ✔
- What's the distance from the balance point? (Signal.) *2.*
  Yes, 2.
  (Add to show:) [98:3C]

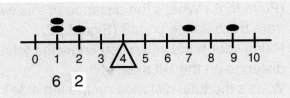

6 2

d. (Point to **7**.) Raise your hand when you know the distance of this weight from the balance point. ✔
- What's the distance from the balance point? (Signal.) *3.*
  Yes, 3.
  (Add to show:) [98:3D]

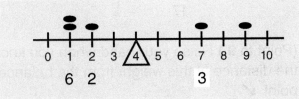

6 2 3

e. (Point to **9**.) Raise your hand when you know the distance of this weight from the balance point. ✔
- What's the distance from the balance point? (Signal.) *5.*
  Yes, 5.
  (Add to show:) [98:3E]

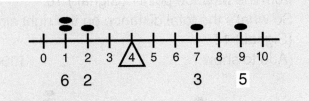

6 2 3 5

f. Now we figure the total distance on each side.
- (Point left.) What's the total distance of 6 and 2? (Signal.) *8.*
- (Point right.) What's the total distance of 3 and 5? (Signal.) *8.*
- Are the distances the same? (Signal.) *Yes.*
- So will the beam balance? (Signal.) *Yes.*
  Remember, figure out the distance of each weight from the balance point. Then add to find the total distance.

g. (Display:) [98:3F]

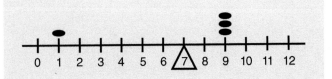

- (Point left.) Raise your hand when you know the total distance on the left side of the balance beam. ✔
- What's the total distance on the left? (Signal.) *6.*
  (Add to show:) [98:3G]

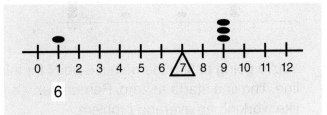

6

- (Point right.) What's the distance of these weights from the balance point? (Signal.) *6.*
- So what's the total distance on the right? (Signal.) *6.*
  (Add to show:) [98:3H]

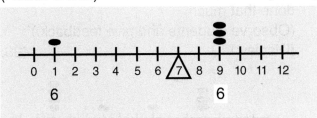

6 6

- Are the total distances the same on both sides? (Signal.) *Yes.*
- So will the beam balance at 7? (Signal.) *Yes.*

h. (Display:) [98:3I]

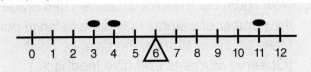

- (Point left.) Raise your hand when you know the distance of these weights from the balance point. ✔
- (Point to **3**.) What's the distance of this weight? (Signal.) *3.*
- (Point to **4**.) What's the distance of this weight? (Signal.) *2.*
  (Add to show:) [98:3J]

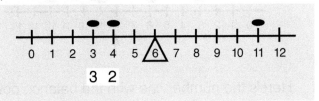

3 2

- What's the total distance on the left? (Signal.) *5.*
- (Point right.) Raise your hand when you know the distance of this point from the balance point. ✔
- What's the total distance on this side? (Signal.) *5.*
- Are the total distances the same? (Signal.) *Yes.*
- So will the beam balance at 6? (Signal.) *Yes.*

a. Open your workbook to Lesson 98 and find part 1. ✔
(Teacher reference:)

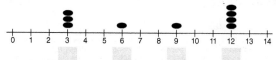

You're going to find the balance point for this line. The line starts at zero. Remember, it's just like working an average problem.
To find the balance point, we figure out the total distance of the weights from zero and divide by the number of weights.

b. Figure out the totals for the weights and write the numbers in the boxes. Stop when you've done that much.
(Observe students and give feedback.)
(Display:) [98:3K]

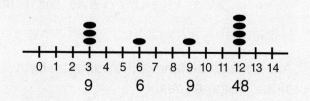

Here are the distances for the weights.

c. Now figure out the total distance. Then divide by the number of weights to find the balance point. Make an arrow to show the balance point.
(Observe students and give feedback.)

d. Check your work.
• What's the total distance? (Signal.) 72.
• What did you divide by? (Signal.) 9.
• What's the balance point for these weights? (Signal.) 8.
(Change to show:) [98:3L]

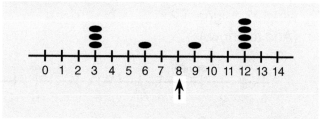

Here's the number line with the balance point at 8.

e. Now we'll figure out the distance from the balance point on each side.
• (Point to 3.) Raise your hand when you know the **total** distance of these 3 weights from the balance point. ✔
• (Point to 3.) What's the total distance of these 3 weights from the balance point? (Signal.) 15.
• (Point to 6.) What's the distance of this weight from the balance point? (Signal.) 2.
• Raise your hand when you know the total distance on the left side. ✔
• What's the total distance on the left side? (Signal.) 17.
(Add to show:) [98:3M]

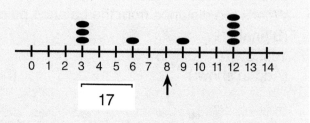

f. (Point to 9.) Raise your hand when you know the distance of this weight from the balance point. ✔
• What's the distance of this weight from the balance point? (Signal.) 1.
• (Point to 12.) Raise your hand when you know the total distance of these 4 weights from the balance point. ✔
• What's the total distance of these 4 weights from the balance point? (Signal.) 16.
• So what's the total distance on the right side? (Signal.) 17.
(Add to show:) [98:3N]

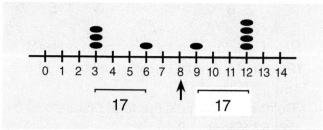

• Are the total distances the same? (Signal.) Yes.
• So will the beam balance at 8? (Signal.) Yes.

EXERCISE 4: FRACTIONS
FOR MIXED NUMBERS　　　　　　　　REMEDY

a. (Display:)　　　　　　　　　　　　　　[98:4A]

$$9\frac{1}{3} =$$

We're going to rewrite this mixed number as a fraction that is more than 1.
We change the whole number into a fraction with a denominator of 3. Then we add the fractions.

- Read the mixed number. (**Signal.**) *9 and 1/3.*
- What's the denominator for the fraction that equals 9? (**Signal.**) *3.*
- Raise your hand when you know the fraction that equals 9. ✔
- What fraction? (**Signal.**) *27/3.*
- What do you add to 27/3? (**Signal.**) *1/3.*
- What's the fraction that equals 9 and 1/3? (**Signal.**) *28/3.*
(Add to show:)　　　　　　　　　　　　[98:4B]

$$9\frac{1}{3} = \frac{28}{3}$$

$$\frac{27}{3} + \frac{1}{3}$$

━━━━ **WORKBOOK PRACTICE** ━━━━

a. Find part 2 in your workbook. ✔
(Teacher reference:)　　　　　　　　**R Part E**

a. $4\frac{2}{7} =$ 　　　　　　b. $10\frac{5}{9} =$

c. $6\frac{2}{10} =$ 　　　　　　d. $12\frac{1}{4} =$

- Read mixed number A. (**Signal.**) *4 and 2/7.*
- First you write the fraction that equals 4. What's the denominator of that fraction? (**Signal.**) *7.*
- Then add 2/7 and complete the equation. (Observe students and give feedback.)

b. Check your work.
- What's the fraction for 4? (**Signal.**) *28/7.*
- What did you add to 28/7? (**Signal.**) *2/7.*
- What fraction does 4 and 2/7 equal? (**Signal.**) *30/7.*
(Display:)　　　　　　　　　　　　　　[98:4C]

a.　$$4\frac{2}{7} = \frac{30}{7}$$

$$\frac{28}{7} + \frac{2}{7}$$

Here's what you should have.
c. Read mixed number B. (**Signal.**) *10 and 5/9.*
- First write the fraction that equals 10. Then add 5/9 and complete the equation. (Observe students and give feedback.)
d. Check your work.
- What's the fraction for 10? (**Signal.**) *90/9.*
- What did you add to 90/9? (**Signal.**) *5/9.*
- What fraction does 10 and 5/9 equal? (**Signal.**) *95/9.*
(Display:)　　　　　　　　　　　　　　[98:4D]

b.　$$10\frac{5}{9} = \frac{95}{9}$$

$$\frac{90}{9} + \frac{5}{9}$$

Here's what you should have.
e. Read mixed number C. (**Signal.**) *6 and 2/10.*
- First write the fraction that equals 6. Then add 2/10 and complete the equation. (Observe students and give feedback.)
f. Check your work.
- What's the fraction for 6? (**Signal.**) *60/10.*
- What did you add to 60/10? (**Signal.**) *2/10.*
- What fraction does 6 and 2/10 equal? (**Signal.**) *62/10.*
(Display:)　　　　　　　　　　　　　　[98:4E]

c.　$$6\frac{2}{10} = \frac{62}{10}$$

$$\frac{60}{10} + \frac{2}{10}$$

Here's what you should have.
g. Read mixed number D. (**Signal.**) *12 and 1/4.*
- First write the fraction that equals 12. Then add 1/4 and complete the equation. (Observe students and give feedback.)

h. Check your work.
- What's the fraction for 12? (Signal.) *48/4.*
- What did you add to 48/4? (Signal.) *1/4.*
- What fraction does 12 and 1/4 equal? (Signal.) *49/4.*
  (Display:) [98:4F]

$$d.\ 12\frac{1}{4} = \frac{49}{4}$$

$$\frac{48}{4} + \frac{1}{4}$$

Here's what you should have.

EXERCISE 5: UNIT CONVERSION
MULTIPLICATION/DIVISION

a. You're going to work measurement problems that don't show the facts you'll use.
   You should know a lot of the facts. Remember, all the facts start with the number 1.
b. Say the fact for feet and inches. (Signal.) *1 foot is 12 inches.*
- Say the fact for hours in a day. (Signal.) *1 day is 24 hours.*
- Say the fact for inches in a foot. (Signal.) *1 foot is 12 inches.*
- Say the fact for ounces in a pound. (Signal.) *1 pound is 16 ounces.*
- Say the fact for months in a year. (Signal.) *1 year is 12 months.*
  (Repeat until firm.)
c. If you don't know a fact, or you get confused, you can find the fact in a table on the inside back cover of your textbook.
- Find the table. ✔
  This table shows many measurement facts.
d. Let's say the problem names miles and yards.
- Which units? (Signal.) *Miles and yards.*
- Find the fact in the table. Raise your hand when you have found it. ✔
- Everybody, what's the fact that names miles and yards? (Signal.) *1 mile is 1760 yards.*
e. Listen: Change 48 tablespoons into cups.
- What two units are named? (Signal.) *Tablespoons and cups.*
- Raise your hand when you know the fact for tablespoons and cups. ✔
- Everybody, say the fact for tablespoons and cups. (Signal.) *1 cup is 16 tablespoons.*

f. Listen: Change 3 pounds into ounces.
- What are the unit names? (Signal.) *Pounds and ounces.*
- Raise your hand when you know the measurement fact. ✔
- Everybody, say the measurement fact for pounds and ounces. (Signal.) *1 pound is 16 ounces.*
g. Listen: Change 7 pints into gallons.
- What are the units named? (Signal.) *Pints and gallons.*
- Raise your hand when you know the measurement fact. ✔
- Everybody, say the measurement fact for pints and gallons. (Signal.) *1 gallon is 8 pints.*

TEXTBOOK PRACTICE

a. Open your textbook to Lesson 98 and find part 1. ✔
   (Teacher reference:)

a. Change 6 pints into cups.	d. Change 120 months into years.	**Part 1**
b. Change 3 years into weeks.	e. Change 2 days into hours.	a.
c. Change 3 days into weeks.	f. Change 252 inches into yards.	

b. Problem A: Change 6 pints into cups.
   If you don't know the measurement fact for pints and cups, look it up in the table. Then work the problem. ✔
- Everybody, how many cups is 6 pints? (Signal.) *12 cups.*
c. Problem B: Change 3 years into weeks.
- Work the problem. If you don't know the measurement fact, look it up. Then work the problem. ✔
- Everybody, how many weeks is 3 years? (Signal.) *156 weeks.*
d. Work the rest of the problems in part 1. (Observe students and give feedback.)
e. Check your work.
- Problem C: Change 3 days into weeks.
- How many weeks is 3 days? (Signal.) *3/7 week.*
f. Problem D: Change 120 months into years.
- How many years is 120 months? (Signal.) *10 years.*
g. Problem E: Change 2 days into hours.
- How many hours is 2 days? (Signal.) *48 hours.*
h. Problem F: Change 252 inches into yards.
- How many yards is 252 inches? (Signal.) *7 yards.*

EXERCISE 6: PROBABILITY
EQUATIONS
REMEDY

a. Last time, you learned about taking trials. You also learned the equation for taking trials. You start with the fraction that tells the probability of something happening.
- If there are 4 things in a bag and 3 are red, what's the probability fraction for the red things? (Signal.) *3/4.*
(Display:)   [98:6A]

$$\frac{3}{4} t = r$$

And the equation for trials is 3/4 T = R. That means that for 3/4 of the trials, you'd expect to pull out a red object.

b. (Display:)   [98:6B]

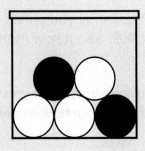

- Raise your hand when you can say the probability for black marbles. ✔
- What's the probability for black marbles? (Signal.) *2/5.*
(Add to show:)   [98:6C]

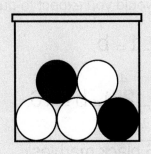

- Raise your hand when you can say the probability for white marbles. ✔
- What's the probability for white marbles? (Signal.) *3/5.*

(Add to show:)   [98:6D]

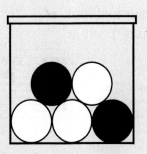

c. (Point to **2/5.**) Say the equation for 2/5 of the trials. (Signal.) *2/5 T = B.*
(Add to show:)   [98:6E]

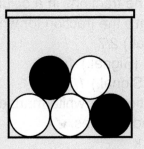

- (Point to **3/5.**) Say the equation for 3/5 of the trials. (Signal.) *3/5 T = W.*
(Add to show:)   [98:6F]

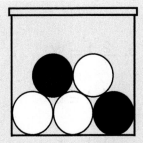

- Which is more likely on any trial—pulling out a black marble or pulling out a white marble? (Signal.) *Pulling out a white marble.*

a. Find part 2 in your textbook. ✔
(Teacher reference:)

a. white
black
red

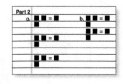

You're going to write equations for trials.
Remember, you first write the probability fraction.

b. Problem A. Write the probability fractions
for white marbles, black marbles, and red
marbles. Write the fractions in a column. ✔
- Everybody, what's the fraction for white
  marbles? (Signal.) *2/7.*
- Black marbles? (Signal.) *4/7.*
- Red marbles? (Signal.) *1/7.*

c. Complete equations for white marbles, black
marbles, and red marbles. ✔

d. Check your work.
- Read the equation for white marbles. (Signal.)
  2/7 T = W.
- Black marbles. (Signal.) *4/7 T = B.*
- Red marbles. (Signal.) *1/7 T = R.*
  (Display:) [98:6G]

> a. $\frac{2}{7} t = w$
>
> $\frac{4}{7} t = b$
>
> $\frac{1}{7} t = r$

Here's what you should have.
- Circle the equation for the color that has the
  highest probability. ✔
- Which color has the highest probability?
  (Signal.) *Black.*
- Which color has the lowest probability?
  (Signal.) *Red.*

e. You're going to use the equation for black
marbles to solve problems. I'll give you the
number for trials or the number of black
marbles. You'll put that number in the equation
and solve for the other letter.

f. (Display:) [98:6H]

> **b.** If you took 28 trials, how many black
> marbles would you expect to pull out?

Problem B: If you took 28 trials, how many
black marbles would you expect to pull out?
- Work the problem. Start with the letter
  equation. The unit name in the answer is black
  marbles.
  (Observe students and give feedback.)

g. Check your work.
You started with the equation 4/7 T = B.
You worked the problem 4/7 times 28.
- What's the whole answer? (Signal.) *16 black
  marbles.*
  (Add to show:) [98:6I]

> **b.** If you took 28 trials, how many black
> marbles would you expect to pull out?
>
> $$\frac{4}{7} t = b$$
>
> $$\frac{4}{\cancel{7}} (\cancel{28}) = b = 16$$
>
> 16 black marbles

Here's what you should have.
If you took 28 trials, you'd expect to pull out
16 black marbles.

h. (Display:) [98:6J]

> **c.** If you took trials until you pulled out
> 4 black marbles, about how many
> trials would you take?

Problem C: If you took trials until you pulled
out 4 black marbles, about how many trials
would you take?
- Work the problem.
  (Observe students and give feedback.)

i. Check your work.
   You started with the equation 4/7 T = B.
   You replaced B with 4, multiplied by the
   reciprocal, and worked the problem 4 times 7/4.
• About how many trials would you take?
   (Signal.) *7 trials.*
   (Add to show:)   [98:6K]

> c. If you took trials until you pulled out
>   4 black marbles, about how many
>   trials would you take?
>
> $$\frac{4}{7}t=b$$
>
> $$\left(\frac{7}{4}\right)\frac{4}{7}t=\overset{1}{\cancel{4}}\left(\frac{7}{\cancel{4}}\right)$$
>
> $$t=7$$
>
> $$\boxed{7\text{ trials}}$$

Here's what you should have.
You'd expect to pull out 4 black marbles in
about 7 trials.

j. (Display:)   [98:6L]

> d. You take 98 trials. About how many
>   black marbles would you expect to
>   pull out?

Problem D: You take 98 trials. About how many
black marbles would you expect to pull out?
• Work the problem.
   (Observe students and give feedback.)
k. Check your work.
   You started with the equation 4/7 T = B.
   You worked the problem 4/7 times 98.
• What's the whole answer? (Signal.)
   56 black marbles.

(Add to show:)   [96:8M]

> d. You take 98 trials. About how many
>   black marbles would you expect to
>   pull out?
>
> $$\frac{4}{7}t=b$$
>
> $$\frac{4}{\cancel{7}}\overset{14}{(\cancel{98})}=b=56$$
>
> $$\boxed{56\text{ black marbles}}$$

Here's what you should have.
l. (Display:)   [98:6N]

> e. You take trials until you pull out
>   80 black marbles. About how many
>   trials would you expect to take?

Problem E: You take trials until you pull out 80
black marbles. About how many trials would
you expect to take?
• Work the problem.
   (Observe students and give feedback.)
m. Check your work.
   You started with the equation 4/7 T = B.
   You worked the problem 80 × 7/4.
• What's the answer? (Signal.) *140 trials.*
   (Add to show:)   [98:6O]

> e. You take trials until you pull out
>   80 black marbles. About how many
>   trials would you expect to take?
>
> $$\frac{4}{7}t=b$$
>
> $$\left(\frac{7}{4}\right)\frac{4}{7}t=\overset{20}{\cancel{80}}\left(\frac{7}{\cancel{4}}\right)$$
>
> $$t=140$$
>
> $$\boxed{140\text{ trials}}$$

Here's what you should have.

Exercise 7: Volume

Complex Figures   [REMEDY]

a. Find part 3 in your textbook. ✔
   (Teacher reference:)

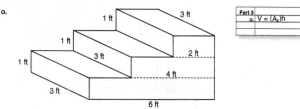

Here's a stairway with three stairs. You figure
out the volume for each stair. Then you add
the volumes to find the total volume.

b. Figure out the volume of the bottom stair.
   (Observe students and give feedback.)
• Check your work. You multiplied 3 × 6 for the
  area of the base. What's the number for the
  area? (Signal.) *18.*
• What's the height? (Signal.) *1.*
• What's the volume? (Signal.) *18 cubic feet.*
  (Display:)   [98:7A]

   V = (A_b) h
   V = (18) 1
   V = 18 cu ft

Here's what you should have.

c. Figure out the volume of the second stair.
   (Observe students and give feedback.)
• Check your work. You multiplied 3 × 4 for the
  area of the base. What's the number for area?
  (Signal.) *12.*
• What's the height? (Signal.) *1.*
• What's the volume? (Signal.) *12 cubic feet.*
  (Add to show:)   [98:7B]

   V = (A_b) h   V = (A_b) h
   V = (18) 1   V = (12) 1
   V = 18 cu ft   V = 12 cu ft

Here's what you should have.

d. Figure out the volume of the top stair.
   (Observe students and give feedback.)
• Check your work. You multiplied 3 × 2 for the
  area of the base. What's the number for the
  area? (Signal.) *6.*
• What's the height? (Signal.) *1.*
• What's the volume? (Signal.) *6 cubic feet.*
  (Add to show:)   [98:7C]

   V = (A_b) h   V = (A_b) h   V = (A_b) h
   V = (18) 1   V = (12) 1   V = (6) 1
   V = 18 cu ft   V = 12 cu ft   V = 6 cu ft

Here's what you should have.

e. Now you'll add the three volumes and find the
   total volume of concrete for the stairs.
• Say the numbers you'll add. (Signal.)
  18 + 12 + 6.
• Work the problem. ✔
• What's the volume of the three stairs?
  (Signal.) *36 cubic feet.*
  (Display:)   [98:7D]

   V = (A_b) h   V = (A_b) h   V = (A_b) h   1
   V = (18) 1   V = (12) 1   V = (6) 1   1 8
   V = 18 cu ft   V = 12 cu ft   V = 6 cu ft   1 2
                                                    + 6
                                                    3 6 cu ft

Here's what you should have.

Assign Independent Work, Textbook parts 4–10
and Workbook parts 3 and 4.

Optional extra math-fact practice worksheets
are available on ConnectED.

Lesson

EXERCISE 1: MENTAL MATH

a. Time for some mental math.
- Listen: 2 times what number equals 8? (Pause.) (Signal.) *4.*
- So 20 times what number equals 80? (Pause.) (Signal.) *4.*
- 2 times what number equals 12? (Pause.) (Signal.) *6.*
- So 20 times what number equals 120? (Pause.) (Signal.) *6.*
- 20 times what number equals 140? (Pause.) (Signal.) *7.*
- 20 times what number equals 180? (Pause.) (Signal.) *9.*
- (Repeat until firm.)

b. Listen: 20 times what number equals 60? (Pause.) (Signal.) *3.*
- 30 times what number equals 60? (Pause.) (Signal.) *2.*
- 30 times what number equals 90? (Pause.) (Signal.) *3.*
- 30 times what number equals 180? (Pause.) (Signal.) *6.*
- (Repeat until firm.)

c. Listen: 45 plus 4. (Pause.) (Signal.) *49.*
- 45 + 6. (Pause.) (Signal.) *51.*
- 54 + 6. (Pause.) (Signal.) *60.*
- 54 + 8. (Pause.) (Signal.) *62.*
- 74 + 8. (Pause.) (Signal.) *82.*
- 78 + 8. (Pause.) (Signal.) *86.*
- 118 + 8. (Pause.) (Signal.) *126.*
- (Repeat until firm.)

d. Listen: 40 plus 18. (Pause.) (Signal.) *58.*
- 120 + 18. (Pause.) (Signal.) *138.*
- 330 + 17. (Pause.) (Signal.) *347.*
- 30 + 55. (Pause.) (Signal.) *85.*
- 30 + 59. (Pause.) (Signal.) *89.*
- (Repeat until firm.)

EXERCISE 2: FRACTION SIMPLIFICATION
MULTIPLICATION
R Test 11

a. (Display:) [99:2A]

$$\frac{40}{100} (50) =$$

- Read the problem. (Signal.) *40/100 × 50.*
You learned that the first thing you do to simplify is to cross out zeros. This problem has zeros in the fraction and a zero in the number 50.

b. So first we cross out zeros in the fraction.
- How many zeros can we cross out in each part of the fraction? (Signal.) *1.*
(Add to show:) [99:2B]

$$\frac{4\cancel{0}}{10\cancel{0}} (50) =$$

We have 4/10 times 50. We can cross out the zeros in 50/10.
(Add to show:) [99:2C]

$$\frac{4\cancel{0}}{10\cancel{0}} (5\cancel{0}) =$$

Now we multiply 4 times 5.
- What's the answer? (Signal.) *20.*
(Add to show:) [99:2D]

$$\frac{4\cancel{0}}{10\cancel{0}} (5\cancel{0}) = 20$$

c. Remember, first you cross out all the zeros you can cross out. Then multiply.

WORKBOOK PRACTICE

a. Open your workbook to Lesson 99 and find part 1. ✔
(Teacher reference:)

Test 11
R Part A

a. $\frac{20}{100}(35) =$

b. $\frac{80}{100}(30) =$

c. $\frac{50}{3}\left(\frac{9}{10}\right) =$

You're going to simplify the fractions and then work the problem. Remember, first you cross out the zeros. Then you simplify the fractions further.

b. Read problem A. (Signal.) *20/100 × 35.*
• Simplify the fractions and work the problem. (Observe students and give feedback.)
c. Check your work.
You crossed out zeros. So you had 2/10.
• Then you simplified 2/10. What does it equal? (Signal.) *1/5.*
• Then you simplified 35/5. What does it equal? (Signal.) *7.*
• What does 20/100 times 35 equal? (Signal.) *7.*
(Display:)   [99:2E]

a. $\frac{\overset{1}{\cancel{20}}}{\underset{5}{\cancel{100}}}(\overset{7}{\cancel{35}}) = 7$

Here's what you should have.
d. Read problem B. (Signal.) *80/100 × 30.*
• Simplify the fractions and work the problem. (Observe students and give feedback.)
e. Check your work.
You crossed out zeros in 80/100. So you had 8/10. Then you crossed out zeros in 30/10.
• Then you multiplied 8 times 3. What's the answer? (Signal.) *24.*
(Display:)   [99:2F]

b. $\frac{8\cancel{0}}{10\cancel{0}}(3\cancel{0}) = 24$

Here's what you should have.
f. Read problem C. (Signal.) *50/3 × 9/10.*
• Simplify the fractions and work the problem. (Observe students and give feedback.)

g. Check your work.
You crossed out zeros. That left you with 5/3 times 9.
• You simplified 9/3. What does that equal? (Signal.) *3.*
• What does 50/3 times 9/10 equal? (Signal.) *15.*
(Display:)   [99:2G]

c. $\frac{5\cancel{0}}{\cancel{3}}\left(\frac{\overset{3}{\cancel{9}}}{\cancel{10}}\right) = 15$

Here's what you should have.

EXERCISE 3: BALANCE BEAMS
AVERAGE DISTANCE   REMEDY

a. Find part 2 in your workbook. ✔
(Teacher reference:)   R Part F

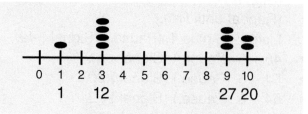

You'll figure out the balance point for this line.
b. Figure out the totals for the weights and write the numbers in the boxes. Stop when you've done that much.
(Observe students and give feedback.)
(Display:)   [99:3A]

Here are the distances for the weights. Now add those distances to get the total distance. ✔
• Everybody, what's the total distance? (Signal.) *60.*
c. Now figure out the balance point and make an arrow to show the balance point. Pencils down when you're finished.
(Observe students and give feedback.)
You worked the problem 60 ÷ 10 to find the balance point.
• What's the balance point for these weights? (Signal.) *6.*

(Add to show:) [99:3B]

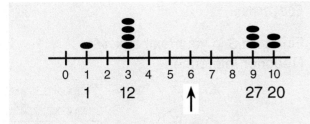

Here's the number line with the balance point at 6.

EXERCISE 4: FRACTIONS
FOR MIXED NUMBERS   [REMEDY]

a. Open your textbook to Lesson 99 and find part 1. ✔
(Teacher reference:)

a. $7\frac{3}{8}$ b. $1\frac{6}{11}$ c. $9\frac{4}{5}$ d. $4\frac{2}{9}$

You'll copy the mixed numbers. Below, you'll write the addition for changing each mixed number into a fraction and complete the equation.

b. Read mixed number A. (Signal.) *7 and 3/8.*
• Copy mixed number A. Below, add the fraction for 7 and the fraction 3/8.
(Observe students and give feedback.)
• Everybody, what's the fraction for 7? (Signal.) *56/8.*
• What did you add to 56/8? (Signal.) *3/8.*
• What's the fraction that equals 7 and 3/8? (Signal.) *59/8.*
(Display:) [99:4A]

a. $7\frac{3}{8} = \boxed{\frac{59}{8}}$

$$\frac{56}{8} + \frac{3}{8}$$

Here's what you should have.
c. Read mixed number B. (Signal.) *1 and 6/11.*
• Copy the number. Below, add the fraction for 1 and the fraction 6/11.
(Observe students and give feedback.)
• Everybody, what's the fraction for 1? (Signal.) *11/11.*
• What did you add to 11/11? (Signal.) *6/11.*
• What's the fraction that equals 1 and 6/11? (Signal.) *17/11.*

(Display:) [99:4B]

b. $1\frac{6}{11} = \boxed{\frac{17}{11}}$

$$\frac{11}{11} + \frac{6}{11}$$

Here's what you should have.
d. Work the rest of the problems in part 1.
(Observe students and give feedback.)
e. Check your work.
Mixed number C: 9 and 4/5.
You worked the problem 45/5 + 4/5.
• What fraction equals 9 and 4/5? (Signal.) *49/5.*
(Display:) [99:4C]

c. $9\frac{4}{5} = \boxed{\frac{49}{5}}$

$$\frac{45}{5} + \frac{4}{5}$$

Here's what you should have.
f. Mixed number D: 4 and 2/9.
You worked the problem 36/9 + 2/9.
• What fraction equals 4 and 2/9? (Signal.) *38/9.*
(Display:) [99:4D]

d. $4\frac{2}{9} = \boxed{\frac{38}{9}}$

$$\frac{36}{9} + \frac{2}{9}$$

Here's what you should have.

EXERCISE 5: UNIT CONVERSION
MULTIPLICATION/DIVISION   [REMEDY]

a. Find the measurement table in the back of your textbook. ✔
You're going to work problems that refer to measurement facts. If you don't know a fact, look it up in this table.
b. Listen: Change 45 cups into gallons.
• Raise your hand when you know the measurement fact. ✔
• Everybody, what's the measurement fact for cups and gallons? (Signal.) *1 gallon is 16 cups.*
c. Listen: Change 3 miles into yards.
• Raise your hand when you know the measurement fact. ✔
• Everybody, what's the measurement fact for miles and yards? (Signal.) *1 mile is 1760 yards.*

a. Find part 2 in your textbook. ✔
(Teacher reference:)

a. Change 30 minutes into hours. d. Change 3 miles into yards.
b. Change 7 pounds into ounces. e. Change 62 days into weeks.
c. Change 690 feet into yards. f. Change 4 gallons into cups.

Remember, don't use the facts in the back unless you need to.

b. Problem A: Change 30 minutes into hours.
• Raise your hand when you know the fact for hours and minutes. ✔
• Everybody, say the fact. (Signal.) *1 hour is 60 minutes.*
• Raise your hand when you can say the problem you'll work. ✔
• Say the problem you'll work. (Signal.) *30 ÷ 60.*
• Write the problem as a fraction and simplify. Pencils down when you're finished. ✔
Check your work.
30 minutes is 30/60 of an hour. That's 1/2 hour.

c. Work the rest of the problems in part 2. (Observe students and give feedback.)

d. Check your work.
• Problem B: Change 7 pounds into ounces.
• Raise your hand if you didn't need to use the table. ✔
• Say the problem you worked. (Signal.) *16 × 7.*
• How many ounces is 7 pounds? (Signal.) *112 ounces.*

e. C: Change 690 feet into yards.
• Raise your hand if you didn't need to use the table. ✔
• Say the problem you worked. (Signal.) *690 ÷ 3.*
• How many yards is 690 feet? (Signal.) *230 yards.*

f. D: Change 3 miles into yards.
• Raise your hand if you didn't need to use the table. ✔
• Say the problem you worked. (Signal.) *1760 × 3.*
• How many yards is 3 miles? (Signal.) *5280 yards.*

g. E: Change 62 days into weeks.
• Raise your hand if you didn't need to use the table. ✔
• Say the problem you worked. (Signal.) *62 ÷ 7.*
• How many weeks is 62 days? (Signal.) *8 and 6/7 weeks.*

h. F: Change 4 gallons into cups.
• Raise your hand if you didn't need to use the table. ✔
• Say the problem you worked. (Signal.) *16 × 4.*
• How many cups is 4 gallons? (Signal.) *64 cups.*

EXERCISE 6: PROBABILITY
EQUATIONS
REMEDY

a. Find part 3 in your textbook. ✔
(Teacher reference:)

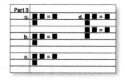

Write the equation for:

a. green
b. purple
c. yellow
d. If you took trials until you pulled out 9 green marbles, about how many trials would you expect to take?
e. If you took 96 trials, about how many yellow marbles would you expect to pull out?
f. If you took 16 trials, about how many purple marbles would you expect to pull out?
g. If you took trials until you pulled out 10 purple marbles, about how many trials would you expect to take?

You're going to write equations for pulling out green marbles, purple marbles, and yellow marbles. Remember, T stands for trials.

b. Problem A. Write the equation for green. ✔
• Everybody, read the equation. (Signal.) *3/8 T = G.*
• B: Write the equation for purple. ✔
• Everybody, read the equation. (Signal.) *1/8 T = P.*
• C: Write the equation for yellow. ✔
• Everybody, read the equation. (Signal.) *4/8 T = Y.*
(Display:)   [99:6A]

$$a. \ \frac{3}{8}t = g$$

$$b. \ \frac{1}{8}t = p$$

$$c. \ \frac{4}{8}t = y$$

Here's what you should have.

c. Problem D: If you took trials until you pulled out 9 green marbles, about how many trials would you expect to take?
• Work the problem.
(Observe students and give feedback.)
• Check your work.
You started with the equation for green 3/8 T = G.
• You worked the problem 9 times 8/3. What's the answer? (Signal.) *24 trials.*

d. Problem E: If you took 96 trials, about how many yellow marbles would you expect to pull out?
• Work the problem.
(Observe students and give feedback.)
• Check your work.
You started with the equation for yellow 4/8 T = Y.

- You worked the problem 4/8 times 96. What's the answer? (Signal.) *48 yellow marbles.*
e. Problem F: If you took 16 trials, about how many purple marbles would you expect to pull out?
- Work the problem.
  (Observe students and give feedback.)
- Check your work.
  You started with the equation for purple 1/8 T = P.
- You worked the problem 1/8 times 16. What's the answer? (Signal.) *2 purple marbles.*
f. Problem G: If you took trials until you pulled out 10 purple marbles, about how many trials would you expect to take?
- Work the problem.
  (Observe students and give feedback.)
- Check your work.
  You started with the equation for purple 1/8 T = P.
- You worked the problem 10 times 8 over 1. What's the answer? (Signal.) *80 trials.*

EXERCISE 7: LONG DIVISION

2-DIGIT QUOTIENT

R Test 11

a. (Display:)   [99:7A]

$$72\overline{)3948}$$

This problem has a two-digit answer.
I'll underline the digits that we use to find the first digit of the answer.
- 39 divided by 72. Can you work that problem? (Signal.) *No.*
  39 is too small.
- 394 divided by 72. Can you work that problem? (Signal.) *Yes.*
b. So I underline 394, and I'll write the first digit of the answer right above the 4.
- How many digits will I underline? (Signal.) *3.*
- I'll write the answer above which digit? (Signal.) *4.*
  (Add to show:)   [99:7B]

$$72\overline{)3948}$$

- Say the estimation problem. (Signal.) *39 ÷ 7.*
- Raise your hand when you know the answer. ✔
- What's the answer? (Signal.) *5.*
  (Add to show:)   [99:7C]

$$72\overline{)\overset{5}{3948}}$$

- Say the multiplication problem for 5. (Signal.) *72 × 5.*
c. Work that problem in part A. ✔
- Everybody, what's 72 times 5? (Signal.) *360.*
  (Add to show:)   [99:7D]

$$72\overline{)\overset{5}{3948}} \\ -360$$

Now we subtract. The answer is 34.
(Add to show:)   [99:7E]

$$72\overline{)\overset{5}{3948}} \\ \underline{-360} \\ 34$$

We've worked the problem for the first digit of the answer.
d. What's the first thing we do to work the problem for the ones digit? (Signal.) *Bring down the ones digit.*
(Add to show:)   [99:7F]

$$72\overline{)\overset{5}{3948}} \\ \underline{-360} \\ 348$$

You know how to work the rest of the problem.
e. (Display:)   [99:7G]

$$28\overline{)745}$$

- Read the problem. (Signal.) *745 ÷ 28.*
- Say the first problem you'll work for dividing by 28. (Signal.) *74 ÷ 28.*
  So I underline 74.
  (Add to show:)   [99:7H]

$$28\overline{)745}$$

f. Your turn: Copy the problem and work the problem for the first digit. Stop when you have the first digit of the answer and a remainder below. ✔

(Add to show:) [99:7I]

$$28\overline{)745}$$ quotient 2, -56, 18

Here's what you should have.
The first digit of the answer is 2.
You subtracted 56. The remainder is 18.

g. Now you'll work the problem for the ones digit of the answer.
- What's the first thing you do? (Signal.) *Bring down the ones digit.*
- Do it and complete the problem.
(Observe students and give feedback.)

h. Check your work. You worked the problem 185 ÷ 28.
- Say the estimation problem. (Signal.) *19 ÷ 3.*
- What's the answer? (Signal.) *6.*
(Add to show:) [99:7J]

$$28\overline{)745}$$ quotient 26, -56, 185

Here's what you should have for that part.

i. Say the multiplication for the ones digit. (Signal.) *28 × 6.*
- What's the answer? (Signal.) *168.*
- Say the subtraction problem for the remainder. (Signal.) *185 − 168.*
- What's the remainder? (Signal.) *17.*
(Add to show:) [99:7K]

$$28\overline{)745}$$ quotient $26\frac{17}{28}$, -56, 185, -168, 17

Here's what you should have for the whole problem.

j. (Display:) [99:7L]

$$41\overline{)2296}$$

- Read the problem. (Signal.) *2296 ÷ 41.*
- Raise your hand when you know the first division problem you'll work. ✔
- Say the first division problem. (Signal.) *229 ÷ 41.*
- Copy the problem. Underline 229 and work the problem for the first digit of the answer. Stop when you have a remainder for the first digit of the answer.
(Observe students and give feedback.)

k. Check your work.
- What's the first digit of the answer? (Signal.) *5.*
- What's the remainder? (Signal.) *24.*
(Add to show:) [99:7M]

$$41\overline{)2296}$$ quotient 5, -205, 24

Here's what you should have.

l. Work the problem for the second digit of the answer.
(Observe students and give feedback.)
You brought down 6 and worked the problem 246 divided by 41. The answer is 6 with no remainder.
- What's the answer to the whole problem? (Signal.) *56.*
(Add to show:) [99:7N]

$$41\overline{)2296}$$ quotient 56, -205, 246, -246, 0

Here's what you should have.

Assign Independent Work, Textbook parts 4–9 and Workbook parts 3–5.

Optional extra math-fact practice worksheets are available on ConnectED.

Lesson 100

EXERCISE 1: MENTAL MATH

a. Time for some mental math.
- Listen: 3 times what number equals 9? (Pause.) (Signal.) *3.*
- So 30 times what number equals 90? (Pause.) (Signal.) *3.*
- 30 times what number equals 150? (Pause.) (Signal.) *5.*
- 30 times what number equals 240? (Pause.) (Signal.) *8.*
- 30 times what number equals 180? (Pause.) (Signal.) *6.*
- (Repeat until firm.)

b. Listen: 40 times what number equals 80? (Pause.) (Signal.) *2.*
- 40 times what number equals 160? (Pause.) (Signal.) *4.*
- 40 times what number equals 240? (Pause.) (Signal.) *6.*
- 40 times what number equals 400? (Pause.) (Signal.) *10.*
- (Repeat until firm.)

c. Listen: 76 plus 4. (Pause.) (Signal.) *80.*
- 76 + 6. (Pause.) (Signal.) *82.*
- 24 + 6. (Pause.) (Signal.) *30.*
- 24 + 16. (Pause.) (Signal.) *40.*
- 74 + 16. (Pause.) (Signal.) *90.*
- (Repeat until firm.)

d. Listen: 48 plus 8. (Pause.) (Signal.) *56.*
- 118 + 8. (Pause.) (Signal.) *126.*
- 40 + 18. (Pause.) (Signal.) *58.*
- 120 + 18. (Pause.) (Signal.) *138.*
- 130 + 15. (Pause.) (Signal.) *145.*
- 130 + 55. (Pause.) (Signal.) *185.*
- 130 + 59. (Pause.) (Signal.) *189.*
- (Repeat until firm.)

EXERCISE 2: BALANCE BEAMS
AVERAGE DISTANCE

REMEDY

a. Open your workbook to Lesson 100 and find part 1. ✔
(Teacher reference:)

R Part G

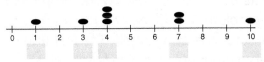

- Figure out the totals for the weights and write the numbers in the boxes. Then figure out the total distance for the weights. Stop when you've done that much.
(Observe students and give feedback.)
(Display:) [100:2A]

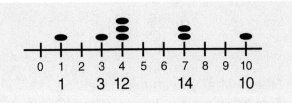

Here are the totals for the weights. You added those numbers to get the total distance.
- What's the total distance? (Signal.) *40.*
- Now figure out the balance point and make an arrow to show it. Pencils down when you're finished.
(Observe students and give feedback.)
- Everybody, what did you divide 40 by? (Signal.) *8.*
- What's the balance point for these weights? (Signal.) *5.*
(Add to show:) [100:2B]

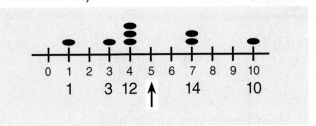

Here's the number line with the balance point at 5.

EXERCISE 3: FRACTION OPERATIONS

SIMPLIFY MULTIPLICATION

a. Find part 2 in your workbook. ✔
(Teacher reference:)

a. $\frac{15}{100}\left(\frac{20}{21}\right) =$ b. $\frac{20}{25}\left(\frac{10}{4}\right) =$

c. $\frac{90}{100}(45) =$ d. $\frac{24}{100}\left(\frac{200}{3}\right) =$

You're going to simplify these fractions as much as possible and then work the problem.

b. Read problem A. (Signal.) *15/100 × 20/21.*
- Cross out zeros and simplify the first fraction. (Observe students and give feedback.)

c. Check your work.
You crossed out zeros. That left 15/10 times 2/21. You simplified 15/10.
- What fraction does it equal? (Signal.) *3/2.*
(Display:)   [100:3A]

a. $\overset{3}{\cancel{15}} \left(\frac{\cancel{20}}{21} \right) = $
$\underset{2}{\cancel{100}}$

Here's what you should have.
- Two fractions can be simplified. Raise your hand when you know what they are. ✔
- What fractions can be simplified? (Call on a student. Idea: *2/2 and 3/21.*)
- Everybody, what does 2/2 equal? (Signal.) *1.*
- Raise your hand when you know what 3/21 equals. ✔
- What does 3/21 equal? (Signal.) *1/7.*
- Simplify those fractions and write the answer to the whole problem. ✔
- Everybody, what's the answer to the whole problem? (Signal.) *1/7.*
(Add to show:)   [100:3B]

a. $\overset{1}{\cancel{\overset{3}{\cancel{15}}}} \left(\frac{\overset{1}{\cancel{20}}}{\underset{7}{\cancel{21}}} \right) = \frac{1}{7}$
$\underset{\cancel{2}}{\cancel{100}}$

Here's what you should have.

d. Read problem B. (Signal.) *20/25 × 10/4.*
- Simplify and work the problem. (Observe students and give feedback.)

e. Check your work.
- Did you cross out zeros? (Signal.) *No.*
You simplified 20/25.
- What does it equal? (Signal.) *4/5.*

(Display:)   [100:3C]

b. $\overset{4}{\cancel{\underset{5}{\cancel{25}}}^{20}} \left(\frac{10}{4} \right) = $

- You simplified 4/4. What does that equal? (Signal.) *1.*
- Then you simplified 10/5. What does it equal? (Signal.) *2.*
(Add to show:)   [100:3D]

b. $\overset{1}{\cancel{\overset{4}{\cancel{20}}}} \left(\frac{\overset{2}{\cancel{10}}}{\cancel{4}} \right) = 2$
$\underset{\cancel{5}}{\cancel{25}}$

Yes, 20/25 times 10/4 equals 2.

f. Read problem C. (Signal.) *90/100 × 45.*
The answer to this problem is a mixed number.
- Do the simplification and work the problem. (Observe students and give feedback.)

g. Check your work.
You crossed out zeros. That left you with 9/10.
- You simplified 45/10. What's the simplified fraction? (Signal.) *9/2.*
You multiplied 9 times 9. That's 81.
- The answer is 81/2. What mixed number does that equal? (Signal.) *40 and 1/2.*
(Display:)   [100:3E]

c. $\frac{\cancel{90}}{\underset{2}{\cancel{100}}} (\overset{9}{\cancel{45}}) = \frac{81}{2} = 40\frac{1}{2}$

Here's what you should have.
You may have simplified in a different order, but you should have the same answer—40 and 1/2.

h. Read problem D. (Signal.) *24/100 × 200/3.*
- Work the problem. (Observe students and give feedback.)

i. Check your work.
You crossed out zeros.
- That left you with 24 times 2/3. What does 24/3 equal? (Signal.) *8.*
- You multiplied 8 times 2. What's the answer? (Signal.) *16.*
(Add to show:)   [100:3F]

d. $\frac{24}{\cancel{100}} \left(\frac{\cancel{200}}{\overset{8}{\cancel{3}}} \right) = 16$

Here's what you should have.

EXERCISE 4: UNIT CONVERSION
MULTIPLICATION/DIVISION

a. Find the measurement table in the back of your textbook. ✔
   You're going to work problems that refer to measurement facts. If you don't know a fact, look it up in this table.

b. Listen: Change 12 tablespoons into cups.
- Raise your hand when you know the measurement fact. ✔
- Everybody, what's the measurement fact for tablespoons and cups? (Signal.) *1 cup is 16 tablespoons.*

c. Listen: Change 2 miles into feet.
- Raise your hand when you know the measurement fact. ✔
- Everybody, what's the measurement fact for miles and feet? (Signal.) *1 mile is 5280 feet.*

══════ TEXTBOOK PRACTICE ══════

a. Open your textbook to Lesson 100 and find part 1. ✔
   (Teacher reference:)

   a. Change 300 seconds into minutes.
   b. Change 660 years into decades.
   c. Change 4 gallons into cups.
   d. Change 62 weeks into days.
   e. Change 45 ounces into pounds.

Part 1	
a.	

   Remember, don't use the facts in the back unless you need to.

b. Problem A: Change 300 seconds into minutes.
- Work the problem. ✔
- What's the fact for seconds and minutes? (Call on a student. Idea: *1 minute is 60 seconds.*)
- Everybody, read the problem you worked. (Signal.) *300 ÷ 60.*
- How many minutes is 300 seconds? (Signal.) *5 minutes.*

c. Work the rest of the problems in part 1. (Observe students and give feedback.)

d. Check your work.
- Problem B: Change 660 years into decades.
- Raise your hand if you didn't need to use the table. ✔

- Read the problem you worked. (Signal.) *660 ÷ 10.*
- How many decades is 660 years? (Signal.) *66 decades.*

e. Problem C: Change 4 gallons into cups.
- Raise your hand if you didn't need to use the table. ✔
- Read the problem you worked. (Signal.) *16 × 4.*
- How many cups is 4 gallons? (Signal.) *64 cups.*

f. D: Change 62 weeks into days.
- Raise your hand if you didn't need to use the table. ✔
- Read the problem you worked. (Signal.) *62 × 7.*
- How many days is 62 weeks? (Signal.) *434 days.*

g. E: Change 45 ounces into pounds.
- Raise your hand if you didn't need to use the table. ✔
- Read the problem you worked. (Signal.) *45 ÷ 16.*
- How many pounds is 45 ounces? (Signal.) *2 and 13/16 pounds.*

EXERCISE 5: PROBABILITY
EQUATIONS

a. Find part 2 in your textbook. ✔
   (Teacher reference:)

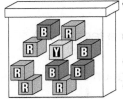

   a. Write the probability equation for pulling a blue cube from the box without looking.
   b. If you took 25 trials, about how many blue cubes would you expect to pull?
   c. Write the probability equation for the yellow cube.
   d. If you took trials until you pulled out 7 yellow cubes, about how many trials would you expect to take?
   e. Write the probability equation for red cubes.
   f. If you took 18 trials, about how many red cubes would you expect to pull out?

   These are cubes that are in a box. You're going to figure out the probabilities for different colors.

b. Problem A: Write the probability equation for pulling a blue cube from the box without looking. ✔
- Everybody, read the equation for pulling out a blue cube. (Signal.) *4/10 T = B.*

c. Problem B: If you took 25 trials, about how many blue cubes would you expect to pull out?
- Work the problem. ✔
- Everybody, what's the whole answer? (Signal.) *10 blue cubes.*

(Display:)   [100:5A]

a. $\frac{4}{10} t = b$

b. $\frac{\overset{2}{\cancel{4}}}{\underset{5}{\cancel{10}}} (\overset{5}{\cancel{25}}) = b = 10$

[10 blue cubes]

Here's what you should have.

d. Work problems C and D.
   (Observe students and give feedback.)
e. Check your work.
• Problem C. Read the equation for yellow
   cubes. (Signal.) *1/10 T = Y.*
• Problem D. What's the whole answer?
   (Signal.) *70 trials.*
   (Display:)   [100:5B]

c. $\frac{1}{10} t = y$

d. $\left(\frac{10}{1}\right) \frac{1}{10} t = 7 \left(\frac{10}{1}\right)$

$t = 70$

[70 trials]

Here's what you should have.

f. Work problems E and F.
   (Observe students and give feedback.)
g. Check your work.
• Problem E. Read the equation for red cubes.
   (Signal.) *5/10 T = R.*
• Problem F. What's the whole answer? (Signal.)
   9 red cubes.
   (Display:)   [100:5C]

e. $\frac{5}{10} t = r$

f. $\frac{\overset{1}{\cancel{5}}}{\underset{2}{\cancel{10}}} (\overset{9}{\cancel{18}}) = r = 9$

[9 red cubes]

Here's what you should have.

EXERCISE 6: LONG DIVISION
2-DIGIT QUOTIENT

a. (Display:)   [100:6A]

$5\,3\overline{\smash{\big)}\,1\,6\,9\,9}$

• Read the problem. (Signal.) *1699 ÷ 53.*
• Copy the problem in part A. ✔
• Raise your hand when you know the first
   division problem you'll work. ✔
• Say the first division problem. (Signal.) *169 ÷ 53.*
• Underline 169 and work the problem for the
   first digit of the answer. Stop when you have a
   remainder for the first digit of the answer.
b. Check your work.
• What's the first digit of the answer? (Signal.) *3.*
• What's the remainder? (Signal.) *10.*
   (Add to show:)   [100:6B]

$$\begin{array}{r} 3 \\ 5\,3\overline{\smash{\big)}\,1\,6\,9\,9} \\ \underline{-1\,5\,9} \\ 1\,0 \end{array}$$

Here's what you should have.

c. Work the problem for the second digit of the
   answer.
   (Observe students and give feedback.)
   You brought down 9 and worked the problem
   109 ÷ 53. The answer is 2 with a remainder
   of 3.
• What's the answer to the whole problem?
   (Signal.) *32 and 3/53.*
   (Add to show:)   [100:6C]

$$\begin{array}{r} 3\,2\frac{3}{53} \\ 5\,3\overline{\smash{\big)}\,1\,6\,9\,9} \\ \underline{-1\,5\,9} \\ 1\,0\,9 \\ \underline{-1\,0\,6} \\ 3 \end{array}$$

Here's what you should have.

d. (Display:) [100:6D]

$$48\overline{)983}$$

- New problem. Read it. (Signal.) *983 ÷ 48.*
- Raise your hand when you know the first division problem you'll work. ✔
- Say the first division problem. (Signal.) *98 ÷ 48.*

e. Copy the problem. Underline 98 and work the problem for the first digit of the answer. Stop when you have a remainder for the first digit of the answer.
   (Observe students and give feedback.)

f. Check your work.
- What's the first digit of the answer? (Signal.) *2.*
   (Add to show:) [100:6E]

$$48\overline{)983} \\ \underline{-96} \\ 2$$ with a 2 above the line

Here's what you should have.

g. Work the problem for the second digit of the answer.
   (Observe students and give feedback.)
   You brought down 3 and worked the problem 23 ÷ 48. The answer is zero with a remainder of 23.
- What's the answer to the whole problem? (Signal.) *20 and 23/48.*
   (Add to show:) [100:6F]

$$48\overline{)983} \\ \underline{-96} \\ 23 \\ \underline{-\ 0} \\ 23$$ with $20\frac{23}{48}$ above the line

Here's what you should have.

EXERCISE 7: SENTENCES
PARTS OF A GROUP

a. My turn: Listen: 3 out of every 5 boys. What's the fraction for 3 out of 5? 3/5.
- Your turn: What's the fraction for 3 out of 5? (Signal.) *3/5.*
- What's the fraction for 4 out of 10? (Signal.) *4/10.*
- What's the fraction for 7 out of 8? (Signal.) *7/8.*
- What's the fraction for 3 out of 11? (Signal.) *3/11.*
   (Repeat until firm.)

b. Here's a sentence: 3 out of every 5 eggs are cracked.
- Say the sentence. (Signal.) *3 out of every 5 eggs are cracked.*
   My turn to say the sentence with a fraction: 3/5 of the eggs are cracked.

c. Listen: 2 out of every 9 windows were open.
- Say the sentence with a fraction for 2 out of 9. (Signal.) *2/9 of the windows were open.*

d. Listen: 7 out of every 8 girls are playing.
- Say the sentence with a fraction for 7 out of 8. (Signal.) *7/8 of the girls are playing.*
   (Repeat until firm.)

e. (Display:) [100:7A]

   5 out of every 8 books were sold.

   I'm going to write an equation for this sentence.
- Read the sentence. (Signal.) *5 out of every 8 books were sold.*
- Say the sentence with a fraction for 5 out of 8. (Signal.) *5/8 of the books were sold.*
   (Add to show:) [100:7B]

   5 out of every 8 books were sold.

$$\frac{5}{8}b = s$$

   Here's the sentence with a fraction and letters for books and sold.
- Read the equation. (Signal.) *5/8 B = S.*

a. Find part 3 in your textbook. ✔
   (Teacher reference:)

   a. 5 out of every 9 envelopes are sealed.
   b. 9 out of every 15 cars were moving.
   c. 2 out of every 3 people were visitors.
   d. 7 out of every 10 cats are sleeping.

   For each sentence, you'll write an equation with a fraction and two letters.

b. Read sentence A. (Signal.) *5 out of every 9 envelopes are sealed.*

• Write the equation. ✔

• Everybody, read the equation. (Signal.) *5/9 E = S.*

c. Write equations for the rest of the items. (Observe students and give feedback.)

d. Check your work.

• Sentence B: 9 out of every 15 cars were moving. Read the equation. (Signal.) *9/15 C = M.*

• C: 2 out of every 3 people were visitors. Read the equation. (Signal.) *2/3 P = V.*

• D: 7 out of every 10 cats are sleeping. Read the equation. (Signal.) *7/10 C = S.*

EXERCISE 8: INDEPENDENT WORK
MIXED NUMBERS/BAR GRAPHS

a. Find part 4 in your textbook. ✔
   (Teacher reference:)

   a. $3\frac{2}{7}$ b. $10\frac{1}{4}$ c. $6\frac{3}{5}$

   As part of your independent work, you're going to write equations that show the fraction each mixed number equals.
   Remember, show the two fractions you add to figure out the answer.

b. Find part 5. ✔
   (Teacher reference:)

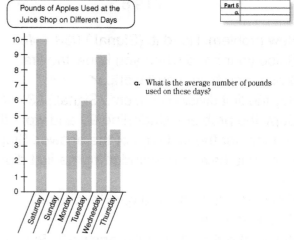

a. What is the average number of pounds used on these days?

   The graph shows the pounds of apples used at a juice shop on different days.

• Raise your hand when you know the number you'll divide by. ✔

• What number will you divide by? (Signal.) *6.* Yes, on Sunday they used zero pounds, but you count Sunday as one of the 6 days.

Assign Independent Work, Textbook parts 4–11 and Workbook part 3.

Optional extra math-fact practice worksheets are available on ConnectED.

Mastery Test 10

Note: Mastery Tests are administered to the entire group. Each student will need a pencil and the *Student Assessment Book.* Try to arrange students so they cannot look at other students' responses.

Note: This test will probably take more than 1 class period.

Note: Students may refer to the table of Measurement Facts at the back of the *Textbook* for part 1 of the test.

Teacher Presentation

a. Find Test 10 in your test booklet. ✔
• Pencils down. ✔
  Listen carefully and do your best work.
b. Touch part 1. ✔
  To work the problems in part 1, you need to use measurement facts.
• Find the table of facts at the back of your textbook. ✔
  If you don't know a fact, look it up in the table.
c. Work parts 1 through 12 of the test by yourself. Read the directions carefully and do your best work. Put your pencil down when you've finished the test. If you don't finish it today, you can finish it in the next math class.
  (Observe but do not give feedback.)

Scoring Notes

a. Collect test booklets. Use the *Answer Key* and Passing Criteria Table to score the tests.

Part	Score	Possible Score	Passing Score
\multicolumn{4}{c}{**Passing Criteria Table—Mastery Test 10**}
1	2 for each item (problem, answer with unit name)	8	6
2	2 for each item (fraction problem, mixed number answer)	6	5
3	3 for each item (letter equation, substitution, answer with unit name)	6	5
4	3 for each balance beam (weights, answers)	6	5
5	3 for each item (problem, renaming, answer with unit name)	6	5
6	2 for each item (product, sign in equation)	6	5
7	3 for each item (set up, computation, answer with unit name)	12	10
8	2 for each item (addition problem, answer)	6	5
9	4 for each item (letter equation, substitution, answers)	12	10
10	2 for each problem (fraction problem, proper mixed-number answer)	6	5
11	3 for each item (column totals, grand total, arrow)	6	5
12	2 for each item (division problem, answer with unit name)	8	7
	Total	88	

b. Complete the Mastery Test 10 Remedy Summary Sheet to determine whether group remedies are needed. Reproducible Remedy Summary Sheets are at the back of the *Answer Key* and the back of the *Teacher's Guide.*

• If ¼ or more of the students did not pass a test part, present the remedy for that part before beginning Lesson 101. The Remedy Table follows and also appears at the end of the Mastery Test 10 *Answer Key.* Remedies worksheets follow Mastery Test 10 in the *Student Assessment Book.*

Part	Test Items	Remedy Lesson	Remedy Ex.	Remedies Worksheet	Textbook
1	Unit Conversion (Multiplication/Division)	94	5	—	—
		97	5	—	Part 2
		99	5	—	Part 2
2	Mixed-Number Operations (Unlike Denominators)	91	4	Part A	—
		92	3	Part B	—
3	Probability (Word Problems)	98	6	—	Part 2
		99	6	—	Part 3
4	Balance Beam (Weights on one side)	94	3	Part C	—
		95	4	Part D	—
5	Mixed-Number Operations (Word Problems)	94	8	—	Part 3
		95	7	—	Part 3
6	Fraction Multiplication (Product/Factor Comparison)	92	6	—	Part 3
		93	4	—	Part 1
7	Volume (Complex Figures)	97	7	—	Part 4
		98	7	—	Part 3
8	Fractions (For Mixed Numbers)	98	4	Part E	
		99	4	—	Part 1
9	Word Problems (Parts of a Group)	89	7	—	Part 4
		91	7	—	Part 3
		93	7	—	Part 3
10	Mixed-Number Operations (Simplified Answer)	96	5	—	Part 2
		97	6	—	Part 3
11	Balance Beam (Balance Point)	99	3	Part F	—
		100	2	Part G	—
12	Fraction Word Problems (Division)	90	6	—	Part 2
		91	6	—	Part 2

Remedy Table—Mastery Test 10

Retest

Retest individual students on any part failed.

Lessons 101-105 Planning Page

	Lesson 101	Lesson 102	Lesson 103	Lesson 104	Lesson 105
Student Learning Objectives	**Exercises** 1. Solve for missing numbers and add using mental math **2. Multiply and divide decimals by multiples by 10** 3. Simplify fractions 4. Solve long division problems with 2-digit quotients 5. Complete equations for probability 6. Solve mixed-number word problems 7. Solve word problems for parts of a group Complete work independently	**Exercises** 1. Solve for missing numbers and add using mental math 2. Multiply and divide decimals by multiples of 10 3. Solve long division problems with 2-digit quotients **4. Show probability with fractions** 5. Solve mixed-number word problems **6. Read decimal numbers written as words** 7. Solve word problems for parts of a group Complete work independently	**Exercises** 1. Add and solve for missing numbers using mental math **2. Find average on a line plot** **3. Use the distributive property of multiplication to evaluate expressions** 4. Show probability with fractions 5. Read decimal numbers written as words 6. Convert mixed numbers to fractions 7. Multiply and divide decimals by multiples of 10 Complete work independently	**Exercises** 1. Solve for missing numbers and add using mental math **2. Find total value for columns on a line plot** 3. Use the distributive property of multiplication to evaluate expressions **4. Show probability with percents** 5. Read decimal numbers written as words 6. Convert mixed numbers to fractions 7. Multiply and divide decimals by multiples of 10 Complete work independently	**Exercises** 1. Solve for missing numbers and add using mental math 2. Find total value for columns on a line plot 3. Show probability with percents 4. Use the distributive property of multiplication to evaluate expressions 5. Convert mixed numbers to fractions 6. Multiply and divide decimals by multiples of 10 7. Complete equations to show probability Complete work independently
Common Core State Standards for Mathematics					
5.OA 1	✔	✔	✔	✔	✔
5.NBT 1–2	✔	✔	✔	✔	✔
5.NBT 3		✔	✔	✔	✔
5.NBT 5–7	✔	✔	✔	✔	✔
5.NF 1	✔	✔	✔	✔	
5.NF 2	✔	✔	✔	✔	✔
5.NF 4	✔	✔	✔	✔	✔
5.NF 5			✔		✔
5.NF 6			✔	✔	
5.MD 1	✔	✔		✔	✔
5.MD 5	✔	✔			
5.G 1		✔		✔	
5.G 2	✔				✔
Teacher Materials	Presentation Book 2, Board Displays CD or chalk board				
Student Materials	Textbook, Workbook, pencil, lined paper, ruler				
Additional Practice	• Student Practice Software: Block 5: Activity 1 (5.MD 1), Activity 2 (5.NF 1 and 5.NF 2), Activity 3 (5.MD 3 and 5.MD 4), Activity 4 (5.NF 4 and 5.NF 6), Activity 5, Activity 6 (5.MD 3 and 5.MD 4) • Provide needed fact practice with Level D or E Math Fact Worksheets.				
Mastery Test					

Lesson

EXERCISE 1: MENTAL MATH

a. Time for some mental math.
- Listen: 4 times what number equals 12? (Signal.) *3.*
- So 40 times what number equals 120? (Signal.) *3.*
- 40 times what number equals 160? (Signal.) *4.*
- 30 times what number equals 180? (Signal.) *6.*
- 30 times what number equals 240? (Signal.) *8.*
  (Repeat until firm.)

b. Listen: 40 times what number equals 240? (Signal.) *6.*
- 40 times what number equals 160? (Signal.) *4.*
- 40 times what number equals 320? (Signal.) *8.*
- 40 times what number equals 400? (Signal.) *10.*
  (Repeat until firm.)

c. Listen: 96 plus 4. (Pause.) (Signal.) *100.*
- 96 + 6. (Pause.) (Signal.) *102.*
- 124 + 6. (Pause.) (Signal.) *130.*
- 124 + 16. (Pause.) (Signal.) *140.*
- 174 + 16. (Pause.) (Signal.) *190.*
  (Repeat until firm.)

d. Listen: 148 plus 8. (Pause.) (Signal.) *156.*
- 118 + 8. (Pause.) (Signal.) *126.*
- 140 + 18. (Pause.) (Signal.) *158.*
- 160 + 18. (Pause.) (Signal.) *178.*
- 160 + 15. (Pause.) (Signal.) *175.*
- 140 + 15. (Pause.) (Signal.) *155.*
- 140 + 25. (Pause.) (Signal.) *165.*
  (Repeat until firm.)

EXERCISE 2: DECIMAL MULTIPLICATION/DIVISION
MULTIPLES OF 10 `REMEDY`

a. (Display:) [101:2A]

> 34.53

- Read this number. (Signal.) *34 and 53 hundredths.*
  You know how to move the decimal point when we multiply or divide by multiples of 10.

b. If you multiply by 10, do you move the decimal point to the left or to the right? (Signal.) *To the right.*
- If you multiply by 10, how many places do you move the decimal point? (Signal.) *1 place.*
- If you multiply by 100, how many places do you move the decimal point? (Signal.) *2 places.*
- So what does 34.53 times 100 equal? (Signal.) *3453.*
- What does 34.53 times 10 equal? (Signal.) *345 and 3 tenths.*
  (Repeat until firm.)

c. (Display:) [101:2B]

> 31.5

- Read this number. (Signal.) *31 and 5 tenths.*
- Which way do we move the decimal point when we divide? (Signal.) *To the left.*
- What's 31.5 divided by 10? (Signal.) *3 and 15 hundredths.*
- What's 31.5 divided by 100? (Signal.) *315 thousandths.*
  (Repeat until firm.)

a. Open your workbook to Lesson 101 and find part 1. ✔
(Teacher reference:) R Part C

| a. | .17 | b. | 3.14 | c. | 460 | d. | .152 |

I'll tell you what you multiply or divide each number by. You'll copy the digits below and write the new decimal value.

b. Read number A. (Signal.) *17 hundredths.*
• Multiply 17 hundredths by 100. Write the new value below. ✔
• Everybody, what's 17 hundredths times 100? (Signal.) *17.*
c. Read number B. (Signal.) *3 and 14 hundredths.*
• Divide 3 and 14 hundredths by 10. Write the new value. ✔
• Everybody, what's 3 and 14 hundredths divided by 10? (Signal.) *314 thousandths.*
d. Read number C. (Signal.) *460.*
• Divide 460 by 1000. ✔
• Everybody, what's 460 divided by 1000? (Signal.) *460 thousandths.*
e. Read number D. (Signal.) *152 thousandths.*
• Multiply 152 thousandths by 10. ✔
• Everybody, what's 152 thousandths times 10? (Signal.) *1 and 52 hundredths.*
(Display:) [101:2C]

a. .17	**b.** 3.14	**c.** 460	**d.** .152
17	.314	.460	1.52

Here's what you should have for each value.

EXERCISE 3: FRACTION OPERATIONS
SIMPLIFY MULTIPLICATION
REMEDY

a. Open your textbook to Lesson 101 and find part 1. ✔
(Teacher reference:)

a. $\frac{50}{100}\left(\frac{20}{3}\right) =$ b. $\frac{35}{100}\left(\frac{10}{28}\right) =$ c. $\frac{20}{50}\left(\frac{10}{3}\right) =$

b. Problem A: 50/100 × 20/3.
• Copy the problem. Cross out all the zeros you can. Stop when you've done that much.
(Observe students and give feedback.)

c. Check your work.
(Display:) [101:3A]

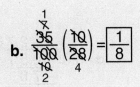

a. $\frac{5\cancel{0}}{1\cancel{0}\cancel{0}}\left(\frac{2\cancel{0}}{3}\right) =$

Here's what you should have.
You crossed out zeros in 50/100. You crossed out zeros in 20/100.
• Raise your hand when you know if you can do any more simplification. ✔
• Can you do more simplification? (Signal.) *No.*
• Multiply and write the fraction and mixed number. ✔
You worked the problem 5 times 2/3. That's 10/3.
• What does that equal? (Signal.) *3 and 1/3.*
(Add to show:) [101:3B]

a. $\frac{5\cancel{0}}{1\cancel{0}\cancel{0}}\left(\frac{2\cancel{0}}{3}\right) = \frac{10}{3} = \boxed{3\frac{1}{3}}$

Here's what you should have.
d. Problem B: 35/100 × 10/28.
• Copy and work the problem. Start by crossing out zeros. ✔
e. Check your work.
(Display:) [101:3C]

b. $\frac{\overset{7}{\cancel{35}}}{\underset{2}{\underset{\cancel{10}}{\cancel{1\cancel{0}\cancel{0}}}}}\left(\frac{\cancel{10}}{\underset{4}{\cancel{28}}}\right) = \boxed{\frac{1}{8}}$

Here's what you should have.
You crossed out zeros in 10/100.
Then you simplified 35/10. That's 7/2.
Then you simplified 7/28. That's 1/4.
• What's the answer to the problem? (Signal.) *1/8.*
f. Problem C: 20/50 × 10/3.
• Copy and work the problem. ✔
g. Check your work.
(Display:) [101:3D]

c. $\frac{2\cancel{0}}{5\cancel{0}}\left(\frac{\overset{2}{\cancel{10}}}{3}\right) = \frac{4}{3} = \boxed{1\frac{1}{3}}$

Here's what you should have.
You crossed out zeros in 20/50.
Then you simplified 10/5. That's 2.
So the answer is 4/3, which is 1 and 1/3.

EXERCISE 4: LONG DIVISION
2-DIGIT QUOTIENT

REMEDY

a. Find part 2 in your textbook. ✔
(Teacher reference:)

a. $39\overline{)1057}$ b. $91\overline{)4924}$

b. Problem A: 1057 ÷ 39.
• Copy the problem. Underline the digits for the first division problem you work. Stop when you've done that much.
(Observe students and give feedback.)
• Everybody, say the first division problem you'll work. (Signal.) *105 ÷ 39.*
• Work that problem. Then work the problem for the ones digit of the answer.
(Observe students and give feedback.)
c. Check your work.
• Say the first division problem you worked. (Signal.) *105 ÷ 39.*
• What's the first digit of the answer? (Signal.) *2.* You brought down the 7.
• Say the second division problem you worked. (Signal.) *277 ÷ 39.*
• What's the second digit of the answer? (Signal.) *7.*
• What's the answer to the whole problem? (Signal.) *27 and 4/39.*
(Display:)   [101:4A]

$$
\begin{array}{r}
27\frac{4}{39} \\
39\overline{)1057} \\
-78 \\
\hline
277 \\
-273 \\
\hline
4
\end{array}
$$
a.

Here's what you should have.
d. Problem B: 4924 ÷ 91.
• Copy the problem. Underline the digits for the first division problem you work. Work that problem. Then work the problem for the ones digit of the answer.
(Observe students and give feedback.)
e. Check your work.
• Say the first division problem you worked. (Signal.) *492 ÷ 91.*
• What's the first digit of the answer? (Signal.) *5.* You brought down the 4.
• Say the second division problem you worked. (Signal.) *374 ÷ 91.*

• What's the second digit of the answer? (Signal.) *4.*
• What's the answer to the whole problem? (Signal.) *54 and 10/91.*
(Display:)   [101:4B]

$$
\begin{array}{r}
54\frac{10}{91} \\
91\overline{)4924} \\
-455 \\
\hline
374 \\
-364 \\
\hline
10
\end{array}
$$
b.

Here's what you should have.

EXERCISE 5: PROBABILITY
EQUATIONS

a. Find part 3 in your textbook. ✔
(Teacher reference:)

a. How many pink rabbits would you expect to pull out if you took 12 trials?
b. About how many trials would you expect to take to pull 4 blue rabbits from the hat?
c. How many yellow rabbits would you expect to pull out if you took 30 trials?

The diagram shows the objects that are in a hat.
• What are those objects? (Signal.) *Rabbits.* You're going to figure out the probability for pulling out a pink rabbit, a blue rabbit, or a yellow rabbit. Remember, first write the probability equation. Then work the problem for a particular color.
b. Problem A: How many pink rabbits would you expect to pull out if you took 12 trials?
• Work the problem.
(Observe students and give feedback.)
c. Check your work.
You started with the probability equation: 2/6 T = P.
• What's the answer? (Signal.) *4 pink rabbits.*
(Display:)   [101:5A]

$$
\text{a. } \frac{2}{6}t = p
$$
$$
\frac{2}{6}(\overset{2}{\cancel{12}}) = p = 4
$$

4 pink rabbits

Here's what you should have.
d. Work the rest of the problems in part 3.
(Observe students and give feedback.)

e. Check your work.
- Problem B: About how many trials would you expect to take to pull 4 blue rabbits from the hat?
- What's the answer? (Signal.) *24 trials.*
  (Display:) [101:5B]

$$\mathbf{b.} \quad \frac{1}{6}t = b$$
$$\left(\frac{6}{1}\right)\frac{1}{6}t = 4\left(\frac{6}{1}\right)$$
$$t = 24$$

$$\boxed{24 \text{ trials}}$$

Here's what you should have.

f. Problem C: How many yellow rabbits would you expect to pull out if you took 30 trials?
- What's the answer? (Signal.) *15 yellow rabbits.*
  (Display:) [101:5C]

$$\mathbf{c.} \quad \frac{3}{6}t = y$$
$$\frac{3}{\cancel{6}}(\cancel{30}) = y = 15$$

$$\boxed{15 \text{ yellow rabbits}}$$

Here's what you should have.

EXERCISE 6: MIXED-NUMBER OPERATIONS
WORD PROBLEMS
REMEDY

a. Find part 4 in your textbook. ✔
   (Teacher reference:)

a. The cow gained $16\frac{1}{2}$ pounds last week and $14\frac{7}{8}$ pounds this week. How much weight did the cow gain during the two-week period?

b. Last fall, the maple tree was $12\frac{1}{4}$ feet tall. This fall, the tree is $15\frac{5}{6}$ feet tall. How much did the tree grow during the year?

These word problems tell about fractions with different denominators. You rewrite the problem with fractions that have the same denominator.

b. Problem A: The cow gained 16 and 1/2 pounds last week and 14 and 7/8 pounds this week. How much weight did the cow gain during the two-week period?
- Say the addition problem you start with. (Signal.) *16 and 1/2 + 14 and 7/8.*
- Write the problem. Rewrite the fractions so they have the same denominator and work the problem.
  (Observe students and give feedback.)
- Everybody, how much weight did the cow gain during the two-week period? (Signal.) *31 and 3/8 pounds.*
  (Display:) [101:6A]

$$\mathbf{a.} \quad 16\,\overset{1}{\frac{1}{2}}\left(\frac{4}{4}\right) = \quad \frac{4}{8}$$
$$+\,14\,\frac{7}{8} \qquad = +\,\frac{7}{8}$$
$$\rule{4cm}{0.4pt}$$
$$30\,\overset{1}{\frac{11}{8}}$$
$$\boxed{31\,\frac{3}{8} \text{ pounds}}$$

Here's what you should have.

c. Problem B: Last fall, the maple tree was 12 and 1/4 feet tall. This fall, the tree is 15 and 5/6 feet tall. How much did the tree grow during the year?
- Raise your hand when you know the problem you'll start with. ✔
- Say the problem you'll start with. (Signal.) *15 and 5/6 – 12 and 1/4.*
- Work the problem.
  (Observe students and give feedback.)
  You rewrote the fractions as 12ths. 10/12 minus 3/12.
- Everybody, how much did the tree grow? (Signal.) *3 and 7/12 feet.*
  (Display:) [101:6B]

$$\mathbf{b.} \quad 15\,\frac{5}{6}\left(\frac{2}{2}\right) = \quad \frac{10}{12}$$
$$-\,12\,\frac{1}{4}\left(\frac{3}{3}\right) = -\,\frac{3}{12}$$
$$\rule{4cm}{0.4pt}$$
$$\boxed{3\,\frac{7}{12} \text{ feet}}$$

Here's what you should have.

EXERCISE 7: SENTENCES
PARTS OF A GROUP
REMEDY

a. (Display:)   [101:7A]

> 5 out of every 7 cups are green.

- Read the sentence. (Signal.) *5 out of every 7 cups are green.*
- Say the fraction for 5 out of every 7. (Signal.) *5/7.*
  Yes, 5/7 of the cups are green.
- In part A, write the equation with a fraction and letters. ✔
- Everybody, read the equation. (Signal.)
  5/7 C = G.
  (Add to show:)   [101:7B]

> 5 out of every 7 cups are green.
>
> $$\frac{5}{7}\,c = g$$

Here's what you should have.

b. You're going to use equations like the one you just wrote to work problems.
  (Display:)   [101:7C]

> **a.** 4 out of every 7 pictures were sold. There were 28 pictures. How many were sold?

Listen: 4 out of every 7 pictures were sold. There were 28 pictures. How many were sold?
- Make an equation for the first sentence. Stop when you've done that much. ✔
- Everybody, read the equation. (Signal.)
  4/7 P = S.
  (Add to show:)   [101:7D]

> **a.** 4 out of every 7 pictures were sold. There were 28 pictures. How many were sold?
>
> $$\frac{4}{7}\,p = s$$

Here's what you should have.

c. Put in the number the problem gives. Then work the problem. ✔
  You worked the problem 4/7 times 28.
- Read the whole answer. (Signal.) *16 sold pictures.*
  Yes, 16 pictures were sold.

d. (Display:)   [101:7E]

> **b.** 2 out of every 5 apartments were rented. There were 16 rented apartments. How many apartments were there in all?

- New problem: 2 out of every 5 apartments were rented. There were 16 rented apartments. How many apartments were there in all?
- Write an equation for the first sentence. Then work the problem.
  (Observe students and give feedback.)

e. Check your work.
- Read the equation you started with. (Signal.)
  2/5 A = R.
  You replaced R with 16. You worked the problem 16 times 5/2.
- What's the whole answer? (Signal.) *40 apartments.*
  Yes, there were 40 apartments in all.

Assign Independent Work, Textbook parts 5–10 and Workbook parts 2–4.

Optional extra math-fact practice worksheets are available on ConnectED.

Lesson 102

EXERCISE 1: MENTAL MATH

a. Time for some mental math.
- Listen: 50 times what number equals 150? (Signal.) *3.*
- So 50 times what number equals 200? (Signal.) *4.*
- 50 times what number equals 350? (Signal.) *7.*
- 40 times what number equals 240? (Signal.) *6.*
- 30 times what number equals 240? (Signal.) *8.*
(Repeat until firm.)

b. Listen: 40 times what number equals 400? (Signal.) *10.*
- 40 times what number equals 160? (Signal.) *4.*
- 40 times what number equals 320? (Signal.) *8.*
- 40 times what number equals 360? (Signal.) *9.*
(Repeat until firm.)

c. Listen: 96 plus 6. (Pause.) (Signal.) *102.*
- 96 + 8. (Pause.) (Signal.) *104.*
- 125 + 6. (Pause.) (Signal.) *131.*
- 125 + 16. (Pause.) (Signal.) *141.*
- 175 + 16. (Pause.) (Signal.) *191.*
(Repeat until firm.)

d. Listen: 148 plus 8. (Pause.) (Signal.) *156.*
- 118 + 8. (Pause.) (Signal.) *126.*
- 140 + 18. (Pause.) (Signal.) *158.*
- 170 + 18. (Pause.) (Signal.) *188.*
- 170 + 15. (Pause.) (Signal.) *185.*
- 190 + 15. (Pause.) (Signal.) *205.*
- 190 + 25. (Pause.) (Signal.) *215.*
(Repeat until firm.)

EXERCISE 2: DECIMAL MULTIPLICATION/DIVISION
MULTIPLES OF 10
REMEDY

a. (Display:)   [102:2A]

> 6.41

- Read this number. (Signal.) *6 and 41 hundredths.*
- If you multiply this number by 10, what's the answer? (Signal.) *64 and 1 tenth.*
- If you multiply this number by 100, what's the answer? (Signal.) *641.*

b. (Display:)   [102:2B]

> 83

- If you divide this number by 10, what's the answer? (Signal.) *8 and 3 tenths.*
If you divide this number by 100, what's the answer? (Signal.) *83 hundredths.*

■■■■■ WORKBOOK PRACTICE ■■■■■

a. Open your workbook to Lesson 102 and find part 1. ✔
(Teacher reference:)   **R** **Part D**

a. .125 (Multiply by 100.)		d. .091 (Multiply by 100.)
b. 11.8 (Divide by 100.)		e. 17.06 (Multiply by 10.)
c. 152 (Divide by 10.)		f. 116 (Divide by 1000.)

You're going to multiply or divide decimal values by 10, 100, or 1000. You'll write the new decimal value below each problem.

b. Problem A. Read the number. (Signal.)
125 thousandths.
- Read the directions in parentheses. (Signal.)
Multiply by 100.
- Figure out where the new decimal point will be and write the answer below. ✔
- Everybody, what's the new decimal value? (Signal.) *12 and 5 tenths.*
(Display:) [102:2C]

$$a. \quad .125$$
$$12.5$$

Here's what you should have.
c. Read number B. (Signal.) *11 and 8 tenths.*
- Read the directions. (Signal.) *Divide by 100.*
- Write the new decimal value. ✔
- Everybody, what's the new decimal value? (Signal.) *118 thousandths.*
d. Work the rest of the problems.
(Observe students and give feedback.)
e. Check your work.
- Problem C. You divided 152 by 10. What's the new decimal value? (Signal.) *15 and 2 tenths.*
- D. You multiplied 91 thousandths by 100. What's the new decimal value? (Signal.) *9 and 1 tenth.*
- E. You multiplied 17 and 6 hundredths by 10. What's the new decimal value? (Signal.) *170 and 6 tenths.*
- F. You divided 116 by 1000. What's the new decimal value? (Signal.) *116 thousandths.*

EXERCISE 3: LONG DIVISION
2-DIGIT QUOTIENT

a. Open your textbook to Lesson 102 and find part 1. ✔
(Teacher reference:)

a. 18⟌679   b. 62⟌1736

b. Problem A: 679 ÷ 18.
- Copy the problem. Underline the digits for the first division problem you work. Work that problem. Then work the problem for the ones digit of the answer.
(Observe students and give feedback.)

c. Check your work.
- Say the first division problem you worked. (Signal.) *67 ÷ 18.*
- What's the first digit of the answer? (Signal.) *3.* You brought down the 9.
- Say the second division problem you worked. (Signal.) *139 ÷ 18.*
- What's the second digit of the answer? (Signal.) *7.*
- What's the answer to the whole problem? (Signal.) *37 and 13/18.*
(Display:) [102:3A]

$$
a. \quad 18\overline{)679} \; {}^{37\frac{13}{18}}
$$
$$
\begin{array}{r}
37\frac{13}{18} \\
18\overline{)679} \\
-54 \\
\hline
139 \\
-126 \\
\hline
13
\end{array}
$$

Here's what you should have.
d. Problem B: 1736 ÷ 62.
- Copy the problem. Underline the digits for the first division problem you work. Work that problem. Then work the problem for the ones digit of the answer.
(Observe students and give feedback.)
e. Check your work.
- Say the first division problem you worked. (Signal.) *173 ÷ 62.*
- What's the first digit of the answer? (Signal.) *2.* You brought down the 6.
- Say the second division problem you worked. (Signal.) *496 ÷ 62.*
- What's the second digit of the answer? (Signal.) *8.*
- What's the answer to the whole problem? (Signal.) *28.*
(Display:) [102:3B]

$$
\begin{array}{r}
28 \\
62\overline{)1736} \\
-124 \\
\hline
496 \\
-496 \\
\hline
0
\end{array}
$$

Here's what you should have.

EXERCISE 4: PROBABILITY

SPINNERS

a. (Display:)   [102:4A]

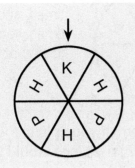

Here's a picture of a spinner. When you take a trial with a spinner, you spin the wheel. When the spinner stops, you look at the part the arrow points to. That's the winning part.

- What is the letter of the winning part in this picture? (Signal.) *K.*
  Yes, K.

b. The fraction for the winning part is based on the spinner.

- Raise your hand when you know how many equal parts this spinner has. ✔
- How many equal parts? (Signal.) *6.*
- How many parts have the letter K? (Signal.) *1.*
- So what's the probability fraction for the spinner stopping at K? (Signal.) *1/6.*

c. Look at the H's on the spinner. Raise your hand when you know the probability fraction for H. ✔

- What's the probability fraction for H? (Signal.) *3/6.*
- Raise your hand when you know the probability fraction for P. ✔
- What's the probability fraction for P? (Signal.) *2/6.*

d. Listen: Each time you spin the wheel, you take a trial.
  (Display:)   [102:4B]

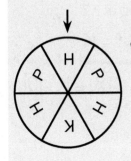

a. If you spin the wheel 30 times, what's the expected number for stopping at H?

Problem A: If you spin the wheel 30 times, what's the expected number for stopping at H?

- Write the probability equation for H in part A. ✔
- Everybody, read the probability equation for H. (Signal.) *3/6 T = H.*
- Put in the number the problem gives for trials and work the problem. Remember, each spin is a trial.
  (Observe students and give feedback.)

e. Check your work. You worked the problem 3/6 times 30.

- What's the number for H? (Signal.) *15.*
  So if you spin the wheel 30 times, you'd expect it to stop at H on 15 of those spins.
  (Add to show:)   [102:4C]

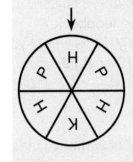

a. If you spin the wheel 30 times, what's the expected number for stopping at H?

$$\frac{3}{6} t = H$$

$$\frac{3}{\cancel{6}} (\cancel{30}) = H = \boxed{15}$$

Here's what you should have.

f. (Display:)   [102:4D]

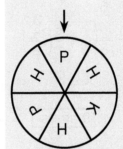

b. About how many times would you expect to spin the wheel for it to stop at P 3 times?

New problem: About how many times would you expect to spin the wheel for it to stop at P 3 times.

- Write the probability equation for P. ✔
- Everybody, read the probability equation for P. (Signal.) *2/6 T = P.*
  The problem asks: About how many times would you expect to spin the wheel for it to stop at P 3 times?
- Is the number 3 the number of times you spin the wheel or the number of times it stops at P? (Signal.) *The number of times it stops at P.*
- So is 3 the number for trials or the number for P? (Signal.) *The number for P.*
- Write the equation with a number for P. ✔

(Add to show:) [102:4E]

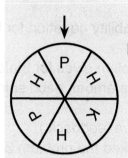

b. About how many times would you expect to spin the wheel for it to stop at P 3 times?

$$\frac{2}{6} t = p$$

$$\frac{2}{6} t = 3$$

Here are the equations you should have.

g. Now figure out the number of trials you'd expect to take for the wheel to stop at P 3 times. (Observe students and give feedback.)
• Everybody, about how many times would you expect to spin the wheel? (Signal.) *9 times.*
  (Add to show:) [102:4F]

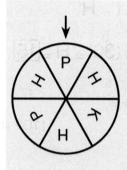

b. About how many times would you expect to spin the wheel for it to stop at P 3 times?

$$\frac{2}{6} t = p$$

$$\left(\frac{6}{2}\right) \frac{2}{6} t = 3\left(\frac{\overset{3}{\cancel{6}}}{\cancel{2}}\right)$$

$$t = 9$$

$$\boxed{9 \text{ times}}$$

Here's what you should have.

h. (Display:) [102:4G]

c. If you spin the wheel 18 times, about how many times would you expect it to stop at K?

New problem: If you spin the wheel 18 times, about how many times would you expect it to stop at K?
• Write the probability equation. Put in the number the problem gives. Then stop. (Observe students and give feedback.)
  (Add to show:) [102:4H]

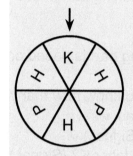

c. If you spin the wheel 18 times, about how many times would you expect it to stop at K?

$$\frac{1}{6} t = K$$

$$\frac{1}{6} (18) = K$$

Here are the equations you should have.

i. Now figure out the expected number for K. (Observe students and give feedback.)
• Everybody, if you spin the wheel 18 times, what's the expected number for stopping at K? (Signal.) *3.*
  (Add to show:) [102:4I]

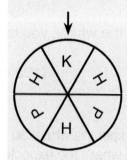

c. If you spin the wheel 18 times, about how many times would you expect it to stop at K?

$$\frac{1}{6} t = K$$

$$\frac{1}{\cancel{6}} (\cancel{18}\overset{3}{}) = K = \boxed{3}$$

Here's what you should have.

EXERCISE 5: MIXED-NUMBER OPERATIONS
WORD PROBLEMS

[REMEDY]

a. Find part 2 in your textbook. ✔
(Teacher reference:)

a. Tony finished the canoe race in $4\frac{1}{3}$ days. Henry finished the race in $6\frac{3}{4}$ days. How many days faster was Tony than Henry?

b. A fuel tank had $15\frac{5}{8}$ gallons of gas in it. $7\frac{9}{16}$ more gallons were poured into the tank. How many gallons ended up in the tank?

To work these problems, you have to rewrite the fractions so they have the same denominator.

b. Problem A: Tony finished the canoe race in 4 and 1/3 days. Henry finished the race in 6 and 3/4 days. How many days faster was Tony than Henry?

• Raise your hand when you know if you add or subtract to work the problem. ✔
• Do you add or subtract? (Signal.) *Subtract.*
• Work the problem.
(Observe students and give feedback.)
• Everybody, how many days faster was Tony than Henry? (Signal.) *2 and 5/12 days.*
(Display:) [102:5A]

$$a.\quad 6\frac{3}{4}\left(\frac{3}{3}\right)=\frac{9}{12}$$
$$-4\frac{1}{3}\left(\frac{4}{4}\right)=-\frac{4}{12}$$
$$\boxed{2\frac{5}{12}\text{ days}}$$

Here's what you should have.

c. Problem B: A fuel tank had 15 and 5/8 gallons of gas in it. 7 and 9/16 more gallons were poured into the tank. How many gallons ended up in the tank?

• Raise your hand when you know if you add or subtract to work the problem. ✔
• Do you add or subtract? (Signal.) *Add.*
• Work the problem.
(Observe students and give feedback.)
• Everybody, how many gallons ended up in the tank? (Signal.) *23 and 3/16 gallons.*

(Display:) [102:5B]

$$b.\quad 15\overset{1}{\frac{5}{8}}\left(\frac{2}{2}\right)=\frac{10}{16}$$
$$+\ 7\frac{9}{16}\qquad=+\frac{9}{16}$$
$$22\overset{1}{\frac{19}{16}}$$
$$\boxed{23\frac{3}{16}\text{ gallons}}$$

Here's what you should have.

EXERCISE 6: NUMBER NAMES
DECIMAL VALUES

a. (Display:) [102:6A]

> a. Five and three tenths
>
> b. Seventeen hundredths
>
> c. Two hundred four thousandths

These items describe decimal numbers. If you read each item correctly, you should be able to write the number.

b. (Point to **A.**) Raise your hand when you can say this number. ✔
• What number? (Signal.) *5 and 3 tenths.*
c. (Point to **B.**) Raise your hand when you can say this number. ✔
• What number? (Signal.) *17 hundredths.*
d. (Point to **C.**) Raise your hand when you can say this number. ✔
• What number? (Signal.) *2 hundred 4 thousandths.*
e. Write the three decimal numbers in part B. ✔
(Add to show:) [102:6B]

a. Five and three tenths	5.3
b. Seventeen hundredths	.17
c. Two hundred four thousandths	.204

Here's what you should have.

TEXTBOOK PRACTICE

a. Find part 3 in your textbook. ✔
(Teacher reference:)

a. One hundred thirteen thousandths

b. Eleven and five tenths

c. Sixty and ninety-nine hundredths

d. One and seven thousandths

e. Forty-five hundredths

Part 3	
a.	

These items describe the decimal values you'll write.

• Write decimal value A. ✔
Check your work. Item A says: 1 hundred 13 **thousandths.**
(Display:)   [102:6C]

> a. .113

Here's what you should have.

b. Write the rest of the decimal values.
(Observe students and give feedback.)

c. Check your work.
(Display:)   [102:6D]

> b. 11.5
>
> c. 60.99
>
> d. 1.007
>
> e. .45

Here's what you should have.

• Read B. (Signal.) *11 and 5 tenths.*
• Read C. (Signal.) *60 and 99 hundredths.*
• Read D. (Signal.) *1 and 7 thousandths.*
• Read E. (Signal.) *45 hundredths.*

EXERCISE 7: WORD PROBLEMS
PARTS OF A GROUP

a. Find part 4 in your textbook. ✔
(Teacher reference:)

a. 5 out of every 7 packages were frozen. There were 35 frozen packages. How many packages were there in all?

b. 2 out of every 9 bugs were spiders. There were 54 bugs. How many spiders were there?

c. 4 out of every 5 workers wore safety glasses. There were 100 workers. How many wore safety glasses?

d. 3 out of every 10 students wore jeans. 36 students wore jeans. How many students were there?

These are word problems that tell about a fraction.

b. Problem A: 5 out of every 7 packages were frozen. There were 35 frozen packages. How many packages were there in all?

• Write a fraction equation with two letters. Stop when you've done that much. ✔
• Everybody read the equation. (Signal.)
5/7 P = F.
• Work the problem and write the answer. ✔
• You worked the problem 35 times 7/5. What's the whole answer? (Signal.) *49 packages.*
(Display:)   [102:7A]

> a. $\dfrac{5}{7} p = f$
>
> $\left(\dfrac{7}{5}\right) \dfrac{5}{7} p = \overset{7}{\cancel{35}} \left(\dfrac{7}{5}\right)$
>
> $p = 49$
>
> 49 packages

Here's what you should have.

c. Problem B: 2 out of every 9 bugs were spiders. There were 54 bugs. How many spiders were there?
- Work problem B.
  (Observe students and give feedback.)
- Everybody, read the equation you started with. (Signal.) *2/9 B = S.*
- You worked the problem 2/9 times 54. What's the whole answer? (Signal.) *12 spiders.*
  (Display:)   [102:7B]

$$\text{b. } \frac{2}{9} b = s$$

$$\frac{2}{\cancel{9}} (\overset{6}{\cancel{54}}) = s = 12$$

| 12 spiders |

Here's what you should have.
d. Work the rest of the problems in part 4.
  (Observe students and give feedback.)
e. Check your work.
  Problem C: 4 out of every 5 workers wore safety glasses. There were 100 workers. How many wore safety glasses?
- Read the equation you started with. (Signal.) *4/5 W = S.*
- You solved for S. What does S equal? (Signal.) *80.*
- So how many workers wore safety glasses? (Signal.) *80 workers.*
f. Problem D: 3 out of every 10 students wore jeans. 36 students wore jeans. How many students were there?
- Read the equation you started with. (Signal.) *3/10 S = J.*
- How many students were there? (Signal.) *120 students.*

EXERCISE 8: INDEPENDENT WORK
FRACTION OPERATIONS—SIMPLIFY MULTIPLICATION

a. Find part 5 in your textbook. ✔
  (Teacher reference:)

a. $\frac{45}{100} \left(\frac{50}{9}\right) =$ b. $\frac{24}{80} \left(\frac{30}{21}\right) =$

c. $\frac{25}{100} \left(\frac{8}{10}\right) =$ d. $\frac{7}{4} \left(\frac{16}{28}\right) =$

For these problems, you'll simplify and then multiply. Remember, first cross out zeros in the numerator and denominator. Then see if other fractions can be simplified.
- Work problem A.
  (Observe students and give feedback.)
  Check your work. You started with 45/100 × 50/9. First you crossed out zeros in 100 and 50, and you simplified 45/10 and 9/9.
- The fraction answer is 5/2. What's the mixed-number answer? (Signal.) *2 and 1/2.*
  You'll work the rest of the problems as part of your independent work.

Assign Independent Work, the rest of Textbook part 5 and parts 6–12 and Workbook parts 2 and 3.

Optional extra math-fact practice worksheets are available on ConnectED.

Lesson 103

EXERCISE 1: MENTAL MATH

a. Time for some mental math. Tell me the answers.
- Listen: 80 plus 20. (Pause.) (Signal.) *100.*
- 80 + 24. (Pause.) (Signal.) *104.*
- 60 + 24. (Pause.) (Signal.) *84.*
- 60 + 28. (Pause.) (Signal.) *88.*
  (Repeat until firm.)

b. Listen: 70 plus 50. (Pause.) (Signal.) *120.*
- 70 + 56. (Pause.) (Signal.) *126.*
- 90 + 50. (Pause.) (Signal.) *140.*
- 90 + 56. (Pause.) (Signal.) *146.*
- 80 + 30. (Pause.) (Signal.) *110.*
  (Repeat until firm.)

c. Listen: 20 times some number equals 100. What number? (Signal.) *5.*
- 50 × some number = 100. What number? (Signal.) *2.*
- 25 × some number = 100. What number? (Signal.) *4.*
- 10 × some number = 100. What number? (Signal.) *10.*
  (Repeat until firm.)

d. Listen: 5 times some number equals 100. What number? (Signal.) *20.*
- 4 × some number = 100. What number? (Signal.) *25.*
- 2 × some number = 100. What number? (Signal.) *50.*
  (Repeat until firm.)

EXERCISE 2: LINE PLOT
AVERAGE `REMEDY`

a. (Display:) [103:2A]

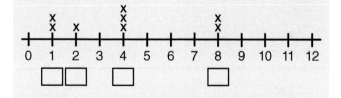

Here's a new kind of problem. It's called a line plot. It shows the Xs above some of the numbers.

b. We're going to find the average. The steps for finding the average are the same as the steps for finding the balance point of weights. First we find the totals for the columns. We add the totals. Then we divide by the number of Xs.
- (Point to **1**.) Listen: The Xs show how many you add. To find the total for the ones, we add 1 + 1.
- What's the total for the ones? (Signal.) *2.* (Add to show:) [103:2B]

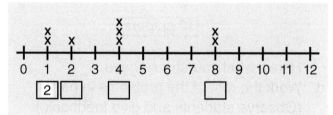

c. (Point to **2**.) What's the total for 2? (Signal.) *2.* (Add to show:) [103:2C]

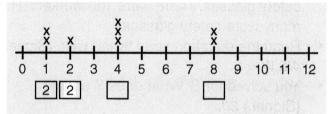

- (Point to **4**.) Say the addition for the fours. (Signal.) *4 + 4 + 4.*
- What's the total for 4? (Signal.) *12.* (Add to show:) [103:2D]

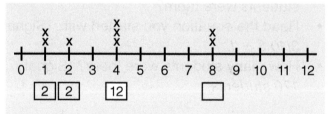

- (Point to **8**.) Say the addition for the eights. (Signal.) *8 + 8.*
- What's the total for 8? (Signal.) *16.* (Add to show:) [103:2E]

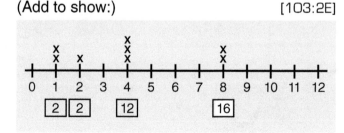

d. Add all the totals. Raise your hand when you know the answer. ✔
• Everybody, what's the total of the totals? (Signal.) *32.*
e. Count all the Xs to see what number we divide by. ✔
• What number do we divide by? (Signal.) *8.*
f. Work the problem.
(Observe students and give feedback.)
• Everybody, what's 32 divided by 8? (Signal.) *4.* So the average is 4.
g. Remember, to find the average, you add the totals for columns and divide by the number of Xs that are on the line plot.

━━━━━ **WORKBOOK PRACTICE** ━━━━━

a. Open your workbook to Lesson 103 and find part 1. ✔
(Teacher reference:) R Part E

b. Problem A. It's like the problem we just worked. Write the total under each number that has Xs. Stop when you've done that much. ✔
(Display:) [103:2F]

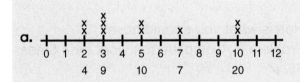

Here's what you should have.
• Now add the totals and divide by the number of Xs.
(Observe students and give feedback.)
• Everybody, what's the total for all the Xs? (Signal.) *50.*
• What did you divide by? (Signal.) *10.*
• What's the average? (Signal.) *5.*
c. Problem B. Write the total for each column and add. Count the number of Xs and figure out the average.
(Observe students and give feedback.)

(Display:) [103:2G]

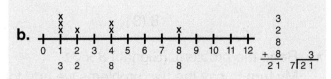

Here's what you should have.
There are 7 Xs. So you worked the problem 21 divided by 7.
• What's the average? (Signal.) *3.*

EXERCISE 3: DISTRIBUTIVE PROPERTY

a. When you work multiplication or division problems, you work problems for the tens and problems for the ones.
(Display:) [103:3A]

5 (32)

Here's 5 times 32.
We can write 32 as 30 plus 2.
(Add to show:) [103:3B]

5 (32)

5 (30 + 2)

Now we add 5×30 plus 5×2.
(Add to show:) [103:3C]

5 (32)

5 (30 + 2)

5 (30) + 5 (2)

• Say the two problems we add. (Signal.) *5 × 30 plus 5 × 2.*
• What's 5 times 30? (Signal.) *150.*
• What's 5 times 2? (Signal.) *10.*
• What's 150 plus 10? (Signal.) *160.*
(Add to show:) [103:3D]

5 (32)

5 (30 + 2)

5 (30) + 5 (2)

150 + 10 = 160

So 5 times 32 equals 160.

b. (Display:) [103:3E]

> 8 (31)

- Read the problem. (Signal.) *8 × 31.*
  My turn to say the two problems we add for 8 times 31: 8 × 30 + 8 × 1.
- Say the two problems we add. (Signal.) *8 × 30 + 8 × 1.*

c. (Display:) [103:3F]

> 4 (76)

- Read the problem. (Signal.) *4 × 76.*
- Say the two problems we add for 4 times 76. (Signal.) *4 × 70 + 4 × 6.*

d. (Display:) [103:3G]

> 3 (51)

- Read the problem. (Signal.) *3 × 51.*
- Say the two problems we add for 3 times 51. (Signal.) *3 × 50 + 3 × 1.*

e. (Display:) [103:3H]

> 9 (28)

- Read the problem. (Signal.) *9 × 28.*
- Say the two problems we add for 9 times 28. (Signal.) *9 × 20 + 9 × 8.*
  (Repeat until firm)

━━━━━━ **WORKBOOK PRACTICE** ━━━━━━

a. Find part 2 in your workbook. ✔
  (Teacher reference:)

  a. 3(54)
  ░░░ + ░░░
  ░░░ + ░░░ = ░░░

  b. 7(91)
  ░░░ + ░░░
  ░░░ + ░░░ = ░░░

  c. 4(85)
  ░░░ + ░░░
  ░░░ + ░░░ = ░░░

b. Problem A. You're going to complete the work for 3 times 54.
- Say the two problems you add. (Signal.) *3 × 50 + 3 × 4.*
- Write those problems below. ✔

(Display:) [103:3I]

> a. 3 (54)
> 3 (50) + 3 (4)

Here's what you should have: 3 × 50 + 3 × 4.
- Write the answers to the problems. Then complete the bottom equation. ✔
- Everybody, what's 3 times 50? (Signal.) *150.*
- What's 3 times 4? (Signal.) *12.*
- What's 150 plus 12? (Signal.) *162.*
(Display:) [103:3J]

> a. 3 (54)
> 3 (50) + 3 (4)
> 150 + 12 = 162

Here's what you should have.

c. Problem B. You're going to complete the work for 7 times 91.
- Say the two problems you add. (Signal.) *7 × 90 + 7 × 1.*
- Write and work those problems below. Then complete the bottom equation. ✔
- Everybody, what's 7 times 90? (Signal.) *630.*
- What's 7 times 1? (Signal.) *7.*
- What's 630 plus 7? (Signal.) *637.*
(Display:) [103:3K]

> b. 7 (91)
> 7 (90) + 7 (1)
> 630 + 7 = 637

Here's what you should have.

d. Problem C. Complete the work for 4 times 85. ✔
- Everybody, what's 4 times 80? (Signal.) *320.*
- What's 4 times 5? (Signal.) *20.*
- What's 320 plus 20? (Signal.) *340.*
(Display:) [103:3L]

> c. 4 (85)
> 4 (80) + 4 (5)
> 320 + 20 = 340

Here's what you should have.

Exercise 4: Probability

Spinners

a. Open your textbook to Lesson 103 and find part 1. ✔
    (Teacher reference:)

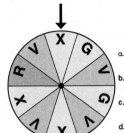

a. If you spin the wheel 20 times, what's the expected number for the spinner stopping at X?
b. How many times would you expect to spin the wheel until it stops at V two times?
c. If you spin the wheel 40 times, what's the expected number for the spinner stopping at R?
d. How many times would you expect to spin the wheel until it stops at G three times?

• Raise your hand when you know the probability fraction for G. ✔
• What's the probability fraction for G? (Signal.) *2/10.*

b. Problem A: If you spin the wheel 20 times, what's the expected number for the spinner stopping at X?
• Work the problem.
    (Observe students and give feedback.)
    You started with the probability equation 3/10 T = X.
• If you spin the spinner 20 times, what's the expected number for stopping at X? (Signal.) *6.*
    (Display:)   [103:4A]

$$\text{a.} \quad \frac{3}{10} t = X$$
$$\frac{3}{10}(20) = X = \boxed{6}$$

Here's what you should have.

c. Work problem B.
    (Observe students and give feedback.)
• Problem B. Read the equation you started with. (Signal.) *4/10 T = V.*
• You multiplied both sides by 10/4. How many times would you expect to spin the wheel until it stops at V 2 times? (Signal.) *5 times.*

(Display:)   [103:4B]

$$\text{b.} \quad \frac{4}{10} t = V$$
$$\left(\frac{10}{4}\right)\frac{4}{10} t = 2\left(\frac{10}{4}\right)$$
$$t = 5$$

$$\boxed{5 \text{ times}}$$

Here's what you should have.

d. Work the rest of the problems.
    (Observe students and give feedback.)

e. Check your work.
    Problem C: If you spin the wheel 40 times, what's the expected number for the spinner stopping at R?
• What's the answer? (Signal.) *4.*
    (Display:)   [103:4C]

$$\text{c.} \quad \frac{1}{10} t = R$$
$$\frac{1}{10}(40) = R = \boxed{4}$$

Here's what you should have.

f. Problem D: How many times would you expect to spin the wheel until it stops at G 3 times?
• What's the answer? (Signal.) *15 times.*
    (Display:)   [103:4D]

$$\text{d.} \quad \frac{2}{10} t = G$$
$$\left(\frac{10}{2}\right)\frac{2}{10} t = 3\left(\frac{10}{2}\right)$$
$$t = 15$$

$$\boxed{15 \text{ times}}$$

Here's what you should have.

EXERCISE 5: NUMBER NAMES
DECIMAL VALUES

a. Find part 2 in your textbook. ✔
(Teacher reference:)

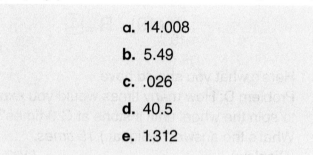

a. Fourteen and eight thousandths
b. Five and forty-nine hundredths
c. Twenty-six thousandths
d. Forty and five tenths
e. One and three hundred twelve thousandths

The decimal numbers that you'll write are shown with words.

b. Raise your hand when you can say number A. ✔
• What's number A? (Signal.) *14 and 8 thousandths.*

c. Raise your hand when you can say number B. ✔
• What's number B? (Signal.) *5 and 49 hundredths.*

d. Write the decimal values for all the items. (Observe students and give feedback.)

e. Check your work.
(Display:) [103:5A]

> a. 14.008
>
> b. 5.49
>
> c. .026
>
> d. 40.5
>
> e. 1.312

Here's what you should have.

f. Read each number.
• (Point to **A.**) Get ready. (Signal.) *14 and 8 thousandths.*
• (Point to **B.**) Get ready. (Signal.) *5 and 49 hundredths.*
• (Point to **C.**) Get ready. (Signal.) *26 thousandths.*
• (Point to **D.**) Get ready. (Signal.) *40 and 5 tenths.*
• (Point to **E.**) Get ready. (Signal.) *1 and 312 thousandths.*

EXERCISE 6: FRACTIONS
FOR MIXED NUMBERS

a. (Display:) [103:6A]

$$2\frac{1}{5} =$$

I'll show you a fast way to change mixed numbers into fractions that are more than 1. You write the denominator of the fraction.
• What's the denominator? (Signal.) *5.*
(Add to show:) [103:6B]

$$2\frac{1}{5} = \frac{}{5}$$

Now you multiply and add.

b. You multiply 5 times 2, then add 1.
• What do you multiply? (Signal.) *5 × 2.*
• Then what do you add? (Signal.) *1.*
(Repeat until firm.)

• What's 5 times 2? (Signal.) *10.*
• What do you add to 10? (Signal.) *1.*
• What's 10 plus 1? (Signal.) *11.*
(Repeat until firm.)
So the fraction is 11/5.
(Add to show:) [103:6C]

$$2\frac{1}{5} = \frac{11}{5}$$

• Read the fraction that equals 2 and 1/5. (Signal.) *11/5.*

c. (Display:) [103:6D]

$$3\frac{5}{9} =$$

- New mixed number. Read it. (Signal.) *3 and 5/9.*
  You're going to say the multiplication, then the addition.
- What's the denominator? (Signal.) *9.*
  (Repeat until firm.)
  (Add to show:)   [103:6E]

$$3\frac{5}{9} = \frac{}{9}$$

- Say the multiplication. (Signal.) *9 × 3.*
- What's the answer? (Signal.) *27.*
- Say the addition. (Signal.) *27 + 5.*
  (Repeat until firm.)
- Raise your hand when you know the answer. ✔
- What's 27 plus 5? (Signal.) *32.*
  (Add to show:)   [103:6F]

$$3\frac{5}{9} = \frac{32}{9}$$

- Read the fraction that equals 3 and 5/9.
  (Signal.) *32/9.*
  Remember, start with the denominator of the fraction and say the multiplication; then say the addition.
- d. (Display:)   [103:6G]

$$6\frac{2}{3} =$$

- New mixed number. Read it. (Signal.) *6 and 2/3.*
- What's the denominator? (Signal.) *3.*
  (Repeat until firm.)
  (Add to show:)   [103:6H]

$$6\frac{2}{3} = \frac{}{3}$$

- Say the multiplication. (Signal.) *3 × 6.*
- What's the answer? (Signal.) *18.*
- Say the addition. (Signal.) *18 + 2.*
- What's 18 plus 2? (Signal.) *20.*
  (Repeat until firm.)
  (Add to show:)   [103:6I]

$$6\frac{2}{3} = \frac{20}{3}$$

- Read the fraction that equals 6 and 2/3.
  (Signal.) *20/3.*
- e. (Display:)   [103:6J]

$$2\frac{7}{10} =$$

- New mixed number. Read it. (Signal.) *2 and 7/10.*
- What's the denominator? (Signal.) *10.*
  (Repeat until firm.)
  (Add to show:)   [103:6K]

$$2\frac{7}{10} = \frac{}{10}$$

- Say the multiplication. (Signal.) *10 × 2.*
- What's the answer? (Signal.) *20.*
- Say the addition. (Signal.) *20 + 7.*
- What's 20 plus 7? (Signal.) *27.*
  (Repeat until firm.)
  (Add to show:)   [103:6L]

$$2\frac{7}{10} = \frac{27}{10}$$

- Read the fraction that equals 2 and 7/10.
  (Signal.) *27/10.*
  Remember, start with the denominator of the fraction and say the multiplication; then say the addition.

EXERCISE 7: DECIMAL MULTIPLICATION/DIVISION
MULTIPLES OF 10

a. Find part 3 in your textbook. ✔
   (Teacher reference:)

a.	.725 (Multiply by 100.)	d.	52.18 (Multiply by 10.)
b.	115.8 (Divide by 10.)	e.	.031 (Multiply by 100.)
c.	7.025 (Multiply by 1000.)	f.	35.4 (Divide by 100.)

You're going to multiply or divide decimal values by 10, 100, or 1000. You'll write the new value.

b. Problem A. Read the decimal number.
   (Signal.) *725 thousandths.*
   • Read the directions. (Signal.) *Multiply by 100.*
   • Figure out where the new decimal point will be and write the number. ✔
   • Everybody, what's 725 thousandths times 100? (Signal.) *72 and 5/10.*
   (Display:)   [103:7A]

> **a.** 72.5

Here's what you should have.
You moved the decimal point 2 places to the right.

c. Work problem B. ✔
   • Everybody, what's 115 and 8 tenths divided by 10? (Signal.) *11 and 58 hundredths.*
   You moved the decimal point one place to the left.

d. Work problem C. ✔
   • Everybody, what's 7 and 25 thousandths times 1000? (Signal.) *7025.*
   You moved the decimal point 3 places to the right.

e. Work problem D. ✔
   • Everybody, what's 52 and 18 hundredths times 10? (Signal.) *521 and 8 tenths.*
   You moved the decimal point one place to the right.

f. Work problem E. ✔
   • Everybody, what's 31 thousandths times 100? (Signal.) *3 and 1 tenth.*
   You moved the decimal point 2 places to the right.

g. Work problem F. ✔
   • Everybody, what's 35 and 4 tenths divided by 100? (Signal.) *354 thousandths.*
   You moved the decimal point 2 places to the left.

Assign Independent Work, Textbook parts 4–13.

Optional extra math-fact practice worksheets are available on ConnectED.

Lesson 104

EXERCISE 1: MENTAL MATH

a. Time for some mental math. Tell me the answers.
- 50 times some number equals 100. What number? (Signal.) *2.*
- 10 × some number = 100. What number? (Signal.) *10.*
- 25 × some number = 100. What number? (Signal.) *4.*
- 20 × some number = 100. What number? (Signal.) *5.*
  (Repeat until firm.)

b. Listen: 2 times some number equals 100. What number? (Signal.) *50.*
- 4 × some number = 100. What number? (Signal.) *25.*
- 5 × some number = 100. What number? (Signal.) *20.*
  (Repeat until firm.)

c. Listen: 50 plus 20. (Pause.) (Signal.) *70.*
- 50 + 24. (Pause.) (Signal.) *74.*
- 70 + 24. (Pause.) (Signal.) *94.*
- 70 + 28. (Pause.) (Signal.) *98.*
- 30 + 50. (Pause.) (Signal.) *80.*
- 30 + 56. (Pause.) (Signal.) *86.*
  (Repeat until firm.)

d. Listen: 80 plus 50. (Pause.) (Signal.) *130.*
- 80 + 56. (Pause.) (Signal.) *136.*
- 130 + 30. (Pause.) (Signal.) *160.*
- 130 + 34. (Pause.) (Signal.) *164.*
- 130 + 44. (Pause.) (Signal.) *174.*
  (Repeat until firm.)

e. Listen: 110 plus 70. (Pause.) (Signal.) *180.*
- 110 + 77. (Pause.) (Signal.) *187.*
- 120 + 77. (Pause.) (Signal.) *197.*
  (Repeat until firm.)

EXERCISE 2: LINE PLOT
COLUMN TOTALS

a. (Display:) [104:2A]

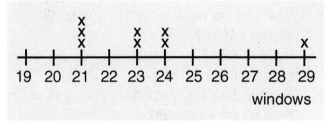

This line plot shows numbers for windows. (Point to **windows.**) So you find the total number of windows.

b. The Xs show how many you add.
- (Point to **24.**) What's this number? (Signal.) *24.*
- How many 24s do you add in this column? (Signal.) *2.*
  (Add to show:) [104:2B]

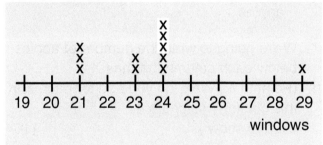

- How many 24s do you add now? (Signal.) *5.* So you work the problem: 24 + 24 + 24 + 24 + 24.
- Say the addition for this number. (Signal.) *24 + 24 + 24 + 24 + 24.*

c. (Point to **21.**) What's this number? (Signal.) *21.*
- Say the addition for this column. (Signal.) *21 + 21 + 21.*

d. (Point to **29.**) What's this number? (Signal.) *29.*
- Say the total for this column. (Signal.) *29.*

e. (Point to **23.**) What's this number? (Signal.) *23.*
- Say the addition for this column. (Signal.) *23 + 23.*

f. (Display:)   [104:2C]

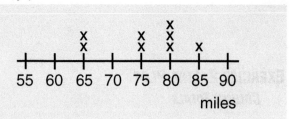

The numbers on this line plot are for miles.
- What unit do we add for this line plot?
  (Signal.) *Miles.*
- (Point to **65.**) Say the addition for this column.
  (Signal.) *65 + 65.*
- (Point to **80.**) Say the addition for this column.
  (Signal.) *80 + 80 + 80.*
- (Point to **75.**) Say the addition for this column.
  (Signal.) *75 + 75.*

g. (Display:)   [104:2D]

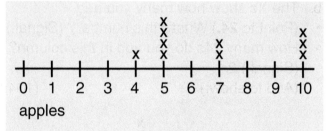

We're going to write the number of apples
below each column that has Xs.

h. (Point to **4.**) What do you add for each X in
   this column? (Signal.) *4.*
   (Add to show:)   [104:2E]

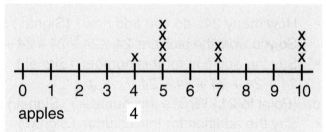

i. (Point to **5.**) Say the addition for this column.
   (Signal.) *5 + 5 + 5 + 5.*
- What's the number of apples for this column?
  (Signal.) *20.*
  (Add to show:)   [104:2F]

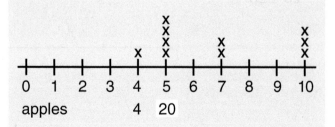

j. (Point to **7.**) Say the addition for this column.
   (Signal.) *7 + 7.*
- What's the number of apples for this column?
  (Signal.) *14.*
  (Add to show:)   [104:2G]

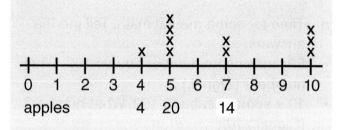

k. (Point to **10.**) Say the addition problem.
   (Signal.) *10 + 10 + 10.*
- What's the number of apples for this column?
  (Signal.) *30.*
  (Add to show:)   [104: 2H]

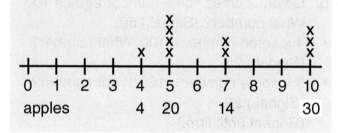

======================= **WORKBOOK PRACTICE** =======================

a. Open your workbook to Lesson 104 and find
   part 1. ✔
   (Teacher reference:)

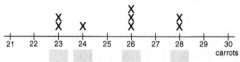

- Everybody, what do the numbers on this line
  plot tell about? (Signal.) *Carrots.*
  The Xs are shown for the columns. You'll write
  numbers below columns that have Xs.

b. Touch 23. ✔
- Say the addition for that column. (Signal.)
  23 + 23.
- Touch 26. ✔
- Say the addition for that column. (Signal.)
  26 + 26 + 26.
- Figure out the total number of carrots for each
  column. Write the totals in the boxes.
  (Observe students and give feedback.)

c. Check your work.
- Say the addition for 23 carrots. (Signal.)
  23 + 23.
- What's the total? (Signal.) *46.*

d. What's the total for 24 carrots? (Signal.) *24.*
• Say the addition for 26 carrots. (Signal.)
  26 + 26 + 26.
• What's the total? (Signal.) *78.*
e. Say the addition for 28 carrots. (Signal.)
  28 + 28.
• What's the total? (Signal.) *56.*

EXERCISE 3: DISTRIBUTIVE PROPERTY

a. You're going to work problems that multiply tens and ones.
  (Display:)   [104:3A]

  4 (89)

• Read the problem. (Signal.) *4 × 89.*
  The problems you add are: 4 × 80 plus 4 × 9.
• Say the problems you add. (Signal.)
  4 × 80 + 4 × 9.
b. (Display:)   [104:3B]

  5 (32)

• Say the problems you add. (Signal.)
  5 × 30 + 5 × 2.
c. (Display:)   [104:3C]

  7 (16)

• Say the problems you add. (Signal.)
  7 × 10 + 7 × 6.

━━━━━━━ WORKBOOK PRACTICE ━━━━━━━

a. Find part 2 in your workbook. ✔
  (Teacher reference:)

  a. 5(68)
     ▢ + ▢
     ▢ + ▢ = ▢

  b. 7(31)
     ▢ + ▢
     ▢ + ▢ = ▢

  c. 3(68)
     ▢ + ▢
     ▢ + ▢ = ▢

You're going to show the multiplication problems you add. Then you'll complete the bottom equation.

b. Read problem A. (Signal.) *5 × 68.*
• Say the problems you add. (Signal.)
  5 × 60 + 5 × 8.
• Write the problems. Then stop.
  (Observe students and give feedback.)
  (Display:)   [104:3D]

  a. 5 (68)
     5 (60) + 5 (8)

Here's what you should have.
• Complete the bottom equation. Write what 5 × 60 and 5 × 8 equal. Then add and write the answer.
  (Observe students and give feedback.)
c. Check your work.
• What's 5 times 60? (Signal.) *300.*
• What's 5 times 8? (Signal.) *40.*
• What's 300 plus 40? (Signal.) *340.*
  (Add to show:)   [104:3E]

  a. 5 (68)
     5 (60) + 5 (8)
     300 + 40 = 340

Here's what you should have.
d. Work problem B.
  (Observe students and give feedback.)
e. Check your work.
  You worked the problems: 7 × 30 plus 7 × 1.
• What's 7 times 30? (Signal.) *210.*
• What's 7 times 1? (Signal.) *7.*
• What's 210 plus 7? (Signal.) *217.*
  (Display:)   [104:3F]

  b. 7 (31)
     7 (30) + 7 (1)
     210 + 7 = 217

Here's what you should have.
f. Work problem C.
  (Observe students and give feedback.)

g. Check your work.
You worked the problems: 3 × 60 plus 3 × 8.
- What's 3 times 60? (Signal.) *180.*
- What's 3 times 8? (Signal.) *24.*
- What's 180 plus 24? (Signal.) *204.*
(Display:)   [104:3G]

> **c.** 3 (68)
>
> 3 (60) + 3 (8)
>
> 180 + 24 = 204

Here's what you should have.

EXERCISE 4: PROBABILITY

SPINNERS WITH PERCENTS

`REMEDY`

a. (Display:)   [104:4A]

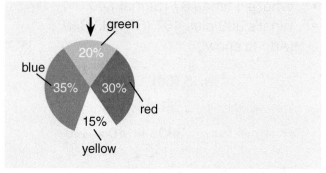

This wheel shows percents. When you work
spinner problems, you write the percents as
fractions.

b. What percent of this wheel is green?
(Signal.) *20%.*
- What's the fraction for 20 percent?
(Signal.) *20/100.*
(Add to show:)   [104:4B]

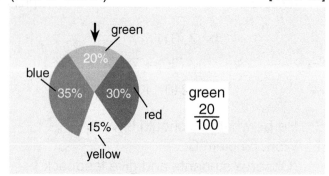

c. Say the probability equation for the spinner
stopping at green. (Signal.) *20/100 T = G.*
- Say the probability equation for the spinner
stopping at yellow. (Signal.) *15/100 T = Y.*
- Say the probability equation for the spinner
stopping at red. (Signal.) *30/100 T = R.*
- Say the probability equation for the spinner
stopping at blue. (Signal.) *35/100 T = B.*
(Repeat until firm.)

d. (Add to show:)   [104:4C]

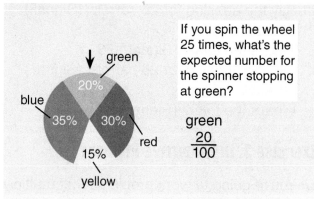

Here's a problem: If you spin the wheel 25 times,
what's the expected number for the spinner
stopping at green?
- In part A, write the probability equation for the
spinner stopping at green. ✔
(Add to show:)   [104:4D]

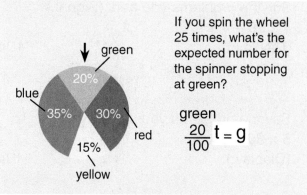

Here's what you should have: 20/100 T = G.

e. Work the problem. Remember to simplify the
fractions. ✔
You started with the equation: 20/100 T = G.
You replaced T with 25 and simplified the
fractions.
- What's the answer? (Signal.) *5.*
(Add to show:)   [104:4E]

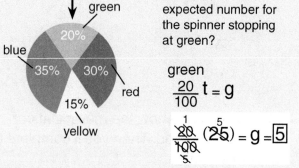

So if you spin the wheel 25 times, you would
expect it to stop at green on 5 of those spins.

f. (Change to show:) [104:4F]

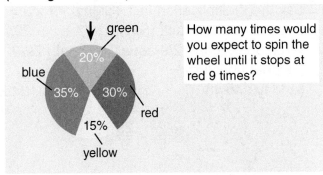

How many times would you expect to spin the wheel until it stops at red 9 times?

New problem: How many times would you expect to spin the wheel until it stops at red 9 times?

- Write the equation. Put in a number for one of the letters. Stop when you've done that much. (Observe students and give feedback.)
- Everybody, read the equation you started with. (Signal.) *30/100 T = R.*
  You replaced R with 9.
- Work the problem. Remember to simplify the fractions.
  (Observe students and give feedback.)
  You figured out how many trials you'd expect to take. That's the number of times you'd spin the wheel.
- What's the answer? (Signal.) *30 times.*

g. (Change to show:) [104:4G]

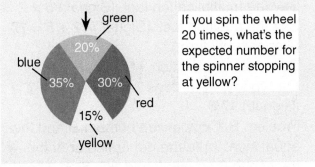

If you spin the wheel 20 times, what's the expected number for the spinner stopping at yellow?

New problem: If you spin the wheel 20 times, what's the expected number for stopping at yellow?

- Write the probability equation. Replace one of the letters with a number. Work the problem. Remember to simplify the fractions.
  (Observe students and give feedback.)
  You started with the equation: 15/100 T = Y.
  You replaced T with 20.
- You figured out the expected number for the spinner to stop at yellow. What's the answer? (Signal.) *3.*
  So if you took 20 trials, you'd expect the spinner to stop at yellow on 3 of those trials.

EXERCISE 5: NUMBER NAMES
DECIMAL VALUES

a. Open your textbook to Lesson 104 and find part 1. ✔
  (Teacher reference:)

  a. Three and five hundredths
  b. Three and five tenths
  c. Three hundred four and five tenths
  d. Thirteen thousandths
  e. Fourteen hundredths

Part 1	
a.	

  The decimal numbers that you'll write are shown with words.

b. Raise your hand when you can say number A. ✔
- What's number A? (Signal.) *3 and 5 hundredths.*
c. Raise your hand when you can say number B. ✔
- What's number B? (Signal.) *3 and 5 tenths.*
d. Write the decimal values for all the items.
  (Observe students and give feedback.)
e. Check your work.
  (Display:) [104:5A]

  a. 3.05
  b. 3.5
  c. 304.5
  d. .013
  e. .14

  Here's what you should have.
f. Read each number.
- (Point to **A.**) Get ready. (Signal.) *3 and 5 hundredths.*
- (Point to **B.**) Get ready. (Signal.) *3 and 5 tenths.*
- (Point to **C.**) Get ready. (Signal.) *304 and 5 tenths.*
- (Point to **D.**) Get ready. (Signal.) *13 thousandths.*
- (Point to **E.**) Get ready. (Signal.) *14 hundredths.*

EXERCISE 6: FRACTIONS
FOR MIXED NUMBERS

REMEDY

a. (Display:) [104:6A]

$$4\frac{1}{9} =$$

You're going to change mixed numbers into fractions.

- Read the mixed number. (Signal.) *4 and 1/9.* Remember, start with the denominator of the fraction and say the multiplication; then say the addition.
- What's the denominator of the fraction? (Signal.) *9.*
(Add to show:) [104:6B]

$$4\frac{1}{9} = \frac{}{9}$$

- Say the multiplication fact. (Signal.) *9 × 4 = 36.*
- Say the addition fact. (Signal.) *36 + 1 = 37.*
(Repeat until firm.)
(Add to show:) [104:6C]

$$4\frac{1}{9} = \frac{37}{9}$$

- What does 4 and 1/9 equal? (Signal.) *37/9.*
b. (Display:) [104:6D]

$$10\frac{3}{7} =$$

- New mixed number. Read it. (Signal.) *10 and 3/7.*
- Say the multiplication fact. (Signal.) *7 × 10 = 70.*
- Say the addition fact. (Signal.) *70 + 3 = 73.*
- What does 10 and 3/7 equal? (Signal.) *73/7.*
(Repeat until firm.)
(Add to show:) [104:6E]

$$10\frac{3}{7} = \frac{73}{7}$$

Yes, 73/7.
c. (Display:) [104:6F]

$$3\frac{2}{9} =$$

- New mixed number. Read it. (Signal.) *3 and 2/9.*
- Say the multiplication fact. (Signal.) *9 × 3 = 27.*
- Say the addition fact. (Signal.) *27 + 2 = 29.*
- What does 3 and 2/9 equal? (Signal.) *29/9.*
(Repeat until firm.)
(Add to show:) [104:6G]

$$3\frac{2}{9} = \frac{29}{9}$$

Yes, 29/9.

════════ **TEXTBOOK PRACTICE** ════════

a. Find part 2 in your textbook. ✔
(Teacher reference:)

a. $2\frac{5}{6} = \blacksquare$ b. $5\frac{1}{8} = \blacksquare$ c. $4\frac{4}{5} = \blacksquare$
d. $3\frac{2}{3} = \blacksquare$ e. $10\frac{1}{8} = \blacksquare$

b. Problem A. Copy mixed number A and the equal sign. Write the denominator of the fraction that equals 2 and 5/6. Stop when you've done that much. ✔
(Display:) [104:6H]

$$a. \ 2\frac{5}{6} = \frac{}{6}$$

Here's what you should have.
- Say the multiplication fact. (Signal.) *6 × 2 = 12.*
- Say the addition fact. (Signal.) *12 + 5 = 17.*
(Repeat until firm.)
- Complete the fraction. ✔
- Everybody, what does 2 and 5/6 equal? (Signal.) *17/6.*
c. Problem B. Copy mixed number B and the equal sign. Write the denominator of the fraction that equals 5 and 1/8. Stop when you've done that much. ✔
(Display:) [104:6I]

$$b. \ 5\frac{1}{8} = \frac{}{8}$$

Here's what you should have.
- Say the multiplication fact. (Signal.) *8 × 5 = 40.*
- Say the addition fact. (Signal.) *40 + 1 = 41.*
- Complete the fraction. ✔
- Everybody, what does 5 and 1/8 equal? (Signal.) *41/8.*

d. Write equations for the rest of the items in part 2. (Observe students and give feedback.)

e. Check your work.

- Problem C: 4 and 4/5.
- Say the multiplication fact. (Signal.) *5 × 4 = 20.*
- Say the addition fact. (Signal.) *20 + 4 = 24.*
- What does 4 and 4/5 equal? (Signal.) *24/5.*

f. Problem D: 3 and 2/3.

- Say the multiplication fact. (Signal.) *3 × 3 = 9.*
- Say the addition fact. (Signal.) *9 + 2 = 11.*
- What does 3 and 2/3 equal? (Signal.) *11/3.*

g. Problem E: 10 and 1/8.

- Say the multiplication fact. (Signal.) *8 × 10 = 80.*
- Say the addition fact. (Signal.) *80 + 1 = 81.*
- What does 10 and 1/8 equal? (Signal.) *81/8.*

EXERCISE 7: DECIMAL MULTIPLICATION/DIVISION
MULTIPLES OF 10

a. (Display:)   [104:7A]

> .0 5 2   (Multiply by 100.)
>
> .6 1   (Divide by 10.)
>
> 8.3   (Multiply by 1000.)

You're going to multiply or divide decimal values by 10, 100, or 1000. You'll write the new values.

For some numbers, you'll write zeros.

b. (Point to **.052**) Read the decimal number. (Signal.) *52 thousandths.*

- Read the direction. (Signal.) *Multiply by 100.*
- Raise your hand when you know the new decimal number. ✔
- What's the new decimal number? (Signal.) *5 and 2 tenths.*

Yes, 5 and 2 tenths. We don't write a zero in front of the 5.

(Add to show:)   [104:7B]

> .0 5 2   (Multiply by 100.)
> 5.2
>
> .6 1   (Divide by 10.)
>
> 8.3   (Multiply by 1000.)

c. (Point to **.61**) Read the decimal number. (Signal.) *61 hundredths.*

- Read the direction. (Signal.) *Divide by 10.*
- Raise your hand when you know the new decimal number. ✔
- What's the new decimal number? (Signal.) *61 thousandths.*

I have to write a zero in front of the 6.

(Add to show:)   [104:7C]

> .0 5 2   (Multiply by 100.)
> 5.2
>
> .6 1   (Divide by 10.)
> .0 6 1
>
> 8.3   (Multiply by 1000.)

d. (Point to **8.3**) Read the decimal number. (Signal.) *8 and 3 tenths.*

- Read the direction. (Signal.) *Multiply by 1000.*
- Raise your hand when you know the new number. ✔
- What's the new number? (Signal.) *8300.*

I have to write two zeros.

(Add to show:)   [104:7D]

> .0 5 2   (Multiply by 100.)
> 5.2
>
> .6 1   (Divide by 10.)
> .0 6 1
>
> 8.3   (Multiply by 1000.)
> 8 3 0 0

a. Find part 3 in your textbook. ✔
(Teacher reference:)

a. 5.2 (Multiply by 100.)   d. .78 (Multiply by 1000.)

b. .3 (Divide by 100.)   e. 8.34 (Multiply by 10.)

c. .78 (Divide by 10.)   f. 8.34 (Divide by 100.)

b. Problem A. Read the decimal number.
(Signal.) *5 and 2 tenths.*
• Read the direction. (Signal.) *Multiply by 100.*
• Figure out where the new decimal point will be
and write the number. ✔
• Everybody, what's 5 and 2 tenths times 100?
(Signal.) *520.*
(Display:)   [104:7E]

> **a. 520**

Here's what you should have.
You moved the decimal point 2 places to the
right and wrote a zero.

c. Work problem B. ✔
• Everybody, what's 3 tenths divided by 100?
(Signal.) *3 thousandths.*
(Display:)   [104:7F]

> **b. .003**

Here's what you should have.
You moved the decimal point 2 places to the
left and wrote 2 zeros.

d. Work problem C. ✔
• Everybody, what's 78 hundredths divided by
10? (Signal.) *78 thousandths.*
(Display:)   [104:7G]

> **c. .078**

Here's what you should have.
You moved the decimal point 1 place to the
left and wrote a zero.

e. Work problem D. ✔
• Everybody, what's 78 hundredths multiplied by
1000? (Signal.) *780.*
(Display:)   [104:7H]

> **d. 780**

Here's what you should have.
You moved the decimal point 3 places to the
right and wrote a zero.

f. Work problem E. ✔
• Everybody, what's 8 and 34 hundredths
multiplied by 10? (Signal.) *83 and 4 tenths.*
(Display:)   [104:7I]

> **e. 83.4**

Here's what you should have.
You moved the decimal point 1 place to the
right.

g. Work problem F. ✔
(Display:)   [102:7J]

> **f. .0834**

Here's what you should have.
You moved the decimal point 2 places to the
left and wrote a zero.
The answer is 834 ten-thousandths.

> Assign Independent Work, Textbook parts 4–10
> and Workbook parts 3 and 4.

Optional extra math-fact practice worksheets
are available on ConnectED.

Lesson 105

EXERCISE 1: MENTAL MATH

a. Time for some mental math. Tell me the answers.

- 10 times some number equals 100. What number? (Signal.) *10.*
- 50 × some number = 100. What number? (Signal.) *2.*
- 25 × some number = 100. What number? (Signal.) *4.*
- 20 × some number = 100. What number? (Signal.) *5.*
- (Repeat until firm.)

b. Listen: 5 times some number equals 100. What number? (Signal.) *20.*

- 4 × some number = 100. What number? (Signal.) *25.*
- 2 × some number = 100. What number? (Signal.) *50.*
- (Repeat until firm.)

c. Listen: 90 plus 20. (Pause.) (Signal.) *110.*

- 90 + 24. (Pause.) (Signal.) *114.*
- 30 + 24. (Pause.) (Signal.) *54.*
- 80 + 28. (Pause.) (Signal.) *108.*
- 80 + 50. (Pause.) (Signal.) *130.*
- 80 + 56. (Pause.) (Signal.) *136.*
- (Repeat until firm.)

d. Listen: 60 plus 50. (Pause.) (Signal.) *110.*

- 60 + 56. (Pause.) (Signal.) *116.*
- 160 + 30. (Pause.) (Signal.) *190.*
- 160 + 34. (Pause.) (Signal.) *194.*
- 160 + 44. (Pause.) (Signal.) *204.*
- (Repeat until firm.)

e. Listen: 320 plus 70. (Pause.) (Signal.) *390.*

- 320 + 77. (Pause.) (Signal.) *397.*
- 330 + 77. (Pause.) (Signal.) *407.*
- (Repeat until firm.)

EXERCISE 2: LINE PLOT
COLUMN TOTALS

a. (Display:)   [105:2A]

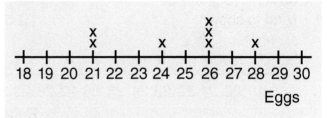

- What's the name under this line plot? (Signal.) *Eggs.*
- So what do the numbers tell about? (Signal.) *Eggs.*

b. (Point to **26**.) Say the addition for this column. (Signal.) *26 + 26 + 26.*

- (Point to **21**.) Say the addition for this column. (Signal.) *21 + 21.*
- (Repeat until firm.)

Remember, the numbers tell the number of eggs you add.

c. (Change to show:)   [105:2B]

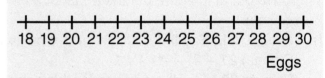

Listen: We're going to show eggs that are in baskets. Some baskets have 21 eggs. Some baskets have 23 eggs.

I'll show a B for each basket. You'll tell me how many eggs are in that basket.

d. (Add to show:)   [105:2C]

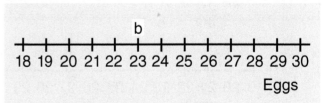

- What's the number for this column? (Signal.) *23.*
- So how many eggs are in this basket? (Signal.) *23.*

e. (Add to show:) [105:2D]

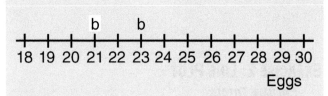

- (Point to **21**.) How many eggs are in this basket? (Signal.) *21.*

f. (Add to show:) [105:2E]

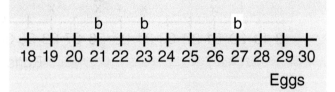

- (Point to **27**.) How many eggs are in this basket? (Signal.) *27.*

g. (Add to show:) [105:2F]

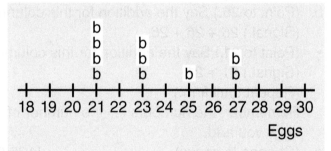

You're going to figure out the eggs for all the baskets.

h. (Point to **21**.) Say the addition for this column. (Signal.) *21 + 21 + 21 + 21.*

i. (Point to **23**.) Say the addition for this column. (Signal.) *23 + 23 + 23.*

j. (Point to **21**.) In part A, figure out the number of eggs in this column. ✔

- Everybody, how many eggs? (Signal.) *84.*
 (Add to show:) [105:2G]

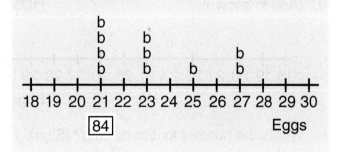

k. (Point to **23**.) Figure out the number of eggs in this column. ✔

- Everybody, how many eggs? (Signal.) *69.*
 (Add to show:) [105:2H]

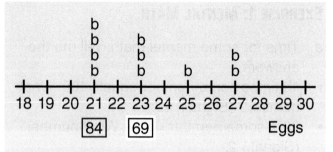

l. (Point to **25**.) How many eggs are in this column? (Signal.) *25.*
 (Add to show:) [105:2I]

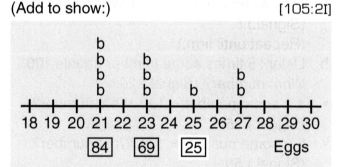

m. (Point to **27**.) Figure out the number of eggs in this column. ✔

- Everybody, how many eggs? (Signal.) *54.*
 (Add to show:) [105:2J]

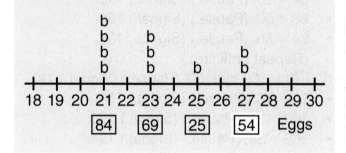

a. Open your workbook to Lesson 105 and find part 1. ✔
(Teacher reference:)

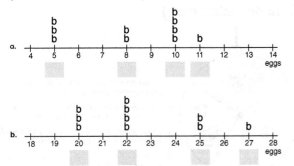

- Everybody, what's the name under the line plot? (Signal.) *Eggs.*
b. Are there any baskets that have 5 eggs? (Signal.) *Yes.*
- What number do you add for every basket in that column? (Signal.) *5.*
c. Are there any baskets that have 6 eggs? (Signal.) *No.*
- Are there any baskets that have 7 eggs? (Signal.) *No.*
d. Are there any baskets that have 8 eggs? (Signal.) *Yes.*
- What do you add for every basket in that column? (Signal.) *8.*
e. Figure out the number of eggs in columns that have Bs. Write the numbers below.
(Observe students and give feedback.)
f. Check your work.
- What's the total for baskets that have 5 eggs? (Signal.) *15.*
- What's the total for baskets that have 8 eggs? (Signal.) *16.*
- What's the total for baskets that have 10 eggs? (Signal.) *40.*
- What's the total for baskets that have 11 eggs? (Signal.) *11.*
g. Problem B. What's the name under the line plot? (Signal.) *Eggs.*
- So what do these numbers tell about? (Signal.) *Eggs.*
- Do any baskets have 20 eggs? (Signal.) *Yes.*
- How many baskets? (Signal.) *3.*
- Do any baskets have 21 eggs? (Signal.) *No.*
- Do any baskets have 22 eggs? (Signal.) *Yes.*
- How many baskets? (Signal.) *4.*

h. Write the total number of eggs below each column that has Bs.
(Observe students and give feedback.)
i. Check your work.
- What's the total for baskets that have 20 eggs? (Signal.) *60.*
- What's the total for baskets that have 22 eggs? (Signal.) *88.*
- What's the total for baskets that have 25 eggs? (Signal.) *50.*
- What's the total for baskets that have 27 eggs? (Signal.) *27.*

EXERCISE 3: PROBABILITY
PERCENTS   `REMEDY`

a. (Display:)   [105:3A]

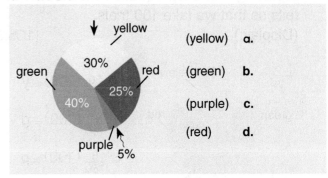

Last time, you worked probability problems for a spinner.
b. In part A, write the probability equation for each color.
(Observe students and give feedback.)
c. Check your work.
(Change to show:)   [105:3B]

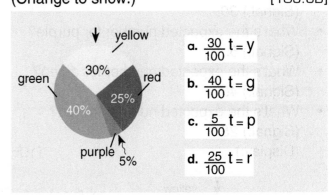

Here's what you should have.
- If you spin the wheel one time, which color is the most likely? (Signal.) *Green.*
- What's the probability of the wheel stopping at green? (Signal.) *40/100.*
- Which color is least likely? (Signal.) *Purple.*
- What's the probability of the wheel stopping at purple? (Signal.) *5/100.*

d. Listen: If you spin the wheel 100 times, the percents show the number of times you would expect the wheel to stop at different colors. If you spin the wheel 100 times, you would expect it to stop at yellow about 30 times because yellow is 30 percent of the wheel.

- If you spin the wheel 100 times, how many times would you expect it to stop at red? (Signal.) *25 times.*
- If you spin the wheel 100 times, how many times would you expect it to stop at green? (Signal.) *40 times.*
- If you spin the wheel 100 times, how many times would you expect it to stop at purple? (Signal.) *5 times.*

e. We can show the numbers are correct by replacing T in each equation with 100. That tells us that we take 100 trials.
(Display:)   [105:3C]

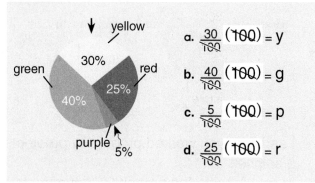

Each equation shows that the percent number is the expected number of times the spinner will stop.

- What's the expected number for yellow? (Signal.) *30.*
- What's the expected number for purple? (Signal.) *5.*
- What's the expected number for green? (Signal.) *40.*
- What's the expected number for red? (Signal.) *25.*

(Display:)   [105:3D]

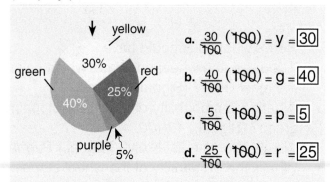

Remember, percent numbers are the expected numbers for 100 trials.

---

a. Open your textbook to Lesson 105 and find part 1. ✔
- Pencils down. ✔
(Teacher reference:)

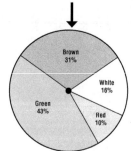

a. If you take 100 trials, about how many times would you expect the spinner to stop at white?
b. If you take 100 trials, about how many times would you expect the spinner to stop at brown?
c. If you take 100 trials, about how many times would you expect the spinner to stop at green?
d. If you take 100 trials, about how many times would you expect the spinner to stop at red?

I'll read each question. You'll tell me the answer.

- Question A: If you take 100 trials, about how many times would you expect the spinner to stop at white? (Signal.) *16 times.*
- B: If you take 100 trials, about how many times would you expect the spinner to stop at brown? (Signal.) *31 times.*
- C: If you take 100 trials, about how many times would you expect the spinner to stop at green? (Signal.) *43 times.*
- D: If you take 100 trials, about how many times would you expect the spinner to stop at red? (Signal.) *10 times.*

EXERCISE 4: DISTRIBUTIVE PROPERTY

a. Find part 2 in your textbook. ✔
(Teacher reference:)

Sample Problem   6(25)
6(20) + 6(5)
120 + 30 = 150

a. 9(42)   b. 2(57)   c. 7(83)

These are multiplication problems. You're going to show the numbers you add for tens and ones.
The sample problem is already worked: 6 × 25. The problems below show the parts that you add: 6 × 20 plus 6 × 5.
Below, you complete the equation: 120 + 30 = 150.

b. Problem A: 9 × 42.
• Work the problem.
 (Observe students and give feedback.)
 (Display:) [105:4A]

$$\mathbf{a.}\ 9\ (42)$$
$$9\ (40) + 9\ (2)$$
$$360\ +\ 18 = \boxed{378}$$

Here's what you should have.
Below the problem, you wrote: 9 × 40 plus 9 × 2.
That's 360 plus 18.
• What's the answer? (Signal.) *378.*
c. Work problem B.
 (Observe students and give feedback.)
 (Display:) [105:4B]

$$\mathbf{b.}\ 2\ (57)$$
$$2\ (50) + 2\ (7)$$
$$100\ +\ 14 = \boxed{114}$$

Here's what you should have.
2 times 57 is 2 times 50 plus 2 times 7. That's 100 plus 14.
• What's the answer? (Signal.) *114.*
d. Work problem C.
 (Observe students and give feedback.)
 (Display:) [105:4C]

$$\mathbf{c.}\ 7\ (83)$$
$$7\ (80) + 7\ (3)$$
$$560\ +\ 21 = \boxed{581}$$

Here's what you should have.
7 times 83 is 7 times 80 plus 7 times 3. That's 560 plus 21.
• What's the answer? (Signal.) *581.*

EXERCISE 5: FRACTIONS
FOR MIXED NUMBERS
`REMEDY`

a. (Display:) [105:5A]

$$8\tfrac{1}{4} =$$

• Read this mixed number. (Signal.) *8 and 1/4.*

b. We're going to write the fraction for this mixed number. Remember, we multiply then add.
• Raise your hand when you know the number we start with for the multiplication. ✔
• What number? (Signal.) *4.*
 (Add to show:) [105:5B]

$$8\tfrac{1}{4} = \dfrac{}{4}$$

• Start with 4 and say the first fact. (Signal.) *4 × 8 = 32.*
• Say the next fact. (Signal.) *32 + 1 = 33.*
 (Repeat until firm.)
• What does 8 and 1/4 equal? (Signal.) *33/4.*
 (Add to show:) [105:5C]

$$8\tfrac{1}{4} = \dfrac{33}{4}$$

c. (Display:) [105:5D]

$$2\tfrac{3}{5} =$$

• Read this mixed number. (Signal.) *2 and 3/5.*
d. We're going to write the fraction for this mixed number.
• Raise your hand when you know the number we start with. ✔
• What number? (Signal.) *5.*
 (Add to show:) [105:5E]

$$2\tfrac{3}{5} = \dfrac{}{5}$$

• Start with 5 and say the first fact. (Signal.) *5 × 2 = 10.*
• Say the next fact. (Signal.) *10 + 3 = 13.*
 (Repeat until firm.)
• What does 2 and 3/5 equal? (Signal.) *13/5.*
 (Add to show:) [105:5F]

$$2\tfrac{3}{5} = \dfrac{13}{5}$$

a. Find part 3 in your textbook. ✔
   (Teacher reference:)

a. $1\frac{3}{7}$ b. $20\frac{1}{2}$ c. $3\frac{4}{5}$ d. $8\frac{1}{3}$ e. $4\frac{7}{10}$

b. Problem A: 1 and 3/7.
• Copy the mixed number and complete the equation.
   (Observe students and give feedback.)
c. Check your work.
• What's the denominator of the fraction? (Signal.) *7.*
• Say the multiplication fact. (Signal.) *7 × 1 = 7.*
• Say the addition fact. (Signal.) *7 + 3 = 10.*
• What's the fraction that equals 1 and 3/7? (Signal.) *10/7.*
d. Work the rest of the problems in part 3.
   (Observe students and give feedback.)
e. Check your work.
• Mixed number B: 20 and 1/2.
• Say the multiplication fact. (Signal.) *2 × 20 = 40.*
• Say the addition fact. (Signal.) *40 + 1 = 41.*
• What fraction equals 20 and 1/2? (Signal.) *41/2.*
f. C: 3 and 4/5.
• Say the multiplication fact. (Signal.) *5 × 3 = 15.*
• Say the addition fact. (Signal.) *15 + 4 = 19.*
• What fraction equals 3 and 4/5? (Signal.) *19/5.*
g. D: 8 and 1/3.
• Say the multiplication fact. (Signal.) *3 × 8 = 24.*
• Say the addition fact. (Signal.) *24 + 1 = 25.*
• What fraction equals 8 and 1/3? (Signal.) *25/3.*
h. E: 4 and 7/10.
• Say the multiplication fact. (Signal.) *10 × 4 = 40.*
• Say the addition fact. (Signal.) *40 + 7 = 47.*
• What fraction equals 4 and 7/10? (Signal.) *47/10.*

EXERCISE 6: DECIMAL MULTIPLICATION/DIVISION
MULTIPLES OF 10

a. Find part 4 in your textbook. ✔
   (Teacher reference:)

a. .13   (Divide by 100.)   d. .099   (Multiply by 1000.)
b. 2.03 (Divide by 10.)   e. 48.2   (Divide by 100.)
c. 9.9   (Divide by 100.)   f. .102   (Multiply by 100.)

You'll follow the direction for each item and write the new decimal number.
b. Read decimal number A. (Signal.) *13 hundredths.*
• Read the direction. (Signal.) *Divide by 100.*
• Write the new number. ✔
   (Display:)   [105:6A]

   a. .0013

Here's what you should have. You added 2 zeros. The new number is 13 ten-thousandths.
c. Read decimal number B. (Signal.) *2 and 3 hundredths.*
• Read the direction. (Signal.) *Divide by 10.*
• Write the new number. ✔
• Everybody, read the new number. (Signal.) *203 thousandths.*
   (Display:)   [105:6B]

   b. .203

Here's what you should have.
d. Work the rest of the problems in part 4.
   (Observe students and give feedback.)
e. Check your work.
   Read each new decimal value.
• C. (Signal.) *99 thousandths.*
• D. (Signal.) *99.*
• E. (Signal.) *482 thousandths.*
• F. (Signal.) *10 and 2 tenths.*
   (Display:)   [105:6C]

   c. .099

   d. 99

   e. .482

   f. 10.2

Here's what you should have for C through F.

EXERCISE 7: PROBABILITY

SPINNERS WITH PERCENTS

a. Find part 5 in your textbook. ✔
(Teacher reference:)

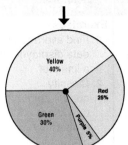

a. If you spin the wheel 10 times, what's the expected number for the wheel stopping at yellow?

b. You spin the wheel until it lands on purple 4 times. How many trials would you expect to take?

c. If you spin the wheel 30 times, what's the expected number for the wheel stopping at green?

d. If you spin the wheel 12 times, what's the expected number for the wheel stopping at red?

Here's a spinner.

* Problem A: If you spin the wheel 10 times, what's the expected number for the wheel stopping at yellow?
* Say the probability equation for yellow. (Signal.) *40/100 T = Y.*

b. Problem B: You spin the wheel until it lands on purple 4 times. How many trials would you expect to take?
* Say the probability equation for purple. (Signal.) *5/100 T = P.*

c. Work problem A. Start with the probability equation.
(Observe students and give feedback.)
Check your work.

* If you spin the wheel 10 times, what's the expected number for the wheel stopping at yellow? (Signal.) *4.*
(Display:) [105:7A]

$$a. \quad \frac{40}{100} t = y$$

$$\frac{40}{100} \overset{1}{(10)} = y = \boxed{4}$$

Here's what you should have.

d. Work problem B.
(Observe students and give feedback.)
Check your work.

* You spin the wheel until it lands on purple 4 times. How many trials would you expect to take? (Signal.) *80 trials.*

(Display:) [105:7B]

$$b. \quad \frac{5}{100} t = p$$

$$\left(\frac{100}{5}\right) \frac{5}{100} t = 4 \left(\frac{\overset{20}{100}}{5}\right)$$

$$t = 80$$

$$\boxed{80 \text{ trials}}$$

Here's what you should have.

e. Work problem C.
(Observe students and give feedback.)
Check your work.

* If you spin the wheel 30 times, what's the expected number for the wheel stopping at green? (Signal.) *9.*
(Display:) [105:7C]

$$c. \quad \frac{30}{100} t = g$$

$$\frac{30}{100} (30) = g = \boxed{9}$$

Here's what you should have.

f. Work problem D.
(Observe students and give feedback.)
Check your work.

* If you spin the wheel 12 times, what's the expected number for the wheel stopping at red? (Signal.) *3.*
(Display:) [105:7D]

$$d. \quad \frac{25}{100} t = r$$

$$\frac{\overset{1}{25}}{\underset{4}{100}} \overset{3}{(12)} = r = \boxed{3}$$

Here's what you should have.

Assign Independent Work, Textbook parts 6–14 and Workbook part 2.

Optional extra math-fact practice worksheets are available on ConnectED.

Lessons 106–110 Planning Page

	Lesson 106	Lesson 107	Lesson 108	Lesson 109	Lesson 110
Student Learning Objectives	**Exercises** 1. Add and divide using mental math 2. Find total values for columns on a line plot 3. **Use the distributive property of multiplication to check division** 4. Convert mixed numbers to fractions 5. **Find surface area of rectangular prisms** 6. **Convert metric units** Complete work independently	**Exercises** 1. Find total values for columns on a line plot 2. **Solve word problems that involve unit conversion** 3. **Solve mixed-number multiplication problems** 4. Use the distributive property of multiplication to check division 5. Find surface area of rectangular prisms 6. Convert metric units Complete work independently	**Exercises** 1. **Find average of data displayed on a line plot** 2. **Classify quadrilaterals based on properties/ hierarchy** 3. Solve mixed-number multiplication problems 4. Use the distributive property of multiplication to check division 5. Solve word problems that involve unit conversion 6. Find surface area of rectangular prisms 7. Convert metric units Complete work independently	**Exercises** 1. Find average of data displayed on a line plot 2. Classify quadrilaterals based on properties/ hierarchy 3. Solve mixed-number multiplication problems 4. Use the distributive property of multiplication to check division 5. Solve word problems that involve unit conversion 6. Find surface area of rectangular prisms Complete work independently	**Exercises** 1. Find average of data displayed on a line plot 2. Classify quadrilaterals based on properties/ hierarchy 3. Solve mixed-number word problems 4. Use the distributive property of multiplication to check division 5. Solve word problems that involve unit conversion 6. Find surface area of rectangular prisms 7. **Multiply by 10 with an exponent** 8. Complete work independently
Common Core State Standards for Mathematics					
5.OA 1	✔	✔	✔	✔	✔
5.NBT 1	✔		✔	✔	✔
5.NBT 2	✔		✔	✔	✔
5.NBT 5–6	✔	✔	✔	✔	✔
5.NBT 7	✔		✔	✔	
5.NF 1–2	✔	✔	✔	✔	✔
5.NF 4	✔	✔	✔	✔	✔
5.NF 5				✔	
5.NF 6	✔	✔	✔	✔	✔
5.MD 1	✔	✔	✔	✔	✔
5.G 1	✔				
5.G 2				✔	
5.G 3–4			✔	✔	✔
Teacher Materials	Presentation Book 2, Board Displays CD or chalk board				
Student Materials	Textbook, Workbook, pencil, lined paper				
Additional Practice	• Student Practice Software: Block 5: Activity 1 (5.MD 1), Activity 2 (5.NF 1 and 5.NF 2), Activity 3 (5.MD 3 and 5.MD 4), Activity 4 (5.NF 4 and 5.NF 6), Activity 5, Activity 6 (5.MD 3 and 5.MD 4) • Provide needed fact practice with Level D or E Math Fact Worksheets.				
Mastery Test					Student Assessment Book (Present Mastery Test 11 following Lesson 110.)

350 Lessons 106–110 Planning Page

Connecting Math Concepts

EXERCISE 1: MENTAL MATH

a. Time for some mental math. Tell me the answers.
- 100 divided by 4. (Pause.) (Signal.) *25.*
- 100 ÷ 5. (Pause.) (Signal.) *20.*
- 100 ÷ 10. (Pause.) (Signal.) *10.*
- (Repeat until firm.)

b. Listen: 100 divided by 20. (Pause.) (Signal.) *5.*
- 100 ÷ 25. (Pause.) (Signal.) *4.*
- 100 ÷ 50. (Pause.) (Signal.) *2.*
- 100 ÷ 2. (Pause.) (Signal.) *50.*
- (Repeat until firm.)

c. Listen: 290 plus 20. (Pause.) (Signal.) *310.*
- 290 + 24. (Pause.) (Signal.) *314.*
- 230 + 24. (Pause.) (Signal.) *254.*
- 280 + 28. (Pause.) (Signal.) *308.*
- 280 + 50. (Pause.) (Signal.) *330.*
- 280 + 59. (Pause.) (Signal.) *339.*
- (Repeat until firm.)

d. Listen: 60 plus 70. (Pause.) (Signal.) *130.*
- 60 + 76. (Pause.) (Signal.) *136.*
- 160 + 80. (Pause.) (Signal.) *240.*
- 160 + 84. (Pause.) (Signal.) *244.*
- 160 + 14. (Pause.) (Signal.) *174.*
- (Repeat until firm.)

e. Listen: 120 plus 14. (Pause.) (Signal.) *134.*
- 120 + 24. (Pause.) (Signal.) *144.*
- 320 + 24. (Pause.) (Signal.) *344.*
- (Repeat until firm.)

EXERCISE 2: LINE PLOT
COLUMN TOTALS

a. Open your workbook to Lesson 106 and find part 1. ✔
(Teacher reference:)

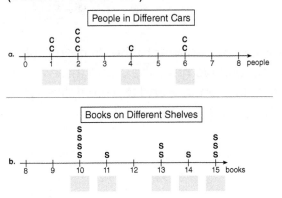

b. Problem A. This line plot shows the number of people in different cars.
- What's the name under the line plot? (Signal.) *People.*
- So what do these numbers tell about? (Signal.) *People.*

c. The Cs tell about cars. Some cars have one person in them.
- How many cars have 1 person? (Signal.) *2 cars.*
- Say the addition for that column. (Signal.) *1 + 1.*

d. Look at the column for 2. ✔
- How many cars have 2 people? (Signal.) *3 cars.*
- Say the addition for that column. (Signal.) *2 + 2 + 2.*

e. How many cars have 4 people? (Signal.) *1 car.*
- How many cars have 6 people? (Signal.) *2 cars.*

f. Write the total for each column.
(Observes students and give feedback.)

g. Check your work.
- What's the total for cars that have one person? (Signal.) *2.*
- What's the total for cars that have 2 people? (Signal.) *6.*
- What's the total for cars that have 4 people? (Signal.) *4.*
- What's the total for cars that have 6 people? (Signal.) *12.*

h. Problem B. This line plot shows the number of books on different shelves.
- What's the name under the line plot? (Signal.) *Books.*
- What does the letter S stand for? (Signal.) *Shelves.*

i. How many shelves have 10 books? (Signal.) *4 shelves.*
- Write the total for shelves that have 10 books. Then figure out the numbers for shelves that have 11 books, 13 books, 14 books, and 15 books.
(Observe students and give feedback.)

j. Check your work.
- What's the total for shelves that have 10 books? (Signal.) *40.*
- What's the total for shelves that have 11 books? (Signal.) *11.*
- What's the total for shelves that have 13 books? (Signal.) *26.*
- What's the total for shelves that have 14 books? (Signal.) *14.*
- What's the total for shelves that have 15 books? (Signal.) *45.*

EXERCISE 3: DISTRIBUTIVE PROPERTY
CHECKING DIVISION ANSWERS

a. (Display:) [106:3A]

$$4\overline{)6_28} \quad 5\overline{)3\,6_10} \quad 9\overline{)4\,5\,0}$$

with quotients: 17, 72, 50

You're going to read these division facts.
- (Point to **4.**) Read this fact. (Signal.) *68 ÷ 4 = 17.*
- (Point to **5.**) Read this fact. (Signal.) *360 ÷ 5 = 72.*
- (Point to **9.**) Read this fact. (Signal.) *450 ÷ 9 = 50.*

b. You're going to say a multiplication fact for each division fact.
- (Point to **4.**) Start with 4 and say the multiplication fact. (Signal.) *4 × 17 = 68.*
- (Point to **5.**) Say the multiplication fact. (Signal.) *5 × 72 = 360.*
- (Point to **9.**) Say the multiplication fact. (Signal.) *9 × 50 = 450.*

c. My turn to say the multiplication for the parts you add.
(Point to **4.**) 4 times 10 plus 4 times 7.
- Say the multiplication for the parts you add. (Signal.) *4 × 10 plus 4 × 7.*
- (Point to **5.**) Say the multiplication for the parts you add. (Signal.) *5 × 70 plus 5 × 2.*
- (Point to **9.**) Say the multiplication for the parts you add. (Signal.) *9 × 50 plus 9 × zero.*
(Repeat until firm.)

d. (Display:) [106:3B]

$$3\overline{)6\,9}$$

- Read the problem. (Signal.) *69 ÷ 3.*
- Say the division problem for the tens digit. (Signal.) *6 ÷ 3.*
- What's the answer? (Signal.) *2.*
(Add to show:) [106:3C]

$$3\overline{)6\,9}$$ with 2 above

- Say the division problem for the ones digit. (Signal.) *9 ÷ 3.*
- What's the answer? (Signal.) *3.*
(Add to show:) [106:3D]

$$3\overline{)6\,9}$$ with 23 above

- Read the problem and the whole answer. (Signal.) *69 ÷ 3 = 23.*

e. You can check your work by doing the multiplication for the tens and ones. The multiplication is 3 times 23.
- Say the multiplication for the parts you add. (Signal.) *3 × 20 plus 3 × 3.*
(Add to show:) [106:3E]

$$3\overline{)6\,9}$$ with 23 above   3 (23)
                                       3 (20) + 3 (3)

- What's 3 times 20? (Signal.) *60.*
- What's 3 times 3? (Signal.) *9.*
- What's 60 plus 9? (Signal.) *69.*
(Add to show:) [106:3F]

$$3\overline{)6\,9}$$ with 23 above   3 (23)
                                       3 (20) + 3 (3)
                                 60 + 9 = 69

We know our answer to the division problem is 23 because we have the same three numbers for the division fact and the multiplication fact: 3, 23, and 69.

f. We can show the multiplication and division of the parts with two rectangles.
(Change to show:) [106:3G]

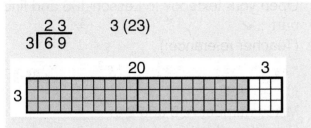

(Point to **rectangle.**) The shaded rectangle shows the multiplication for the tens—3 × 20. The white square shows the multiplication for the ones—3 × 3.
• What's the multiplication for the tens? (Signal.) *3 × 20.*
• What's the multiplication for the ones? (Signal.) *3 × 3.*
(Add to show:) [106:3H]

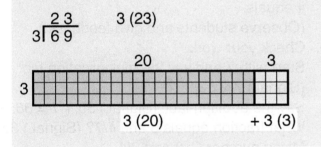

• What's 3 times 20? (Signal.) *60.*
• What's 3 times 3? (Signal.) *9.*
• What's 60 plus 9? (Signal.) *69.*
(Add to show:) [106:3I]

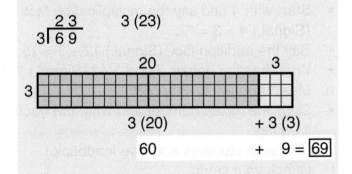

Here's the multiplication for the two parts that are added. The total for both parts is 69.
We can show 69 divided by 3. That's one row in the diagram.

(Change to show:) [106:3J]

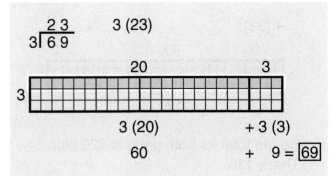

• Raise your hand when you know the number of squares in a row. ✔
• What's the number of squares in a row? (Signal.) *23.*
That's the answer to the problem 69 ÷ 3.
g. (Display:) [106:3K]

4 (32)

New problem: 4 × 32.
• Say the problem for the tens. (Signal.) *4 × 30.*
• Say the problem for the ones. (Signal.) *4 × 2.*
(Repeat until firm.)
• We make the rectangle for the tens and the ones.
(Add to show:) [106:3L]

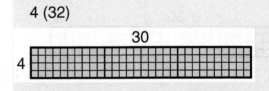

Here's the rectangle for 4 × 30.
(Add to show:) [106:3M]

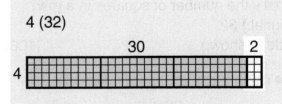

Here's the rectangle for 4 × 2.
The part for the tens has 4 times 30 squares.
• How many is that? (Signal.) *120.*
(Add to show:) [106:3N]

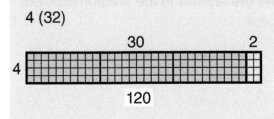

The part for the ones has 4 times 2 squares.
• How many is that? (Signal.) *8.*

(Add to show:)   [106:3O]

4 (32)

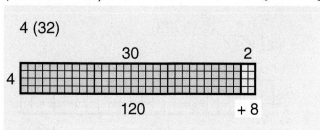

So the total for both parts is 120 plus 8.
That's 128.
So 4 times 32 equals 128.

(Add to show:)   [106:3P]

4 (32)

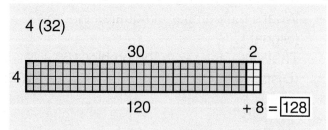

- Say the problem that divides by 4. **(Signal.)**
  128 ÷ 4.
  There are 4 rows, so the number of squares
  in a row is the answer to the division problem
  128 ÷ 4.

(Change to show:)   [106:3Q]

4 (32)

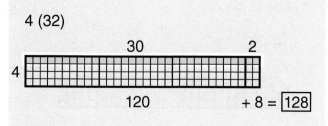

- Raise your hand when you know the number
  of squares in a row. ✔
- What's the number of squares in a row?
  (Signal.) *32.*

(Add to show:)   [106:3R]

4 (32)

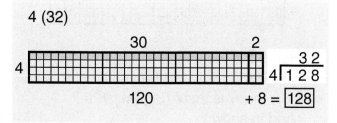

That's the answer to the division problem
128 ÷ 4.

EXERCISE 4: FRACTIONS
FOR MIXED NUMBERS

a. Open your textbook to Lesson 106 and find
    part 1. ✔
    (Teacher reference:)

a. $3\frac{2}{9}$ b. $5\frac{1}{7}$ c. $3\frac{3}{4}$ d. $6\frac{2}{5}$

b. Mixed number A: 3 and 2/9.
- Copy the mixed number and write the fraction
  it equals.
  (Observe students and give feedback.)
c. Check your work.
- Start with 9 and say the multiplication fact.
  (Signal.) *9 × 3 = 27.*
- Say the addition fact. **(Signal.)** *27 + 2 = 29.*
- What fraction equals 3 and 2/9? **(Signal.)** *29/9.*
d. Mixed number B: 5 and 1/7.
- Copy the mixed number and write the fraction
  it equals.
  (Observe students and give feedback.)
e. Check your work.
- Start with 7 and say the multiplication fact.
  (Signal.) *7 × 5 = 35.*
- Say the addition fact. **(Signal.)** *35 + 1 = 36.*
- What fraction equals 5 and 1/7? **(Signal.)** *36/7.*
f. Mixed number C: 3 and 3/4.
- Copy the mixed number and write the fraction
  it equals.
  (Observe students and give feedback.)
g. Check your work.
- Start with 4 and say the multiplication fact.
  (Signal.) *4 × 3 = 12.*
- Say the addition fact. **(Signal.)** *12 + 3 = 15.*
- What fraction equals 3 and 3/4? **(Signal.)** *15/4.*
h. Mixed number D: 6 and 2/5.
- Copy the mixed number and write the fraction
  it equals.
  (Observe students and give feedback.)
i. Check your work.
- Start with 5 and say the multiplication fact.
  (Signal.) *5 × 6 = 30.*
- Say the addition fact. **(Signal.)** *30 + 2 = 32.*
- What fraction equals 6 and 2/5? **(Signal.)** *32/5.*

EXERCISE 5: SURFACE AREA
OF RECTANGULAR PRISMS

a. (Display:)   [106:5A]

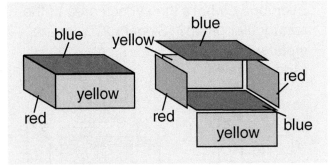

You know about the volume of a rectangular prism. You're going to learn about the surface area of a rectangular prism.
Listen: A rectangular prism has 6 faces.
• Say the fact. (Signal.) *A rectangular prism has 6 faces.*
The picture on the right shows the faces separated. The faces are in pairs.
The faces in each pair have the same color.
The faces in each pair have the same **area.**
The top and the bottom have the same area.
The front and the back have the same area.
The left and right faces have the same area.

b. Which face has the same area as the front? (Signal.) *The back.*
• Which face has the same area as the top? (Signal.) *The bottom.*
• Which face has the same area as the left face? (Signal.) *The right face.*
  (Repeat until firm.)

c. (Add to show:)   [106:5B]

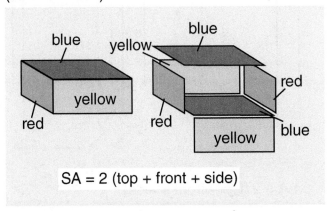

SA = 2 (top + front + side)

Here's the equation that we'll use to figure out surface area: Surface area equals 2 times the quantity, top plus front plus side.
• Say the equation. (Signal.) *Surface area equals 2 times the quantity, top plus front plus side.*
  (Repeat until firm.)

d. If we have 2 times the top, we'll have the top and the bottom. If we have 2 times the front face, we'll have the front and back. If we have 2 times a side, we'll have the left and right sides.
When we find the surface area, the units are square units.
• What are the units of surface area? (Signal.) *Square units.*
(Display:)   [106:5C]

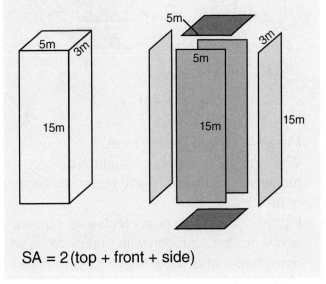

SA = 2 (top + front + side)

Here's a rectangular prism. We'll put in numbers for the area of the top, the area of the front, and the area of a side and add. Then we'll multiply by 2 to find the area of all the faces.
Once more: We put in the area for the top, the area for the front, and the area for a side.
• Look at the top. ✔
• Say the multiplication for the area. (Call on a student. Idea: *5 × 3, or 3 × 5.*)
• Look at the front. ✔
• Say the multiplication for the area. (Call on a student. Idea: *5 × 15, or 15 × 5.*)
• Look at the right side. ✔
• Say the multiplication for the area. (Call on a student. Idea: *3 × 15, or 15 × 3.*)

e. Copy the equation in part A. Figure out the area of the top, the front, and the right side, and write the equation with those numbers. (Observe students and give feedback.)
• Everybody, what's the number for the top face? (Signal.) *15.*
• What's the number for the front face? (Signal.) *75.*
• What's the number for the right face? (Signal.) *45.*

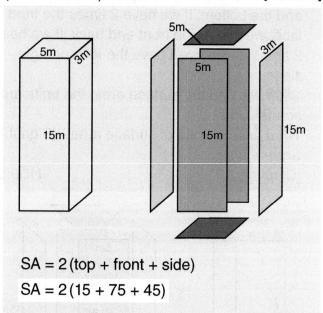

SA = 2 (top + front + side)
SA = 2 (15 + 75 + 45)

Here's what you should have.

f. We're going to add these numbers, then multiply by 2. That will give us the surface area of all 6 faces.

• Figure out what 15 plus 75 plus 45 equals, and write the equation with that number. ✔

• Everybody, what's the total for the 3 faces? (Signal.) *135.*
(Add to show:) [106:5E]

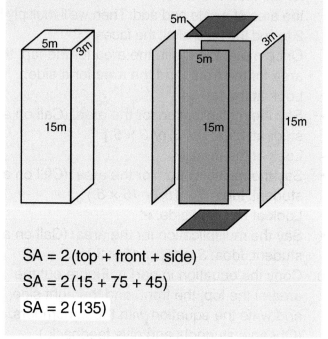

SA = 2 (top + front + side)
SA = 2 (15 + 75 + 45)
SA = 2 (135)

Here's what you should have.

g. Now you'll multiply 2 times 135 to find the area of all 6 faces. The unit name in the answer is square meters. Work the problem and write the answer. ✔

• Everybody, what's the surface area of the rectangular prism? (Signal.) *270 square meters.*
(Add to show:) [106:5F]

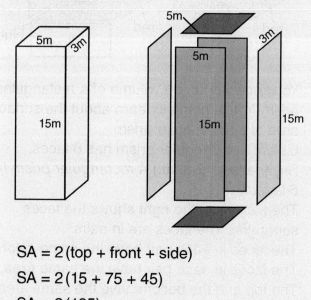

SA = 2 (top + front + side)
SA = 2 (15 + 75 + 45)
SA = 2 (135)
SA = 270 sq m

Here's what you should have.

h. (Display:) [106:5G]

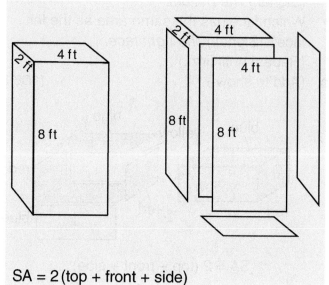

SA = 2 (top + front + side)

- New problem. Read the equation for finding the surface area of a rectangular prism. (Signal.) *Surface area equals 2 times the quantity, top plus front plus side.*
- Say the multiplication for the top. (Call on a student. Idea: *2 × 4, or 4 × 2.*)
- Say the multiplication for the front. (Call on a student. Idea: *8 × 4, or 4 × 8.*)
- Say the multiplication for the left side. (Call on a student. Idea: *2 × 8, or 8 × 2.*)
- Everybody, figure out the area of the top, the front, and the left side, and write the equation with those numbers.
  (Observe students and give feedback.)
- Everybody, what's the number for the area of the top face? (Signal.) *8.*
- What's the number for the front face? (Signal.) *32.*
- What's the number for the left face? (Signal.) *16.*
  (Add to show:) [106:5H]

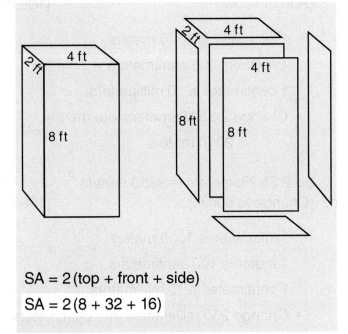

SA = 2 (top + front + side)

SA = 2 (8 + 32 + 16)

Here's what you should have.

i. Add the 3 areas and write the equation with the total. ✔

- Everybody, what's the total for the 3 faces? (Signal.) *56.*

(Add to show:) [106:5I]

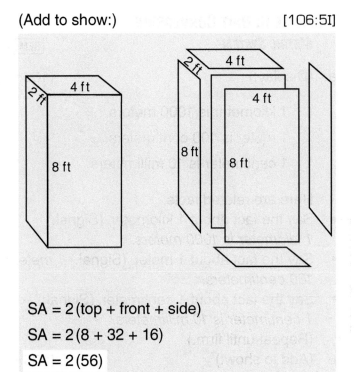

SA = 2 (top + front + side)

SA = 2 (8 + 32 + 16)

SA = 2 (56)

Here's what you should have.
We multiply by 2 to find the area of all 6 faces.

- Do it and write the answer with a unit name.
  (Observe students and give feedback.)
- Everybody, what's the area of all 6 faces? (Signal.) *112 square feet.*
  (Add to show:) [106:5J]

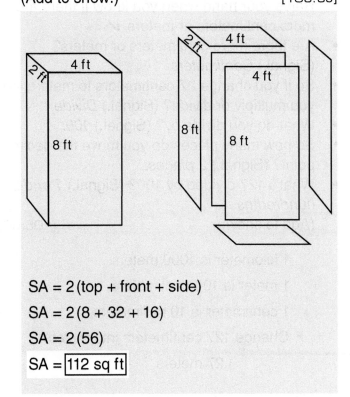

SA = 2 (top + front + side)

SA = 2 (8 + 32 + 16)

SA = 2 (56)

SA = 112 sq ft

You've worked some hard problems.

EXERCISE 6: UNIT CONVERSION
METRIC SYSTEM

REMEDY

a. (Display:)   [106:6A]

> 1 kilometer is 1000 meters.
>
> 1 meter is 100 centimeters.
>
> 1 centimeter is 10 millimeters.

Here are related facts.
- Say the fact about 1 kilometer. (Signal.) *1 kilometer is 1000 meters.*
- Say the fact about 1 meter. (Signal.) *1 meter is 100 centimeters.*
- Say the fact about 1 centimeter. (Signal.) *1 centimeter is 10 millimeters.*
  (Repeat until firm.)

b. (Add to show:)   [106:6B]

> 1 kilometer is 1000 meters.
>
> 1 meter is 100 centimeters.
>
> 1 centimeter is 10 millimeters.
>
> • Change 127 centimeters into meters.

Change 127 centimeters into meters.
- Raise your hand when you know if there are more centimeters or meters. ✔
- Are there more centimeters or meters? (Signal.) *Centimeters.*
- So if you change 27 centimeters to meters, do you multiply or divide? (Signal.) *Divide.*
- What do you divide by? (Signal.) *100.*
- So how many places do you move the decimal point? (Signal.) *2 places.*
- What's 127 divided by 100? (Signal.) *1 and 27 hundredths.*
  (Add to show:)   [106:6C]

> 1 kilometer is 1000 meters.
>
> 1 meter is 100 centimeters.
>
> 1 centimeter is 10 millimeters.
>
> • Change 127 centimeters into meters.
>
>   1.27 meters

- How many meters is 127 centimeters? (Signal.) *1.27 meters.*

c. (Change to show:)   [106:6D]

> 1 kilometer is 1000 meters.
>
> 1 meter is 100 centimeters.
>
> 1 centimeter is 10 millimeters.
>
> • Change 2.35 kilometers into meters.

New problem: Change 2.35 kilometers into meters.
- Raise your hand when you know if there are more kilometers or meters. ✔
- Are there more kilometers or meters? (Signal.) *Meters.*
- So if you change 2.35 kilometers into meters, do you multiply or divide? (Signal.) *Multiply.*
- What do you multiply by? (Signal.) *1000.* Yes, 1 kilometer is 1000 meters.
- How many places do you move the decimal point? (Signal.) *3 places.*
- What is 2.35 times 1000? (Signal.) *2350.*
  (Add to show:)   [106:6E]

> 1 kilometer is 1000 meters.
>
> 1 meter is 100 centimeters.
>
> 1 centimeter is 10 millimeters.
>
> • Change 2.35 kilometers into meters.
>
>   2350 meters

So 2.35 kilometers is 2350 meters.

d. (Change to show:)   [106:6F]

> 1 kilometer is 1000 meters.
>
> 1 meter is 100 centimeters.
>
> 1 centimeter is 10 millimeters.
>
> • Change 250 millimeters into centimeters.

New problem: Change 250 millimeters into centimeters.
- In part A, write the answer with a unit name. ✔

- Everybody, are there more millimeters or centimeters? (Signal.) *Millimeters.*
- What did you divide by? (Signal.) *10.*
- What's the answer? (Signal.) *25 centimeters.*
  (Add to show:) [106:6G]

> 1 kilometer is 1000 meters.
>
> 1 meter is 100 centimeters.
>
> 1 centimeter is 10 millimeters.
>
> - Change 250 millimeters into centimeters.
>
>   25.0 centimeters

Here's what you should have.
If you just wrote 25 centimeters, that's also correct.

═══════════ **TEXTBOOK PRACTICE** ═══════════

a. Find part 2 in your textbook. ✔
   (Teacher reference:)

a. Change 43.7 meters into centimeters.
b. Change .7 kilometers into meters.
c. Change 456 meters into kilometers.
d. Change .004 centimeters into millimeters.
e. Change .8 centimeters into meters.

Part 2
a.

Facts:
- 1 kilometer is 1000 meters.
- 1 meter is 100 centimeters.
- 1 centimeter is 10 millimeters.

Problem A: Change 43.7 meters into centimeters.
- Write the answer with a unit name. ✔
- Everybody, how many centimeters is 43.7 meters? (Signal.) *4370 centimeters.*

b. Work the rest of the items.
   (Observe students and give feedback.)
c. Check your work.
- Item B: Change .7 kilometers into meters.
  What's the answer? (Signal.) *700 meters.*
- Item C: Change 456 meters into kilometers.
  What's the answer? (Signal.) *456 thousandths kilometers.*
- Item D: Change .004 centimeters into millimeters.
  What's the answer? (Signal.) *4 hundredths millimeters.*
- Item E: Change .8 centimeters into meters.
  What's the answer? (Signal.) *8 thousandths meters.*

> Assign Independent Work, Textbook parts 3–10 and Workbook parts 2 and 3.

Optional extra math-fact practice worksheets are available on ConnectED.

Lesson 107

EXERCISE 1: LINE PLOT
COLUMN TOTALS

a. Open your workbook to Lesson 107 and find part 1. ✔
• Pencils down. ✔
(Teacher reference:)

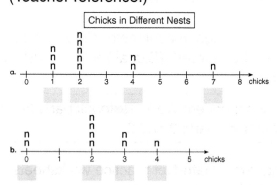

• Touch line plot A. ✔
This line plot shows the number of chicks in different nests.
• What's the name under the line plot? (Signal.) *Chicks.*
• So what do the numbers show? (Signal.) *Chicks.*
• What does the letter N stand for? (Signal.) *Nest.*
• How many nests have one chick? (Signal.) *3 nests.*

b. Write the total for nests that have one chick. Then figure out the totals for nests that have 2 chicks, 4 chicks, and 7 chicks. Pencils down when you're finished.
(Observe students and give feedback.)

c. Check your work.
• What's the total for nests that have 1 chick? (Signal.) *3.*
• What's the total for nests that have 2 chicks? (Signal.) *10.*
• What's the total for nests that have 4 chicks? (Signal.) *8.*
• What's the total for nests that have 7 chicks? (Signal.) *7.*

(Display:) [107:1A]

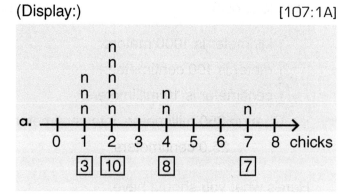

Here's what you should have.

d. Write the totals for the nests in line plot B. Pencils down when you're finished.
(Observe students and give feedback.)

e. Check your work.
• What's the total for nests that have zero chicks? (Signal.) *Zero.*
• What's the total for nests that have 2 chicks? (Signal.) *8.*
• What's the total for nests that have 3 chicks? (Signal.) *6.*
• What's the total for nests that have 4 chicks? (Signal.) *4.*

(Display:) [107:1B]

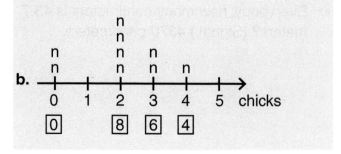

Here's what you should have.

EXERCISE 2: UNIT CONVERSION
WORD PROBLEMS
REMEDY

a. Some word problems name two units that are related, like hours and minutes or feet and inches. The question in the problem names the unit you end up with. So you change everything to that unit.
(Display:) [107:2A]

> **a.** The counter was 3 feet long. It was made 48 inches longer. How many feet long is the counter now?

The counter was 3 feet long. It was made 48 inches longer. How many feet long is the counter now?

- What unit does the **question** name? (Signal.) *Feet.*
  So we have to change a unit to feet.
- Raise your hand when you know which unit we change to feet. ✔
- Which unit? (Signal.) *Inches.*
  Yes, we'll change inches into feet.
- Raise your hand when you know if we multiply or divide to change inches into feet. ✔
- Do we multiply or divide? (Signal.) *Divide.*
- Raise your hand when you know how many feet equals 48 inches. ✔
- How many feet is 48 inches? (Signal.) *4 feet.*
  I'll read the first part of the problem again: The counter was 3 feet long. It was made 48 inches longer.
- How many feet long was the counter at first? (Signal.) *3 feet.*
- How many feet were added? (Signal.) *4 feet.*
- So how many feet long is the counter now? (Signal.) *7 feet.*

b. Remember, the question names the unit that you need in the answer. So you change any measurement that is not that unit.

c. (Display:) [107:2B]

> **b.** The board was 50 inches long. A worker sawed off 2 feet. How many inches long is the board now?

The board was 50 inches long. A worker sawed off 2 feet. How many inches long is the board now?

- What unit does the question name? (Signal.) *Inches.*
- So which unit in the problem do you have to change to inches? (Signal.) *Feet.*
  The worker sawed off 2 feet.
- Raise your hand when you know how many inches equals 2 feet. ✔
- How many inches equals 2 feet? (Signal.) *24 inches.*

d. Now you can work the problem.
- In part A, figure out the answer and write it with a unit name. ✔
  You worked the problem 50 minus 24.
- How long is the board now? (Signal.) *26 inches.*

e. (Display:) [107:2C]

> **c.** Mr. Brown went on a 2-week vacation in June. In December he went on a 10-day vacation. How many vacation days did he have during that year?

Mr. Brown went on a 2-week vacation in June. In December he went on a 10-day vacation. How many vacation days did he have during that year?

- What's the unit named in the question? (Signal.) *Days.*
- Change the measurement that is not days into days. Then work the problem.
  (Observe students and give feedback.)
  You changed 2 weeks into days.
- How many days? (Signal.) *14 days.*
  You worked the problem 14 plus 10.
- How many vacation days did he have during that year? (Signal.) *24 days.*

EXERCISE 3: MIXED-NUMBER OPERATIONS

MULTIPLICATION

a. (Display:)   [107:3A]

$$2\frac{3}{4} \times \frac{1}{2}$$

Here's a new kind of problem.
- Read it. (Signal.) *2 and 3/4 × 1/2.*
  We work the problem by changing the mixed number into a fraction.
- Raise your hand when you know the fraction that equals 2 and 3/4. ✔
- What's the fraction for 2 and 3/4? (Signal.) *11/4.*
  (Add to show:)   [107:3B]

$$2\frac{3}{4} \times \frac{1}{2}$$
$$\frac{11}{4} \times \frac{1}{2} =$$

So the new problem we work is 11/4 times 1/2.

b. Say the multiplication fact for the numerators. (Signal.) *11 × 1 = 11.*
- Say the multiplication fact for the denominators. (Signal.) *4 × 2 = 8.*
- What does 11/4 times 1/2 equal? (Signal.) *11/8.*
  (Add to show:)   [107:3C]

$$2\frac{3}{4} \times \frac{1}{2}$$
$$\frac{11}{4} \times \frac{1}{2} = \frac{11}{8}$$

- Raise your hand when you know the mixed number that equals 11/8. ✔
- What's the mixed number for 11/8? (Signal.) *1 and 3/8.*
  (Add to show:)   [107:3D]

$$2\frac{3}{4} \times \frac{1}{2}$$
$$\frac{11}{4} \times \frac{1}{2} = \frac{11}{8} = 1\frac{3}{8}$$

So 2 and 3/4 × 1/2 = 1 and 3/8.

c. (Display:)   [107:3E]

$$\frac{1}{4} \times 2\frac{3}{7}$$

- New problem. Read it. (Signal.) *1/4 × 2 and 3/7.*
- Raise your hand when you know the fraction that equals 2 and 3/7. ✔
- What fraction? (Signal.) *17/7.*
  (Add to show:)   [107:3F]

$$\frac{1}{4} \times 2\frac{3}{7}$$
$$\frac{1}{4} \times \frac{17}{7} =$$

- Read the problem we work. (Signal.) *1/4 × 17/7.*
- Raise your hand when you know the answer. ✔
- Everybody, what's 1/4 times 17/7?
  (Signal.) *17/28.*
  (Add to show:)   [107:3G]

$$\frac{1}{4} \times 2\frac{3}{7}$$
$$\frac{1}{4} \times \frac{17}{7} = \frac{17}{28}$$

═══════════ **WORKBOOK PRACTICE** ═══════════

a. Find part 2 in your workbook. ✔
  (Teacher reference:)

  a. $\frac{3}{5} \times 1\frac{3}{4}$ b. $3\frac{1}{3} \times \frac{2}{7}$

  c. $6\frac{4}{5} \times \frac{1}{9}$ d. $\frac{3}{4} \times 2\frac{1}{4}$

b. Read problem A. (Signal.) *3/5 × 1 and 3/4.*
- Write the problem with a fraction for the mixed number and work it.
  (Observe students and give feedback.)
- Everybody, you changed 1 and 3/4 into a fraction. What fraction? (Signal.) *7/4.*
- You worked the problem 3/5 times 7/4. That's 21/20.
- What's the mixed number that equals 21/20? (Signal.) *1 and 1/20.*

c. Read problem B. (Signal.) *3 and 1/3 × 2/7.*
- Write the equation below with a fraction for the mixed number. Figure out the answer.
  (Observe students and give feedback.)
- Everybody, what fraction did you write for 3 and 1/3? (Signal.) *10/3.*
- Read the equation. (Signal.) *10/3 × 2/7 = 20/21.*
d. Read problem C. (Signal.) *6 and 4/5 × 1/9.*
- Write the problem with a fraction for the mixed number and work it.
  (Observe students and give feedback.)
- Everybody, what fraction did you write for 6 and 4/5? (Signal.) *34/5.*
- You multiplied 34/5 × 1/9. What's the answer? (Signal.) *34/45.*
e. Read problem D. (Signal.) *3/4 × 2 and 1/4.*
- Write the problem with a fraction for the mixed number and work it.
  (Observe students and give feedback.)
- Everybody, what fraction did you write for 2 and 1/4? (Signal.) *9/4.*
- You multiplied 3/4 × 9/4. What's the fraction answer? (Signal.) *27/16.*
- What's the mixed-number answer? (Signal.) *1 and 11/16.*

EXERCISE 4: DISTRIBUTIVE PROPERTY
CHECKING DIVISION ANSWERS

a. (Display:)   [107:4A]

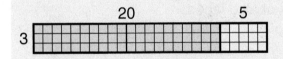

Here's a rectangle divided into a part for the tens and a part for the ones. The whole rectangle shows 3 × 25.
- What does the whole rectangle show? (Signal.) *3 × 25.*
- How many rows are there? (Signal.) *3.*
- Start with 3 and say the multiplication for the tens. (Signal.) *3 × 20.*
- What's the answer? (Signal.) *60.*
  (Add to show:)   [107:4B]

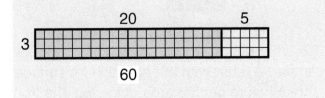

- Start with 3 and say the multiplication for the ones. (Signal.) *3 × 5.*
- What's the answer? (Signal.) *15.*
  (Add to show:)   [107:4C]

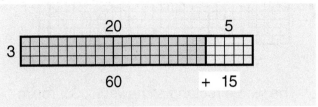

- How many squares are in the whole rectangle? (Signal.) *75.*
  (Add to show:)   [107:4D]

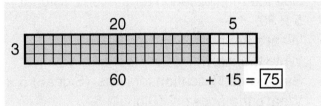

So the simple multiplication equation is 3 times 25 equals 75.
- Say the simple multiplication equation. (Signal.) *3 × 25 = 75.*
b. Now we'll work the division problem.
- How many rows are there? (Signal.) *3.*
- So what number do you divide by? (Signal.) *3.*
- Say the division problem for the whole rectangle. (Signal.) *75 ÷ 3.*
  The answer is the number of squares in the top row.
- What's the answer? (Signal.) *25.*
  (Change to show:)   [107:4E]

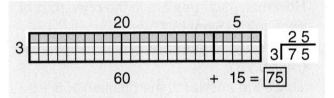

- How many shaded squares are in the tens part? (Signal.) *20.*
- How many shaded squares are in the ones part? (Signal.) *5.*
- How many shaded squares are in the whole row? (Signal.) *25.*
  So 75 divided by 3 equals 25. The diagram shows the multiplication and the division.

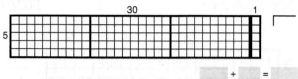

WORKBOOK PRACTICE

a. Find part 3 in your workbook. ✔
(Teacher reference:)

The whole rectangle shows 5 × 31. You're going to figure out the total number of squares in the rectangle then work the division problem.

- Say the multiplication for the tens. (Signal.) 5 × 30.
- What's the answer? (Signal.) 150.
- Write the answer below that part. ✔
- Say the multiplication for ones. (Signal.) 5 × 1.
- What's the answer? (Signal.) 5.

b. Write that answer and complete the addition equation for both parts. ✔

- Everybody, how many squares are in the whole rectangle? (Signal.) 155.

c. Now we'll work the division problem.

- How many rows are there? (Signal.) 5.
- So what number do you divide by? (Signal.) 5.
- Say the division problem for the whole rectangle. (Signal.) 155 ÷ 5.
  The answer is the number of squares in each row.
- How many squares are in the tens part of each row? (Signal.) 30.
- How many squares are in the ones part of each row? (Signal.) 1.
- How many squares are in each whole row? (Signal.) 31.
  That's the answer to the problem 155 ÷ 5.
- Write the problem and the answer next to the rectangle. Then shade the top row to show the answer. Remember, each row shows the division for the tens part and the division for the ones part. ✔

EXERCISE 5: SURFACE AREA
OF RECTANGULAR PRISMS

a. Last time, you worked problems to find the surface area of all the faces on a rectangular prism.

- How many faces does a rectangular prism have? (Signal.) 6.
  The faces are in pairs. Both faces of a pair have the same area.
- Which face has the same area as the bottom face? (Signal.) The top face.
- Which face has the same area as the back face? (Signal.) The front face.
- Which face has the same area as the left face? (Signal.) The right face.
  (Repeat until firm.)

b. (Display:) [107:5A]

SA = 2 (top + front + side)

Here's the equation for finding the area of all 6 faces: Surface area equals 2 times the quantity, top plus front plus side.

- Read the equation for finding the surface area of a rectangular prism. (Signal.) *Surface area equals 2 times the quantity, top plus front plus side.*
  Yes, we find the area of the top face, the front face, and a side face and add. Then we multiply the answer by 2. That gives us the area of all 6 faces.

c. (Display:) [107:5B]

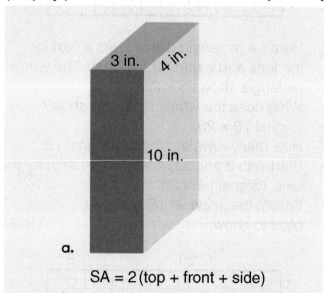

SA = 2 (top + front + side)

- In part A, start with the equation for surface area. Figure out the area of the top, the front, and the right side, and write the equation with those numbers. ✔

- Everybody, what's the number for the area of the top face? (Signal.) *12.*
- What's the number for the front face? (Signal.) *30.*
- What's the number for the right face? (Signal.) *40.*
  (Add to show:)   [107:5C]

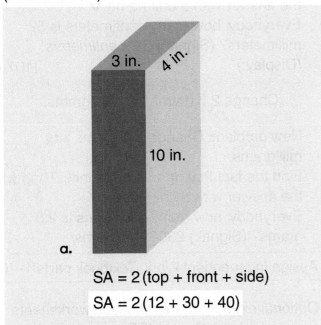

$$SA = 2\,(\text{top} + \text{front} + \text{side})$$
$$SA = 2\,(12 + 30 + 40)$$

Here's what you should have.

d. Add the 3 areas and write the total in the equation. ✔
- Everybody, what's the total of the 3 faces? (Signal.) *82.*
e. We multiply by 2 to find the area of all 6 faces. Do it and write the answer with a unit name. (Observe students and give feedback.)
- Everybody, what's the area of all 6 faces? (Signal.) *164 square inches.*
f. (Display:)   [107:5D]

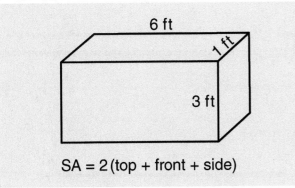

$$SA = 2\,(\text{top} + \text{front} + \text{side})$$

- New problem. Figure out the area of the top, the front, and the right side, and write the equation with those numbers. ✔
- Everybody, what's the number for the area of the top face? (Signal.) *6.*

- What's the number for the area of the front face? (Signal.) *18.*
- What's the number for the area of the right face? (Signal.) *3.*
  (Add to show:)   [107:5E]

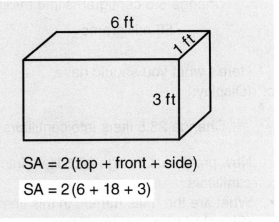

$$SA = 2\,(\text{top} + \text{front} + \text{side})$$
$$SA = 2\,(6 + 18 + 3)$$

Here's what you should have.

g. Add the 3 areas and write the total in the equation. ✔
- Everybody, what's the total for the 3 faces? (Signal.) *27.*
h. We multiply by 2 to find the area of all 6 faces. Do it and write the answer with a unit name. (Observe students and give feedback.)
- Everybody, what's the area of all 6 faces? (Signal.) *54 square feet.*

EXERCISE 6: UNIT CONVERSION
METRIC SYSTEM   REMEDY

a. Find the table of metric units on the inside back cover of your textbook. ✔
  I'll show problems. If you don't know the measurement fact for a problem, look it up in the table.
b. (Display:)   [107:6A]

Change 5.8 centigrams into milligrams.

Change 5.8 centigrams into milligrams.
- What are the units named in this item? (Signal.) *Centigrams and milligrams.*
- Find the fact that has both names. ✔
- Everybody, read the fact. (Signal.) *1 centigram is 10 milligrams.*
- Are there more centigrams or milligrams? (Signal.) *Milligrams.*
- So if you end up with milligrams, do you multiply or divide? (Signal.) *Multiply.*
- What do you multiply by? (Signal.) *10.*
  (Repeat until firm.)

- Write the answer and the unit name in part A. ✔
- Everybody, how many milligrams is 5.8 centigrams? (Signal.) *58 milligrams*. (Add to show:) [107:6B]

> Change 5.8 centigrams into milligrams.
>
> 58 milligrams

Here's what you should have.

c. (Display:) [107:6C]

> Change 28.5 liters into centiliters.

New problem: Change 28.5 liters into centiliters.
- What are the units named in this item? (Signal.) *Liters and centiliters*.
- Find the fact that names both units. Then write the answer with a unit name. ✔
- Everybody, how many centiliters is 28.5 liters? (Signal.) *2850 centiliters*.

d. (Display:) [107:6D]

> Change 802 meters into kilometers.

New problem: Change 802 meters into kilometers.
- Find the fact that names both units. Then write the answer with a unit name. ✔
- Everybody, how many kilometers is 802 meters? (Signal.) *802 thousandths kilometers*.

e. (Display:) [107:6E]

> Change 52 millimeters into centimeters.

New problem: Change 52 millimeters into centimeters.
- Find the fact that names both units. Then write the answer with a unit name. ✔
- Everybody, how many centimeters is 52 millimeters? (Signal.) *5.2 centimeters*.

f. (Display:) [107:6F]

> Change 2.3 grams into milligrams.

New problem: Change 2.3 grams into milligrams.
- Find the fact that names both units. Then write the answer with a unit name. ✔
- Everybody, how many milligrams is 2.3 grams? (Signal.) *2300 milligrams*.

Assign Independent Work, Textbook parts 1–10.

Optional extra math-fact practice worksheets are available on ConnectED.

Lesson

EXERCISE 1: LINE PLOT
AVERAGE

a. (Display:) [108:1A]

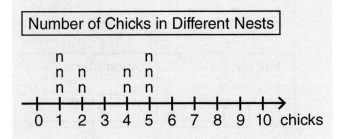

Number of Chicks in Different Nests

This line plot shows the number of chicks in different nests.

• What does this line plot show? (Signal.) *The number of chicks in different nests.*
• What do the numbers for the columns show? (Signal.) *Chicks.*
• What do the Ns stand for? (Signal.) *Nests.*
  (Repeat until firm.)
b. (Point to **1**.) Raise your hand when you know the total number of chicks for this column. ✔
• What's the total? (Signal.) *3.*
  Yes, there was one chick in each nest.
  (Add to show:) [108:1B]

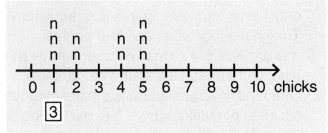

c. (Point to **2**.) Raise your hand when you know the total number of chicks for this column. ✔
• What's the total? (Signal.) *4.*
  (Add to show:) [108:1C]

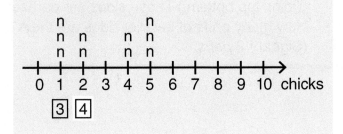

d. (Point to **4**.) Raise your hand when you know the total number of chicks for this column. ✔
• What's the total? (Signal.) *8.*
  (Add to show:) [108:1D]

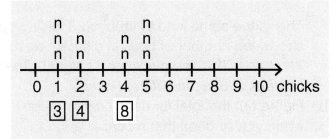

e. (Point to **5**.) Raise your hand when you know the total number of chicks for this column. ✔
• What's the total? (Signal.) *15.*
  (Add to show:) [108:1E]

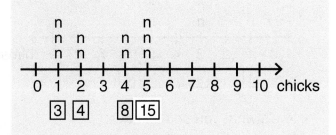

f. Now we find the average number of chicks in each nest. We first add the totals, then divide by the number of nests.
• Work the problem in part A. Remember, divide by the number of nests.
  (Observe students and give feedback.)
g. Check your work.
  You found the total number of chicks for all the columns.
• What's the total number of chicks? (Signal) *30.*
• What did you divide by? (Signal.) *10.*
  Yes, there are 10 nests.
• What's the average number of chicks in each nest? (Signal.) *3 chicks.*
• Yes, 3 chicks. Make sure you have the unit name **chicks** in your answer. ✔

WORKBOOK PRACTICE

a. Open your workbook to Lesson 108 and find part 1. ✔
(Teacher reference:)

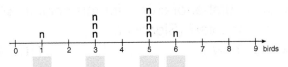

a. This line plot shows the number of birds in different nests.

This is the same kind of problem. The line plot shows the number of birds in different nests.

• What does this line plot show? (Signal.) *The number of birds in different nests.*

b. Figure out the total for each column. Stop when you've done that much.
(Observe students and give feedback.)
(Display:) [108:1F]

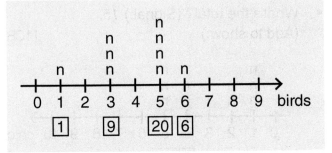

Here's what you should have.

c. Now you'll figure out the average number of birds in each nest.

• Add the totals to find the number of birds. Then divide by the number of nests.
(Observe students and give feedback.)

d. Check your work.

• What's the total number of birds? (Signal.) *36.*

• What number did you divide by? (Signal.) *9.*

• What's the average number of birds in each nest? (Signal.) *4 birds.*

• Make sure you have the unit name **birds** in your answer. ✔

EXERCISE 2: QUADRILATERALS

HIERARCHY

a. (Display:) [108:2A]

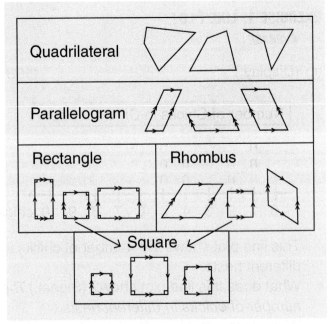

This table shows quadrilaterals.

• Say **quadrilaterals.** (Signal.) *Quadrilaterals.*
(Repeat until firm.)

b. A quadrilateral is a figure with four straight sides and four angles.

• How many straight sides? (Signal.) *4 straight sides.*

• How many angles? (Signal.) *4 angles.*

c. (Point to quadrilaterals, no parallel sides.) The only name for these figures is quadrilateral. They have four sides and four angles.
You can see the names of other figures that have four straight sides and four angles.

d. One type of quadrilateral is a parallelogram. (Point to parallelograms.) Say **parallelogram.** (Signal.) *Parallelogram.*
All parallelograms have two pairs of parallel sides. The little arrows show the pairs of sides that are parallel.

e. (Point: left, right.) These sides are parallel. (Point: top bottom.) These sides are parallel.

• How many pairs of parallel sides are there? (Signal.) *2 pairs.*

f. So all the figures below are parallelograms. The types of parallelograms are: Rectangle, rhombus, and square.
- Say **rhombus.** (Signal.) *Rhombus.*
- Say **rectangle, rhombus, and square.** (Signal.) *Rectangle, rhombus, and square.*

g. Does a rectangle have two pairs of parallel sides? (Signal.) *Yes.*
- Does a square have two pairs of parallel sides? (Signal.) *Yes.*
  You can see the pictures of the rhombus. It also has two pairs of parallel sides.

h. What are a rectangle, a rhombus, and a square? (Signal.) *Parallelograms.*
- Are they also quadrilaterals? (Signal.) *Yes.*
- What makes them quadrilaterals? (Call on a student. Idea: *They have 4 straight sides and 4 angles.*)
- Everybody, are rectangles, rhombuses, and squares also parallelograms? (Signal.) *Yes.*
- What makes them parallelograms? (Signal.) *They have 2 pairs of parallel sides.*
  Yes, they have two pairs of parallel sides.

i. (Point to rectangles.) Does a rectangle have four straight sides? (Signal.) *Yes.*
- How many angles does a rectangle have? (Signal.) *4.*
- How many degrees are in each angle? (Signal.) *90.*
  Yes, all angles of a rectangle are 90 degrees. You can see the 90-degree markers.

j. Is a rectangle a quadrilateral? (Signal.) *Yes.*
- How do you know? (Call on a student. Idea: *It has 4 straight sides and 4 angles.*)
  Yes, it has four straight sides and four angles.

k. Is a rectangle a parallelogram? (Signal.) *Yes.*
- How do you know? (Call on a student. Idea: *It has 2 pairs of parallel sides.*)
- How is a rectangle different from other parallelograms? (Call on a student. Idea: *It has four 90-degree angles.*)
  Yes, all angles of a rectangle are 90 degrees.

l. (Point to squares.) An arrow goes from the rectangles to the squares because a square is a special kind of rectangle.
- How is a square different from other rectangles? (Call on a student. Idea: *All sides are the same length.*)
  Yes, all sides of a square are the same length.

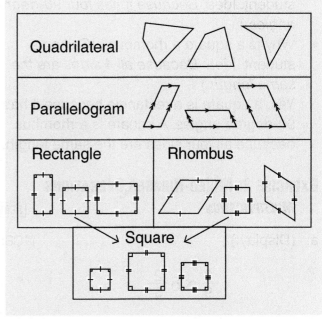

The little lines show the sides that are the same length.

m. (Point to rhombus.) This parallelogram is a rhombus. Say **rhombus.** (Signal.) *Rhombus.* All sides of a rhombus are the same length.
- Say the rule about the sides of a rhombus. (Signal.) *All sides of a rhombus are the same length.*
  A rhombus is a parallelogram, so a rhombus has two pairs of sides that are parallel.
- Is a rhombus a parallelogram? (Signal.) *Yes.*
- Why is it a parallelogram? (Call on a student. Idea: *It has two pairs of parallel sides.*)
  All sides of a rhombus are the same length. So a square is a rhombus.
  You can see an arrow going from the rhombus to the square. The square has two pairs of parallel sides and the sides are the same length, so a square is a rhombus.

n. (Point to top row.) Once more: What is the name for any figure that has four straight sides and four angles? (Signal.) *Quadrilateral.*
- What is the quadrilateral called if it has two pairs of parallel sides? (Signal.) *Parallelogram.*
- What is any parallelogram called if all the angles are 90 degrees? (Signal.) *Rectangle.*
- What is any parallelogram called if all the sides are the same length? (Signal.) *Rhombus.*
- What is a rhombus called if all the angles are 90 degrees? (Signal.) *Square.*
  (Repeat until firm.)

o. Why is a square a rectangle? (Call on a student. Idea: *Because it has four 90-degree angles.*)

- Why is a square a rhombus? (Call on a student. Idea: *Because all 4 sides are the same length.*)

Yes, a square is a rectangle because it has 90-degree angles. A square is a rhombus because all four sides are the same length.

EXERCISE 3: MIXED-NUMBER OPERATIONS

MULTIPLICATION

REMEDY

a. (Display:) [108:3A]

$$\frac{2}{9} \times 5\frac{2}{5}$$

Last time, you worked problems that multiply a fraction and a mixed number. Remember, simplify before you multiply.

- Read the problem. (Signal.) *2/9 × 5 and 2/5.* We can't simplify before we change the mixed number into a fraction.
- Raise your hand when you know the fraction that equals 5 and 2/5. ✔
- What fraction? (Signal.) *27/5.*

(Add to show:) [108:3B]

$$\frac{2}{9} \times 5\frac{2}{5}$$

$$\frac{2}{9} \times \frac{27}{5} =$$

- Read the problem. (Signal.) *2/9 × 27/5.*

b. You can simplify 27 over 9. What does it equal? (Signal.) *3.*

Yes, 3.

(Add to show:) [108:3C]

$$\frac{2}{9} \times 5\frac{2}{5}$$

$$\frac{2}{\cancel{9}} \times \frac{\cancel{27}^{3}}{5} =$$

- Can we simplify anything else? (Signal.) *No.*
- Raise your hand when you know the fraction we get when we multiply. ✔
- What fraction? (Signal.) *6/5.*

(Add to show:) [108:3D]

$$\frac{2}{9} \times 5\frac{2}{5}$$

$$\frac{2}{\cancel{9}} \times \frac{\cancel{27}^{3}}{5} = \frac{6}{5}$$

- Raise your hand when you know the mixed-number answer. ✔
- What's the answer? (Signal.) *1 and 1/5.*

(Add to show:) [108:3E]

$$\frac{2}{9} \times 5\frac{2}{5}$$

$$\frac{2}{\cancel{9}} \times \frac{\cancel{27}^{3}}{5} = \frac{6}{5} = \boxed{1\frac{1}{5}}$$

Yes, 1 and 1/5.

c. (Display:) [108:3F]

$$3\frac{1}{9} \times \frac{3}{7}$$

- New problem. Read it. (Signal.) *3 and 1/9 × 3/7.* First we change the mixed number into a fraction. Then we simplify if we can.
- Raise your hand when you know the fraction for 3 and 1/9. ✔
- What fraction? (Signal.) *28/9.*

(Add to show:) [108:3G]

$$3\frac{1}{9} \times \frac{3}{7}$$

$$\frac{28}{9} \times \frac{3}{7} =$$

- Copy the problem and work it in part A. You can simplify twice before you multiply. (Observe students and give feedback.)

d. You simplified 28 over 7. What does that equal? (Signal.) *4.*

- You simplified 3 over 9. What does that equal? (Signal.) *1/3.*

(Add to show:) [108:3H]

$$3\frac{1}{9} \times \frac{3}{7}$$

$$\frac{\cancel{28}^{4}}{\cancel{9}_{3}} \times \frac{\cancel{3}^{1}}{\cancel{7}} = \frac{4}{3} = \boxed{1\frac{1}{3}}$$

Here's what you should have.

- What's the fraction answer? (Signal.) *4/3.*
- What's the mixed-number answer? (Signal.) *1 and 1/3.*

WORKBOOK PRACTICE

a. Find part 2 in your workbook. ✔
(Teacher reference:) R Part B

a. $\frac{3}{4} \times 2\frac{2}{3}$ b. $1\frac{1}{10} \times \frac{5}{6}$

b. Problem A: 3/4 × 2 and 2/3.
• Change the mixed number into a fraction, simplify, and do the multiplication.
(Observe students and give feedback.)

c. Check your work.
• You changed 2 and 2/3 into a fraction. What fraction? (Signal.) *8/3.*
You simplified 3/3 and 8/4. You worked the problem 1 times 2.
• What's the answer? (Signal.) *2.*
(Display:) [108:3I]

$$\text{a.}\quad \frac{3}{4} \times 2\frac{2}{3}$$
$$\frac{\cancel{3}}{4} \times \frac{\cancel{8}}{\cancel{3}} = 2$$

Here's what you should have.

d. Problem B: 1 and 1/10 × 5/6.
• Change the mixed number into a fraction, simplify, and do the multiplication.
(Observe students and give feedback.)

e. Check your work.
• You changed 1 and 1/10 into a fraction. What fraction? (Signal.) *11/10.*
• You simplified 5 over 10. What does it equal? (Signal.) *1/2.*
• You multiplied 11 times 1 and 2 times 6. What's the answer? (Signal.) *11/12.*
(Display:) [108:3J]

$$\text{b.}\quad 1\frac{1}{10} \times \frac{5}{6}$$
$$\frac{11}{\cancel{10}_{2}} \times \frac{\cancel{5}^{1}}{6} = \frac{11}{12}$$

Here's what you should have.

EXERCISE 4: DISTRIBUTIVE PROPERTY
CHECKING DIVISION ANSWERS

a. Find part 3 in your workbook. ✔
(Teacher reference:)

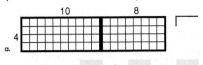

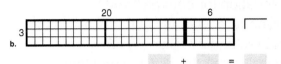

Rectangle A. You're going to figure out the total number of squares in the rectangle, then work the division problem.
• Say the multiplication for squares in the tens part. (Signal.) *4 × 10.*
• What's the answer? (Signal.) *40.*
• Write the answer below that part. ✔

b. Say the multiplication for the ones part. (Signal.) *4 × 8.*
• What's the answer? (Signal.) *32.*
• Write the number for the ones and complete the equation for both parts. ✔
• Everybody, how many squares are in the whole rectangle? (Signal.) *72.*

c. Now we'll work the division problem.
• How many rows are there? (Signal.) *4.*
• So what number do you divide by? (Signal.) *4.*
• Say the division problem for finding the number of squares in the top row. (Signal.) *72 ÷ 4.*
• Write that problem next to the rectangle. ✔
• Everybody, how many squares are in the tens part of that row? (Signal.) *10.*
• How many squares are in the ones part of that row? (Signal.) *8.*
• How many squares are in the whole row? (Signal.) *18.*
That's the answer to the problem 72 divided by 4.
• Write the answer. ✔
Remember, each row shows the division for the tens part and the division for the ones part.

d. Rectangle B. You're going to figure out the total number of squares in the rectangle, then work the division problem.
- Say the multiplication for the squares in the tens part. (Signal.) *3 × 20.*
- Write the answer below that part. ✔
- Everybody, say the multiplication for the squares in the ones part. (Signal.) *3 × 6.*
- Write the number for that part and complete the equation for both parts. ✔
- Everybody, how many squares are in the whole rectangle? (Signal.) *78.*
e. Now we'll work the division problem.
- How many rows are there? (Signal.) *3.*
- So what number do you divide by? (Signal.) *3.*
- Say the division problem for finding the number of squares in the top row. (Signal.) *78 ÷ 3.*
- Write the problem next to the rectangle. Then stop. ✔
- Everybody, how many squares are in the tens part of the top row? (Signal.) *20.*
- How many squares are in the ones part of that row? (Signal.) *6.*
- How many squares are in the whole row? (Signal.) *26.*
  That's the answer to the problem 78 divided by 3.
- Write the answer. ✔
f. Remember, each row shows the division for the tens part and the division for the ones part.

EXERCISE 5: UNIT CONVERSION
WORD PROBLEMS   REMEDY

a. Open your textbook to Lesson 108 and find part 1. ✔
   (Teacher reference:)

a. A rope was 90 feet long. 15 yards were cut from the rope. How many feet long is the rope now?

b. The carpenter had 3 boards. One was 3 yards long. One was 45 inches long. The last board was 2 yards long. What's the total number of yards for all three boards?

c. There were 27 gallons of milk in the tank. Then 11 quarts of milk were added. How many gallons of milk are in the tank now?

Part 1	
a.	

Each problem names two related units. The question tells which unit name you need for all of the measurements.

b. Problem A: A rope was 90 feet long. 15 yards were cut from the rope. How many feet long is the rope now?
- What unit is named in the question? (Signal.) *Feet.*
- One measurement refers to yards. Change that measurement to feet and work the problem.
  (Observe students and give feedback.)
- Check your work. You changed 15 yards into feet.
  How many feet is 15 yards? (Signal.) *45 feet.*
- You worked the problem 90 minus 45. How long is the rope now? (Signal.) *45 feet.*
c. Problem B: The carpenter had 3 boards. One was 3 yards long. One was 45 inches long. The last board was 2 yards long. What's the total number of yards for all three boards?
- What unit does the question name? (Signal.) *Yards.*
  So you have to change one of the measurements into yards. When you work this problem, you'll end up with a mixed number. Remember to simplify.
- Work the problem.
  (Observe students and give feedback.)
- Everybody, you changed 45 inches into yards. How many yards is 45 inches? (Signal.) *1 and 1/4 yards.*
- You worked the problem 3 plus 1 and 1/4 plus 2. What's the answer? (Signal.) *6 and 1/4 yards.*
d. Problem C: There were 27 gallons of milk in the tank. Then 11 quarts of milk were added. How many gallons of milk were in the tank now?
- What unit does the question name? (Signal.) *Gallons.*
- Work the problem.
  (Observe students and give feedback.)
- Everybody, you changed 11 quarts into gallons.
  What's the mixed number? (Signal.) *2 and 3/4.*
- You worked the problem 27 plus 2 and 3/4. How much milk was in the tank now? (Signal.) *29 and 3/4 gallons.*

EXERCISE 6: SURFACE AREA
OF RECTANGULAR PRISMS

a. (Display:) [108:6A]

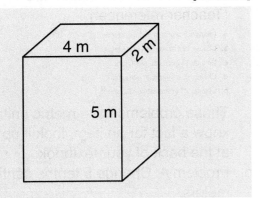

We're going to find the surface area of this rectangular prism.
- How many faces does the prism have? (Signal.) *6.*

b. We use an equation that figures out the area of 3 faces, then multiplies by 2.
- Raise your hand when you can say the equation. ✔
- Say the equation. (Signal.) *Surface area equals 2 times the quantity, top plus front plus side.*

(Add to show:) [108:6B]

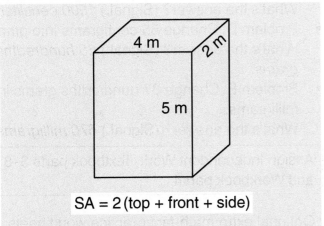

SA = 2 (top + front + side)

c. Which face do we figure out first? (Signal.) *The top face.*
- One number for the top is 4. What's the other number for the top? (Signal.) *2.*
- Say the multiplication for the top. (Signal.) *4 × 2.* (Repeat until firm.)

d. Which face do we figure out next? (Signal.) *The front face.*
- One number for the front is 5. What's the other number for the front? (Signal.) *4.*

e. Which face do we figure out next? (Signal.) *A side face.*
- One number for a side is 2. What's the other number for the side? (Signal.) *5.*

f. In part B, write the equation for the surface area. Below, write the equation with a number for the top, the front, and a side. Then figure out the surface area.
(Observe students and give feedback.)
- What's the number for the top face? (Signal.) *8.*
- What's the number for the front face? (Signal.) *20.*
- What's the number for the side face? (Signal.) *10.*

(Add to show:) [108:6C]

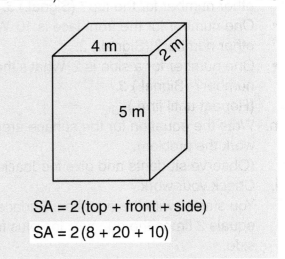

SA = 2 (top + front + side)

SA = 2 (8 + 20 + 10)

- You added 8 plus 20 plus 10. What's the total for those faces? (Signal.) *38.*
- Then you multiplied 38 times 2. What's the answer? (Signal.) *76.*
- What's the unit name? (Signal.) *Square meters.*

(Add to show:) [108:6D]

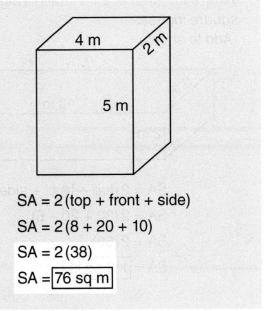

SA = 2 (top + front + side)

SA = 2 (8 + 20 + 10)

SA = 2 (38)

SA = 76 sq m

Here's what you should have.

g. (Display:) [108:6E]

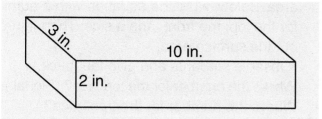

You're going to figure out the surface area of this rectangular prism.

• One number for the top face is 10. What's the other number for the top? (Signal.) *3.*
• One number for the front face is 10. What's the other number? (Signal.) *2.*
• One number for a side is 2. What's the other number? (Signal.) *3.*
(Repeat until firm.)

h. Write the equation for the surface area and work the problem.
(Observe students and give feedback.)

i. Check your work.
You started with the equation: Surface area equals 2 times the quantity, top plus front plus side.

• What's the number for the top? (Signal.) *30.*
• What's the number for the front? (Signal.) *20.*
• What's the number for a side? (Signal.) *6.*
You added 30 plus 20 plus 6.
• What's the total for those faces? (Signal.) *56.*
You multiplied 56 by 2.
• What's the area of all 6 faces? (Signal.) *112 square inches.*
(Add to show:) [108:6F]

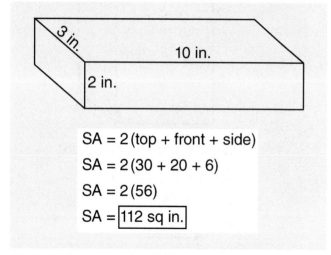

$$SA = 2\,(\text{top} + \text{front} + \text{side})$$
$$SA = 2\,(30 + 20 + 6)$$
$$SA = 2\,(56)$$
$$SA = \boxed{112 \text{ sq in.}}$$

Here's what you should have.

EXERCISE 7: UNIT CONVERSION
METRIC SYSTEM

a. Find part 2 in your textbook. ✔
(Teacher reference:)

a. Change .5 centimeters into meters.
b. Change 228 milligrams into grams.
c. Change 15 liters into centiliters.
d. Change 65 centigrams into grams.
e. Change .67 grams into milligrams.

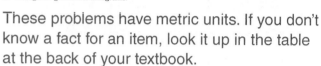

These problems have metric units. If you don't know a fact for an item, look it up in the table at the back of your textbook.

b. Problem A: Change 5 tenths centimeters into meters.
• Work the problem. ✔
• Everybody, how many meters is .5 centimeters? (Signal.) *5 thousandths meters.*
c. Problem B: Change 228 milligrams into grams.
• Work the problem. ✔
• Everybody, how many grams is 228 milligrams? (Signal.) *228 thousandths grams.*
d. Work the rest of the items in part 2.
(Observe students and give feedback.)
e. Check your work.
• Problem C: Change 15 liters into centiliters. What's the answer? (Signal.) *1500 centiliters.*
• Problem D: Change 65 centigrams into grams. What's the answer? (Signal.) *65 hundredths grams.*
• Problem E: Change 67 hundredths grams into milligrams.
What's the answer? (Signal.) *670 milligrams.*

Assign Independent Work, Textbook parts 3–8 and Workbook part 4.

Optional extra math-fact practice worksheets are available on ConnectED.

Lesson 109

EXERCISE 1: LINE PLOT
   AVERAGE

a. Open your workbook to Lesson 109 and find part 1. ✔
   (Teacher reference:)

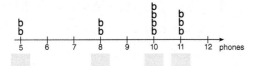

a. This line plot shows the number of phones in different buildings.

b. This line plot shows the number of chairs in different rooms.

Problem A. The sentence tells about the line plot. This line plot shows the number of phones in different buildings.

• Do the numbers for the columns show phones or buildings? (Signal.) *Phones.*
   The letter B stands for building.
   You're going to find the average number of phones in each building. To do that, you first figure out the total number of phones for each column. Then you add the totals for phones. Then you divide by the number of **buildings.**

b. Figure out the total for each column. Stop when you've done that much.
   (Observe students and give feedback.)
   (Display:)   [109:1A]

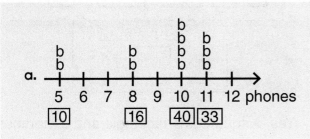

Here's what you should have.

c. Now find the average number of phones. Add the totals for phones. Then divide by the number of buildings.
   (Observe students and give feedback.)

d. Check your work.
• What's the total for phones? (Signal.) *99.*
• What did you divide by? (Signal.) *11.*
   Yes, there are 11 buildings, so you divide by 11.
• What's the average number of phones in each building? (Signal.) *9 phones.*
• Make sure you have the unit name **phones** in your answer. ✔

e. Problem B. This line plot shows the number of chairs in different rooms.
• Do the numbers show chairs or rooms? (Signal.) *Chairs.*
   Each R tells about a room.
• Touch the zero column. ✔
• What's the total for the zero column? (Signal.) *Zero.*
   Yes, zero plus zero equals zero.

f. W rite the totals for all the columns. Stop when you've done that much.
   (Observe students and give feedback.)
   (Display:)   [109:1B]

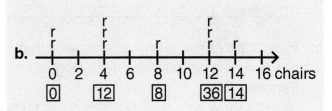

Here's what you should have.
• Now find the average number of chairs in each room.
   (Observe students and give feedback.)

g. Check your work.
• What's the total for chairs? (Signal.) *70.*
• What did you divide by? (Signal.) *10.*
   Yes, there are 10 rooms.
• What's the average number of chairs in a room? (Signal.) *7 chairs.*
• Make sure you have the unit name **chairs** in your answer. ✔
   You're working some hard problems.

EXERCISE 2: QUADRILATERALS

HIERARCHY

R Test 12

a. (Display:)   [109:2A]

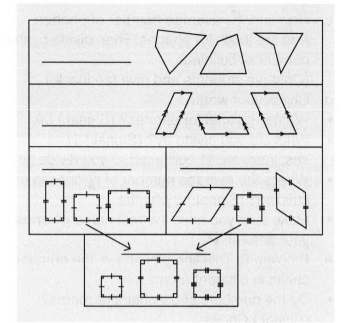

• What is the name for any figure that has four straight sides and four angles? (Signal.) *Quadrilateral.*
Yes, a quadrilateral has four straight sides and four angles.
(Add to show:)   [109:2B]

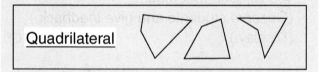

b. What's the name of any quadrilateral that has two pairs of parallel sides? (Signal.) *Parallelogram.*
(Add to show:)   [109:2C]

c. What's the name of any parallelogram that has four 90-degree angles? (Signal.) *Rectangle.*
(Add to show:)   [109:2D]

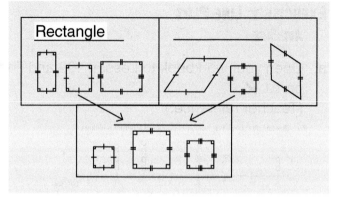

d. What's the name of any parallelogram that has four sides of the same length? (Signal.) *Rhombus.*
(Add to show:)   [109:2E]

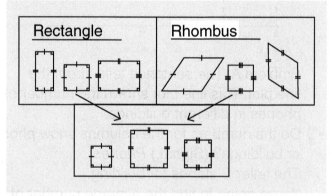

e. What's the name of a rhombus that has four 90-degree angles? (Signal.) *Square.*
(Repeat until firm.)
(Add to show:)   [109:2F]

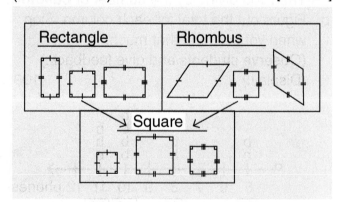

Yes, a square is a rectangle and a rhombus.

a. Find part 2 in your workbook. ✔
(Teacher reference:) R Test 12: Part A

a. _____

b. _____

c. _____ d. _____

e. _____

f. What is any parallelogram with four 90-degree angles? _____

g. What is any quadrilateral with two pairs of parallel sides? _____

h. What is any parallelogram with all sides the same length? _____

i. What is any rectangle with all sides the same length? _____

(Display:) [109:2G]

square, parallelogram, rhombus,
quadrilateral, rectangle

Here are the names that you'll put in the chart.

b. Read each name. (Point to each name
as students say:) *Square, parallelogram,
rhombus, quadrilateral, rectangle.*

• Put the names in the correct places in the
chart. Make sure you spell them correctly.
(Observe students and give feedback.)

c. Check your work.

• What's the name for A? (Signal.) *Quadrilateral.*

• What's the name for B? (Signal.)
Parallelogram.

• What's the name for C? (Signal.) *Rectangle.*

• What's the name for D? (Signal.) *Rhombus.*

• What's the name for E? (Signal.) *Square.*

d. Write answers to the questions below the table.
(Observe students and give feedback.)

e. Check your work.

• Item F: What is any parallelogram with four
90-degree angles? (Signal.) *Rectangle.*

• G: What is any quadrilateral with two pairs of
parallel sides? (Signal.) *Parallelogram.*

• H: What is any parallelogram with all sides the
same length? (Signal.) *Rhombus.*

• I. What is any rectangle with all sides the
same length? (Signal.) *Square.*

EXERCISE 3: MIXED-NUMBER OPERATIONS
MULTIPLICATION   REMEDY

a. Open your textbook to Lesson 109 and find
part 1. ✔
(Teacher reference:)

Copy each problem. Write the new problem below. Simplify. Then multiply.

a. $2\frac{3}{10} \times \frac{20}{7}$ b. $\frac{9}{5} \times 6\frac{2}{3}$ c. $\frac{1}{7} \times 2\frac{4}{5}$ d. $3\frac{3}{8} \times \frac{4}{9}$

These are multiplication problems that have
a mixed number and a fraction. I'll read the
directions: **Copy each problem. Write the
new problem below. Simplify. Then multiply.**

• You change the mixed number into a fraction.
Then what do you do? (Signal.) *Simplify.*
Yes, you simplify before you multiply.

b. Problem A: 2 and 3/10 × 20/7.

• Copy the problem. Below, write the new
problem and work it. Then write the answer as
a mixed number.
(Observe students and give feedback.)

c. Check your work.

• What fraction did you write for 2 and 3/10?
(Signal.) *23/10.*

• You simplified 20 tenths. What does it equal?
(Signal.) *2.*

• Then you multiplied. What's the answer?
(Signal.) *46/7.*

• What's the mixed number answer? (Signal.)
6 and 4/7.

d. Problem B: 9/5 × 6 and 2/3.

• Copy and work the problem.
(Observe students and give feedback.)

e. Check your work.

• What fraction did you write for 6 and 2/3?
(Signal.) *20/3.*

• You simplified 9 over 3. What does it equal?
(Signal.) *3.*

• You simplified 20 over 5. What does it equal?
(Signal.) *4.*

• Then you multiplied 3 times 4. What's the
answer? (Signal.) *12.*

f. Work the rest of the problems in part 1. If the
answer is more than one, write it as a whole
number or a mixed number.
(Observe students and give feedback.)

g. Check your work.
- Problem C: 1/7 × 2 and 4/5.
- What's the answer to the problem? (Signal.) *2/5.*
(Display:) [109:3A]

$$c. \quad \frac{1}{7} \times 2\frac{4}{5}$$
$$\frac{1}{\cancel{7}} \times \frac{\cancel{14}^{2}}{5} = \boxed{\frac{2}{5}}$$

Here's what you should have.
h. Problem D: 3 and 3/8 × 4/9.
- What's the mixed number answer to the problem? (Signal.) *1 and 1/2.*
(Display:) [109:3B]

$$d. \quad 3\frac{3}{8} \times \frac{4}{9}$$
$$\frac{\cancel{27}^{3}}{\cancel{8}_{2}} \times \frac{\cancel{4}^{1}}{\cancel{9}} = \frac{3}{2} = \boxed{1\frac{1}{2}}$$

Here's what you should have.

EXERCISE 4: DISTRIBUTIVE PROPERTY
CHECKING DIVISION ANSWERS

a. Find part 2 in your textbook. ✔
(Teacher reference:)

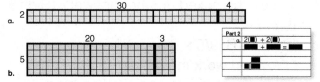

You're going to work the multiplication and division problem for these rectangles.
b. Problem A. You'll figure out the area of the whole rectangle.
- Write the multiplication for the tens part and the ones part. Stop when you've done that much. ✔
- Everybody, what's the multiplication for the tens part? (Signal.) *2 × 30.*
- What's the multiplication for the ones part? (Signal.) *2 × 4.*
(Display:) [109:4A]

a. 2 (30) + 2 (4)

c. Below, write the answer for the tens and ones and figure out the area of the whole rectangle. ✔
- Everybody, read the addition equation. (Signal.) *60 + 8 = 68.*

(Add to show:) [109:4B]

a. 2 (30) + 2 (4)
   60 + 8 = 68

Here's what you should have.
- What's the area of the whole rectangle? (Signal.) *68 square units.*
d. Now you'll do the division the fast way.
- Write the division problem for the whole rectangle. Remember, the number of rows tells what you divide by. Stop when you've written the problem. ✔
- Everybody, read the division problem. (Signal.) *68 ÷ 2.*
The answer is the number of squares in each row.
•. Raise your hand when you know the number of squares in the top row. ✔
- Everybody, how many squares are in the top row? (Signal.) *34.*
- Write 34 as the answer to the division problem. ✔
e. Problem B. Write the multiplication for the tens part and the ones part. Stop when you've done that much. ✔
- Everybody, what's the multiplication for the tens part? (Signal.) *5 × 20.*
- What's the multiplication for the ones part? (Signal.) *5 × 3.*
(Display:) [109:4C]

b. 5 (20) + 5 (3)

Here's what you should have.
f. Below, write the answer for the tens and ones and figure out the area of the whole rectangle. ✔
- Everybody, what's the area of the whole rectangle? (Signal.) *115 square units.*
g. Now you'll do the division the fast way.
- Write the division problem for the whole rectangle. Remember, the number of rows tells what you divide by. Stop when you've written the problem. ✔
- Everybody, read the division problem. (Signal.) *115 ÷ 5.*
The answer is the number of squares in each row.
h. Figure out the number of squares in the top row and write it as the answer to the division problem. ✔
- Everybody, what's 115 divided by 5? (Signal.) *23.*
Yes, the top row has 2 tens and 3 ones.

EXERCISE 5: UNIT CONVERSION
WORD PROBLEMS

a. Find part 3 in your textbook. ✔
(Teacher reference:)

   a. A dog weighed 46 pounds. A cat weighed 96 ounces. How many pounds did the two animals weigh together?

   b. The loaf of bread lasted 3 days. The mustard lasted 12 weeks. How many more days did the mustard last than the bread?

   c. The cat was 39 months old. The dog was 3 years old. How many months older was the cat than the dog?

b. Problem A: A dog weighed 46 pounds. A cat weighed 96 ounces. How many pounds did the two animals weigh together?

• What unit does the question name? (Signal.) *Pounds.*
• Change all the measurements into pounds and work the problem.
(Observe students and give feedback.)
Check your work.
You changed 96 ounces into pounds.
• How many pounds? (Signal.) *6 pounds.*
• Read the problem you worked. (Signal.) *46 + 6.*
• What's the whole answer? (Signal.) *52 pounds.*
Yes, together both animals weighed 52 pounds.

c. Problem B: The loaf of bread lasted 3 days. The mustard lasted 12 weeks. How many more days did the mustard last than the bread?

• What unit does the question name? (Signal.) *Days.*
• Change all the measurements into days and work the problem.
(Observe students and give feedback.)
You changed 12 weeks into days.
• Did you multiply or divide? (Signal.) *Multiply.*
• How many days is 12 weeks? (Signal.) *84 days.*
Then you worked the problem 84 minus 3.
• How many days longer did the mustard last? (Signal.) *81 days.*

d. Problem C: The cat was 39 months old. The dog was 3 years old. How many months older was the cat than the dog?

• Work the problem.
(Observe students and give feedback.)
You changed 3 years into months.
• How many months is 3 years? (Signal.) *36 months.*
You worked the problem 39 minus 36.
• How many months older was the cat? (Signal.) *3 months.*

EXERCISE 6: SURFACE AREA
OF RECTANGULAR PRISMS

a. Find part 4 in your textbook.
(Teacher reference:)

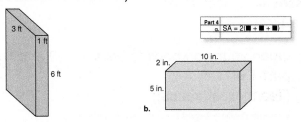

You're going to find the surface area of these rectangular prisms.

b. Problem A. Write the equation. Then stop. ✔
• Everybody, read the equation. (Signal.)
Surface area equals 2 times the quantity, top plus front plus side.

c. Work the problem. Remember to write the unit name in the answer.
(Observe students and give feedback.)

d. Check your work. You figured out the area of 3 faces.
• What's the number for the top? (Signal.) *3.*
• What's the number for the front? (Signal.) *6.*
• What's the number for the left? (Signal.) *18.*
• What's 3 plus 6 plus 18? (Signal.) *27.*
• You multiplied 27 by 2. What's the area of all 6 faces? (Signal.) *54 square feet.*

e. Problem B. Work the problem. Remember to write the unit name in the answer.
(Observe students and give feedback.)

f. Check your work.
You figured out the area of 3 faces.
• What's the number for the top? (Signal.) *20.*
• What's the number for the front? (Signal.) *50.*
• What's the number for the right? (Signal.) *10.*
• What's 20 plus 50 plus 10? (Signal.) *80.*
• You multiplied 80 by 2. What's the area of all 6 faces? (Signal.) *160 square inches.*

> Assign Independent Work, Textbook parts 5–11 and Workbook parts 3 and 4.

> Optional extra math-fact practice worksheets are available on ConnectED.

Lesson 110

EXERCISE 1: LINE PLOT
AVERAGE

a. Open your workbook to Lesson 110 and find part 1. ✔
(Teacher reference:)

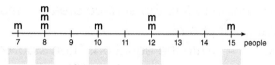

a. This line plot shows the number of people in different minibuses.

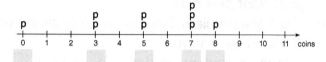

b. This line plot shows the number of coins in different pockets.

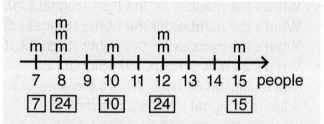

You're going to figure out the average for each problem.

b. Problem A. Read the sentence that tells about the line plot. (Signal.) *This line plot shows the number of people in different minibuses.*
• Do the numbers show people or minibuses? (Signal.) *People.*

c. Figure out the total number for each of the columns that has minibuses. Stop when you've written the totals.
(Observe students and give feedback.)

d. Check your work.
(Display:)   [110:1A]

m
m m   m
m m m   m m   m
7 8 9 10 11 12 13 14 15 people
[7] [24]   [10]   [24]   [15]

Here's what you should have.

e. Now figure out the average number of people in each minibus.
(Observe students and give feedback.)

f. Check your work.
• What's the total for people? (Signal.) *80.*
• What did you divide by? (Signal.) *8.*
Yes, there are 8 minibuses.

• What's the average number of people in each minibus? (Signal.) *10 people.*
Yes, 80 divided by 8 is 10.
• Make sure you have the unit name **people** in your answer. ✔

g. Problem B. Work the problem and find the average number of coins in each pocket.
(Observe students and give feedback.)

h. Check your work.
• What's the total for coins? (Signal.) *45.*
• What did you divide by? (Signal.) *9.*
Yes, there are 9 pockets.
• What's the average number of coins in each pocket? (Signal.) *5 coins.*

EXERCISE 2: QUADRILATERALS
HIERARCHY

a. (Display:)   [110:2A]

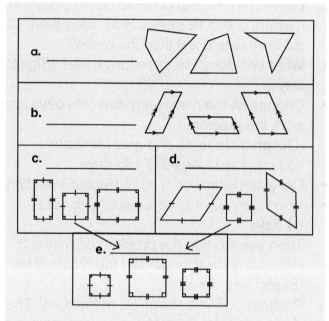

You learned the names for these figures.
What's the name for all the figures? (Signal.) *Quadrilateral.*
• (Point to **A.**) So what name goes in space A? (Signal.) *Quadrilateral.*
• (Point to **B.**) What name is B? (Signal.) *Parallelogram.*
• (Point to **C.**) What name is C? (Signal.) *Rectangle.*
• (Point to **D.**) What name is D? (Signal.) *Rhombus.*
• (Point to **E.**) What name is E? (Signal.) *Square.*
(Repeat until firm.)

a. Open your textbook to Lesson 110 and find part 1. ✔
(Teacher reference:)

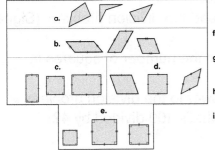

f. What is any rectangle with all sides the same length?

g. What is any quadilateral with two pairs of parallel sides?

h. What is any parallelogram with four 90-degree angles?

i. What is any parallelogram with all sides the same length?

(Display:)   [110:2B]

> parallelogram, square, quadrilateral, rectangle, rhombus

Here are the names that go in the chart.

- Read the names. (Point to each name as students read:) *Parallelogram, square, quadrilateral, rectangle, rhombus.*

b. Write the names for A through E. Make sure you spell them correctly.
(Observe students and give feedback.)

c. Check your work.

- What's the name for A? (Signal.) *Quadrilateral.*
- What's the name for B? (Signal.) *Parallelogram.*
- What's the name for C? (Signal.) *Rectangle.*
- What's the name for D? (Signal.) *Rhombus.*
- What's the name for E? (Signal.) *Square.*

d. Write answers to questions F through I.
(Observe students and give feedback.)

e. Check your work.

- Question F: What is any rectangle with all sides the same length? (Signal.) *Square.*
- G: What is any quadrilateral with two pairs of parallel sides? (Signal.) *Parallelogram.*
- H: What is any parallelogram with four 90-degree angles? (Signal.) *Rectangle.*
- I: What is any parallelogram with all sides the same length? (Signal.) *Rhombus.*

EXERCISE 3: MIXED-NUMBER OPERATIONS
MULTIPLICATION WORD PROBLEMS

a. (Display:)   [110:3A]

> What's $\frac{2}{3}$ of $2\frac{1}{2}$ pounds?

- Read the question. (Signal.) *What's 2/3 of 2 and 1/2 pounds?*
  2/3 of 2 and 1/2 is 2/3 **times** 2 and 1/2.
- In part A, write the problem and work it. Write the answer as a mixed number with a unit name.
  (Observe students and give feedback.)
- Everybody, what's 2/3 of 2 and 1/2 pounds? (Signal.) *1 and 2/3 pounds.*
  (Add to show:)   [110:3B]

> What's $\frac{2}{3}$ of $2\frac{1}{2}$ pounds?
>
> $\frac{2}{3} \times 2\frac{1}{2}$
>
> $\frac{\cancel{2}^1}{3} \times \frac{5}{\cancel{2}} = \frac{5}{3} = \boxed{1\frac{2}{3} \text{ pounds}}$

Here's what you should have.

b. (Display:)   [110:3C]

> What's $\frac{5}{7}$ of $3\frac{1}{5}$ gallons?

- Read the question. (Signal.) *What's 5/7 of 3 and 1/5 gallons?*
- Write the problem and work it. ✔
- Everybody, what's 5/7 of 3 and 1/5 gallons? (Signal.) *2 and 2/7 gallons.*
  (Add to show:)   [110:3D]

> What's $\frac{5}{7}$ of $3\frac{1}{5}$ gallons?
>
> $\frac{5}{7} \times 3\frac{1}{5}$
>
> $\frac{\cancel{5}^1}{7} \times \frac{16}{\cancel{5}} = \frac{16}{7} = \boxed{2\frac{2}{7} \text{ gallons}}$

Here's what you should have.

c. (Display:) [110:3E]

> What's $\frac{1}{2}$ of $1\frac{3}{5}$ hours?

- Read the question. (Signal.) *What's 1/2 of 1 and 3/5 hours?*
- Write the problem and work it. ✔
- Everybody, what's 1/2 of 1 and 3/5 hours? (Signal.) *4/5 hour.*
  (Add to show:) [110:3F]

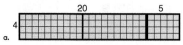

> What's $\frac{1}{2}$ of $1\frac{3}{5}$ hours?
>
> $$\frac{1}{2} \times 1\frac{3}{5}$$
>
>
>
> $$\frac{1}{\cancel{2}} \times \frac{\overset{4}{\cancel{8}}}{5} = \boxed{\frac{4}{5}} \text{ hours}$$

Here's what you should have.

EXERCISE 4: DISTRIBUTIVE PROPERTY
CHECKING DIVISION ANSWERS

a. Find part 2 in your textbook. ✔
   (Teacher reference:)

   You're going to work the multiplication and division problem for this rectangle. You'll figure out the area of the whole rectangle.
b. Write the multiplication for the tens part and the ones part. ✔
- Everybody, what's the multiplication for the tens part? (Signal.) *4 × 20.*
- What's the multiplication for the ones part? (Signal.) *4 × 5.*
c. Below, write the answer for the tens and ones and figure out the area of the whole rectangle. ✔
- Everybody, what's the area of the whole rectangle? (Signal.) *100 square units.*

d. Now you'll do the division the fast way.
- Write the division problem for the whole rectangle. Remember, the number of rows tells what you divide by. Stop when you've written the problem. ✔
- Everybody, read the division problem. (Signal.) *100 ÷ 4.*
  The answer is the number of squares in each row.
e. Write the answer to the division problem. ✔
- Everybody, what's 100 divided by 4? (Signal.) *25.*

EXERCISE 5: UNIT CONVERSION
WORD PROBLEMS

a. Find part 3 in your textbook. ✔
   (Teacher reference:)

   > a. Tank X held 12 gallons of oil. Tank Y held 30 pints of oil. How many more pints were there in Tank X than in Tank Y?
   >
   > b. Driveway T was 69 feet long. Driveway R was 12 yards long. How many yards longer was driveway T than driveway R?
   >
   > c. Cindy went hiking for 13 days in May and 2 weeks in September. How many weeks did she hike in all?

   These problems name two units.
- Where do you find the unit you should have in your answer? (Signal.) *In the question.*
  (Repeat until firm.)
  Yes, the question tells the unit for all the measurements.
b. Problem A: Tank X held 12 gallons of oil. Tank Y held 30 pints of oil. How many more pints were there in tank X than tank Y?
- Work the problem.
  (Observe students and give feedback.)
  Check your work. You changed 12 gallons into pints.
- How many pints were in tank X? (Signal.) *96 pints.*
  You worked the problem 96 minus 30.
- How many more pints were in tank X? (Signal.) *66 pints.*

c. Problem B: Driveway T was 69 feet long. Driveway R was 12 yards long. How many yards longer was driveway T than driveway R?

- Work the problem.
  (Observe students and give feedback.)
  Check your work. You changed 69 feet into yards.
- How many yards? (Signal.) *23 yards.*
  Then you worked the problem 23 minus 12.
- How many yards longer was driveway T? (Signal.) *11 yards.*

d. Problem C: Cindy went hiking for 13 days in May and 2 weeks in September. How many weeks did she hike in all?

- Work the problem.
  (Observe students and give feedback.)
  Check your work. You changed 13 days into weeks.
- How many weeks is that? (Signal.) *1 and 6/7 weeks.*
  Then you worked the problem 1 and 6/7 plus 2.
- How many weeks did she hike in all? (Signal.) *3 and 6/7 weeks.*

EXERCISE 6: SURFACE AREA
OF RECTANGULAR PRISMS

a. Find part 4 in your textbook.
(Teacher reference:)

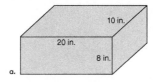 

- Find the surface area of the rectangular prism. Remember to start with the equation for surface area.
  (Observe students and give feedback.)
b. Check your work. You figured out the area of 3 faces.
- What's the number for the top? (Signal.) *200.*
- What's the number for the front? (Signal.) *160.*
- What's the number for the right? (Signal.) *80.*
- What's 200 plus 160 plus 80? (Signal.) *440.*
- You multiplied 440 by 2. What's the area of all 6 faces? (Signal.) *880 square inches.*

EXERCISE 7: EXPONENTS
BASE 10

a. (Display:)   [110:7A]

a. 12×10^2	**e.** $\dfrac{12}{10^3}$
b. 12×10^3	
c. 12×10	**f.** $.012 \times 10^2$
d. $\dfrac{12}{10^2}$	**g.** $.012 \times 10^3$

Read these values.

- (Point to **A.**) (Signal.) *12 × 10 to the second.*
- (Point to **B.**) (Signal.) *12 × 10 to the third.*
- (Point to **C.**) (Signal.) *12 × 10.*
- (Point to **D.**) (Signal.) *12 ÷ 10 to the second.*
- (Point to **E.**) (Signal.) *12 ÷ 10 to the third.*
- (Point to **F.**) (Signal.) *12 thousandths times 10 to the second.*
- (Point to **G.**) (Signal.) *12 thousandths times 10 to the third.*

b. (Add to show:)   [110:7B]

a. 12×10^2	**e.** $\dfrac{12}{10^3}$
b. 12×10^3	
c. 12×10	**f.** $.012 \times 10^2$
d. $\dfrac{12}{10^2}$	**g.** $.012 \times 10^3$

$\boxed{120}$

- (Point to **120.**) Read this number. (Signal.) *120.*
  One of the operations shown above equals 120.
- Raise your hand when you know the letter of the operation that equals 120. ✔
- Which operation equals 120? (Signal.) *C.*
c. (Point to **A.**) 12 times 10 to the second does not have the right number of zeros.
- How many zeros does 120 have? (Signal.) *1.*
- How many zeros does 12 × 10 to the second have? (Signal.) *2.*
  That's not the right number of zeros.
- Everybody, what does 12 × 10 to the second equal? (Signal.) *1 thousand 200.*

d. (Point to **G**.) How do you know that G does not equal 120? (Call on a student. Idea: *You would have to multiply by 10 to the fourth to end up with 120.*)

- Raise your hand when you know what 12 thousandths times 10 to the third equals. ✔
- What does it equal? (Signal.) *12.*
  (Change to show:) [110:7C]

a. 12×10^2		**e.** $\dfrac{12}{10^3}$	
b. 12×10^3			
c. 12×10		**f.** $.012 \times 10^2$	
d. $\dfrac{12}{10^2}$		**g.** $.012 \times 10^3$	

$\boxed{.12}$

- Raise your hand when you know the letter of the operation that equals 12 hundredths. ✔
- Which operation equals 12 hundredths? (Signal.) *D.*

e. (Point to **E**.) How do you know that E does not equal 12 hundredths? (Call on a student. Idea: *There would be 3 decimal places in the answer, not 2.*)

f. (Point to **F**.) How do you know that F does not equal 12 hundredths? (Call on a student. Idea: *You would move the decimal point 2 places and get 1 and 2 tenths, not 12 hundredths.*)

EXERCISE 8: INDEPENDENT WORK
MIXED-NUMBER OPERATIONS—MULTIPLICATION

a. Find part 5 in your textbook. ✔
   (Teacher reference:)

a. $1\frac{2}{5} \times \frac{3}{4}$ b. $\frac{5}{8} \times 1\frac{3}{10}$ c. $\frac{6}{5} \times 3\frac{1}{2}$

You'll work these problems independently. You'll change the mixed number into a fraction. Remember to simplify before you multiply.

Assign Independent Work, Textbook parts 5–12.

Optional extra math-fact practice worksheets are available on ConnectED.

Mastery Test 11

Teacher Presentation

a. Find Test 11 in your test booklet. ✔
- Pencils down. ✔
  Listen carefully and do your best work.
b. Touch part 1. ✔
  To work the problems in part 1, you need to use measurement facts.
- Find the table of facts at the back of your textbook. ✔
  If you don't know a fact, look it up in the table.
c. Touch part 2. ✔
  You also need measurement facts to work part 2.
d. Work parts 1 through 11 of the test by yourself. Read the directions carefully and do your best work. Put your pencil down when you've finished the test.
  (Observe but do not give feedback.)

Scoring Notes

a. Collect test booklets. Use the *Answer Key* and Passing Criteria Table to score the tests.

Part	Score	Possible Score	Passing Score
	Passing Criteria Table—Mastery Test 11		
1	1 for each item	3	3
2	3 for each addition/ subtraction problem (both numbers in problem, operation, answer with unit name)	9	7
3	1 for each item	2	2
4	2 for each item (addition/subtraction problem, answer with unit name)	6	5
5	1 for each item	3	3
6	1 for each item—a and c 2 for each item—b and d (substitution, answer)	6	5
7	2 for each item (fraction problem, simplified answer)	6	5
8	1 for each item	4	3
9	3 for each item (letter equation, substitution, answer)	6	5
10	2 for the item (division problem, mixed-number answer)	2	2
11	3 for each item (each digit in the answer, remainder)	6	5
	Total	53	

b. Complete the Mastery Test 11 Remedy Summary Sheet to determine whether group remedies are needed. Reproducible Remedy Summary Sheets are at the back of the *Answer Key* and the back of the *Teacher's Guide.*

• If ¼ or more of the students did not pass a test part, present the remedy for that part before beginning Lesson 111. The Remedy Table follows and also appears at the end of the Mastery Test 11 *Answer Key.* Remedies worksheets follow Mastery Test 11 in the *Student Assessment Book.*

Remedy Table—Mastery Test 11					
Part	Test Items	Lesson	Ex.	Remedies Worksheet	Textbook
1	Unit Conversion (Metric System)	106	6	—	Part 2
		107	6		Lined paper*
2	Unit Conversion (Word Problems)	107	2	—	—
		108	5	—	Part 1
3	Fraction Operations (Simplify Multiplication)	99	2	Part A	—
		101	3	—	Part 1
4	Mixed-Number Operations (Word Problems)	101	6	—	Part 4
		102	5	—	Part 2
5	Fractions (For Mixed Numbers)	104	6	—	Part 2
		105	5	—	Part 3
6	Probability (Spinner with Percents)	104	4	—	—
		105	3	—	Part 1
7	Mixed-Number Operations (Multiplication)	108	3	Part B	—
		109	3	—	Part 1
8	Decimal Multiplication/ Division (Multiples of 10)	101	2	Part C	—
		102	2	Part D	—
9	Word Problems (Parts of a Group)	101	7	—	—
		102	7	—	Part 4
10	Line Plot (Average)	103	2	Part E	—
11	Long Division (2-Digit Quotient)	99	7	—	—
		101	4	—	Part 2

*For this remedy, students refer to the table of Measurement Facts at the back of the textbook.

Retest

Retest individual students on any part failed.

	Lesson 111	Lesson 112	Lesson 113	Lesson 114	Lesson 115
Student Learning Objectives	**Exercises** 1. Find average of data displayed on a line plot 2. Classify quadrilaterals based on properties/hierarchy 3. Use the distributive property of multiplication to check division 4. Solve mixed-number multiplication problems 5. Solve word problems that involve fractions 6. Find surface area of rectangular prisms 7. Multiply by 10 with an exponent 8. Complete work independently	**Exercises** 1. **Work with line plots that have fractional units** 2. Classify quadrilaterals based on properties/hierarchy 3. Use the distributive property of multiplication to check division 4. Solve mixed-number multiplication problems 5. Multiply by 10 with an exponent 6. Solve word problems that involve fractions Complete work independently	**Exercises** 1. **Solve division word problems as fractions** 2. **Use diagrams to represent decimal addition** 3. Solve word problems that involve fractions 4. Work with line plots that have fractional units 5. Compare fractions with unlike denominators 6. Solve mixed-number multiplication problems Complete work independently	**Exercises** 1. **Compare fractions with unlike denominators** 2. Use diagrams to represent decimal addition 3. Solve word problems that involve fractions 4. **Find average of data displayed on line plots that have fractional units** 5. Solve division word problems as fractions 6. Solve probability problems with mental math 7. Solve mixed-number word problems Complete work independently	**Exercises** 1. Compare fractions with unlike denominators 2. Solve probability problems with mental math 3. Find average of data displayed on line plots that have fractional units 4. **Use diagrams to represent fraction multiplication** 5. Solve division word problems with fractions 6. Use diagrams to represent decimal addition 7. Solve division word problems as fractions Complete work independently
Common Core State Standards for Mathematics					
5.OA 1	✔	✔	✔	✔	✔
5.NBT 1–2	✔	✔	✔		✔
5.NBT 5	✔	✔	✔	✔	✔
5.NBT 6	✔	✔	✔	✔	
5.NBT 7	✔		✔	✔	✔
5.NF 1–2	✔		✔	✔	✔
5.NF 3			✔	✔	✔
5.NF 4	✔	✔	✔	✔	✔
5.NF 5			✔	✔	✔
5.NF 6	✔	✔	✔	✔	✔
5.NF 7			✔	✔	✔
5.MD 1	✔	✔	✔	✔	✔
5.MD 2		✔	✔	✔	✔
5.G 2	✔				
5.G 3–4	✔	✔			✔
Teacher Materials	Presentation Book 2, Board Displays CD or chalk board				
Student Materials	Textbook, Workbook, pencil, lined paper				
Additional Practice	• Student Practice Software: Block 5: Activity 1 (5.MD 1), Activity 2 (5.NF 1 and 5.NF 2), Activity 3 (5.MD 3 and 5.MD 4), Activity 4 (5.NF 4 and 5.NF 6), Activity 5, Activity 6 (5.MD 3 and 5.MD 4) • Provide needed fact practice with Level D or E Math Fact Worksheets.				
Mastery Test					

Lesson 111

EXERCISE 1: LINE PLOT
AVERAGE

a. Open your textbook to Lesson 111 and find part 1. ✔
(Teacher reference:)

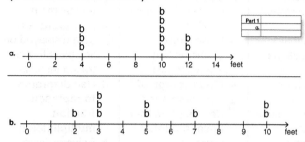

- Look at line plot A. ✔
This line plot shows the lengths of different boards. The numbers show feet. Some are 4 feet long, some are 10 feet long, and some are 12 feet long. Remember, the numbers show feet. The Bs show how many boards are each length.

b. How many boards are 4 feet long? (Signal.) 3.
- How many boards are 10 feet long? (Signal.) 5.
- How many boards are 12 feet long? (Signal.) 2.

c. Find the average length of these boards. Write your answer as a simplified mixed number.
(Observe students and give feedback.)

d. Check your work.
You found the total for each column.
(Display:)   [111:1A]

$$
\begin{array}{r}
1\,2 \\
5\,0 \\
+\,2\,4 \\
\hline
\end{array}
$$

Here's the problem you worked to find the total number of feet.
- What's the answer? (Signal.) 86.
- What did you divide 86 by? (Signal.) 10.
- What's the average length of the boards? (Signal.) 8 and 3/5 feet.

e. Line plot B also shows boards of different lengths. Figure out the total number of feet. Then find the average length of the boards.
(Observe students and give feedback.)

f. Check your work.
(Display:)   [111:1B]

$$
\begin{array}{r}
2 \\
9 \\
1\,0 \\
7 \\
+\,2\,0 \\
\hline
\end{array}
$$

Here's the problem you worked to find the total number of feet.
- What's the answer? (Signal.) 48.
- What did you divide 48 by? (Signal.) 9.
- What's the average length of the boards? (Signal.) 5 and 1/3 feet.
Yes, 5 and 1/3 feet. That's 5 feet 4 inches.

EXERCISE 2: QUADRILATERALS
HIERARCHY   [REMEDY]

a. Find part 2 in your textbook. ✔
(Teacher reference:)

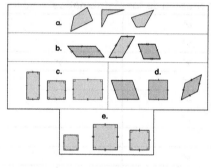

f. What is any quadrilateral that has two pairs of parallel sides?

g. What is any figure that has four straight sides and four angles?

h. What is any parallelogram that has four 90-degree angles and four sides the same length?

i. What is any parallelogram that has four sides the same length?

j. What is any parallelogram that has four 90-degree angles?

(Display:)   [111:2A]

parallelogram, square, quadrilateral, rectangle, rhombus

Here are the names for the figures in the chart.
- Read the names. (Point to each name as students read:) *Parallelogram, square, quadrilateral, rectangle, rhombus.*

b. Write the names for A through E. Stop when you've written the names.
(Observe students and give feedback.)

c. Check your work.
- What name is A? (Signal.) *Quadrilateral.*
- What name is B? (Signal.) *Parallelogram.*
- What name is C? (Signal.) *Rectangle.*
- What name is D? (Signal.) *Rhombus.*
- What name is E? (Signal.) *Square.*
d. Write answers to questions F through J. (Observe students and give feedback.)
e. Check your work.
- Question F: What is any quadrilateral that has two pairs of parallel sides? (Signal.) *Parallelogram.*
- G: What is any figure that has four straight sides and four angles? (Signal.) *Quadrilateral.*
- H: What is any parallelogram that has four 90-degree angles and four sides the same length? (Signal.) *Square.*
- I: What is any parallelogram that has four sides the same length? (Signal.) *Rhombus.*
- J: What is any parallelogram that has four 90-degree angles? (Signal.) *Rectangle.*

EXERCISE 3: DISTRIBUTIVE PROPERTY
CHECKING DIVISION ANSWERS

a. Find part 3 in your textbook. ✔
(Teacher reference:)

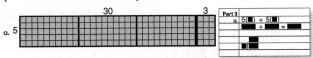

You're going to work the multiplication and division problem for this rectangle. You'll figure out the area of the whole rectangle. Then you'll figure out the number of squares in one row.
b. Write the multiplication for the tens part and the ones part. ✔
- Everybody, what's the multiplication for the tens part? (Signal.) *5 × 30.*
- What's the multiplication for the ones part? (Signal.) *5 × 3.*
c. Below, write the answer for the tens and ones and figure out the area of the whole rectangle. ✔
- Everybody, what's the area of the whole rectangle? (Signal.) *165 square units.*

d. Now you'll do the division the fast way.
- Write the division problem. Remember, the number of rows tells what you divide by. Stop when you've written the problem. ✔
- Everybody, read the division problem. (Signal.) *165 ÷ 5.*
The answer is the number of squares in each row.
e. Figure out the number of squares in the top row and write it as the answer to the division problem. ✔
- You worked the problem 165 divided by 5. How many squares are in each row? (Signal.) *33.*

EXERCISE 4: MIXED-NUMBER OPERATIONS
MULTIPLICATION WORD PROBLEMS   REMEDY

a. (Display:)   [111:4A]

> What's $\frac{3}{2}$ of $4\frac{1}{2}$ years?

- Read the question. (Signal.) *What's 3/2 of 4 and 1/2 years?*
3/2 of 4 and 1/2 is 3/2 **times** 4 and 1/2.
- In part A, write the problem and work it. Write the answer as a mixed number with a unit name.
(Observe students and give feedback.)
- Everybody, what's 3/2 of 4 and 1/2 years? (Signal.) *6 and 3/4 years.*
(Add to show:)   [111:4B]

> What's $\frac{3}{2}$ of $4\frac{1}{2}$ years?
>
> $\frac{3}{2} \times 4\frac{1}{2}$
>
> $\frac{3}{2} \times \frac{9}{2} = \frac{27}{4} = \boxed{6\frac{3}{4} \text{ years}}$

Here's what you should have.
b. (Display:)   [111:4C]

> What's $\frac{1}{3}$ of $2\frac{1}{4}$ pounds?

- Read the question. (Signal.) *What's 1/3 of 2 and 1/4 pounds?*
- Write the problem and work it. ✔

- Everybody, what's 1/3 of 2 and 1/4 pounds? (Signal.) *3/4 pound.*
  (Add to show:) [111:4D]

  What's $\frac{1}{3}$ of $2\frac{1}{4}$ pounds?

  $$\frac{1}{3} \times 2\frac{1}{4}$$

  $$\frac{1}{\cancel{3}} \times \frac{\cancel{9}^{3}}{4} = \boxed{\frac{3}{4} \text{ pound}}$$

  Here's what you should have.

c. (Display:) [111:4E]

  What's $\frac{3}{5}$ of $2\frac{1}{3}$ miles?

- Read the question. (Signal.) *What's 3/5 of 2 and 1/3 miles?*
- Write the problem and work it. ✔
- Everybody, what's 3/5 of 2 and 1/3 miles? (Signal.) *1 and 2/5 miles.*
  (Add to show:) [111:4F]

  What's $\frac{3}{5}$ of $2\frac{1}{3}$ miles?

  $$\frac{3}{5} \times 2\frac{1}{3}$$

  $$\frac{\cancel{3}^{1}}{5} \times \frac{7}{\cancel{3}} = \frac{7}{5} = \boxed{1\frac{2}{5} \text{ miles}}$$

  Here's what you should have.

EXERCISE 5: SENTENCES
FRACTION COMPARISON

REMEDY

a. (Display:) [111:5A]

  The truck was $\frac{3}{4}$ as long as the plane.

  This sentence compares two things.
  The truck was 3/4 as long as the plane.
- What does the sentence name first? (Signal.) *The truck.*
- What does the sentence tell about the truck? (Signal.) *It was 3/4 as long as the plane.*
b. I'm going to write an equation for this sentence.
  The first part of the sentence says, **the truck was.** I write **T equals** for that part.

(Add to show:) [111:5B]

  The truck was $\frac{3}{4}$ as long as the plane.

  $$t =$$

  The sentence tells that the truck was **3/4 as long as the plane.** I write 3/4 P for that part.
  (Add to show:) [111:5C]

  The truck was $\frac{3}{4}$ as long as the plane.

  $$t = \frac{3}{4}p$$

- Read the equation for the sentence. (Signal.) *T = 3/4 P.*
c. (Display:) [111:4D]

  The dog weighed $\frac{1}{4}$ as much as the goat.

  New sentence: The dog weighed 1/4 as much as the goat.
- What does the sentence name first? (Signal.) *The dog.*
- What do I write for **the dog weighed?** (Signal.) *D =.*
- What do I write for **1/4 as much as the goat?** (Signal.) *1/4 G.*
  (Repeat until firm.)
  (Add to show:) [111:5E]

  The dog weighed $\frac{1}{4}$ as much as the goat.

  $$d = \frac{1}{4}g$$

- Read the equation for the sentence. (Signal.) *D = 1/4 G.*
d. (Display:) [111:5F]

  Building A is $\frac{7}{5}$ the height of building B.

  New sentence: Building A is 7/5 the height of building B.
- What does the sentence name first? (Signal.) *Building A.*
- What do I write for **building A is?** (Signal.) *A =.*
- What do I write for **7/5 the height of building B?** (Signal.) *7/5 B.*
  (Repeat until firm.)

(Add to show:) [111:5G]

> Building A is $\frac{7}{5}$ the height of building B.
>
> $$A = \frac{7}{5} B$$

• Read the equation. (Signal.) *A = 7/5 B.*

e. (Display:) [111:5H]

> Julia weighed $\frac{3}{5}$ as much as her father.

New sentence: Julia weighed 3/5 as much as her father.

• What does the sentence name first? (Signal.) *Julia.*

• What do I write for **Julia weighed?** (Signal.) *J =.*

• What do I write for **3/5 as much as her father?** (Signal.) *3/5 F.*
(Repeat until firm.)
(Add to show:) [111:5I]

> Julia weighed $\frac{3}{5}$ as much as her father.
>
> $$J = \frac{3}{5} f$$

• Read the equation. (Signal.) *J = 3/5 F.*

TEXTBOOK PRACTICE

a. Find part 4 in your textbook. ✔
(Teacher reference:)

a. The truck was $\frac{9}{2}$ the weight of the car.

b. The wallet has $\frac{3}{4}$ as much money as the piggy bank.

c. The dinner costs $\frac{7}{4}$ as much as the show.

d. The rainfall in August was $\frac{9}{5}$ as much as the rainfall in July.

You're going to write equations for these sentences.

• Read sentence A. (Signal.) *The truck was 9/2 the weight of the car.*

• What does the sentence name first? (Signal.) *The truck.*

• Write the equation for the sentence.
(Observe students and give feedback.)

• Everybody, read the equation you wrote.
(Signal.) *T = 9/2 C.*

(Display:) [111:5J]

> **a.** $t = \dfrac{9}{2} c$

Here's what you should have.

b. Read sentence B. (Signal.) *The wallet has 3/4 as much money as the piggy bank.*

• Write the equation. ✔

• Everybody, read the equation you wrote.
(Signal.) *W = 3/4 P.*
(Display:) [111:5K]

> **b.** $w = \dfrac{3}{4} p$

Here's what you should have.

c. Write equations for the rest of the items in part 4.
(Observe students and give feedback.)

d. Check your work.

• Sentence C: The dinner costs 7/4 as much as the show.

• Read the equation you wrote. (Signal.) *D = 7/4 S.*

e. D: The rainfall in August was 9/5 as much as the rainfall in July.

• Read the equation you wrote. (Signal.) *A = 9/5 J.*

EXERCISE 6: SURFACE AREA
OF RECTANGULAR PRISMS

a. Find part 5 in your textbook.
(Teacher reference:)

 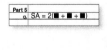

a.

• Find the surface area of the rectangular prism. Remember to start with the equation for surface area.
(Observe students and give feedback.)

b. Check your work. You figured out the area of 3 faces.

• What's the number for the top? (Signal.) *21.*

• What's the number for the front? (Signal.) *28.*

• What's the number for the right? (Signal.) *12.*

• What's 21 plus 28 plus 12? (Signal.) *61.*

• You multiplied 61 by 2. What's the area of all 6 faces? (Signal.) *122 square feet.*

EXERCISE 7: EXPONENTS
BASE 10

a. (Display:)   [111:7A]

> **a.** $.035 \times 10^4$ **e.** $\dfrac{350}{10^2}$
>
> **b.** 3.5×10^2
>
> **c.** $.35 \times 10^2$ **f.** $\dfrac{350}{10^3}$
>
> **d.** $.35 \times 10^4$
>
> $\boxed{35}$

- In part B, write the letter of the operation that equals 35. ✔
- Everybody, what's the letter of the operation that equals 35? (Signal.) *C.*
  Yes, 35 tenths times 10 to the second equals 35.

b. (Point to **A**.) How do you know that A does not equal 35? (Call on a student. Idea: *If you move the decimal point 4 places to the right, you end up with more than 35.*)
- Everybody, what does A equal? (Signal.) *350.*

c. (Point to **B**.) How do you know that B does not equal 35? (Call on a student. Idea: *When you move the decimal point 2 places to the right, you end up with more than 35.*)
- Everybody, what does B equal? (Signal.) *350.*

d. (Change to show:)   [111:7B]

> **a.** $.035 \times 10^4$ **e.** $\dfrac{350}{10^2}$
>
> **b.** 3.5×10^2
>
> **c.** $.35 \times 10^2$ **f.** $\dfrac{350}{10^3}$
>
> **d.** $.35 \times 10^4$
>
> $\boxed{.350}$

- Write the letter of the operation that equals 350 thousandths. ✔
- Everybody, what's the letter of the operation that equals 350 thousandths? (Signal.) *F.*
  Yes, 350 divided by 10 to the third.

e. (Point to **E**.) How do you know E does not equal 350 thousandths? (Call on a student. Idea: *If you move the decimal point 2 places to the left, you end up with hundredths, not thousandths.*)
- Everybody, raise your hand when you know what E equals. ✔
- What does E equal? (Signal.) *3 and 50 hundredths.*
  Yes, $350 \div 10^2$ equals 3 and 50 hundredths.

EXERCISE 8: INDEPENDENT WORK
UNIT CONVERSION—WORD PROBLEMS

a. Find part 6 in your textbook. ✔
(Teacher reference:)

> **a.** Tim is 70 inches tall. Roberto is 6 feet tall. How many inches taller is Roberto than Tim?
>
> **b.** Boat A weighed 2.1 tons. Boat B weighed 1700 pounds. How many pounds do the boats weigh together?
>
> **c.** A tree was 28 inches tall. It grew 3 feet in the next five years. How many feet tall was the tree then?

I'll read problem A: Tim is 70 inches tall. Roberto is 6 feet tall. How many inches taller is Roberto than Tim?
The problem gives two units. So you have to change one of the units. The question tells which unit you'll write in the answer.
- Raise your hand when you know the unit in the question. ✔
- What unit? (Signal.) *Inches.*
  So you have to change the other unit into inches before you work the problem.

> Assign Independent Work, Textbook parts 6–11 and Workbook parts 1 and 2.

> Optional extra math-fact practice worksheets are available on ConnectED.

Lesson 112

EXERCISE 1: LINE PLOT
WITH FRACTIONS

a. (Display:) [112:1A]

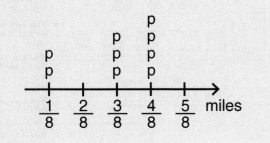

Here's a new kind of problem.
Each column has a fraction—not a whole number.
This line plot shows the number of miles people walk.
Remember, the numbers for the columns are miles.
The Ps are for people.

b. (Point to $\frac{1}{8}$.) 2 people walk 1/8 of a mile.
* How many people walk 2/8 of a mile? (Signal.) *Zero.*
* How many people walk 3/8 of a mile? (Signal.) *3.*
* How many people walk 4/8 of a mile? (Signal.) *4.*

c. We work the problem the same way we work problems with whole numbers. We find the total for each column. We add the totals to find the total number of miles. Then we divide by the number of people to find the average.
* Say the problem for the 1/8 column. (Signal.) *1/8 + 1/8.*
* Say the problem for the 3/8 column. (Signal.) *3/8 + 3/8 + 3/8.*
* Say the problem for the 4/8 column. (Signal.) *4/8 + 4/8 + 4/8 + 4/8.*

d. In part A, figure out the total for each column. Leave your answers as eighths. Stop when you've done that much.
(Observe students and give feedback.)

e. Check your work.
(Add to show:) [112:1B]

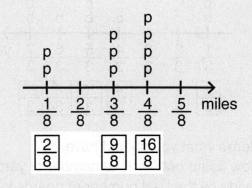

Here are the totals you should have.

f. Now add those fractions to find the total number of miles. Leave your answer as a fraction. ✔
* Everybody, what's the total number of miles? (Signal.) *27/8.*

g. Divide by the number of people to find the average.
(Observe students and give feedback.)

h. Check your work.
You worked the problem 27/8 divided by 9.
* What's the answer? (Signal.) *3/8.*
So the average distance the people walked is 3/8 of a mile.

═══ **WORKBOOK PRACTICE** ═══

a. Open your workbook to Lesson 112 and find part 1. ✔
(Teacher reference:)

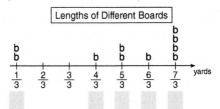

This is like the problem we just worked.
The fractions show the length of different boards. The units are yards.
* What unit do the fractions show? (Signal.) *Yards.*
* What do the Bs show? (Signal.) *Boards.*
(Repeat until firm.)

b. Figure out the total for each column and write it below. Stop when you've done that much.
(Observe students and give feedback.)
(Display:)   [112:1C]

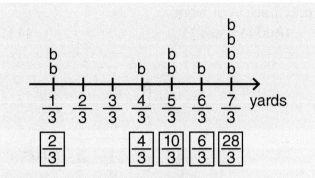

Here's what you should have.

c. Now figure out the total number of yards. Then divide by the total number of boards to find the average length of the boards.
(Observe students and give feedback.)

d. Check your work. You added the totals to find the total number of yards.

• What's the total number of yards? (Signal.) *50/3.*

• What did you divide 50/3 by? (Signal.) *10.*

• You worked the problem 50/3 × 1/10. What's the average length of these boards? (Signal.) *1 and 2/3 yards.*
(Add to show:)   [112:1D]

$$\frac{50}{3} \div 10$$

$$\frac{5\cancel{0}}{3} \times \frac{1}{1\cancel{0}} = \frac{5}{3} = \boxed{1 \frac{2}{3} \text{ yards}}$$

Here's what you should have.

EXERCISE 2: QUADRILATERALS
HIERARCHY

a. Open your textbook to Lesson 112 and find part 1. ✔
(Teacher reference:)

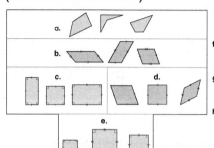

f. What is any parallelogram that has four 90-degree angles?

g. What is any figure that has four straight sides and four angles?

h. What is any parallelogram that has four 90-degree angles and four sides the same length?

i. What is any parallelogram that has four sides the same length?

j. What is any quadrilateral that has two pairs of parallel sides?

(Display:)   [112:2A]

quadrilateral, rhombus, square, parallelogram, rectangle

• Write the names for A through E. Stop when you've written the names. ✔

b. Check your work.

• What's name A? (Signal.) *Quadrilateral.*

• What's name B? (Signal.) *Parallelogram.*

• What's name C? (Signal.) *Rectangle.*

• What's name D? (Signal.) *Rhombus.*

• What's name E? (Signal.) *Square.*

c. Write answers to questions F through J. ✔

d. Check your work.

• Question F: What is any parallelogram that has four 90-degree angles? (Signal.) *Rectangle.*

• G: What is any figure that has four straight sides and four angles? (Signal.) *Quadrilateral.*

• H: What is any parallelogram that has four 90-degree angles and four sides the same length? (Signal.) *Square.*

• I: What is any parallelogram that has four sides the same length? (Signal.) *Rhombus.*

• J: What is any quadrilateral that has two pairs of parallel sides? (Signal.) *Parallelogram.*

EXERCISE 3: DISTRIBUTIVE PROPERTY
CHECKING DIVISION ANSWERS

a. (Display:)   [112:3A]

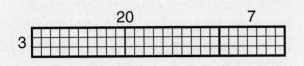

We're going to do the multiplication and division for this rectangle the fast way. But you're going to answer questions about some of the steps we take.

b. Listen: Can we write the division problem before we do the multiplication? (Signal.) *No.*

• Why not? (Call on a student. Idea: *We need to know the total number of squares before we divide.*)
  Yes, we need to know the total squares.

• How can we figure out the total squares the fast way? (Call on a student. Idea: *Do the multiplication for the tens part. Do the multiplication for the ones part. Then add.*)

c. Raise your hand when you know the total number of squares in the rectangle. ✔

• What's the total? (Signal.) *81.*
  Yes, 3 times 20 equals **60.** 3 times 7 equals **21.** So the total number of squares is 81.

d. Now we can work the division problem.

• How do you know what to divide by? (Call on a student. Idea: *We multiplied by the number of rows, so we divide by the same number.*)

e. In part B, write the division problem for this rectangle. ✔

• Everybody, read the division problem. (Signal.) *81 ÷ 3.*

f. You can figure out the answer the fast way.

• How do you do that? (Call on a student. Idea: *Figure out the number of squares in a row.*)

• Everybody, figure out the squares in a row and write the answer to the division problem. ✔

• Everybody, what's the answer to the problem 81 divided by 3? (Signal.) *27.*

• How do you know that the tens digit of the answer is 2? (Call on a student. Idea: *There are two groups of 10 in each row.*)

• How do you know that the ones digit of the answer is 7? (Call on a student. Idea: *There are 7 ones in each row.*)

EXERCISE 4: MIXED-NUMBER OPERATIONS
MULTIPLICATION WORD PROBLEMS   REMEDY

a. (Display:)   [112:4A]

> $5\frac{3}{4}$ cans were filled. Each can held $\frac{2}{3}$ gallon. How many gallons were in the cans?

5 and 3/4 cans were filled. Each can held 2/3 gallon. How many gallons were in the cans? You know the gallons in 1 can, and you have to find the gallons in 5 and 3/4 cans, so you multiply 5 and 3/4 times 2/3.

• Say the multiplication problem. (Signal.) *5 and 3/4 × 2/3.*

b. In part B, write the multiplication problem. Change the mixed number into a fraction, and work the problem. Write the answer as a mixed number and unit name.
  (Observe students and give feedback.)

• Everybody, how many gallons were in the cans? (Signal.) *3 and 5/6 gallons.*
  (Add to show:)   [112:4B]

> $5\frac{3}{4}$ cans were filled. Each can held $\frac{2}{3}$ gallon. How many gallons were in the cans?
>
> $$5\frac{3}{4} \times \frac{2}{3}$$
>
> $$\frac{23}{\overset{\,}{\underset{2}{4}}} \times \frac{\overset{1}{\cancel{2}}}{3} = \frac{23}{6} = \boxed{3\frac{5}{6} \text{ gallons}}$$

Here's what you should have. You changed 5 and 3/4 into 23/4, then multiplied.

a. Find part 2 in your textbook. ✔
(Teacher reference:)

a. A pipe leaked at the rate of $1\frac{2}{5}$ gallons each day. The pipe leaked for 8 days. How many gallons did the pipe leak?

b. A recipe calls for $1\frac{1}{2}$ cups of oil. How much oil is needed for $\frac{3}{4}$ of the amount the recipe calls for?

c. Two students wrote pages of notes. Jay wrote $4\frac{3}{4}$ pages. Fran wrote $1\frac{3}{5}$ times as much as Jay. How many pages did Fran write?

These problems are like the one you just worked.

b. Problem A: A pipe leaked at the rate of 1 and 2/5 gallons each day. The pipe leaked for 8 days. How many gallons did the pipe leak? The pipe leaked for 8 days, so you multiply 1 and 2/5 gallons by 8.

• Write the multiplication problem, change the mixed number into a fraction, and work the problem. Write the answer as a mixed number and unit name.
(Observe students and give feedback.)

• Everybody, how many gallons did the pipe leak in 8 days? (Signal.) *11 and 1/5 gallons.*
(Display:) [112:4C]

$$a. \quad 1\frac{2}{5} \times 8$$
$$\frac{7}{5} \times 8 = \frac{56}{5} = \boxed{11\frac{1}{5}} \text{ gallons}$$

Here's what you should have. You changed 1 and 2/5 into 7/5, then multiplied.

c. Problem B: A recipe calls for 1 and 1/2 cups of oil. How much oil is needed for 3/4 of the amount the recipe calls for?
3/4 of an amount is 3/4 times the amount.

• Work the problem.
(Observe students and give feedback.)

• Everybody, how much oil is needed for 3/4 of what the recipe calls for? (Signal.)
1 and 1/8 cups.
(Display:) [112:4D]

$$b. \quad 1\frac{1}{2} \times \frac{3}{4}$$
$$\frac{3}{2} \times \frac{3}{4} = \frac{9}{8} = \boxed{1\frac{1}{8}} \text{ cups}$$

Here's what you should have.

d. Problem C: Two students wrote pages of notes. Jay wrote 4 and 3/4 pages. Fran wrote 1 and 3/5 times as much as Jay. How many pages did Fran write?

• Write the problem. Change both mixed numbers into fractions and work the problem. Show your answer as a mixed number and unit name.
(Observe students and give feedback.)

• Everybody, how many pages did Fran write? (Signal.) *7 and 3/5 pages.*
(Display:) [112:4E]

$$c. \quad 4\frac{3}{4} \times 1\frac{3}{5}$$
$$\frac{19}{\cancel{4}} \times \frac{\cancel{8}^{2}}{5} = \frac{38}{5} = \boxed{7\frac{3}{5}} \text{ pages}$$

Here's what you should have.

EXERCISE 5: EXPONENTS
BASE 10

a. (Display:) [112:5A]

a. 2.7×10^4	d. $\frac{270}{10^2}$
b. 2.7×10^2	e. $.027 \times 10^1$
c. $\frac{2.7}{10^4}$	f. 27×10^3

$$\boxed{27,000}$$

In part C, write the letters of the two operations that equal 27,000. ✔

• Everybody, what are the letters of operation that equal 27,000? (Signal.) *A and F.*

b. (Point to **A.**) How do you know that this operation equals 27,000? (Call on a student. Idea: *Moving the decimal point 4 places to the right adds 3 zeros after 27.*)

c. (Point to **F.**) How do you know that this operation equals 27,000? (Call on a student. Idea: *Multiplying by 10^3 adds 3 zeros after 27.*)

d. (Point to **C.**) How do you know that C does **not** equal 27,000? (Call on a student. Idea: *You move the decimal point to the left, so the answer is less than 2.7.*)

e. (Change to show:) [112:5B]

> **a.** 2.7×10^4 **d.** $\dfrac{270}{10^2}$
>
> **b.** 2.7×10^2 **e.** $.027 \times 10^1$
>
> **c.** $\dfrac{2.7}{10^4}$ **f.** 27×10^3
>
> $\boxed{2.7}$

- Write the letter of the operation that equals 2 and 7 tenths. ✔
- Everybody, what's the letter of the operation that equals 2.7? (Signal.) *D.*
- How do you know that D equals 2.7? (Call on a student. Idea: *Moving the decimal point 2 places to the left gives 2.70, which equals 2.7.*)
f. (Point to **E.**) How do you know that E does not equal 2.7? (Call on a student. Idea: *If you move the decimal point 1 place to the right, you end up with hundredths not tenths.*)
- Everybody, raise your hand when you know what E equals. ✔
- What does E equal? (Signal.) *27 hundredths.* Yes, 27 thousandths times 10 equals 27 hundredths.

EXERCISE 6: WORD PROBLEMS
FRACTION COMPARISON

a. (Display:) [112:6A]

> $G = \dfrac{2}{3} R$ $\boxed{G = 4}$

- (Point to **G = $\frac{2}{3}$R.**) Read this problem. (Signal.) *G = 2/3R.*
- The problem gives a number for G. What number? (Signal.) *4.*
  So I write the equation with 4.
  (Add to show:) [112:6B]

> $G = \dfrac{2}{3} R$ $\boxed{G = 4}$
>
> $4 = \dfrac{2}{3} R$

- Read the new equation. (Signal.) *4 = 2/3R.* You have to figure out R.
- What do you multiply both sides by? (Signal.) *3/2.*
- Do it and work the problem in part C. (Observe students and give feedback.)
- Everybody, what does R equal? (Signal.) *6.*

(Add to show:) [112:6C]

> $G = \dfrac{2}{3} R$ $\boxed{G = 4}$
>
> $\left(\dfrac{3}{2}\right)\overset{2}{\cancel{4}} = \dfrac{2}{3} R \left(\dfrac{3}{2}\right)$
>
> $\boxed{6 = R}$

Here's what you should have.

b. (Display:) [112:6D]

> $M = 5K$ $\boxed{M = 20}$

This problem gives a number for M.
- Copy the problem and work it. Figure out what K equals.
  (Observe students and give feedback.)
- Everybody, what does K equal? (Signal.) *4.*
  (Add to show:) [112:6E]

> $M = 5K$ $\boxed{M = 20}$
>
> $\left(\dfrac{1}{5}\right)\overset{4}{\cancel{20}} = 5K \left(\dfrac{1}{5}\right)$
>
> $\boxed{4 = K}$

Here's what you should have.

c. (Display:) [112:6F]

> The cat weighs $\dfrac{1}{5}$ as much as the dog.

- Read this sentence. (Signal.) *The cat weighs 1/5 as much as the dog.*
  You learned to write equations for sentences like this one.
- What does the sentence name first? (Signal.) *The cat.*
- What do I write for the cat weighs? (Signal.) *C =.*
- What do I write on the other side of the equals? (Signal.) *1/5 D.*
  (Add to show:) [112:6G]

> The cat weighs $\dfrac{1}{5}$ as much as the dog.
>
> $c = \dfrac{1}{5} d$

- Read the equation. (Signal.) *C = 1/5 D.*

> The cat weighs $\frac{1}{5}$ as much as the dog.
>
> The cat weighs 12 pounds. How many pounds does the dog weigh?
>
> $$c = \frac{1}{5}d$$

Here's a problem: The cat weighs 1/5 as much as the dog. The cat weighs 12 pounds. How many pounds does the dog weigh?
The problem gives a number for the cat. So I write the equation with a number for cat.

> The cat weighs $\frac{1}{5}$ as much as the dog.
>
> The cat weighs 12 pounds. How many pounds does the dog weigh?
>
> $$c = \frac{1}{5}d$$
>
> $$12 = \frac{1}{5}d$$

Now we have the equation 12 = 1/5 D.
d. We solve for the letter D. We have 1/5 D.
• What do we multiply both sides by? (Signal.) *5/1.*

> The cat weighs $\frac{1}{5}$ as much as the dog.
>
> The cat weighs 12 pounds. How many pounds does the dog weigh?
>
> $$c = \frac{1}{5}d$$
>
> $$\left(\frac{5}{1}\right) 12 = \frac{1}{5}d\left(\frac{5}{1}\right)$$

On the right side we have one D. On the other side we have 12 times 5. That's 60.

> The cat weighs $\frac{1}{5}$ as much as the dog.
>
> The cat weighs 12 pounds. How many pounds does the dog weigh?
>
> $$c = \frac{1}{5}d$$
>
> $$\left(\frac{5}{1}\right) 12 = \frac{1}{5}d\left(\frac{5}{1}\right)$$
>
> $$60 = d$$

e. The problem asks: How many pounds does the dog weigh?
• Say the answer with the unit name. (Signal.) *60 pounds.*

> The cat weighs $\frac{1}{5}$ as much as the dog.
>
> The cat weighs 12 pounds. How many pounds does the dog weigh?
>
> $$c = \frac{1}{5}d$$
>
> $$\left(\frac{5}{1}\right) 12 = \frac{1}{5}d\left(\frac{5}{1}\right)$$
>
> $$60 = d$$
>
> 60 pounds

Yes, the unit name is pounds.
The first sentence of the problem says: The cat weighs 1/5 as much as the dog.
• Which animal weighs less—the cat or the dog? (Signal.) *The cat.*
• How many times heavier is the dog than the cat? (Signal.) *5 times.*
The cat weighs 12 pounds and the dog weighs 60 pounds. Those numbers show that the cat weighs 1/5 as much as the dog.

f. (Display:) [112:6M]

> The building is $\frac{4}{5}$ as tall as the tower.
>
> The building is 200 feet tall. How many feet tall is the tower?

New problem: The building is 4/5 as tall as the tower. The building is 200 feet tall. How many feet tall is the tower?
- Raise your hand when you can say the letter equation for the first sentence. ✔
- Say the letter equation. (Signal.) *B = 4/5 T.*
(Add to show:) [112:6N]

> The building is $\frac{4}{5}$ as tall as the tower.
>
> The building is 200 feet tall. How many feet tall is the tower?
>
> $$b = \frac{4}{5} t$$

g. The problem gives a number for building.
- What number? (Signal.) *200.*
(Add to show:) [112:6O]

> The building is $\frac{4}{5}$ as tall as the tower.
>
> The building is 200 feet tall. How many feet tall is the tower?
>
> $$b = \frac{4}{5} t$$
>
> $$200 = \frac{4}{5} t$$

- Raise your hand when you know what we multiply both sides by. ✔
- What do we multiply by? (Signal.) *5/4.*
(Add to show:) [112:6P]

> The building is $\frac{4}{5}$ as tall as the tower.
>
> The building is 200 feet tall. How many feet tall is the tower?
>
> $$b = \frac{4}{5} t$$
>
> $$\left(\frac{5}{4}\right) 200 = \frac{4}{5} t \left(\frac{5}{4}\right)$$

We work the problem 5/4 times 200.

- Write the problem 5/4 times 200 and work it. ✔
- Everybody, what's the answer? (Signal.) *250.*
(Add to show:) [112:6Q]

> The building is $\frac{4}{5}$ as tall as the tower.
>
> The building is 200 feet tall. How many feet tall is the tower?
>
> $$b = \frac{4}{5} t$$
>
> $$\left(\frac{5}{\cancel{4}}\right) \overset{50}{\cancel{200}} = \frac{4}{5} t \left(\frac{5}{4}\right)$$
>
> $$\boxed{250 = t}$$

h. The problem asks: How many feet tall is the tower?
- What's the whole answer? (Signal.) *250 feet.*
(Add to show:) [112:6R]

> The building is $\frac{4}{5}$ as tall as the tower.
>
> The building is 200 feet tall. How many feet tall is the tower?
>
> $$b = \frac{4}{5} t$$
>
> $$\left(\frac{5}{\cancel{4}}\right) \overset{50}{\cancel{200}} = \frac{4}{5} t \left(\frac{5}{4}\right)$$
>
> $$250 = t$$
> $$\boxed{250 \text{ feet}}$$

Yes, 250 feet.

TEXTBOOK PRACTICE

a. Find part 3 in your textbook. ✔
(Teacher reference:)

> a. The camera weighed $\frac{3}{5}$ as much as the book. The camera weighed $\frac{9}{4}$ pounds. How many pounds did the book weigh?
>
> b. Tamir picked $\frac{3}{4}$ as many apples as his mom picked. His mom picked 12 apples. How many apples did Tamir pick?

b. Problem A: The camera weighed 3/5 as much as the book. The camera weighed 9/4 pounds. How many pounds did the book weigh?
- Write the equation with two letters. Below, write the equation with one letter replaced by a number.
(Observe students and give feedback.)

- Everybody, read the equation with two letters. (Signal.) *C = 3/5 B.*
- Read the equation with one letter. (Signal.) *9/4 = 3/5 B.*
  (Display:) [112:6S]

  a.　　$c = \dfrac{3}{5} b$

  　　　$\dfrac{9}{4} = \dfrac{3}{5} b$

  Here's what you should have.
- Raise your hand when you know what you multiply both sides by. ✔
- What do you multiply by? (Signal.) *5/3.*
c. Work the problem and write the answer with a unit name.
d. Check your work.
- What did the camera weigh? (Signal.) *3 and 3/4 pounds.*
  (Display:) [112:6T]

  a.　　$c = \dfrac{3}{5} b$

  　　$\left(\dfrac{5}{\cancel{3}}\right) \dfrac{\cancel{9}^{\,3}}{4} = \dfrac{3}{5} b \left(\dfrac{5}{3}\right)$

  　　　$\dfrac{15}{4} = b$

  　　$\boxed{3 \dfrac{3}{4} \text{ pounds}}$

  Here's what you should have.

e. Problem B: Tamir picked 3/4 as many apples as his mom picked. His mom picked 12 apples. How many apples did Tamir pick?
- Write the letter equation. Replace one of the letters with a number. Stop when you've done that much.
  (Observe students and give feedback.)
- Everybody, read the letter equation. (Signal.) *T = 3/4 M.*
- Read the equation with a number. (Signal.) *T = 3/4 × 12.*
- Work the problem. Show the answer with a unit name.
  (Observe students and give feedback.)
f. Check your work.
  You worked the problem 3/4 times 12.
- What's the whole answer? (Signal.) *9 apples.*
  (Display:) [112:6U]

  b. $T = \dfrac{3}{4} m$

  　$T = \dfrac{3}{\cancel{4}} (\cancel{12}^{\,3}) = 9$

  　$\boxed{9 \text{ apples}}$

  Here's what you should have.

  Assign Independent Work, Textbook parts 4–12.

  Optional extra math-fact practice worksheets are available on ConnectED.

Lesson 113

EXERCISE 1: DIVISION WORD PROBLEMS
AS FRACTIONS

a. Some division problems are easy to work because you can write the answer as a fraction.
(Display:)   [113:1A]

> You divide 3 units into 4 equal parts.

Here's a problem: You divide 3 units into 4 equal parts.
That's 3 divided by 4.
You write 3 divided by 4 as a fraction—3/4.
Each part is 3/4 of a unit.
(Add to show:)   [113:1B]

> You divide 3 units into 4 equal parts.
>
> $$3 \div 4 = \boxed{\dfrac{3}{4}}$$

Here's the equation: 3 divided by 4 equals 3/4.
* Say the equation. (Signal.) *3 ÷ 4 = 3/4.*

b. (Display:)   [113:1C]

> You divide 3 units into 7 equal parts.

New problem: You divide 3 units into 7 equal parts.
* Say the division problem you work. (Signal.) *3 ÷ 7.*
* Say the fraction for 3 divided by 7. (Signal.) *3/7.*
(Add to show:)   [113:1D]

> You divide 3 units into 7 equal parts.
>
> $$3 \div 7 = \boxed{\dfrac{3}{7}}$$

* Say the equation. (Signal.) *3 ÷ 7 = 3/7.*

c. (Display:)   [113:1E]

> You divide 3 units into 2 equal parts.

New problem: You divide 3 units into 2 equal parts.
* Say the division problem you work. (Signal.) *3 ÷ 2.*
* Say the fraction for 3 divided by 2. (Signal.) *3/2.*

(Add to show:)   [113:1F]

> You divide 3 units into 2 equal parts.
>
> $$3 \div 2 = \dfrac{3}{2}$$

* Say the equation. (Signal.) *3 ÷ 2 = 3/2.* 3/2 is more than 1, so we can write it as a mixed number.
* What mixed number? (Signal.) *1 and 1/2.*
(Add to show:)   [113:1G]

> You divide 3 units into 2 equal parts.
>
> $$3 \div 2 = \dfrac{3}{2} = \boxed{1\dfrac{1}{2}}$$

Yes, 3/2 equals 1 and 1/2.

d. (Display:)   [113:1H]

> You divide 5 units into 9 equal parts.

New problem: You divide 5 units into 9 equal parts.
* Say the division problem you work. (Signal.) *5 ÷ 9.*
* Say the fraction for 5 divided by 9. (Signal.) *5/9.*
* Say the equation for 5 divided by 9. (Signal.) *5 ÷ 9 = 5/9.*

e. (Display:)   [113:1I]

> **a.** You divide 5 units into 7 equal parts.

Your turn to write the equations for division problems.
Problem A: You divide 5 units into 7 equal parts.
* Write the equation in part A. ✔
* Everybody, read the equation. (Signal.) *5 ÷ 7 = 5/7.*
(Add to show:)   [113:1J]

> **a.** You divide 5 units into 7 equal parts.
>
> $$\textbf{a.}\ 5 \div 7 = \boxed{\dfrac{5}{7}}$$

Here's what you should have.

f. (Display:) [113:1K]

> **b.** You divide 5 units into 3 equal parts.

Problem B: You divide 5 units into 3 equal parts.

- Write the equation. If the answer is more than 1, show it as a mixed number. ✔
- 5 divided by 3 is a mixed number. What mixed number? (Signal.) *1 and 2/3.*
  (Add to show:) [113:1L]

> **b.** You divide 5 units into 3 equal parts.
>
> **b.** $5 \div 3 = \dfrac{5}{3} = \boxed{1\dfrac{2}{3}}$

Here's what you should have.
5 divided by 3 is 5/3. That equals 1 and 2/3.

g. (Display:) [113:1M]

> **c.** You divide 7 units into 10 equal parts.

Problem C: You divide 7 units into 10 equal parts.

- Write the equation. ✔
- Everybody, read the equation. (Signal.) *7 ÷ 10 = 7/10.*
  (Add to show:) [113:1N]

> **c.** You divide 7 units into 10 equal parts.
>
> **c.** $7 \div 10 = \boxed{\dfrac{7}{10}}$

Here's what you should have.

h. (Display:) [113:1O]

> **d.** You divide 9 units into 2 equal parts.

Problem D: You divide 9 units into 2 equal parts.

- Write the equation. ✔
  9 divided by 2 is a mixed number.
- What mixed number? (Signal.) *4 and 1/2.*
  (Add to show:) [113:1P]

> **d.** You divide 9 units into 2 equal parts.
>
> **d.** $9 \div 2 = \dfrac{9}{2} = \boxed{4\dfrac{1}{2}}$

Here's what you should have.
Remember, you can write division problems as fractions.

EXERCISE 2: DECIMAL OPERATIONS
REPRESENTATION

a. (Display:) [113:2A]

$$\begin{array}{r} .7 \\ + .8 \\ \hline \end{array}$$

- Read the problem. (Signal.) *7 tenths plus 8 tenths.*
  Listen: 7 tenths plus 8 tenths is 15 tenths. When we work the problem, what do we write for 15 tenths? (Signal.) *1 and 5 tenths.*
  (Add to show:) [113:2B]

$$\begin{array}{r} .7 \\ + .8 \\ \hline 1.5 \end{array}$$

b. We can make a diagram to show that the answer is both 15 tenths and 1 and 5 tenths.
   (Add to show:) [113:2C]

Each row shows tenths.
The problem is 7 tenths plus 8 tenths. So we show 7 tenths in the top row.

- How many tenths do we show in the next row? (Signal.) *8 tenths.*
- How many total tenths are there? (Signal.) *15 tenths.*
  (Add to show:) [113:2D]

The shaded parts show the tenths that we add.

c. Listen: The answer to the written problem is 1 and 5 tenths. We can show that the answer is correct if we **move** shaded tenths without adding tenths or taking tenths away. We'll move tenths to fill the top row so it has 10 tenths.

- Raise your hand when you know how many tenths we move to fill the top row. ✔
- How many tenths? (Signal.) *3 tenths.*
  So I cross out 3 and show them in the top row.
  (Change to show:) [113:2E]

$$\begin{array}{r} .7 \\ +\ .8 \\ \hline 1.5 \end{array}$$ +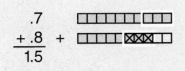

Now the picture shows 1 and 5 tenths. The top row is 1. The other row has 5 tenths.

━━━━━━ **WORKBOOK PRACTICE** ━━━━━━

a. Open your workbook to Lesson 113 and find part 1. ✔
  (Teacher reference:)

$$\begin{array}{r} .5 \\ +\ .9 \\ \hline \end{array}$$ +

You're going to work a problem like the one we just worked.
- Work the column problem first. Raise your hand when you know the answer. ✔
- Everybody, what's the answer? (Signal.) *1 and 4 tenths.*

b. You're going to show the two values that are added. You'll shade 5 tenths in the top row.
- How many tenths are you going to shade in the top row? (Signal.) *5 tenths.*
- How many tenths in the second row? (Signal.) *9 tenths.*
- Lightly shade in the tenths. ✔
  (Display:) [113:2F]

$$\begin{array}{r} .5 \\ +\ .9 \\ \hline 1.4 \end{array}$$ +

Here's what you should have.
- How many tenths did you shade all together? (Signal.) *14 tenths.*

c. The answer to the column problem is 1 and 4 tenths. You're going to show that answer by moving some tenths from the second row to the top row.
- Raise your hand when you know how many tenths you will move. ✔
- How many tenths will you move? (Signal.) *5 tenths.*
- Cross out 5 tenths and shade 5 tenths in the top row. ✔

(Change to show:) [113:2G]

$$\begin{array}{r} .5 \\ +\ .9 \\ \hline 1.4 \end{array}$$ +

- What decimal number does your diagram show now? (Signal.) *1 and 4 tenths.*
- Why did you have to move 5 tenths to the top row? (Call on a student. Ideas: *To show the top row with 10 tenths (1); to show the values in the diagram equal 1 and 4 tenths.*)

EXERCISE 3: WORD PROBLEMS
FRACTION COMPARISON   [REMEDY]

a. Open your textbook to Lesson 113 and find part 1. ✔
  (Teacher reference:)

a. Jan weighed 120 pounds. Jan weighed $\frac{4}{5}$ as much as her mother. How much did her mother weigh?

b. The boat is 28 feet long. The truck is $\frac{7}{4}$ the length of the boat. How long is the truck?

c. Gross Hill is $\frac{3}{8}$ the height of Dillard Hill. Dillard Hill is 240 feet high. How high is Gross Hill?

You're going to work problems that compare two things.

b. Problem A: Jan weighed 120 pounds. Jan weighed 4/5 as much as her mother. How much did her mother weigh?
  The sentence that compares tells how to write the equation.
- Raise your hand when you know the sentence that compares. ✔
- Everybody, read the sentence that compares two things. (Signal.) *Jan weighed 4/5 as much as her mother.*
- Write an equation for that sentence. Below, write an equation with a letter replaced by a number. Raise your hand when you've done that much.
  (Observe students and give feedback.)

c. Check your work.
- Read the letter equation. (Signal.) *J = 4/5 M.*
- Read the equation with a number. (Signal.) *120 = 4/5 M.*
  (Display:) [113:3A]

$$a. \quad J = \frac{4}{5} m$$

$$120 = \frac{4}{5} m$$

Here's what you should have: 120 = 4/5 M.

d. Raise your hand when you know if you multiply both sides by the reciprocal. ✔
- Do you multiply by the reciprocal? (Signal.) *Yes.*
- Work the problem and figure out how much her mother weighs.
  (Observe students and give feedback.)

e. Check your work.
- How much does her mother weigh? (Signal.) *150 pounds.*
  (Add to show:) [113:3B]

$$a. \quad J = \frac{4}{5} m$$

$$\left(\frac{5}{4}\right) \overset{30}{\cancel{120}} = \frac{4}{5} m \left(\frac{5}{4}\right)$$

$$\boxed{150 = m}$$
$$\boxed{150 \text{ pounds}}$$

Here's what you should have.

f. Problem B: The boat is 28 feet long. The truck is 7/4 the length of the boat. How long is the truck?
- Write the equation for the sentence that compares two things. Then write the equation with a letter replaced by a number. ✔

g. Check your work.
- Read the letter equation. (Signal.) *T = 7/4 B.*
- Read the equation with a number. (Signal.) *T = 7/4 × 28.*
- Raise your hand when you know if you multiply both sides by the reciprocal. ✔
- Do you multiply by the reciprocal? (Signal.) *No.*
- Work the problem.
  (Observe students and give feedback.)

h. Check your work.
You replaced B with 28 and worked the problem 7/4 times 28.
- How long is the truck? (Signal.) *49 feet.*
  (Display:) [113:3C]

$$b. \quad t = \frac{7}{4} b$$

$$t = \frac{7}{\cancel{4}} (\overset{7}{\cancel{28}}) = 49$$

$$\boxed{49 \text{ feet}}$$

Here's what you should have.

i. Problem C: Gross Hill is 3/8 the height of Dillard Hill. Dillard Hill is 240 feet high. How high is Gross Hill?
- Write the equation for the sentence that compares two things. ✔
- Read the equation. (Signal.) *G = 3/8 D.*
- Work the problem.
  (Observe students and give feedback.)

j. Check your work. You replaced D with 240 and worked the problem 3/8 times 240.
- How high is Gross Hill? (Signal.) *90 feet.*
  (Display:) [113:3D]

$$c. \quad G = \frac{3}{8} D$$

$$G = \frac{3}{\cancel{8}} (\overset{30}{\cancel{240}}) = 90$$

$$\boxed{90 \text{ feet}}$$

Here's what you should have.

EXERCISE 4: LINE PLOT
WITH FRACTIONS

a. Find part 2 in your textbook. ✔
   (Teacher reference:)

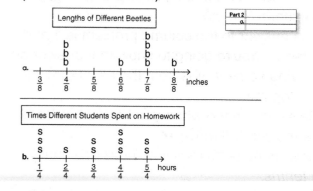

b. Problem A. The fractions on this line plot show the length of beetles in inches.
- How long are the beetles in the first column? (Signal.) *4/8 inch.*
- How long are the beetles in the last column? (Signal.) *8/8 inch.*

c. Figure out the total number of inches. Then divide by the total number of beetles. Show your answer as eighths. Raise your hand when you know the average length of these beetles.
(Observe students and give feedback.)

d. Check your work.
(Display:)   [113:4A]

$$\text{a. } \frac{54}{8} \div 9$$

$$\frac{\overset{6}{\cancel{54}}}{8} \times \frac{1}{\cancel{9}} = \boxed{\frac{6}{8} \text{ inch}}$$

Here's what you should have.
- What's the total for the inches? (Signal.) *54/8.*
- What did you divide by? (Signal.) *9.*
- What's the average length of the beetles? (Signal.) *6/8 inch.*

e. Problem B. The fractions on this line plot show the hours students spent doing homework. Each S shows a student.
- How many hours did each student in the first column spend on homework? (Signal.) *1/4 hour.*
- How many hours did each student in the last column spend on homework? (Signal.) *5/4 hours.*

f. Figure out the total number of hours. Then divide by the total number of students. Raise your hand when you know the average length of time spent on homework.
(Observe students and give feedback.)

g. Check your work.
(Display:)   [113:4B]

$$\text{b. } \frac{33}{4} \div 11$$

$$\frac{\overset{3}{\cancel{33}}}{4} \times \frac{1}{\cancel{11}} = \boxed{\frac{3}{4} \text{ hour}}$$

Here's what you should have.
- What's the total for hours? (Signal.) *33/4.*
- What did you divide by? (Signal.) *11.*
- What's the average length of time? (Signal.) *3/4 hour.*

EXERCISE 5: FRACTION COMPARISON
UNLIKE DENOMINATORS

a. (Display:)   [113:5A]

$$\text{a. } \frac{4}{5} \quad \frac{9}{10} \qquad \text{b. } \frac{10}{12} \quad \frac{3}{4} \qquad \text{c. } \frac{11}{18} \quad \frac{2}{3}$$

We're going to write statements to show if the fractions are equal or if one fraction is greater than the other.
The fractions in each problem have different denominators. We'll find the common denominator for those fractions.

b. Problem A. The fractions are 4/5 and 9/10. The denominators are 5 and 10.
- What's the common denominator? (Signal.) *10.* So we change 4/5 into tenths.
- What fraction of 1 do we multiply by to change fifths into tenths? (Signal.) *2/2.*
(Add to show:)   [113:5B]

$$\text{a. } \frac{4}{5} \quad \frac{9}{10} \qquad \text{b. } \frac{10}{12} \quad \frac{3}{4} \qquad \text{c. } \frac{11}{18} \quad \frac{2}{3}$$

$$\frac{4}{5} \left(\frac{2}{2} \right) = \frac{}{10}$$

- What does 4/5 equal? (Signal.) *8/10.*
(Add to show:)   [113:5C]

$$\text{a. } \frac{4}{5} \quad \frac{9}{10} \qquad \text{b. } \frac{10}{12} \quad \frac{3}{4} \qquad \text{c. } \frac{11}{18} \quad \frac{2}{3}$$

$$\frac{4}{5} \left(\frac{2}{2} \right) = \frac{8}{10}$$

- Which is more—8/10 or 9/10? (Signal.) *9/10.*
- So which fraction is more—4/5 or 9/10? (Signal.) *9/10.*
Now we complete the original statement to show that 9/10 is more.
(Add to show:)   [113:5D]

$$\text{a. } \frac{4}{5} < \frac{9}{10} \qquad \text{b. } \frac{10}{12} \quad \frac{3}{4} \qquad \text{c. } \frac{11}{18} \quad \frac{2}{3}$$

$$\frac{4}{5} \left(\frac{2}{2} \right) = \frac{8}{10}$$

c. Problem B. Read the fractions. (Signal.)
10/12, 3/4.

- Which fraction do you change? (Signal.) *3/4.*
- Raise your hand when you know how many twelfths 3/4 equals. ✔
- What does 3/4 equal? (Signal.) *9/12.*
  (Add to show:)   [113:5E]

a. $\frac{4}{5} < \frac{9}{10}$ b. $\frac{10}{12}$ $\frac{3}{4}$ c. $\frac{11}{18}$ $\frac{2}{3}$

$\frac{4}{5}\left(\frac{2}{2}\right) = \frac{8}{10}$ $\frac{3}{4}\left(\frac{3}{3}\right) = \frac{9}{12}$

So we show that 10/12 is more.
(Add to show:)   [113:5F]

a. $\frac{4}{5} < \frac{9}{10}$ b. $\frac{10}{12} > \frac{3}{4}$ c. $\frac{11}{18}$ $\frac{2}{3}$

$\frac{4}{5}\left(\frac{2}{2}\right) = \frac{8}{10}$ $\frac{3}{4}\left(\frac{3}{3}\right) = \frac{9}{12}$

d. Problem C. Read the fractions. (Signal.)
11/18, 2/3.

- Which fraction do you change? (Signal.) *2/3.*
- Raise your hand when you know how many 18ths 2/3 equals. ✔
- What does 2/3 equal? (Signal.) *12/18.*
  (Add to show:)   [113:5G]

a. $\frac{4}{5} < \frac{9}{10}$ b. $\frac{10}{12} > \frac{3}{4}$ c. $\frac{11}{18}$ $\frac{2}{3}$

$\frac{4}{5}\left(\frac{2}{2}\right) = \frac{8}{10}$ $\frac{3}{4}\left(\frac{3}{3}\right) = \frac{9}{12}$ $\frac{2}{3}\left(\frac{6}{6}\right) = \frac{12}{18}$

That's more than 11/18, so we show that 2/3 is more.
(Add to show:)   [113:5H]

a. $\frac{4}{5} < \frac{9}{10}$ b. $\frac{10}{12} > \frac{3}{4}$ c. $\frac{11}{18} < \frac{2}{3}$

$\frac{4}{5}\left(\frac{2}{2}\right) = \frac{8}{10}$ $\frac{3}{4}\left(\frac{3}{3}\right) = \frac{9}{12}$ $\frac{2}{3}\left(\frac{6}{6}\right) = \frac{12}{18}$

EXERCISE 6: MIXED-NUMBER OPERATIONS
MULTIPLICATION WORD PROBLEMS

a. Find part 3 in your textbook. ✔
   (Teacher reference:)

a. Jaylen collected $5\frac{2}{3}$ pounds of shrimp. Andrea collected 3 times as many pounds of shrimp. How many pounds did Andrea collect?

b. Ali's stand sold $8\frac{3}{4}$ gallons of orange juice this morning. Jan's stand sold $\frac{3}{5}$ as much orange juice. How much did Jan sell this morning?

To work these problems, you change mixed numbers into fractions.

b. Problem A: Jaylen collected 5 and 2/3 pounds of shrimp. Andrea collected 3 times as many pounds of shrimp. How many pounds did Andrea collect?

- Change the mixed number into a fraction and work the problem. Show the answer as a mixed number or a whole number and unit name.
  (Observe students and give feedback.)
c. Check your work.
- How many pounds of shrimp did Andrea collect? (Signal.) *17 pounds.*
  (Display:)   [113:6A]

a. $5\frac{2}{3} \times 3$

   $\frac{17}{3} \times \overset{1}{\cancel{3}} = \boxed{17 \text{ pounds}}$

Here's what you should have.

d. Problem B: Ali's stand sold 8 and 3/4 gallons of orange juice this morning. Jan's stand sold 3/5 as much orange juice. How much did Jan sell this morning?

- Work the problem.
  (Observe students and give feedback.)
e. Check your work.
- How many gallons of orange juice did Jan sell this morning? (Signal.) *5 and 1/4 gallons.*
  (Display:)   [113:6B]

b. $8\frac{3}{4} \times \frac{3}{5}$

   $\overset{7}{\cancel{\frac{35}{4}}} \times \frac{3}{\cancel{5}} = \frac{21}{4} = \boxed{5\frac{1}{4} \text{ gallons}}$

Here's what you should have.

Assign Independent Work, Textbook parts 4–12 and Workbook part 2.

Optional extra math-fact practice worksheets are available on ConnectED.

Lesson 114

EXERCISE 1: FRACTION COMPARISON
UNLIKE DENOMINATORS

REMEDY

a. Open your workbook to Lesson 114 and find part 1. ✔
(Teacher reference:)

R Part B

a. $\frac{30}{20}$ ▢ $\frac{7}{5}$　　b. $\frac{3}{7}$ ▢ $\frac{5}{14}$　　c. $\frac{8}{3}$ ▢ $\frac{42}{15}$

You're going to figure out which fraction is more and complete the statement. To do that, you change the fraction that has the smaller denominator.

b. Problem A: Read the fractions. (Signal.) *30/20, 7/5.*
- Which fraction will you change? (Signal.) *7/5.*
- What denominator will you have in the new fraction? (Signal.) *20.*
- Figure out how many 20ths 7/5 is. Then write the sign to complete the original statement. (Observe students and give feedback.)

c. Check your work.
- What does 7/5 equal? (Signal.) *28/20.*
- Which is more—30/20 or 7/5? (Signal.) *30/20.*
(Display:)　　　　　　　　　　　[114:1A]

$$\textbf{a.}\quad \frac{30}{20} > \frac{7}{5}$$
$$\frac{7}{5}\left(\frac{4}{4}\right) = \frac{28}{20}$$

Here's what you should have.

d. Problem B. Read the fractions. (Signal.) *3/7, 5/14.*
- Change one of the fractions so it has the same denominator as the other fraction. Then complete the original statement. (Observe students and give feedback.)

e. Check your work.
- Which fraction did you change? (Signal.) *3/7.*
- What did you change 3/7 to? (Signal.) *6/14.*
- Which is more—3/7 or 5/14? (Signal.) *3/7.*
(Display:)　　　　　　　　　　　[114:1B]

$$\textbf{b.}\quad \frac{3}{7} > \frac{5}{14}$$
$$\frac{3}{7}\left(\frac{2}{2}\right) = \frac{6}{14}$$

Here's what you should have.

f. Problem C. Read the fractions. (Signal.) *8/3, 42/15.*
- Change one of the fractions so it has the same denominator as the other fraction. Then complete the original statement. (Observe students and give feedback.)

g. Check your work.
- Which fraction did you change? (Signal.) *8/3.*
- What did you change 8/3 to? (Signal.) *40/15.*
- Which is more—8/3 or 42/15? (Signal.) *42/15.*
(Display:)　　　　　　　　　　　[114:1C]

$$\textbf{c.}\quad \frac{8}{3} < \frac{42}{15}$$
$$\frac{8}{3}\left(\frac{5}{5}\right) = \frac{40}{15}$$

Here's what you should have.

EXERCISE 2: DECIMAL OPERATIONS
REPRESENTATION

a. Find part 2 in your workbook. ✔
   (Teacher reference:)

b. Read problem A. (Signal.) *6 tenths + 4 tenths.*
* Write the answer to the decimal problem. Then make the diagram showing the two values: 6 tenths plus 4 tenths. Stop when you have those values lightly shaded. ✔
   (Display:)   [114:2A]

$$
\begin{array}{l}
\textbf{a.} \quad .6 \\
\quad +\,.4 \\
\hline
\quad 1.0
\end{array}
$$

Here's what you should have.
The answer to the decimal problem is 10 tenths. That's one and zero tenths.

c. Now move tenths from the bottom row to show that answer. Cross out the tenths you move in the bottom row.
   (Observe students and give feedback.)
   (Change to show:)   [114:2B]

$$
\begin{array}{l}
\textbf{a.} \quad .6 \\
\quad +\,.4 \\
\hline
\quad 1.0
\end{array}
$$

Here's what you should have.

d. Read problem B. (Signal.) *7 tenths + 5 tenths.*
* Write the answer to the decimal problem. Then show the two values shaded in the diagram. Stop when you've done that much. ✔
* Everybody, how many tenths did you shade for the first value? (Signal.) *7 tenths.*
* How many tenths for the second value? (Signal.) *5 tenths.*
   (Display:)   [114:2C]

$$
\begin{array}{l}
\textbf{b.} \quad .7 \\
\quad +\,.5 \\
\hline
\quad 1.2
\end{array}
$$

Here's what you should have.

e. Now move tenths to show the same answer as the decimal problem.
   (Observe students and give feedback.)
   (Change to show:)   [114:2D]

$$
\begin{array}{l}
\textbf{b.} \quad .7 \\
\quad +\,.5 \\
\hline
\quad 1.2
\end{array}
$$

Here's what you should have.
* What's the answer to the decimal problem? (Signal.) *1 and 2 tenths.*
* How many tenths did you move? (Signal.) *3 tenths.*
* Why did you move 3 tenths? (Call on a student. Ideas: *To make the top row 10 tenths (1); to show that the added values equal 1 and 2 tenths.*)
* What answer does the diagram show? (Signal.) *1 and 2 tenths.*

EXERCISE 3: WORD PROBLEMS
FRACTION COMPARISON   REMEDY

a. Open your textbook to Lesson 114 and find part 1. ✔
   (Teacher reference:)

a. The amount they spent on dinner was $\frac{7}{4}$ the amount they spent on breakfast. They spent $56 on dinner. How much did they spend on breakfast?

b. There are 36 students in the library. The number of students in the computer room is $\frac{3}{4}$ the number of students in the library. How many students are in the computer room?

c. The boat weighed $\frac{3}{7}$ as much as the truck. The truck weighed 14 tons. How much did the boat weigh?

d. Blue Lake has 480 fish. Blue Lake has $\frac{1}{5}$ as many fish as Cloud Lake. How many fish are in Cloud Lake?

These are problems that compare two things.

b. Problem A: The amount they spent on dinner was 7/4 the amount they spent on breakfast. They spent 56 dollars on dinner. How much did they spend on breakfast?
* Write the letter equation for the sentence that compares dinner and breakfast. Then work the problem.
   (Observe students and give feedback.)
* Everybody, read the equation you started with. (Signal.) *D = 7/4 B.*
   You replaced D with 56. You multiplied both sides by 4/7.
* How much did they spend on breakfast? (Signal.) *$32.*

a.
$$d = \frac{7}{4}b$$

$$\left(\frac{4}{\cancel{7}}\right)\overset{8}{\cancel{56}} = \frac{7}{4}b\left(\frac{4}{7}\right)$$

$$32 = b$$

$$\boxed{\$32}$$

Here's what you should have.

c. Work problem B.
   (Observe students and give feedback.)
d. Check your work.
• Problem B: There are 36 students in the library. The number of students in the computer room is 3/4 the number of students in the library. How many students are in the computer room?
• Read the equation you started with. (Signal.)
  C = 3/4 L.
• How many students are in the computer room? (Signal.) *27 students.*
  (Display:)   [114:3B]

b. $C = \frac{3}{4}l$

$$C = \frac{3}{\cancel{4}}(\overset{9}{\cancel{36}}) = 27$$

$$\boxed{27 \text{ students}}$$

Here's what you should have.

e. Work the rest of the problems in part 1.
   (Observe students and give feedback.)
f. Check your work.
   Problem C: The boat weighed 3/7 as much as the truck. The truck weighed 14 tons. How much did the boat weigh?
• Read the equation you started with. (Signal.)
  B = 3/7 T.
• How much did the boat weigh? (Signal.)
  6 tons.
  (Display:)   [114:3C]

c. $b = \frac{3}{7}t$

$$b = \frac{3}{\cancel{7}}(\overset{2}{\cancel{14}}) = 6$$

$$\boxed{6 \text{ tons}}$$

Here's what you should have.

g. Problem D: Blue Lake has 480 fish. Blue Lake has 1/5 as many fish as Cloud Lake. How many fish are in Cloud Lake?
• Read the equation you started with. (Signal.)
  B = 1/5 C.
• How many fish does Cloud Lake have? (Signal.) *2400 fish.*
  (Display:)   [114:3D]

d. $b = \frac{1}{5}c$

$$\left(\frac{5}{1}\right)480 = \frac{1}{5}c\left(\frac{5}{1}\right)$$

$$2400 = c$$

$$\boxed{2400 \text{ fish}}$$

Here's what you should have.

EXERCISE 4: LINE PLOT
WITH FRACTIONS   REMEDY

a. Find part 2 in your textbook. ✔
   (Teacher reference:)

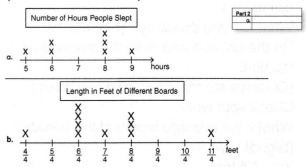

From now on, the only letter shown on line plots will be X. You have to read the problem to figure out what the Xs stand for.

b. Problem A. The line plot shows the number of hours people spent sleeping.
• Do the numbers show hours or people? (Signal.) *Hours.*
• What do the Xs stand for? (Signal.) *People.*
  (Repeat until firm.)
c. Problem B. The line plot shows the length of boards in feet.
• What do the numbers show? (Signal.) *Feet.*
• What does each X stand for? (Signal.)
  A board.
  (Repeat until firm.)

d. Work problem A. Find the average number of hours each person slept.
(Observe students and give feedback.)
e. Check your work.
- What's the total number of hours? (Signal.) *50.*
- What did you divide by? (Signal.) *7.*
- What's the average number of hours these people slept? (Signal.) *7 and 1/7 hours.*
f. Problem B. You'll find the average length of the boards. For 6/4, you'll add four fractions: 6/4 + 6/4 + 6/4 + 6/4.
- Say what you'll add for 6/4. (Signal.) *6/4 + 6/4 + 6/4 + 6/4.*
- Say what you'll add for 8/4. (Signal.) *8/4 + 8/4 + 8/4.*
- Add the totals for all of the columns. Raise your hand when you know the total number of feet.
(Observe students and give feedback.)
g. Check your work.
- What's the total for all the columns? (Signal.) *70/4.*
h. Now you'll find the average.
- Raise your hand when you know what you'll divide by. ✔
- What will you divide by? (Signal.) *10.*
- Do the division and write the average as a fraction.
(Observe students and give feedback.)
i. Check your work.
- What's the average length of the boards? (Signal.) *7/4 feet.*
Yes, 7/4 feet.
j. Look at line plot B. Raise your hand when you know how many boards are above the average length. ✔
- How many boards are **above** the average length? (Signal.) *4 boards.*
- Raise your hand when you know how many boards are below the average length. ✔
- How many boards are **below** the average length? (Signal.) *5 boards.*

EXERCISE 5: DIVISION WORD PROBLEMS
AS FRACTIONS

a. Last time, you wrote equations for division problems. You wrote them as fractions.
(Display:)   [114:5A]

> You divide 3 units into 4 equal parts.

You divide 3 units into 4 equal parts.
- Say the problem for dividing 3 units into 4 parts. (Signal.) *3 ÷ 4.*
- What's the fraction for 3 divided by 4? (Signal.) *3/4.*
(Add to show:)   [114:5B]

> You divide 3 units into 4 equal parts.
>
> $3 \div 4 = \dfrac{3}{4}$

Yes, each part is 3/4 of a unit.
b. (Add to show:)   [114:5C]

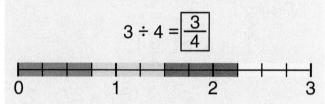

> You divide 3 units into 4 equal parts.
>
> $3 \div 4 = \dfrac{3}{4}$

Here's a number line that shows 3 units divided into 4 parts. You can see the 4 parts in different colors. You can also see that each part is 3/4 of a unit.
c. (Display:)   [114:5D]

> You divide 10 units into 11 equal parts.

You divide 10 units into 11 equal parts.
- Say the division problem. (Signal.) *10 ÷ 11.*
- Say the fraction for 10 divided by 11. (Signal.) *10/11.*
(Add to show:)   [114:5E]

> You divide 10 units into 11 equal parts.
>
> $10 \div 11 = \dfrac{10}{11}$

Here's the equation.

d. (Display:) [114:5F]

> You divide 10 units into 2 equal parts.

You divide 10 units into 2 equal parts.
- Say the division problem. (Signal.) *10 ÷ 2.*
- Say the fraction. (Signal.) *10/2.*
(Add to show:) [114:5G]

> You divide 10 units into 2 equal parts.
>
> $$10 \div 2 = \frac{10}{2}$$

- Is 10/2 more than 1 or less than 1? (Signal.) *More than 1.*
- Does it equal a mixed number or a whole number? (Signal.) *A whole number.*
- What does 10 halves equal? (Signal.) *5.*
(Add to show:) [114:5H]

> You divide 10 units into 2 equal parts.
>
> $$10 \div 2 = \frac{10}{2} = \boxed{5}$$

━━━━━━━━ **TEXTBOOK PRACTICE** ━━━━━━━━

a. Find part 3 in your textbook. ✔
(Teacher reference:)

> a. You divide 5 into 6 equal parts.
> b. You divide 7 into 6 equal parts.
> c. You divide 6 into 2 equal parts.
> d. You divide 12 into 5 equal parts.
> e. You divide 12 into 3 equal parts.
> f. You divide 12 into 17 equal parts.

> Part 3
> a. ■ ÷ ■ = ■

You'll write the equation for each problem. If the fraction is more than one, you'll show it as a mixed number or a whole number.
b. Problem A: You divide 5 into 6 equal parts.
- Write the equation. ✔
- Everybody, read the equation. (Signal.) *5 ÷ 6 = 5/6.*
(Display:) [114:5I]

> a. $5 \div 6 = \boxed{\dfrac{5}{6}}$

Here's what you should have.

c. Problem B: You divide 7 into 6 equal parts.
- Write the equation. ✔
- 7 divided by 6 is a mixed number. What mixed number? (Signal.) *1 and 1/6.*
(Display:) [114:5J]

> b. $7 \div 6 = \dfrac{7}{6} = \boxed{1\dfrac{1}{6}}$

Here's what you should have.
d. Problem C: You divide 6 into 2 equal parts.
- Write the equation. ✔
- 6 divided by 2 is a whole number. What whole number? (Signal.) *3.*
(Display:) [114:5K]

> c. $6 \div 2 = \dfrac{6}{2} = \boxed{3}$

Here's what you should have.
e. Problem D: You divide 12 into 5 equal parts.
- Write the equation. ✔
- 12 divided by 5 is a mixed number. What mixed number? (Signal.) *2 and 2/5.*
(Display:) [114:5L]

> d. $12 \div 5 = \dfrac{12}{5} = \boxed{2\dfrac{2}{5}}$

Here's what you should have.
f. Problem E: You divide 12 into 3 equal parts.
- Write the equation. ✔
- 12 divided by 3 is a whole number. What whole number? (Signal.) *4.*
(Display:) [114:5M]

> e. $12 \div 3 = \dfrac{12}{3} = \boxed{4}$

Here's what you should have.
g. Problem F: You divide 12 into 17 equal parts.
- Write the equation. ✔
- 12 divided by 17 is a fraction. What fraction? (Signal.) *12/17.*
(Display:) [114:5N]

> f. $12 \div 17 = \boxed{\dfrac{12}{17}}$

Here's what you should have.

EXERCISE 6: MENTAL MATH
PROBABILITY

a. (Display:) [114:6A]

A person took 30 trials at pulling cards from a bag.

These are the results:

Red	8
Blue	12
Yellow	10
Total	30

A person took 30 trials at pulling cards from a bag. These are the results.
We don't know the number of objects in the bag. But we can use the results to answer some probability questions.

- How many trials did the person take? (Signal.) *30.*
- On how many trials did the person pull out a red card? (Signal.) *8.*
- On how many trials did the person pull out a blue card? (Signal.) *12.*
- On how many trials did the person pull out a yellow card? (Signal.) *10.*

b. If the person took twice as many trials, would you expect the person to pull out 8 red cards or more than 8? (Signal.) *More than 8.*
Yes, if the person took twice as many trials, you would expect the person to pull out twice as many red cards.
- How many red cards is that? (Signal.) *16.*
- How many blue cards would you expect the person to pull out? (Signal.) *24.*
How many yellow cards would you expect the person to pull out? (Signal.) *20.*
(Repeat until firm.)

c. Listen: The person took 30 trials. If the person took 3 times as many trials, how many trials would the person take? (Signal.) *90.*
- On how many of those trials would you expect the person to pull a red card? (Signal.) *24.*
- On how many trials would you expect the person to pull out a blue card? (Signal.) *36.*
- On how many trials would you expect the person to pull out a yellow card? (Signal.) *30.*
(Repeat until firm.)

d. (Display:) [114:6B]

A person takes 20 trials at pulling a card from a bag.

These are the results:

Green	10
White	3
Blue	7
Total	20

New problem: A person takes 20 trials at pulling cards from a bag. Here are the results.
- On how many trials did the person pull out a green card? (Signal.) *10.*
- On how many trials did the person pull out a white card? (Signal.) *3.*
- On how many trials did the person pull out a blue card? (Signal.) *7.*
- If you took **4 times** as many trials, how many trials would you take? (Signal.) *80.*
(Add to show:) [114:6C]

A person takes 20 trials at pulling a card from a bag.

These are the results:

Green	10		
White	3		
Blue	7		
Total	20		80

- On how many of those trials would you expect to pull out a green card? (Signal.) *40.*
- On how many trials would you expect to pull out a white card? (Signal.) *12.*
- On how many trials would you expect to pull out a blue card? (Signal.) *28.*

(Add to show:) [114:6D]

A person takes 20 trials at pulling a card from a bag.

These are the results:

Green	10	40
White	3	12
Blue	7	28
Total	20	80

Here are the expected numbers.

e. (Change to show:) [114:6E]

A person takes 20 trials at pulling a card from a bag.

These are the results:

Green	10	
White	3	
Blue	7	
Total	20	

Listen: Another person takes 5 times as many trials.

* How many trials would the person take? (Signal.) *100.*
(Add to show:) [114:6F]

A person takes 20 trials at pulling a card from a bag.

These are the results:

Green	10	
White	3	
Blue	7	
Total	20	100

f. Copy the column for 100 trials in part A. Write the number of trials you would expect the person to draw a green card. ✔
* Everybody, on how many trials would you expect green? (Signal.) *50.*
(Add to show:) [114:6G]

A person takes 20 trials at pulling a card from a bag.

These are the results:

Green	10	50
White	3	
Blue	7	
Total	20	100

Yes, 50.

g. Now write the number of trials you expect the person to draw a white card and the number of trials you expect the person to draw a blue card. ✔
h. Check your work.
* On how many trials would you expect the person to pull out a white card? (Signal.) *15.*
* On how many trials would you expect the person to pull out a blue card? (Signal.) *35.*
(Add to show:) [114:6H]

A person takes 20 trials at pulling a card from a bag.

These are the results:

Green	10	50
White	3	15
Blue	7	35
Total	20	100

Here's what you should have.

EXERCISE 7: MIXED-NUMBER OPERATIONS
MULTIPLICATION WORD PROBLEMS

a. Find part 4 in your textbook. ✔
   (Teacher reference:)

a. The otter in the zoo ate $8\frac{2}{5}$ pounds of fish. The dolphin ate $2\frac{1}{2}$ times as much fish. How many pounds of fish did the dolphin eat?

b. Sally picked $1\frac{2}{3}$ baskets of grapes yesterday. She picked $3\frac{1}{5}$ times more baskets today. How many baskets of grapes did she pick today?

To work these problems, you change mixed numbers into fractions.

b. Problem A: The otter in the zoo ate 8 and 2/5 pounds of fish. The dolphin ate 2 and 1/2 times as much fish. How many pounds of fish did the dolphin eat?

• Change each mixed number into a fraction and work the problem. Show the answer as a mixed number or a whole number and unit name.
   (Observe students and give feedback.)

c. Check your work.

• How many pounds of fish did the dolphin eat? (Signal.) *21 pounds.*
   (Display:)   [114:7A]

$$\text{a. } 8\frac{2}{5} \times 2\frac{1}{2}$$
$$\frac{\overset{21}{\cancel{42}}}{5} \times \frac{\overset{1}{\cancel{5}}}{\cancel{2}} = \boxed{21 \text{ pounds}}$$

Here's what you should have.

d. Problem B: Sally picked 1 and 2/3 baskets of grapes yesterday. She picked 3 and 1/5 times more baskets today. How many baskets of grapes did she pick today?

• Work the problem.
   (Observe students and give feedback.)

e. Check your work.

• How many baskets of grapes did she pick today? (Signal.) *5 and 1/3 baskets.*
   (Display:)   [114:7B]

$$\text{b. } 1\frac{2}{3} \times 3\frac{1}{5}$$
$$\frac{\overset{1}{\cancel{5}}}{3} \times \frac{16}{\cancel{5}} = \frac{16}{3} = \boxed{5\frac{1}{3} \text{ baskets}}$$

Here's what you should have.

Assign Independent Work, Textbook parts 5–13.

Optional extra math-fact practice worksheets are available on ConnectED.

Lesson 115

EXERCISE 1: FRACTION COMPARISON
UNLIKE DENOMINATORS

a. Open your workbook to Lesson 115 and find part 1. ✔

(Teacher reference:)

a. $\frac{5}{2}$ ▮ $\frac{31}{12}$ b. $\frac{16}{20}$ ▮ $\frac{3}{4}$ c. $\frac{10}{18}$ ▮ $\frac{2}{3}$

You're going to change one of the fractions in each problem so both fractions have the same denominator. Then you'll complete the original statement.

b. Read the fractions in problem A. (Signal.) *5/2, 31/12.*

• Work the problem.
(Observe students and give feedback.)

• Everybody, what did you change 5/2 to? (Signal.) *30/12.*

• Which is more—5/2 or 31/12? (Signal.) *31/12.*
(Display:)   [115:1A]

$$a. \quad \frac{5}{2} < \frac{31}{12}$$
$$\frac{5}{2}\left(\frac{6}{6}\right) = \frac{30}{12}$$

Here's what you should have.

c. Read the fractions in problem B. (Signal.) *16/20, 3/4.*

• Work the problem. ✔

• Everybody, what did you change 3/4 to? (Signal.) *15/20.*

• Which is more—16/20 or 3/4? (Signal.) *16/20.*
(Display:)   [115:1B]

$$b. \quad \frac{16}{20} > \frac{3}{4}$$
$$\frac{3}{4}\left(\frac{5}{5}\right) = \frac{15}{20}$$

Here's what you should have.

d. Read the fractions in problem C. (Signal.) *10/18, 2/3.*

• Work the problem. ✔

• Everybody, what did you change 2/3 to? (Signal.) *12/18.*

• Which is more—10/18 or 2/3? (Signal.) *2/3.*
(Display:)   [115:1C]

$$c. \quad \frac{10}{18} < \frac{2}{3}$$
$$\frac{2}{3}\left(\frac{6}{6}\right) = \frac{12}{18}$$

Here's what you should have.

EXERCISE 2: MENTAL MATH
PROBABILITY   REMEDY

a. Find part 2 in your workbook.

• Pencils down. ✔
(Teacher reference:)   R Part D

A person took 25 trials. These are the results:	3 times as many trials	100 trials	250 trials
Yellow	1		
Purple	4		
Brown	20		
Total	25		

These are the results of a person pulling colored marbles from a bag.

b. The person took 25 trials.

• What's the number for yellow? (Signal.) *1.*

• What's the number for purple? (Signal.) *4.*

• What's the number for brown? (Signal.) *20.*

c. Here's the information for the second column of numbers: The person took 3 times as many trials.
- If the person took 3 times as many trials, how many trials would the person take? (Signal.) *75.*
- On how many of those trials would you expect the person to pull out a yellow marble? (Signal.) *3.*
- On how many of those trials would you expect the person to pull out a purple marble? (Signal.) *12.*
- On how many of those trials would you expect the person to pull out a brown marble? (Signal.) *60.*

d. Write the expected numbers for the second column. ✔
(Display:)   [115:2A]

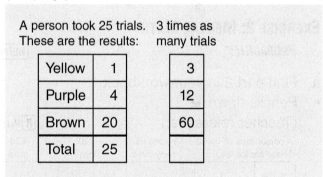

Here's what you should have.
- Add the numbers and write the total for that column. ✔
- Everybody, what's the total number of trials for that column? (Signal.) *75.*

e. Listen: The person takes 100 trials.
- 100 is how many times 25? (Signal.) *4 times.*
So the person takes 4 times as many trials.
- Figure out the number of yellows, purples, and browns you would expect to pull out of the bag, and write the expected numbers.
(Observe students and give feedback.)

f. Check your work.
- How many yellows would you expect? (Signal.) *4.*
- How many purples would you expect? (Signal.) *16.*

- How many browns would you expect? (Signal.) *80.*
(Add to show:)   [115:2B]

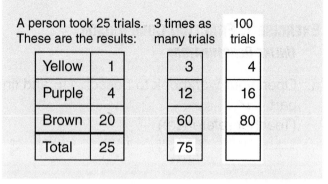

Here's what you should have.
- Add the numbers and write the total for the column. ✔
- Everybody, what's the total number of trials for that column? (Signal.) *100.*

g. Listen: Another person took 250 trials.
- 250 is how many times 25? (Signal.) *10 times.*
- Figure out the expected number of yellows, purples, and browns, and write them in the last column. ✔
- Everybody, what's the expected number for yellows? (Signal.) *10.*
- What's the expected number for purples? (Signal.) *40.*
- What's the expected number for browns? (Signal.) *200.*

h. Now add the numbers and write the total. ✔
- Everybody, what's the total number of trials for that column? (Signal.) *250.*
(Add to show:)   [115:2C]

A person took 25 trials. These are the results:		3 times as many trials	100 trials	250 trials
Yellow	1	3	4	10
Purple	4	12	16	40
Brown	20	60	80	200
Total	25	75	100	250

Here's what you should have.

EXERCISE 3: LINE PLOT
WITH FRACTIONS

a. Find part 3 in your workbook. ✔
(Teacher reference:)

Distances in Miles from School to Different Houses

a. The distance is $\frac{3}{5}$ mile for Dan's house, Laura's house and Jamie's house.

b. The distance is $\frac{4}{5}$ mile for Nan's house and Reggie's house.

c. The distance is 1 mile for Jake's house, Joe's house, and Chet's house.

d. The distance is $\frac{2}{5}$ mile for Lee's house.

e. The distance is $\frac{1}{5}$ mile for Barb's house.

You're going to make a line plot and answer questions. Each point on the line plot will be fifths between zero and 1.

- Write the fractions for 1/5 to 4/5. ✔
(Display:)   [115:3A]

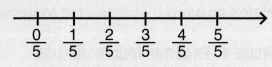

Here's what you should have.

b. The fractions tell how far it is from the school to different houses. The fractions are fifths of a mile.
- Write miles as the unit name under the line plot. ✔

c. Fact A: The distance is 3/5 mile for Dan's house, Laura's house, and Jamie's house.
- What's the distance for those houses? (Signal.) *3/5 mile.*
- How many houses? (Signal.) *3.*
- Make Xs for those houses in the 3/5 column. ✔
(Add to show:)   [115:3B]

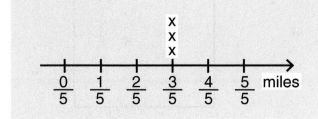

Here's what you should have.

d. Fact B: The distance is 4/5 mile for Nan's house and Reggie's house.
- Make Xs for those houses. ✔
- Everybody, how many Xs did you make? (Signal.) *2.*
- How far are those houses from the school? (Signal.) *4/5 mile.*

e. Fact C: The distance is one mile for Jake's house, Joe's house, and Chet's house.
- What's the fraction for one mile on the line plot? (Signal.) *5/5.*
- How many Xs will you make at 5/5? (Signal.) *3.*
- Make Xs for those houses. ✔

f. Fact D: The distance is 2/5 mile for Lee's house.
- Make an X for that house. ✔
- Everybody, how many Xs did you make? (Signal.) *1.*
- How far is that house from the school? (Signal.) *2/5 mile.*

g. Fact E: The distance is 1/5 mile for Barb's house.
- Make an X for that house. ✔
- Everybody, how many Xs did you make? (Signal.) *1.*
- How far is that house from the school? (Signal.) *1/5 mile.*
(Add to show:)   [115:3C]

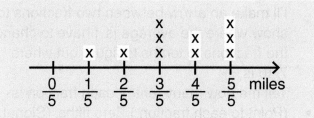

Here's what you should have.

h. How far from the school is the closest house? (Signal.) *1/5 mile.*
- How many houses are that distance from the school? (Signal.) *1.*
- How far from the school are the furthest houses? (Signal.) *5/5 mile.*
- How many houses are that distance from the school? (Signal.) *3.*

i. Find the total for each column. Write each total below the line plot. Stop when you've done that much.
(Observe students and give feedback.)
(Add to show:) [115:3D]

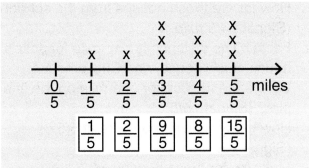

Here's what you should have.

j. Now find the total distance for the houses. Then divide to find the average distance.
(Observe students and give feedback.)

k. Check your work.
You worked the problem 35/5 divided by 10.

•   What's the average distance of the houses? (Signal.) *7/10 mile.*

l. (Change to show:) [115:3E]

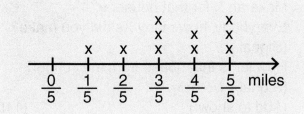

I'll make an arrow between two fractions to show where the average is. I have to change the fractions to tenths to figure out where 7/10 is.
Tell me how many tenths each fraction is.

•   (Point to each fraction.) Zero fifths. (Signal.) *Zero tenths.*

•   One fifth. (Signal.) *2 tenths.*

•   2 fifths. (Signal.) *4 tenths.*

•   3 fifths. (Signal.) *6 tenths.*

•   4 fifths. (Signal.) *8 tenths.*

•   5 fifths. (Signal.) *10 tenths.*
(Repeat until firm.)

(Add to show:) [115:3F]

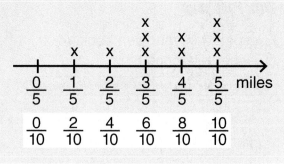

•   The average is 7/10. That's between two fractions on the original line plot. Which two fractions? (Signal.) *3/5 and 4/5.*
(Add to show:) [115:3G]

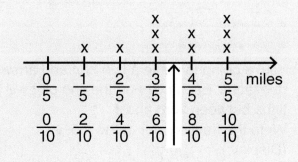

Here's an arrow that shows the average.

EXERCISE 4: FRACTION MULTIPLICATION
REPRESENTATION

a. We're going to divide square units into equal parts and shade some of the parts.
The sides will show fractions that are multiplied.
If we multiply fourths, we divide each side into 4 parts.
If we multiply halves, we divide each side into 2 parts.

b. (Display:) [115:4A]

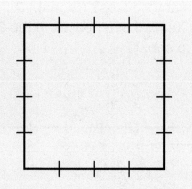

•   Look at the sides. How many parts is each side divided into? (Signal.) *4.*
So each part is 1/4 of the side.

c. (Add to show:) [115:4B]

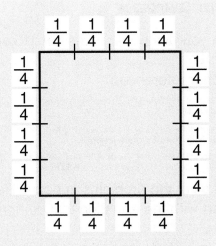

We're going to make a rectangle that is 2/4 times 3/4.
(Add to show:) [115:4C]

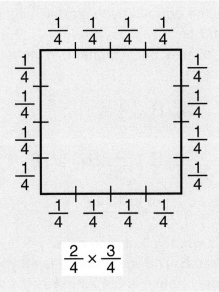

$$\frac{2}{4} \times \frac{3}{4}$$

d. Now we'll show what 2/4 times 3/4 equals.
We just show the parts inside the square unit.
We'll show 2/4 across and 3/4 down.
(Add to show:) [115:4D]

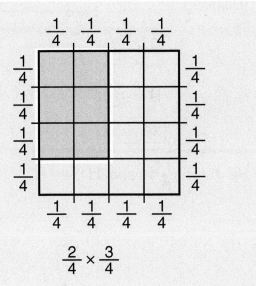

$$\frac{2}{4} \times \frac{3}{4}$$

e. If we count the parts that are shaded, we'll know the fraction that is 2/4 times 3/4.
- Raise your hand when you know how many parts are shaded. ✔
- How many parts are shaded? (Signal.) *6.*
- Raise your hand when you know how many parts there are in the whole square. ✔
- How many parts are in the whole square? (Signal.) *16.*
- What's the fraction for the shaded part? (Signal.) *6/16.*
- So what's 2/4 times 3/4? (Signal.) *6/16.*
(Add to show:) [115:4E]

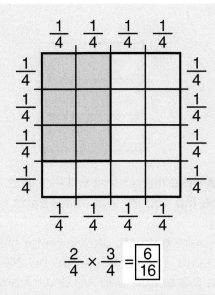

$$\frac{2}{4} \times \frac{3}{4} = \boxed{\frac{6}{16}}$$

So the area of the shaded rectangle is 6/16 of the square unit.

f. You get the same answer when you multiply without simplifying.
The problem for the numerators is 2 times 3.
- What's the answer? (Signal.) *6.*
The problem for the denominators is 4 times 4.
- What's the answer? (Signal.) *16.*
- So what fraction equals 2/4 times 3/4? (Signal.) *6/16.*

WORKBOOK PRACTICE

a. Find part 4 in your workbook. ✔
(Teacher reference:)

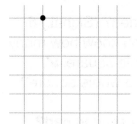 $\frac{2}{3} \times \frac{1}{3} =$

b. Touch the dot. ✔
You're going to start at the dot and draw a square that has 3 parts on each side.
(Display:)   [115:4F]

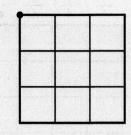

Here's what the square will look like.
- Outline the square that has 3 parts on each side. ✔
c. Now you're going to draw a rectangle inside the square. The rectangle will be 2/3 times 1/3.
- Shade 2/3 across and 1/3 down from the dot. ✔
(Add to show:)   [115:4G]

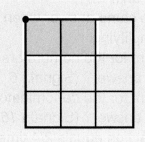

Here's what you should have.
d. You're going to write the fraction for the shaded parts.
- How many parts are shaded? (Signal.) *2.*
- How many parts are in the whole square? (Signal.) *9.*
- So what's the fraction for the shaded part? (Signal.) *2/9.*
e. Work the multiplication problem 2/3 times 1/3. See if you get the same answer. ✔
- Everybody, what's 2/3 times 1/3? (Signal.) *2/9.*
So the area of the shaded rectangle is 2/9 of the square unit.

EXERCISE 5: WORD PROBLEMS
FRACTION COMPARISON

a. Open your textbook to Lesson 115 and find part 1. ✔
(Teacher reference:)

a. Bill was $\frac{4}{3}$ the height of his sister. His sister was 57 inches tall. How tall was Bill?

b. The white house is 48 years old. The gray house is $\frac{3}{4}$ as old as the white house. How old is the gray house?

c. The library had $\frac{9}{2}$ the number of books as the truck. There were 810 books in the library. How many books were in the truck?

b. Work all of the problems in part 1. Remember to start with the equation that compares two things.
(Observe students and give feedback.)
c. Check your work.
- Problem A: Bill was 4/3 the height of his sister. His sister was 57 inches tall. How tall was Bill?
- Read the equation you started with. (Signal.) *B = 4/3 S.*
- How tall was Bill? (Signal.) *76 inches.*
(Display:)   [115:5A]

$$a. \; B = \frac{4}{3} S$$
$$B = \frac{4}{\cancel{3}} \overset{19}{(\cancel{57})} = 76$$
$$\boxed{76 \text{ inches}}$$

Here's what you should have.
d. Problem B: The white house is 48 years old. The gray house is 3/4 as old as the white house. How old is the gray house?
- Read the equation you started with. (Signal.) *G = 3/4 W.*
- How old is the gray house? (Signal.) *36 years.*
(Display:)   [115:5B]

$$b. \; g = \frac{3}{4} W$$
$$g = \frac{3}{\cancel{4}} \overset{12}{(\cancel{48})} = 36$$
$$\boxed{36 \text{ years}}$$

Here's what you should have.

e. Problem C: The library had 9/2 the number of books as the truck. There were 810 books in the library. How many books were in the truck?
- Read the equation you started with. (Signal.) $L = 9/2\ T$.
- How many books were in the truck? (Signal.) *180 books.*
(Display:) [115:5C]

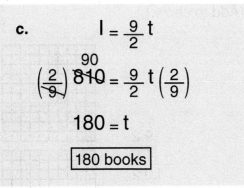

Here's what you should have.

EXERCISE 6: DECIMAL OPERATIONS
REPRESENTATION

a. (Display:) [115:6A]

$$\begin{array}{r} .9 \\ +\ .5 \\ \hline \end{array}$$

- Read the problem. (Signal.) *9 tenths + 5 tenths.*
- What's 9 plus 5? (Signal.) *14.*
(Add to show:) [115:6B]

$$\begin{array}{r} .9 \\ +\ .5 \\ \hline 1\ 4 \end{array}$$

- What's missing in the answer? (Signal.) *The decimal point.*
(Add to show:) [115:6C]

$$\begin{array}{r} .9 \\ +\ .5 \\ \hline 1.4 \end{array}$$

- What's the answer? (Signal.) *1 and 4 tenths.*
(Add to show:) [115:6D]

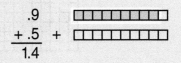

b. You're going to explain how to make a diagram that shows that 9 tenths plus 5 tenths equals 14 tenths, and equals 1 and 4 tenths.
- What's the first value I will show in the diagram? (Signal.) *9 tenths.*
- How will I show 9 tenths? (Call on a student. Idea: *Shade 9 squares in the top row.*) Yes, I shade 9 squares in the top row.
(Add to show:) [115:6E]

- Everybody, how do I show the other value? (Signal.) *Shade 5 tenths in the bottom row.* Yes, I shade 5 tenths in the bottom row.
(Add to show:) [115:6F]

- How many total tenths are shaded in this diagram? (Signal.) *14 tenths.*
c. How can I show that 14 tenths is 1 and 4 tenths? (Call on a student. Idea: *Cross out 1 tenth from the bottom row, and shade 1 tenth in the top row.*)
(Change to show:) [115:6G]

- What does the diagram show now? (Signal.) *1 and 4 tenths.*

d. (Display:) [115:6H]

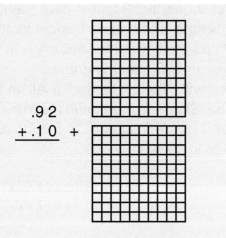

.9 2
+ .1 0   +
─────

We can do the same thing with hundredths.
- Read the problem. (Signal.) *92 hundredths plus 10 hundredths.*
- What's 92 plus 10? (Signal.) *102.*
  (Add to show:) [115:6I]

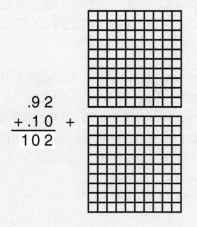

.9 2
+ .1 0   +
─────
10 2

- Where do we put the decimal point in the answer? (Call on a student. Idea: *After the 1.*)
  (Add to show:) [115:6J]

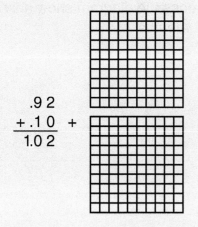

.9 2
+ .1 0   +
─────
1.0 2

- Everybody, what's the answer? (Signal.) *1 and 2 hundredths.*

e. The diagram shows hundredths. The first value is 92 hundredths.
- How do I show the first value? (Call on a student. Idea: *Shade 92 parts.*)
  Yes, I shade 92 parts.
- Everybody, what's the second value? (Signal.) *10 hundredths.*
- How do I show that value? (Signal.) *Shade 10 parts.*
  (Add to show:) [115:6K]

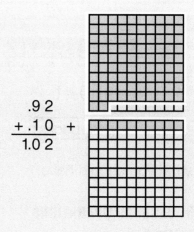

.9 2
+ .1 0   +
─────
1.0 2

f. We have 92 hundredths and 10 hundredths.
- How many hundredths is that in all? (Signal.) *102 hundredths.*
  The answer to the written problem is 1 and 2 hundredths.
- How do I show that we have that number in the diagram? (Call on a student. Idea: *Cross out 8 hundredths and move them to the top square to show 100 hundredths (1).*)
  (Change to show:) [115:6L]

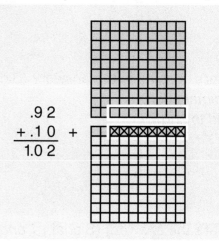

.9 2
+ .1 0   +
─────
1.0 2

- What decimal value does the diagram show now? (Signal.) *1 and 2 hundredths.*
- Does the diagram show the answer to the problem? (Signal.) *Yes.*

EXERCISE 7: DIVISION WORD PROBLEMS
AS FRACTIONS

a. You have worked division problems that involve whole numbers.
- If you divide 6 units into 7 equal parts, what's the fraction for each part? (Signal.) *6/7.*
- If you divide 4 units into 13 equal parts, what's the fraction for each part? (Signal.) *4/13.*
- If you divide 12 units into 7 equal parts, what's the fraction for each part? (Signal.) *12/7.*
(Repeat until firm.)

b. You're going to write fractions that show division problems.
(Display:)   [115:7A]

> **a.** You divide 12 units into 16 equal parts. Write the fraction for each part.

Problem A: You divide 12 units into 16 equal parts.
- In part A, write the fraction for each part. ✔
- Everybody, what's the fraction for each part? (Signal.) *12/16.*

c. (Display:)   [115:7B]

> **b.** You divide 7 units into 10 equal parts. Write the fraction for each part.

Problem B: You divide 7 units into 10 equal parts.
- Write the fraction for each part. ✔
- Everybody, what's the fraction for each part? (Signal.) *7/10.*

d. (Display:)   [115:7C]

> **c.** You divide 4 units into 9 equal parts. Write the fraction for each part.

Problem C: You divide 4 units into 9 equal parts.
- Write the fraction for each part. ✔
- Everybody, what's the fraction for each part? (Signal.) *4/9.*

e. Now we're going to work some problems that tell about cakes. Pencils down. ✔
(Display:)   [115:7D]

> You have 4 cakes. You want to divide them into 9 equal parts. What fraction of a cake is each part?

Listen: You have 4 cakes. You want to divide them into 9 equal parts. What fraction of a cake is each part?

f. Each cake is a unit. You divide 4 units into 9 equal parts.
- Say the division problem. (Signal.) *4 ÷ 9.*
- Say the fraction for 4 divided by 9. (Signal.) *4/9.* That's the answer to the problem.
(Add to show:)   [115:7E]

> You have 4 cakes. You want to divide them into 9 equal parts. What fraction of a cake is each part?
>
> $\frac{4}{9}$ cake

- What fraction of a cake is each part? (Signal.) *4/9.*
Yes, 4/9 of a cake.
- What's the unit name? (Signal.) *Cake.*

g. (Display:)   [115:7F]

> **d.** You have 6 cakes. You want to divide them into 18 parts. What fraction of a cake is each part?

Problem D: You have 6 cakes. You want to divide them into 18 parts. What fraction of a cake is each part?
- Write the fraction. Simplify the fraction if you can, and write the unit name.
(Observe students and give feedback.)
(Add to show:)   [115:7G]

> **d.** You have 6 cakes. You want to divide them into 18 parts. What fraction of a cake is each part?
>
> $\frac{6}{18} = \frac{1}{3}$ cake

Here's what you should have.
6/18 is the fraction you wrote. So each part is 1/3 of a cake.

h. (Display:) [115:7H]

> **e.** You have 6 pies. You want to divide them into 4 equal parts. What is the size of each part?

Problem E: You have 6 pies. You want to divide them into 4 equal parts. What is the size of each part?

- Write the fraction. Then show the answer as a simplified mixed number and a unit name. ✔ You wrote 6/4. That's 3/2.
- What's the mixed number for 3/2? (Signal.) *1 and 1/2.*
  (Add to show:) [115:7I]

> **e.** You have 6 pies. You want to divide them into 4 equal parts. What is the size of each part?
>
> $$\frac{6}{4} = \frac{3}{2} = \boxed{1\frac{1}{2}} \text{ pies}$$

Here's what you should have.
Each part is 1 and 1/2 pies.

i. (Display:) [115:7J]

> **f.** You have 10 pies. You want to divide them equally between 8 people. How much pie will each person receive?

Problem F: You have 10 pies. You want to divide them equally between 8 people. How much pie will each person receive?

- Write the fraction. Then show the answer as a simplified mixed number and a unit name. ✔ You wrote the fraction 10/8. Then you simplified.
- How much pie will each person receive? (Signal.) *1 and 1/4 pies.*
  Yes, each person will receive 1 and 1/4 pies.

Assign Independent Work, Textbook parts 2–12.

Optional extra math-fact practice worksheets are available on ConnectED.

Lessons 116–120 Planning Page

	Lesson 116	Lesson 117	Lesson 118	Lesson 119	Lesson 120
Student Learning Objectives	**Exercises** 1. Compare fractions with unlike denominators 2. Answer questions about line plots that have fractional units 3. Use diagrams to represent fraction multiplication 4. Solve division word problems with fractions 5. Use diagrams to represent parts of a group 6. Solve probability problems with mental math 7. Solve division word problems as fractions Complete work independently	**Exercises** 1. Compare fractions with unlike denominators 2. Use diagrams to represent fraction multiplication 3. Use diagrams to represent operations with fractions and decimals 4. Use diagrams to represent parts of a group 5. Find the average of data displayed on line plots that have fractional units 6. Solve word problems with fractions 7. Interpret expressions Complete work independently	**Exercises** 1. **Use knowledge of fractions and multiplication to explain answers** 2. **Plot lines on a coordinate system by using related function rules** 3. Use diagrams to represent parts of a group 4. Use diagrams to represent decimal addition 5. Compare fractions with unlike denominators 6. **Determine the reasonableness of answers in fraction operations** 7. Interpret expressions Complete work independently	**Exercises** 1. Use knowledge of fractions and multiplication to explain answers 2. Plot lines on a coordinate system by using related function rules 3. Use diagrams to represent parts of a group 4. Represent answers to problems as fractions and decimals 5. Compare fractions with unlike denominators 6. Determine the reasonableness of answers in fraction operations 7. Write an equation to represent a sentence 8. Complete work independently	**Exercises** 1. Plot lines on a coordinate system by using related function rules 2. Use diagrams to represent parts of a group 3. Use knowledge of fractions and multiplication to explain answers 4. Use diagrams to represent decimal multiplication 5. Determine the reasonableness of answers in fraction operations Complete work independently

Common Core State Standards for Mathematics					
5.OA 2		✔	✔	✔	✔
5.OA 3			✔	✔	✔
5.NBT 5–6	✔	✔	✔	✔	✔
5.NBT 7		✔	✔	✔	✔
5.NF 1	✔	✔	✔	✔	
5.NF 2	✔	✔	✔	✔	✔
5.NF 3	✔		✔		
5.NF 4–6	✔	✔	✔	✔	✔
5.NF 7	✔	✔			
5.MD 1	✔				
5.MD 2	✔	✔		✔	✔
5.G 2			✔	✔	✔
5.G 3–4					✔

Teacher Materials	Presentation Book 2, Board Displays CD or chalk board				
Student Materials	Textbook, Workbook, pencil, lined paper, ruler				
Additional Practice	• Student Practice Software: Block 5: Activity 1 (5.MD 1), Activity 2 (5.NF 1 and 5.NF 2), Activity 3 (5.MD 3 and 5.MD 4), Activity 4 (5.NF 4 and 5.NF 6), Activity 5, Activity 6 (5.MD 3 and 5.MD 4) • Provide needed fact practice with Level D or E Math Fact Worksheets.				
Mastery Test					Student Assessment Book (Present Mastery Test 12 following Lesson 120.)

Lesson 116

EXERCISE 1: FRACTION COMPARISON
UNLIKE DENOMINATORS

REMEDY

a. (Display:) [116:1A]

$$\frac{3}{4} \quad \frac{5}{6}$$

Here's a new kind of problem. We're going to compare these fractions. To do that, we have to change both denominators. The denominators are 4 and 6.

- Raise your hand when you know the lowest common denominator for 4 and 6. ✔
- What's the lowest common denominator for 4 and 6? (Signal.) *12.*
  So we change 3/4 into 12ths and 5/6 into 12ths.
  (Add to show:) [116:1B]

$$\frac{3}{4} \quad \frac{5}{6}$$

$$\frac{3}{4}\left(\frac{}{}\right) = \frac{}{12}$$

- Raise your hand when you know how many 12ths 3/4 equals. ✔
- What does 3/4 equal? (Signal.) *9/12.*
  (Add to show:) [116:1C]

$$\frac{3}{4} \quad \frac{5}{6}$$

$$\frac{3}{4}\left(\frac{3}{3}\right) = \frac{9}{12}$$

Now we change 5/6 into 12ths.
(Add to show:) [116:1D]

$$\frac{3}{4} \quad \frac{5}{6}$$

$$\frac{3}{4}\left(\frac{3}{3}\right) = \frac{9}{12}$$

$$\frac{5}{6}\left(\frac{}{}\right) = \frac{}{12}$$

- Raise your hand when you know how many 12ths 5/6 equals. ✔
- What does 5/6 equal? (Signal.) *10/12.*
  (Add to show:) [116:1E]

$$\frac{3}{4} \quad \frac{5}{6}$$

$$\frac{3}{4}\left(\frac{3}{3}\right) = \frac{9}{12}$$

$$\frac{5}{6}\left(\frac{2}{2}\right) = \frac{10}{12}$$

- So which is more—3/4 or 5/6? (Signal.) *5/6.*
  (Add to show:) [116:1F]

$$\frac{3}{4} < \frac{5}{6}$$

$$\frac{3}{4}\left(\frac{3}{3}\right) = \frac{9}{12}$$

$$\frac{5}{6}\left(\frac{2}{2}\right) = \frac{10}{12}$$

We complete the original statement to show that 5/6 is more.

━━━━━ **WORKBOOK PRACTICE** ━━━━━

a. Open your workbook to Lesson 116 and find part 1. ✔
   (Teacher reference:) **R** **Part C**

a. $\frac{3}{8}$ ▢ $\frac{2}{5}$ b. $\frac{7}{6}$ ▢ $\frac{9}{8}$

b. Problem A. Read the fractions. (Signal.)
3/8, 2/5.
This is like the problems we just worked. You have to change both denominators.
- Raise your hand when you know the lowest common denominator for 8ths and 5ths. ✔
- What's the lowest common denominator? (Signal.) *40.*
- Rewrite both fractions. Then complete the original statement.
(Observe students and give feedback.)
- Everybody, what did you change 3/8 to? (Signal.) *15/40.*
- What did you change 2/5 to? (Signal.) *16/40.*
- Which is more—3/8 or 2/5? (Signal.) *2/5.*
(Display:)   [116:1G]

$$a. \quad \frac{3}{8} < \frac{2}{5}$$
$$\frac{3}{8}\left(\frac{5}{5}\right) = \frac{15}{40}$$
$$\frac{2}{5}\left(\frac{8}{8}\right) = \frac{16}{40}$$

Here's what you should have.
c. Problem B. Read the fractions. (Signal.) *7/6, 9/8.*
- Raise your hand when you know the lowest common denominator for 6ths and 8ths. ✔
- What's the lowest common denominator? (Signal.) *24.*
- Rewrite both fractions. Then complete the original statement.
(Observe students and give feedback.)
- Everybody, what did you change 7/6 to? (Signal.) *28/24.*
- What did you change 9/8 to? (Signal.) *27/24.*
- Which is more—7/6 or 9/8? (Signal.) *7/6.*
(Display:)   [116:1H]

$$b. \quad \frac{7}{6} > \frac{9}{8}$$
$$\frac{7}{6}\left(\frac{4}{4}\right) = \frac{28}{24}$$
$$\frac{9}{8}\left(\frac{3}{3}\right) = \frac{27}{24}$$

Here's what you should have.

EXERCISE 2: LINE PLOT
WITH FRACTIONS

a. Find part 2 in your workbook. ✔
(Teacher reference:)

Length of Moths Bernie Collected

Column Totals:

a. Bernie collected 3 moths that were $\frac{10}{10}$ inch long.
b. Bernie collected 1 moth that was $\frac{18}{10}$ inches long.
c. Bernie collected 2 moths that were $\frac{9}{10}$ inch long.
d. Bernie collected 1 moth that was $\frac{8}{10}$ inch long.
e. Bernie collected 1 moth that was $\frac{14}{10}$ inches long.

You're going to make a line plot and answer questions. Each point on the line plot will show tenths of an inch.
- What's the first fraction shown? (Signal.) *7/10.*
- What's the next fraction going to be? (Signal.) *8/10.*
b. Write fractions for the tenths. ✔
(Display:)   [116:2A]

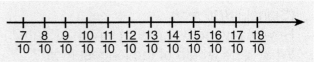

Here's what you should have.
c. The fractions tell the length of the moths that Bernie collected. The unit name is inches.
- Write inches as the unit name. ✔
d. Fact A: Bernie collected 3 moths that were 10/10 inch long.
- Make Xs for those moths. ✔
- Everybody, how many Xs did you make? (Signal.) *3.*
- How long were those moths? (Signal.) *10/10 inch.*
e. Fact B: Bernie collected 1 moth that was 18/10 inches long.
- Make an X for that moth. ✔
- Everybody, how many Xs did you make? (Signal.) *1.*
- How long was that moth? (Signal.) *18/10 inches.*

f. Fact C: Bernie collected 2 moths that were 9/10 inch long.
- Make Xs for those moths. ✔
- Everybody, how many Xs did you make? (Signal.) *2.*
- How long were those moths? (Signal.) *9/10 inch.*

g. Fact D: Bernie collected 1 moth that was 8/10 inch long.
- Make an X for that moth. ✔
- Everybody, how many Xs did you make? (Signal.) *1.*
- How long was that moth? (Signal.) *8/10 inch.*

h. Fact E: Bernie collected 1 moth that was 14/10 inches long.
- Make an X for that moth. ✔
- Everybody, how many Xs did you make? (Signal.) *1.*
- How long was that moth? (Signal.) *14/10 inches.*
(Add to show:)   [116:2B]

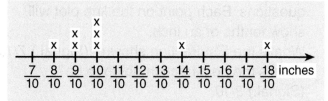

Here's what you should have.

i. What's the length of the shortest moth? (Signal.) *8/10 inch.*
- How many moths are that length? (Signal.) *1.*

j. What's the length of the longest moth? (Signal.) *18/10 inches.*
- How many moths are 18/10 inches long? (Signal.) *1.*

k. How many moths are 10/10 inch long? (Signal.) *3.*
- How many moths are 1 inch long? (Signal.) *3.*

l. Raise your hand when you know how many moths are less than 13/10 inches long. ✔
- How many moths are less than 13/10 inches long? (Signal.) *6.*

m. Raise your hand when you know how many moths are more than 13/10 inches long. ✔
- How many moths are more than 13/10 inches long? (Signal.) *2.*

n. Find the total for each column. Stop when you've done that much.
(Observe students and give feedback.)

(Display:)   [116:2C]

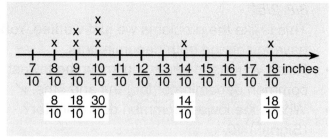

Here's what you should have.

o. Now find the total for all the columns with Xs. Remember, the total is a fraction.
(Observe students and give feedback.)
- Everybody, what's the total for all the columns? (Signal.) *88/10.*
- Raise your hand when you can say the problem you work to find the average. ✔
- Say the problem you'll work. (Signal.) *88/10 ÷ 8.*

p. Figure out the average length. Leave the answer as tenths.
(Observe students and give feedback.)
- Everybody, what's the average length of the moths? (Signal.) *11/10 inches.*

q. Make an arrow to show where the average is. ✔
(Display:)   [116:2D]

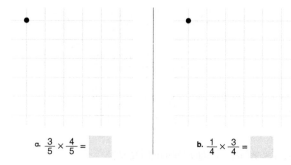

Here's what you should have.

EXERCISE 3: FRACTION MULTIPLICATION
REPRESENTATION

a. Find part 3 in your workbook. ✔
(Teacher reference:)

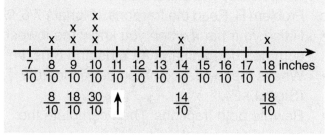

a. $\frac{3}{5} \times \frac{4}{5} =$ �_____   b. $\frac{1}{4} \times \frac{3}{4} =$ �_____

b. Item A. You're going to make a shaded rectangle inside the square.
- Start at the dot and make a square that has 5 parts on each side. ✔
(Display:) [116:3A]

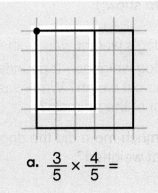

a. $\dfrac{3}{5} \times \dfrac{4}{5} =$

Here's what you should have.
c. Now you're going to draw a rectangle inside the square. The rectangle will show 3/5 times 4/5 of the square. The sides of the square show fifths.
- Make the rectangle 3/5 across and 4/5 down. ✔
(Add to show:) [116:3B]

a. $\dfrac{3}{5} \times \dfrac{4}{5} =$

Here's what you should have.
- Shade in the rectangle. ✔
d. You're going to figure out what fraction of the whole square is the shaded rectangle.
- How many parts are shaded? (Signal.) *12.*
  That's the numerator of the fraction.
- How many parts are in the whole square? (Signal.) *25.*
  That's the denominator of the fraction.
- What's the numerator of the fraction? (Signal.) *12.*
- What's the denominator? (Signal.) *25.*
  (Repeat until firm.)

e. Write the fraction for the shaded rectangle. ✔
- Everybody, what's 3/5 times 4/5? (Signal.) *12/25.*
  So the area of the shaded rectangle is 12/25 of the whole square.
(Add to show:) [116:3C]

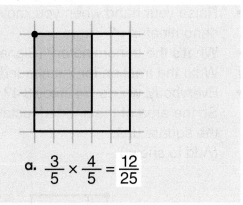

a. $\dfrac{3}{5} \times \dfrac{4}{5} = \dfrac{12}{25}$

Here's what you should have.
f. Item B. You're going to make a shaded rectangle inside the square.
- Start at the dot and make a square that has 4 parts on each side. ✔
(Display:) [116:3D]

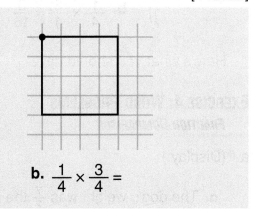

b. $\dfrac{1}{4} \times \dfrac{3}{4} =$

Here's what you should have.
g. Now you're going to draw a rectangle inside the square. The rectangle will show 1/4 times 3/4.
- Make the rectangle 1/4 across and 3/4 down. Then shade the rectangle. ✔
(Add to show:) [116:3E]

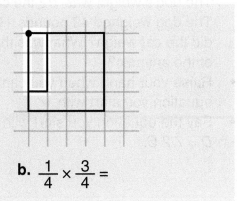

b. $\dfrac{1}{4} \times \dfrac{3}{4} =$

Here's what you should have.

h. You're going to figure out what fraction of the whole square is the shaded rectangle.
- Raise your hand when you know the numerator of the fraction. ✔
- What's the numerator? (Signal.) *3.*
- Raise your hand when you know the denominator. ✔
- What's the denominator? (Signal.) *16.*

i. Write the fraction for the shaded parts. ✔
- Everybody, what's 1/4 times 3/4? (Signal.) *3/16.*
So the area of the shaded rectangle is 3/16 of the square unit.
(Add to show:) [116:3F]

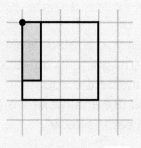

b. $\dfrac{1}{4} \times \dfrac{3}{4} = \dfrac{3}{16}$

Here's what you should have.

EXERCISE 4: WORD PROBLEMS
FRACTION COMPARISON

a. (Display:) [116:4A]

> **a.** The dog's weight was $\frac{7}{2}$ the cat's weight. The dog weighed 42 pounds. How much did the cat weigh? What was the total weight of the animals?

Here's a problem you know how to work. It has a second question.
The dog's weight was 7/2 the cat's weight. The dog weighed 42 pounds. How much did the cat weigh? What was the total weight of the animals?
- Raise your hand when you can say the letter equation you start with. ✔
- Say the equation you start with. (Signal.) *D = 7/2 C.*

(Add to show:) [116:4B]

> **a.** The dog's weight was $\frac{7}{2}$ the cat's weight. The dog weighed 42 pounds. How much did the cat weigh? What was the total weight of the animals?
>
> $$d = \frac{7}{2}\,c$$

b. Yes, the dog's weight equals 7/2 the cat's weight. Start with that equation. In part A, figure out the answer to the first question. Then figure out the answer to the second question.
(Observe students and give feedback.)
- Everybody, what's the answer to the first question? (Signal.) *12 pounds.*
Yes, the cat weighed 12 pounds.
Then you added the weights of 42 and 12.
- What was the total weight of the animals? (Signal.) *54 pounds.*

c. (Change to show:) [116:4C]

> **a.** The dog's weight was $\frac{7}{2}$ the cat's weight. The dog weighed 42 pounds. How much did the cat weigh? What was the total weight of the animals?
>
> How much more did the dog weigh than the cat weighed?

Here's another question: How much more did the dog weigh than the cat weighed?
- Work the problem. Write the answer with a unit name.
(Observe students and give feedback.)
- Everybody, read the problem you worked. (Signal.) *42 – 12.*
- How much more did the dog weigh than the cat? (Signal.) *30 pounds.*

d. (Display:) [116:4D]

> **b.** The bench weighed $\frac{4}{9}$ as much as the table. The table weighed 54 pounds. How much did the bench weigh? How much did both pieces of furniture weigh together? How much more did the table weigh than the bench weighed?

Here's a problem that has three questions: The bench weighed 4/9 as much as the table. The table weighed 54 pounds. How much did the bench weigh? How much did both pieces of furniture weigh together? How much more did the table weigh than the bench weighed?

e. Work the problem and write the answer to the first question. Stop when you've done that much.
   (Observe students and give feedback.)
* Check your work.
* Read the letter equation you started with. (Signal.) *B = 4/9 T.*
* How much did the bench weigh? (Signal.) *24 pounds.*
f. Now figure out the answer to the second question: How much did both pieces of furniture weigh together? ✔
* Check your work.
   You worked the problem 54 plus 24.
* How much did both pieces of furniture weigh? (Signal.) *78 pounds.*
g. Figure out the answer to the third question: How much more did the table weigh than the bench weighed? ✔
* Check your work.
   You worked the problem 54 minus 24.
* How much more did the table weigh? (Signal.) *30 pounds.*

EXERCISE 5: FRACTION REPRESENTATION
PARTS OF A GROUP

a. We're going to make groups from fractions.
   (Display:) [116:5A]

$$\frac{12}{3}$$

* Read this fraction. (Signal.) *12/3.*
* Read the fraction as a division problem. (Signal.) *12 ÷ 3.*
   That tells you that you have 12 divided into 3 equal groups.
* What do you have? (Signal.) *12 divided into 3 equal groups.*
b. I'll show each group as a row of eggs.
* How many rows of eggs do I make? (Signal.) *3.*
   (Add to show:) [116:5B]

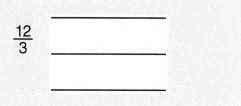

c. The answer to the division problem tells you the number of eggs in each row.
* What's 12 divided by 3? (Signal.) *4.*
* So what's the number of eggs in each row? (Signal.) *4.*
   (Add to show:) [116:5C]

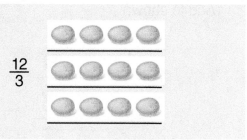

Yes, each row is a group of 4 eggs.

d. (Display:) [116:5D]

$$\frac{14}{7}$$

- New fraction. Read it. (Signal.) *14/7.*
- Say the division problem. (Signal.) *14 ÷ 7.*
  Yes, 14 divided into 7 equal groups.
- How many groups do I make? (Signal.) *7.*
  (Add to show:) [116:5E]

$$\frac{14}{7}$$

e. The answer to the division problem tells you
  the number for each group.
- What's 14 divided by 7? (Signal.) *2.*
- So what's the number in each group?
  (Signal.) *2.*
  I'll show 2 cups in each group.
  (Add to show:) [116:5F]

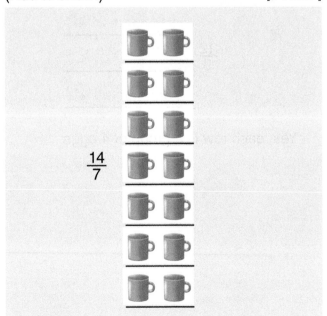

$$\frac{14}{7}$$

- How many groups of cups are there?
  (Signal.) *7.*

a. Find part 4 in your workbook. ✔
  (Teacher reference:)

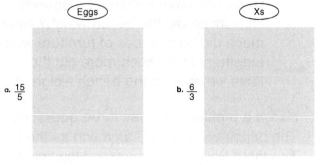

Eggs   Xs

a. $\frac{15}{5}$ b. $\frac{6}{3}$

b. Problem A: 15/5.
- Read the fraction as a division problem.
  (Signal.) *15 ÷ 5.*
- How many groups will you make? (Signal.) *5.*
- Make 5 lines to show the 5 equal groups. ✔
c. The answer to the division problem tells how
  many are in each group.
- What's 15 divided by 5? (Signal.) *3.*
- So how many are in each group? (Signal.) *3.*
d. Make 3 eggs in each group. ✔
  (Display:) [116:5G]

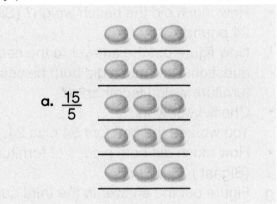

a. $\frac{15}{5}$

Here's what you should have.
- How many groups are there? (Signal.) *5.*
- How many eggs are there all together?
  (Signal.) *15.*
- How many eggs are in each group? (Signal.) *3.*
e. Problem B: 6/3.
- Read the fraction as a division problem.
  (Signal.) *6 ÷ 3.*
- How many equal groups will you make?
  (Signal.) *3.*
- Make lines for the equal groups. ✔
- Everybody, what's 6 divided by 3? (Signal.) *2.*
- So how many are in each group? (Signal.) *2.*
- Make 2 Xs in each group. ✔

(Display:) [116:5H]

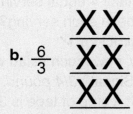

b. $\frac{6}{3}$

Here's what you should have.
- How many Xs are there in all? (Signal.) *6.*
- How many Xs are in each group? (Signal.) *2.*
- How many equal groups are there? (Signal.) *3.*
f. Remember, say the fraction as a division problem.
The problem tells how many groups there are. The answer to the problem tells how many objects are in each group.

EXERCISE 6: MENTAL MATH
PROBABILITY   `REMEDY`

a. Open your textbook to Lesson 116 and find part 1. ✔
(Teacher reference:)

a. | Flowers of Different Colors a Person Pulled from a Box Without Looking |
| --- |

yellow	2
red	7
blue	1

- How many flowers of each color would you expect the person to pull from the box if she took 40 trials?

b. | Marbles of Different Colors a Person Pulled from a Bag Without Looking |
| --- |

white	8
pink	5
orange	7

- How many marbles of each color would you expect the person to pull from the bag if she took 120 trials?

You'll work these problems with mental math.
b. Problem A. The numbers show how many flowers of different colors a person pulled from a box without looking. You can figure out the number of trials the person took. You just add the numbers for all the colors.
- On how many trials did the person pull out a yellow flower? (Signal.) *2.*
- On how many trials did the person pull out a red flower? (Signal.) *7.*
- On how many trials did the person pull out a blue flower? (Signal.) *1.*

c. Raise your hand when you know the number of trials the person took. ✔
- How many trials did the person take? (Signal.) *10.*
- Raise your hand when you know how many flowers of each color you would expect the person to pull from the box if she took **40** trials. ✔
- If a person took 40 trials, what's the expected number for yellow? (Signal.) *8.*
- What's the expected number for red? (Signal.) *28.*
- What's the expected number for blue? (Signal.) *4.*
Yes, the expected numbers are 4 times the numbers for 10 trials.
d. Problem B. The numbers show how many marbles of different colors a person pulled from a bag without looking.
- Figure out the number of trials the person took.
Raise your hand when you know the number of trials. ✔
- Everybody, how many trials did the person take? (Signal.) *20.*
Yes, 8 + 5 + 7 = 20.
e. Raise your hand when you know how many marbles of each color you would expect the person to pull from the bag if she took **120** trials. ✔
- Everybody, if a person took 120 trials, what's the expected number for white? (Signal.) *48.*
- What's the expected number for pink? (Signal.) *30.*
- What's the expected number for orange? (Signal.) *42.*
Yes, the expected numbers are 6 times the numbers for 20 trials.

EXERCISE 7: DIVISION WORD PROBLEMS
AS FRACTIONS

a. Find part 2 in your textbook. ✔
   (Teacher reference:)

> a. Joe has 40 pounds of flour and wants to divide it equally into 32 bags. How many pounds of flour will be in each bag?
>
> b. Lacy has 3 pounds of salad. She wants to make 4 equal servings. How much salad will be in each serving?
>
> c. A strip of tape is 34 inches long. We divide it into 5 equal parts. How many inches long is each part?

Part 2	
a.	

You're going to work problems that involve dividing units into equal parts.

b. Problem A: Joe has 40 pounds of flour and wants to divide it equally into 32 bags. How many pounds of flour will be in each bag?

• Write the fraction. Stop when you've done that much. ✔
   (Display:)   [116:7A]

$$a.\ \frac{40}{32}$$

Here's what you should have: 40/32.

c. Simplify the fraction and write the answer as a mixed number with a unit name. ✔

• Everybody, how much flour will be in each bag? (Signal.) *1 and 1/4 pounds.*
   (Add to show:)   [116:7B]

$$a.\ \frac{40}{32} = \frac{5}{4} = \boxed{1\frac{1}{4}\ \text{pounds}}$$

Here's what you should have.

d. Problem B: Lacy has 3 pounds of salad. She wants to make 4 equal servings. How much salad will be in each serving?

• Work the problem. ✔
• Everybody, how much salad will be in each serving? (Signal.) *3/4 pound.*

e. Problem C: A strip of tape is 34 inches long. We divide it into 5 equal parts. How many inches long is each part?

• Work the problem and write the simplified answer with a unit name. ✔
   You wrote 34/5, which simplifies to 6 and 4/5.

• So how many inches long is each part? (Signal.) *6 and 4/5 inches.*

Assign Independent Work, Textbook parts 3–11.

Optional extra math-fact practice worksheets are available on ConnectED.

Lesson 117

EXERCISE 1: FRACTION COMPARISON
UNLIKE DENOMINATORS

a. Open your workbook to Lesson 117 and find part 1. ✔
    (Teacher reference:)

You're going to complete the statements. You have to rewrite both fractions so they have the lowest common denominator.

b. Problem A. Read the fractions. (Signal.) *6/9, 4/6.*
•   Raise your hand when you know the lowest common denominator. ✔
•   What's the lowest common denominator? (Signal.) *18.*
•   Rewrite both fractions. Then complete the original statement.
    (Observe students and give feedback.)
•   Everybody, what did you change 6/9 to? (Signal.) *12/18.*
•   What did you change 4/6 to? (Signal.) *12/18.*
•   Which is more? (Call on a student. Idea: *Neither; they are the same.*)
•   Everybody, what sign did you write? (Signal.) *Equals.*
    (Display:) [117:1A]

$$a. \quad \frac{6}{9} = \frac{4}{6}$$
$$\frac{6}{9}\left(\frac{2}{2}\right) = \frac{12}{18}$$
$$\frac{4}{6}\left(\frac{3}{3}\right) = \frac{12}{18}$$

Here's what you should have.

c. Problem B. Read the fractions. (Signal.) *3/4, 8/10.*
•   Raise your hand when you know the lowest common denominator. ✔
•   What's the lowest common denominator? (Signal.) *20.*
•   Rewrite both fractions. Then complete the original statement.
    (Observe students and give feedback.)
•   Everybody, what did you change 3/4 to? (Signal.) *15/20.*
•   What did you change 8/10 to? (Signal.) *16/20.*
•   Which is more—3/4 or 8/10? (Signal.) *8/10.*
    (Display:) [117:1B]

$$b. \quad \frac{3}{4} < \frac{8}{10}$$
$$\frac{3}{4}\left(\frac{5}{5}\right) = \frac{15}{20}$$
$$\frac{8}{10}\left(\frac{2}{2}\right) = \frac{16}{20}$$

Here's what you should have.

d. Problem C. Read the fractions. (Signal.) *9/5, 5/3.*
•   Raise your hand when you know the lowest common denominator. ✔
•   What's the lowest common denominator? (Signal.) *15.*
•   Rewrite both fractions. Then complete the original statement.
    (Observe students and give feedback.)
•   Everybody, what did you change 9/5 to? (Signal.) *27/15.*
•   What did you change 5/3 to? (Signal.) *25/15.*
•   Which is more—9/5 or 5/3? (Signal.) *9/5.*
    (Display:) [117:1C]

$$c. \quad \frac{9}{5} > \frac{5}{3}$$
$$\frac{9}{5}\left(\frac{3}{3}\right) = \frac{27}{15}$$
$$\frac{5}{5}\left(\frac{5}{5}\right) = \frac{25}{15}$$

Here's what you should have.

EXERCISE 2: FRACTION MULTIPLICATION
REPRESENTATION

a. Find part 2 in your workbook. ✔
   (Teacher reference:)

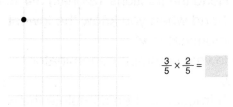

You're going to make a rectangle inside the unit square and complete the equation for 3/5 times 2/5.

b. The fractions show how to make the unit square.

• Both fractions are fifths. So how many parts will be on each side of the square? (Signal.) *5.*
  Then you'll shade the rectangle that shows 3/5 times 2/5.

c. Start at the dot and make the square that has 5 parts on each side. Then make the rectangle for 3/5 times 2/5. Stop when you've done that much. ✔
   (Display:)   [117:2A]

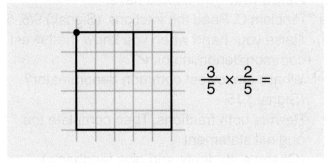

Here's what you should have.

d. Shade the rectangle and complete the equation by showing the fraction for the shaded parts. ✔

• Everybody, what's the fraction for the shaded parts? (Signal.) *6/25.*
  So the area of the shaded rectangle is 6/25 of the square unit.
  That's the same answer you get when you multiply 3/5 times 2/5.

EXERCISE 3: FRACTION AND DECIMAL OPERATIONS
DIAGRAMS

a. (Display:)   [117:3A]

$$\frac{3}{10}\left(\frac{\ \ }{\ \ }\right) = \frac{\ \ }{100}$$

Here's 3/10 on one side of the equal and hundredths on the other side.

• What do we multiply 3/10 by to change it into hundredths? (Signal.) *10/10.*
  (Add to show:)   [117:3B]

$$\frac{3}{10}\left(\frac{10}{10}\right) = \frac{\ \ }{100}$$

Yes, we multiply by 10/10.

• What's 3 times 10? (Signal.) *30.*
  (Add to show:)   [117:3C]

$$\frac{3}{10}\left(\frac{10}{10}\right) = \frac{30}{100}$$

• How many hundredths equals 3/10? (Signal.) *30/100.*
  (Add to show:)   [117:3D]

$$\frac{3}{10}\left(\frac{10}{10}\right) = \frac{30}{100}$$

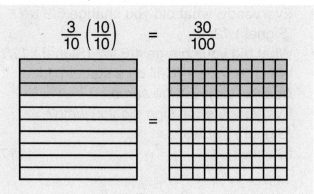

Here are the diagrams to show the equation is true.

b. (Point to tenths diagram.) Here's 3/10. There are 10 parts and 3 of them are colored.

• (Point to hundredths diagram.) Here's 30/100. There are 100 parts and 30 of them are colored.
  The shaded areas are the same size.

c. We can show the same equation with decimal values.
(Add to show:) [117:3E]

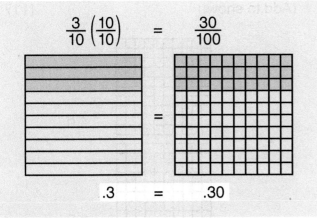

$$\frac{3}{10}\left(\frac{10}{10}\right) = \frac{30}{100}$$

$$.3 = .30$$

• Read the equation. (Signal.) *3 tenths = 30 hundredths.*
Writing a zero after the 3 does not change the value. It is simply multiplying the value by 10/10. We have 10 times more parts and 10 times more parts shaded.

d. Remember, if you write a zero after a decimal value, you don't change the value because you're multiplying by 10/10.

e. (Display:) [117:3F]

$$\frac{4}{10}$$
$$+ \frac{5}{100}$$

Here's a problem that adds tenths and hundredths.
• Read the problem. (Signal.) *4/10 + 5/100.*
• Are the denominators the same? (Signal.) *No.*

f. So we change tenths into hundredths.
(Add to show:) [117:3G]

$$\frac{4}{10}\left(\frac{}{}\right) = \frac{}{100}$$
$$+ \frac{5}{100}$$

• What do we multiply 4/10 by? (Signal.) *10/10.*
(Add to show:) [117:3H]

$$\frac{4}{10}\left(\frac{10}{10}\right) = \frac{}{100}$$
$$+ \frac{5}{100}$$

• What does 4/10 equal? (Signal.) *40/100.*

(Add to show:) [117:3I]

$$\frac{4}{10}\left(\frac{10}{10}\right) = \frac{40}{100}$$
$$+ \frac{5}{100} \qquad + \frac{5}{100}$$

g. Now we add 40/100 and 5/100.
• What's the answer? (Signal.) *45/100.*
(Add to show:) [117:3J]

$$\frac{4}{10}\left(\frac{10}{10}\right) = \frac{40}{100}$$
$$+ \frac{5}{100} \qquad + \frac{5}{100}$$
$$\boxed{\frac{45}{100}}$$

h. We can do the same thing with decimal numbers.
(Add to show:) [117:3K]

$$\frac{4}{10}\left(\frac{10}{10}\right) = \frac{40}{100} \qquad .4$$
$$+ \frac{5}{100} \qquad + \frac{5}{100} \qquad + .0\,5$$
$$\boxed{\frac{45}{100}}$$

• (Point to .4.) Read the problem. (Signal.) *4 tenths + 5 hundredths.*

i. We can change 4 tenths into hundredths.
• What hundredths value equals 4 tenths? (Signal.) *40 hundredths.*
(Add to show:) [117:3L]

$$\frac{4}{10}\left(\frac{10}{10}\right) = \frac{40}{100} \qquad .4\,0$$
$$+ \frac{5}{100} \qquad + \frac{5}{100} \qquad + .0\,5$$
$$\boxed{\frac{45}{100}}$$

So the new problem is 40 hundredths plus 5 hundredths.
• What's the answer? (Signal.) *45 hundredths.*

(Add to show:) [117:3M]

$$\frac{4}{10}\left(\frac{10}{10}\right) = \frac{40}{100} \qquad .40$$
$$+\frac{5}{100} \qquad +\frac{5}{100} \qquad +.05$$
$$\boxed{\frac{45}{100}} \qquad \overline{.45}$$

Yes, 45 hundredths.

(Display:) [117:3N]

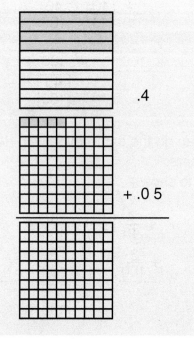

Here's a diagram of the original problem.

j. The first box shows 4 tenths. The second box shows 5 hundredths.

We can't add these decimals because one is tenths and the other is hundredths.

We change the first box to hundredths.

(Add to show:) [117:3O]

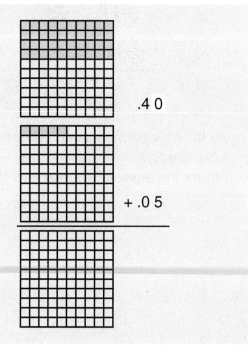

Now we have 40 hundredths plus 5 hundredths.

- Can we add those decimals? (Signal.) *Yes.*
- What's the answer? (Signal.) *45 hundredths.*

(Add to show:) [117:3P]

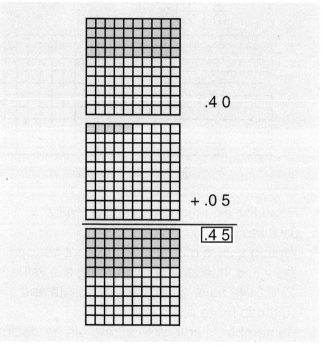

Here's the diagram of the answer.

EXERCISE 4: FRACTION REPRESENTATION
PARTS OF A GROUP

a. (Display:) [117:4A]

$$\frac{18}{3}$$

- Read the fraction as a division problem. (Signal.) *18 ÷ 3.*
- Which number tells how many groups there are? (Signal.) *3.*

(Add to show:) [117:4B]

$$\frac{18}{3}$$

Here are the 3 groups.

b. What's 18 divided by 3? (Signal.) *6.*
• So how many are in each group? (Signal.) *6.*
   (Add to show:)   [117:4C]

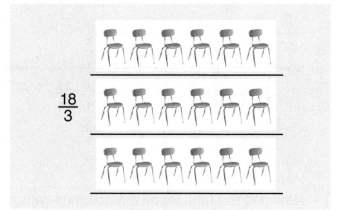

Here's the picture for the fraction 18/3.
There are 3 equal groups with 6 chairs in each
group.

━━━━━━━━━ **WORKBOOK PRACTICE** ━━━━━━━━━

a. Find part 3 in your workbook. ✔
   (Teacher reference:)

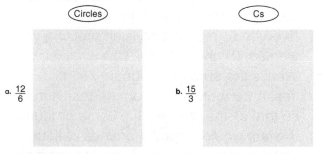

b. Problem A: 12/6.
• Say the division problem. (Signal.) *12 ÷ 6.*
• What's the number of equal groups? (Signal.) *6.*
• What's 12 divided by 6? (Signal.) *2.*
• So how many are in each group? (Signal.) *2.*
   (Repeat until firm.)
c. Show 6 groups, with 2 circles in each group. ✔
   (Display:)   [117:4D]

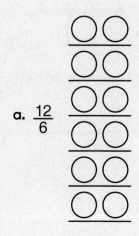

Here's what you should have.

d. Problem B: 15/3.
• Say the division problem. (Signal.) *15 ÷ 3.*
• What's the number of equal groups? (Signal.) *3.*
• What's 15 divided by 3? (Signal.) *5.*
• So how many are in each group? (Signal.) *5.*
   (Repeat until firm.)
e. Show 5 Cs in each group. ✔
   (Display:)   [117:4E]

$$\text{b.} \ \frac{15}{3} \quad \begin{array}{l} \underline{CCCCC} \\ \underline{CCCCC} \\ \underline{CCCCC} \end{array}$$

Here's what you should have.

EXERCISE 5: LINE PLOT
WITH FRACTIONS

a. Find part 4 in your workbook. ✔
   (Teacher reference:)

| Pounds of Flour in Different Jars |

Column
Totals:

a. 3 jars have $\frac{5}{4}$ pounds of flour.

b. 1 jar has 2 pounds of flour.

c. 2 jars have $\frac{10}{4}$ pounds of flour.

d. 3 jars have $\frac{7}{4}$ pounds of flour.

e. 1 jar has $\frac{6}{4}$ pounds of flour.

You're going to make a line plot and answer
questions. Each point on the line plot will
show fractions.
• What's the first fraction shown? (Signal.) *3/4.*
• What's the next fraction going to be?
   (Signal.) *4/4.*
b. Write fractions for the fourths. ✔
   (Display:)   [117:5A]

$$\frac{3}{4} \quad \frac{4}{4} \quad \frac{5}{4} \quad \frac{6}{4} \quad \frac{7}{4} \quad \frac{8}{4} \quad \frac{9}{4} \quad \frac{10}{4} \quad \frac{11}{4}$$

Here's what you should have.
• The fractions tell the pounds of flour in
   different jars. The unit name is pounds. Write
   pounds as the unit name. ✔

c. Fact A: 3 jars have 5/4 pounds of flour.
- Make Xs for those jars. ✔
- Everybody, how many Xs did you make? (Signal.) *3.*
- How much flour is in each of those jars? (Signal.) *5/4 pounds.*

d. Fact B: 1 jar has 2 pounds of flour.
- Make an X at the fraction that equals 2. ✔
- Everybody, what's the fraction for 2? (Signal.) *8/4.*

e. Fact C: 2 jars have 10/4 pounds of flour.
- Make Xs for those jars. ✔
- Everybody, how many Xs did you make? (Signal.) *2.*
- How much flour is in each of those jars? (Signal.) *10/4 pounds.*

f. Fact D: 3 jars have 7/4 pounds of flour.
- Make Xs for those jars. ✔
- Everybody, how many Xs did you make? (Signal.) *3.*
- How much flour is in each of those jars? (Signal.) *7/4 pounds.*

g. Fact E: 1 jar has 6/4 pounds of flour.
- Make an X for that jar. ✔
- Everybody, how many Xs did you make? (Signal.) *1.*
- How much flour is in that jar? (Signal.) *6/4 pounds.*
  (Add to show:) [117:5B]

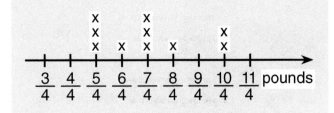

Here's what you should have.

h. What's the least amount of flour in a jar? (Signal.) *5/4 pounds.*
- How many jars have 5/4 pounds of flour? (Signal.) *3.*
- What's the most amount of flour in a jar? (Signal.) *10/4 pounds.*
- How many jars have 10/4 pounds? (Signal.) *2.*
- How many jars have 1 pound of flour? (Signal.) *Zero.*

i. Write the total for each column. Stop when you've done that much. ✔
   (Add to show:) [117:5C]

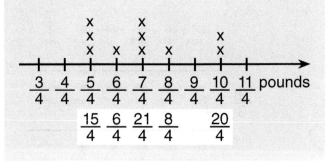

Here's what you should have.

j. Now find the total for all the columns with Xs. Remember, show the total as a fraction.
   (Observe students and give feedback.)
- Everybody, what's the total for all the columns? (Signal.) *70/4.*

k. Now you're going to figure out how much would be in each jar if all the jars had the same amount of flour. That's the average amount. Raise your hand when you know how much would be in each jar. ✔
- Everybody, read the division problem you worked. (Signal.) *70/4 ÷ 10.*
- What's the answer? (Signal.) *7/4.*
   Yes, if we took all the flour and put it in the jars so that all the jars had the same amount, all the jars would have 7/4 pounds of flour.

l. Make an arrow to show where the average is. ✔
   (Add to show:) [117:5D]

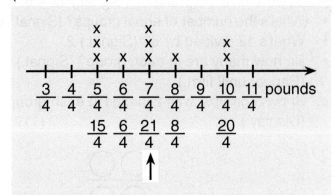

Here's what you should have.

EXERCISE 6: WORD PROBLEMS
FRACTION COMPARISON

a. Open your textbook to Lesson 117 and find part 1. ✔
(Teacher reference:)

- The fir board is 72 inches long. The maple board is $\frac{5}{3}$ the length of the fir board.
a. What's the length of the maple board?
b. What's the total length of both boards?
c. How much longer is the maple board than the fir board?
- In the study room, the number of books was $\frac{5}{2}$ the number of magazines. There were 95 books.
d. How many magazines were there?
e. How many fewer magazines than books were there?
f. What was the total number of books and magazines?

Part 1	
a.	■ = ■ ■

The first problem has three questions—A, B, and C.
- Work the problem. Answer all three questions. Remember the unit names.
(Observe students and give feedback.)
b. Check your work.
- Read the equation you started with. (Signal.) $M = 5/3 \ F$.
Tell me the answers to the questions.
- A: What's the length of the maple board? (Signal.) *120 inches.*
- B: What's the total length of both boards? (Signal.) *192 inches.*
- C: How much longer is the maple board than the fir board? (Signal.) *48 inches.*
c. Work the next problem. Answer questions D, E, and F. Remember the unit names.
(Observe students and give feedback.)
d. Check your work.
- Read the equation you started with. (Signal.) $B = 5/2 \ M$.
Tell me the answers to the questions.
- D: How many magazines were there? (Signal.) *38 magazines.*
- E: How many fewer magazines than books were there? (Signal.) *57 fewer magazines.*
- F: What was the total number of books and magazines? (Signal.) *133 books and magazines.*

EXERCISE 7: EXPRESSIONS
INTERPRETATION

a. (Display:)   [117:7A]

> 6 (C)
>
> 2 (C)
>
> $\frac{1}{2}$ (C)
>
> 10 (C)

Here are expressions that multiply by C. We don't know the number for C, but C is some number.
b. (Point to **6.**) Read this expression. (Signal.) $6 \times C$.
- (Point to **2.**) Read this expression. (Signal.) $2 \times C$.
- (Point to **1/2.**) Read this expression. (Signal.) $1/2 \times C$.
- (Point to **10.**) Read this expression. (Signal.) $10 \times C$.
c. Which is more—6 times C or 2 times C? (Signal.) $6 \times C$.
- Which is more—2 times C or 1/2 times C? (Signal.) *2 times C.*
- Which of these expressions has the largest value? (Signal.) $10 \times C$.
- Which expression has the smallest value? (Signal.) $1/2 \times C$.
d. Raise your hand when you know how many times greater 6 times C is than 2 times C. ✔
- How many times greater? (Signal.) *3 times greater.*
- Raise your hand when you know how many times greater 10 times C is than 2 times C. ✔
- How many times greater? (Signal.) *5 times greater.*

a. Find part 2 in your textbook. ✔
(Teacher reference:)

P 12(1856 − 428)
Q 3(1856 − 428)
R $\frac{1}{4}$(1856 − 428)
S $\frac{1}{2}$(1856 − 428)
T 4(1856 − 428)

Part 2	
a.	b.

  a. Write the letter of the expression that has the greatest value.
  b. Write the letter of the expression that has the smallest value.
  c. Write the letters of the expressions that are more than $\frac{1}{2}$ (1856 − 428).
  d. Write the letter of the expression that is 3 times greater than 4(1856 − 428).

  The quantity (1856 − 428) is 1428.

  e. Figure out what 3(1856 − 428) equals.
  f. Figure out what $\frac{1}{2}$(1856 − 428) equals.

The value inside the parentheses is the same for all these expressions. I'll read each item. You'll write the answer.

b. Item A. Write the letter of the expression that has the greatest value. ✔
• Everybody, which expression has the greatest value? (Signal.) *P.*

c. Item B: Write the letter of the expression that has the smallest value. ✔
• Everybody, which expression has the smallest value? (Signal.) *R.*

d. Item C: Write the letter of the expressions that are more than 1/2 times the quantity, 1856 minus 428. ✔
• Everybody which expressions are more than 1/2 times the quantity, 1856 minus 428? (Signal.) *P, Q, and T.*

e. Item D: Write the letter of the expression that is 3 times greater than 4 times the quantity, 1856 minus 428. ✔
• Everybody, which expression is 3 times greater than 4 times the quantity, 1856 minus 248? (Signal.) *P.*

f. The quantity 1856 minus 428 is 1428.
• Item E: Figure out what 3 times the quantity 1856 minus 428 equals. ✔
• Everybody, what does 3 times the quantity 1856 minus 428 equal? (Signal.) *4284.*

g. Item F: Figure out what 1/2 times the quantity 1856 minus 428 equals. ✔
• Everybody, what does 1/2 times the quantity 1856 minus 428 equal? (Signal.) *714.*

Assign Independent Work, Textbook parts 3–7.

Optional extra math-fact practice worksheets are available on ConnectED.

Lesson

EXERCISE 1: MULTIPLICATION
REASONING

a. (Display:)   [118:1A]

$$\frac{A}{B} = 1$$

Here are letters that stand for numbers. The numerator is A and the denominator is B. The fraction equals one whole unit.

b. One possible pair of numbers for A and B is 2 and 2. Does 2 halves equal one? (Signal.) *Yes.*
• What's another pair of numbers for a fraction that equals 1? (Call on a student. Accept any pair of identical numbers.)

c. (Change to show:)   [118:1B]

$$\frac{A}{B} = 6$$

• If A over B equals 6, what's a possible pair of numbers for a fraction that equals 6? (Call on different students. Accept any pair of numbers for which A is 6 times B—6/1, 12/2, 18/3 … .)

d. (Change to show:)   [118:1C]

$$\frac{A}{B} = 10$$

• If A over B equals 10, what's a possible pair of numbers for a fraction? (Call on different students. Accept any pair of numbers for which A is 10 times B.)

e. (Change to show:)   [118:1D]

$$\frac{A}{B} = 20$$

• If A over B equals 20, what's a possible pair of numbers for a fraction? (Call on different students. Accept any pair of numbers for which A is 20 times B.)

f. If we multiply A over B by 1, will we end up with more than A over B or less than A over B or a value equal to A over B? (Signal.) *Equal to A over B.*

g. (Change to show:)   [118:1E]

$$\frac{A}{B} \left(\frac{6}{6} \right)$$

• If you multiply the numerators, what's the answer? (Signal.) *6A.*
• If you multiply the denominators, what's the answer? (Signal.) *6B.*
(Add to show:)   [118:1F]

$$\frac{A}{B} \left(\frac{6}{6} \right) = \frac{6A}{6B}$$

h. What can you do to show that 6A over 6B equals A over B? (Call on a student. Idea: *Cross out the fraction that equals 1.*)
(Add to show:)   [118:1G]

$$\frac{A}{B} \left(\frac{6}{6} \right) = \frac{\cancel{6}A}{\cancel{6}B}$$

You have one times A over B. That's A over B. If we multiply A over B by one, we always end with a value that equals A over B.

• If we multiply A over B by a value that is more than 1, will we end with A over B? (Signal.) *No.*
• Will we end with more than A over B or less than A over B? (Signal.) *More than A over B.*

i. (Display:)   [118:1H]

$$\frac{A}{B} \left(\frac{7}{6} \right)$$

Let's say that A over B equals a whole number.

• What are we multiplying the whole number by? (Signal.) *7/6.*
• Is 7/6 more than 1, equal to 1, or less than 1? (Signal.) *More than 1.*
• So will the answer be more than A over B or less than A over B? (Signal.) *More than A over B.*

j. (Add to show:) [118:1I]

$$\frac{A}{B}\left(\frac{7}{6}\right)$$
$$12\left(\frac{7}{6}\right) =$$

Let's say that A over B equals 12. We multiply 12 by 7/6. Let's see if the answer is more than 12.

- What's the first thing I do to work the problem? (Call on a student. Idea: *Simplify 12/6.*)
(Add to show:) [118:1J]

$$\frac{A}{B}\left(\frac{7}{6}\right)$$
$$\overset{2}{\cancel{12}}\left(\frac{7}{\cancel{6}}\right) =$$

- What is 2 times 7? (Signal.) *14.*
(Add to show:) [118:1K]

$$\frac{A}{B}\left(\frac{7}{6}\right)$$
$$\overset{2}{\cancel{12}}\left(\frac{7}{\cancel{6}}\right) = \boxed{14}$$

- Is 14 more than 12? (Signal.) *Yes.*
So when we multiply A over B by more than 1, we end up with more than A over B.

k. (Display:) [118:1L]

$$\frac{A}{B}\left(\frac{2}{3}\right)$$

- If we multiply by 2/3 will we end up with more than A over B or less than A over B? (Signal.) *Less than A over B.*
- How do you know you'll end up with less than A over B? (Call on a student. Idea: *Because we are multiplying by less than 1.*)

l. (Add to show:) [118:1M]

$$\frac{A}{B}\left(\frac{2}{3}\right)$$
$$12\left(\frac{2}{3}\right) =$$

Let's say A over B equals 12. We multiply by 2/3.

- How could you show me that the answer has to be less than 12? (Call on a student. Idea: *Simplify 12/3 and multiply.*)
(Add to show:) [118:1N]

$$\frac{A}{B}\left(\frac{2}{3}\right)$$
$$\overset{4}{\cancel{12}}\left(\frac{2}{\cancel{3}}\right) =$$

- What's 4 times 2? (Signal.) *8.*
(Add to show:) [118:1O]

$$\frac{A}{B}\left(\frac{2}{3}\right)$$
$$\overset{4}{\cancel{12}}\left(\frac{2}{\cancel{3}}\right) = \boxed{8}$$

- Is the answer more than 12 or less than 12? (Signal.) *Less than 12.*
So if we multiply A over B by less than 1, we end up with less than A over B.

m. (Display:) [118:1P]

$$\frac{A}{B}\left(-\right) =$$

- If we want to end up with A over B, what would we multiply by? (Call on a student. Idea: *A fraction that equals one.*)

n. Let's say that A over B equals a whole number.

- If we multiply A over B by a whole number, would the answer always be another whole number? (Signal.) *Yes.*
- If we multiplied by a fraction that equals 5, would we end up with a whole number? (Signal.) *Yes.*
- Would that number be more than A over B? (Signal.) *Yes.*
- How many times more than A over B would the answer be? (Signal.) *5 times.*
- If we multiplied by a fraction that equals 10, would we end up with a whole number? (Signal.) *Yes.*
- How many times more than A over B would the answer be? (Signal.) *10 times.*

EXERCISE 2: COORDINATE SYSTEM

RELATED RULES

a. (Display:) [118:2A]

> Add 3

• We're going to make a line on the coordinate system for two rules.

b. Here's the first rule: Add 3.

• Say the first rule. (Signal.) *Add 3.*
(Add to show:) [118:2B]

> Add 3
> 0
> 3
> 6

Here are the first numbers for that rule if we start at zero: Zero, 3, 6.

• Say the next number that adds 3. (Signal.) *9.*
• Say the next number that adds 3. (Signal.) *12.*

c. (Add to show:) [118:2C]

> Add 3 Add 6
> 0
> 3
> 6
> 9
> 12

Here's the second rule: Add 6.
If we start at zero, the first numbers for that rule are zero and 6.

• What's the next number that adds 6? (Signal.) *12.*
• What's the next number that adds 6?
(Signal.) *18.*
• What's the next number that adds 6?
(Signal.) *24.*
(Add to show:) [118:2D]

> Add 3 Add 6
> 0   0
> 3   6
> 6   12
> 9   18
> 12 24

d. We can show both those rules with a line on the coordinate system. The rule for 3 gives the X values. The rule for 6 gives the Y values.

• What does the rule for 3 give? (Signal.)
The X values.
• What does the rule for 6 give? (Signal.)
The Y values.

e. (Add to show:) [118:2E]

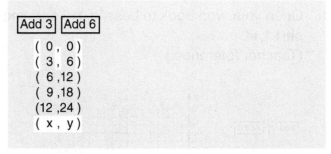

> Add 3 Add 6
> (0, 0)
> (3, 6)
> (6 ,12)
> (9 ,18)
> (12 ,24)
> (x , y)

The X and Y coordinates for the first point are zero comma zero.
The X and Y coordinates for the next point are 3 comma 6.

• What are the coordinates for the next point?
(Signal.) *6 comma 12.*
• What are the coordinates for the next point?
(Signal.) *9 comma 18.*
• What are the coordinates for the last point?
(Signal.) *12 comma 24.*
(Add to show:) [118:2F]

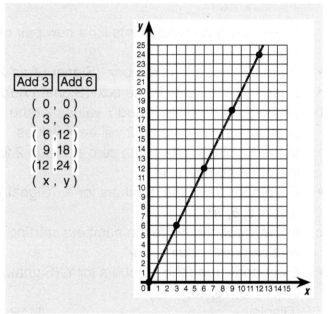

Here's the line with the points shown.
Every point shows a value for X and a value for Y.

f. (Point to **(6,12)**.) What's the X value for this point? (Signal.) *6.*
• What's the Y value for this point? (Signal.) *12.*
• Which is more, the X value or the Y value?
(Signal.) *The Y value.*
• How many times X is Y? (Signal.) *2 times.*

g. (Point to **(12, 24)**.) What's the X value for this point? (Signal.) *12.*
• What's the Y value for this point? (Signal.) *24.*
• Which is more—the X value or the Y value?
(Signal.) *The Y value.*
• How many times X is Y? (Signal.) *2 times.*

a. Open your workbook to Lesson 118 and find part 1. ✔
(Teacher reference:)

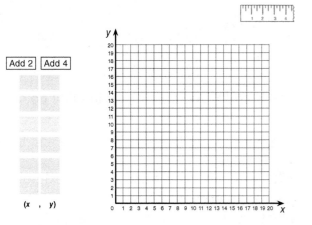

a. What's the *y* value of the points that has the *x* value of 10? _____

b. What's the *x* value of the point that has a *y* value of 12? _____

c. For the points on this line, which is more—the *x* value or the *y* value? _____

d. How many times more? _____

You're going to make points for a new pair of rules.
• Read the first rule in the box. (Signal.) *Add 2.*
• Read the other rule in the box. (Signal.) *Add 4.*
b. You're going to write X and Y values for the points. The numbers for 2 will be X values.
• In the X column, start with zero and add 2 for the other numbers. ✔
• Everybody, read the numbers for X. (Signal.) *0, 2, 4, 6, 8, 10.*
c. In the Y column, write the numbers starting with zero and adding 4. ✔
• Everybody, read the numbers for Y. (Signal.) *0, 4, 8, 12, 16, 20.*
(Display:) [118:2G]

Add 2	Add 4
0	0
2	4
4	8
6	12
8	16
10	20
(x ,	y)

Here's what you should have.
Each row shows X and Y coordinates for a point.
d. The first point is: zero comma zero.
• What's the next point? (Signal.) *2 comma 4.*
• What's the next point? (Signal.) *4 comma 8.*
• What's the next point? (Signal.) *6 comma 12.*

e. Show the parentheses and commas for the points. Then plot the points and make the line.
(Observe students and give feedback.)
(Add to show:) [118:2H]

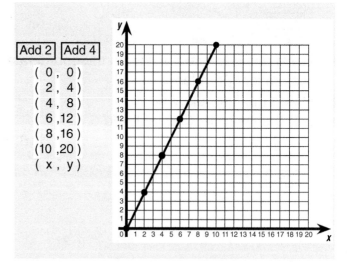

Here's the line you should have.
f. Now answer the questions.
(Observe students and give feedback.)
g. Check your work.
• A: What's the Y value of the point that has the X value of 10? (Signal.) *20.*
• B: What's the X value of the point that has a Y value of 12? (Signal.) *6.*
• C: For the points on the line, which is more— the X value or the Y value? (Signal.) *The Y value.*
• D: How many times more? (Signal.) *2 times.* Yes, Y is always 2 times X for this line.

EXERCISE 3: FRACTION REPRESENTATION
PARTITIONS OF A GROUP

a. (Display:) [118:3A]

$$\frac{18}{6}$$

• Read the fraction. (Signal.) *18/6.*
We can rewrite 18/6 as 1/6 times 18.
(Add to show:) [118:3B]

$$\frac{18}{6} = \frac{1}{6} \times 18$$

• Read the equation. (Signal.) *18/6 = 1/6 × 18.*

b. (Display:) [118:3C]

$$\frac{36}{4} = \frac{1}{4} \times 36$$

We can rewrite 36/4 as 1/4 times 36.
• Read the equation. (Signal.) *36/4 = 1/4 × 36.*

c. (Display:) [118:3D]

$$\frac{15}{5} =$$

• Your turn: In part A, rewrite 15/5 as a fraction times 15.
   (Observe students and give feedback.)
• Everybody, read what 15/5 equals. (Signal.)
   1/5 × 15.

d. (Display:) [118:3E]

$$\frac{38}{2} =$$

• Rewrite 38/2 as a fraction times 38. ✔
• Everybody, read what 38/2 equals. (Signal.)
   1/2 × 38.

e. (Display:) [118:3F]

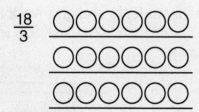

$$\frac{18}{3}$$

Here's a diagram of 18/3.
• It shows how many are in each group. How many? (Signal.) *6.*
   We can show the number in one group as 1/3 times 18.
   (Change to show:) [118:3G]

$$\frac{1}{3} \times 18 = 6$$

• What equation can we write to show the number in two groups? (Signal.) *2/3 × 18 = 12.*

f. (Change to show:) [118:3H]

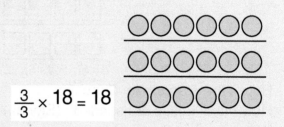

$$\frac{2}{3} \times 18 = 12$$

• What can we write to show the number in three groups? (Signal.) *3/3 × 18 = 18.*
   (Change to show:) [118:3I]

$$\frac{3}{3} \times 18 = 18$$

g. (Display:) [118:3J]

$$\frac{20}{5}$$

We'll make diagrams for 20/5.
• How many objects are there in all? (Signal.) *20.*
• How many groups do I make? (Signal.) *5.*
• How many will be in each group? (Signal.) *4.*
   (Add to show:) [118:3K]

$$\frac{20}{5} = 4$$

• Read the equation. (Signal.) *20/5 = 4.*
   I'll show 4 boxes in each group.
   (Add to show:) [118:3L]

$$\frac{20}{5} = 4$$

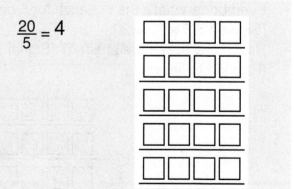

h. The equation for one group is 1/5 times 20 = 4.
- Write the equation for 2 groups. ✔
- Everybody, what's the equation for 2 groups? (Signal.) *2/5 × 20 = 8.*
(Change to show:) [118:3M]

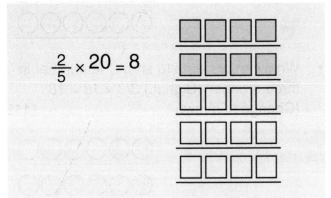

$$\frac{2}{5} \times 20 = 8$$

Here's the diagram for 2 groups.
- How many boxes are shaded? (Signal.) *8.*
i. Write the equation for 4 groups. ✔
- Everybody, what's the equation for 4 groups? (Signal.) *4/5 × 20 = 16.*
(Change to show:) [118:3N]

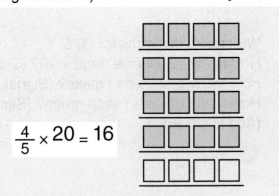

$$\frac{4}{5} \times 20 = 16$$

Here's the diagram for 4 groups.
- How many boxes are shaded? (Signal.) *16.*
j. Write the equation for 5 groups. ✔
- Everybody, what's the equation for 5 groups? (Signal.) *5/5 × 20 = 20.*
- How many boxes are shaded? (Signal.) *20.*
(Change to show:) [118:3O]

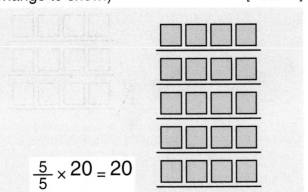

$$\frac{5}{5} \times 20 = 20$$

Here's the diagram for 5/5. All the boxes are shaded.

EXERCISE 4: DECIMAL ADDITION
REPRESENTATION

a. Find part 2 in your workbook. ✔
(Teacher reference:)

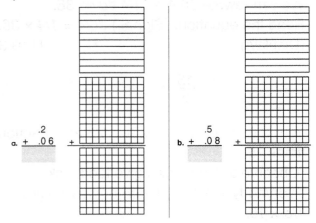

You're going to make the diagrams to show the answer to each problem.
b. Problem A: 2 tenths + 6 hundredths.
- Write the answer to the problem. ✔
- Everybody, what's the answer? (Signal.) *26 hundredths.*
c. Make the diagram for 2 tenths in the top box. Shade rows to show 2 tenths. Make the diagram for 6 hundredths in the second box. (Observe students and give feedback.)
d. Show the answer in the last box. ✔
(Display:) [118:4A]

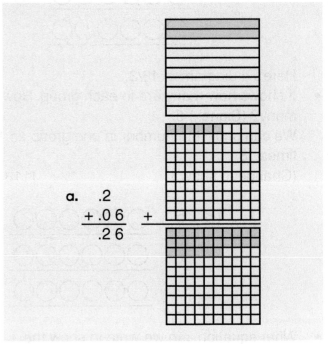

a. .2
  + .0 6   +
   .2 6

Here's one way to show 26 hundredths. Two rows are shaded. That's 20 hundredths. And 6 hundredths in another row are shaded.

e. Problem B: 5 tenths plus 8 hundredths.
• Don't work the problem. Make the diagram for 5 tenths in the first box and 8 hundredths in the second box. Show the answer in the last box. (Observe students and give feedback.)
(Display:) [118:4B]

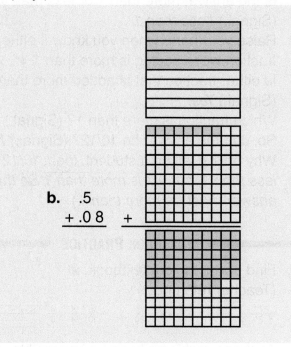

b. .5
+ .0 8 +

Here's what you should have.
The last box shows what 5 tenths and 8 hundredths equal.
• What's the answer? (Signal.) *58 hundredths.*
• Write the answer to the column problem. ✔

EXERCISE 5: FRACTION COMPARISON
UNLIKE DENOMINATORS

a. Open your textbook to Lesson 118 and find part 1. ✔
(Teacher reference:)

a. $\frac{6}{10} \blacksquare \frac{2}{3}$ b. $\frac{9}{4} \blacksquare \frac{14}{6}$ c. $\frac{9}{15} \blacksquare \frac{6}{10}$

b. Problem A. Read the fractions. (Signal.) *6/10, 2/3.*
• Copy the fractions. Below, rewrite both fractions with the lowest common denominator. Then complete the original statement.
(Observe students and give feedback.)
c. Check your work.
You rewrote both fractions with a common denominator.
• What denominator is that? (Signal.) *30.*

(Display:) [118:5A]

a. $\frac{6}{10} < \frac{2}{3}$

$\frac{6}{10}\left(\frac{3}{3}\right) = \frac{18}{30}$

$\frac{2}{3}\left(\frac{10}{10}\right) = \frac{20}{30}$

Here's what you should have.
• What does 6/10 equal? (Signal.) *18/30.*
• What does 2/3 equal? (Signal.) *20/30.*
• Which is more—6/10 or 2/3? (Signal.) *2/3.*
d. Problem B: 9/4, 14/6.
• Copy the fractions and work the problem. (Observe students and give feedback.)
• Everybody, what's the common denominator for 9/4 and 14/6? (Signal.) *12.*
• What did you change 9/4 to? (Signal.) *27/12.*
• What did you change 14/6 to? (Signal.) *28/12.*
• Which is more—9/4 or 14/6? (Signal.) *14/6.*
(Display:) [118:5B]

b. $\frac{9}{4} < \frac{14}{6}$

$\frac{9}{4}\left(\frac{3}{3}\right) = \frac{27}{12}$

$\frac{14}{6}\left(\frac{2}{2}\right) = \frac{28}{12}$

Here's what you should have.
e. Problem C: 9/15, 6/10.
• Copy the fractions and work the problem. (Observe students and give feedback.)
• Everybody, what's the common denominator for 9/15 and 6/10? (Signal.) *30.*
• What did you change 9/15 to? (Signal.) *18/30.*
• What did you change 6/10 to? (Signal.) *18/30.*
• Which is more—9/15 or 6/10? (Call on a student. Idea: *Neither; they are the same value.*)
(Display:) [118:5C]

c. $\frac{9}{15} = \frac{6}{10}$

$\frac{9}{15}\left(\frac{2}{2}\right) = \frac{18}{30}$

$\frac{6}{10}\left(\frac{3}{3}\right) = \frac{18}{30}$

Here's what you should have.

EXERCISE 6: FRACTION OPERATIONS

REASONABLENESS OF ANSWERS

a. (Display:) [118:6A]

$$\frac{1}{2} + \frac{2}{10} = \frac{3}{8}$$

- Read this equation. (Signal.) *1/2 + 2/10 = 3/8.* The answer shown is 3/8. I can tell the answer is wrong without working the problem.
- Raise your hand when you know if 3/8 is more than 1/2 or less than 1/2. ✔
- Tell me about 3/8. (Signal.) *Less than 1/2.*
- Is 1/2 + 2/10 more than 1/2 or less than 1/2? (Signal.) *More than 1/2.*
  So the answer must be more than 1/2. It can't be 3/8 because 3/8 is not more than 1/2.

b. (Display:) [118:6B]

$$\frac{2}{5} + \frac{4}{7} = \frac{5}{10}$$

- Read the equation. (Signal.) *2/5 + 4/7 = 5/10.* The answer shown is 5/10.
- Is that more than 1/2, less than 1/2, or equal to 1/2? (Signal.) *Equal to 1/2.*
- Raise your hand when you know if either fraction we're adding is more than 1/2. ✔
- Is either fraction we're adding more than 1/2? (Signal.) *Yes.*
- Which fraction is more than 1/2? (Signal.) *4/7.*
- If one of the fractions we add is **more** than 1/2, could the answer be 1/2? (Signal.) *No.*
- So, could the answer be 5/10? (Signal.) *No.*
- Why not? (Call on a student. Idea: *4/7 is more than 1/2. So the answer must be more than 1/2.*)

c. (Display:) [118:6C]

$$\frac{9}{7} + \frac{1}{5} = \frac{10}{12}$$

- Read the equation. (Signal.) *9/7 + 1/5 = 10/12.*
- Is the answer more than 1 or less than 1? (Signal.) *Less than 1.*
- Raise your hand when you know if either fraction we're adding is more than 1. ✔
- Is either fraction that is added more than 1? (Signal.) *Yes.*
- Which fraction is more than 1? (Signal.) *9/7.*
- So, can the answer be 10/12? (Signal.) *No.*
- Why not? (Call on a student. Idea: *10/12 is less than 1, but 9/7 is more than 1. So the answer must be more than 1.*)

TEXTBOOK PRACTICE

a. Find part 2 in your textbook. ✔
  (Teacher reference:)

  a. $\frac{7}{8} - \frac{1}{2} = \frac{8}{7}$ b. $\frac{2}{5} + \frac{7}{8} = \frac{9}{20}$ c. $\frac{10}{10} - \frac{4}{4} = \frac{6}{6}$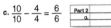

  All these problems have an impossible answer. You're going to write a sentence or two that tells why the answer is impossible.

b. Read item A. (Signal.) *7/8 – 1/2 = 8/7.*
- Write why the answer is impossible. ✔
- (Call on a student.) Read your answer. (Accept answers that express the idea that 7/8 minus 1/2 is less than 1, but the answer shown is more than 1.)

c. Read item B. (Signal.) *2/5 + 7/8 = 9/20.*
- Write why the answer is impossible. ✔
- (Call on a student.) Read your answer. (Accept answers that express the idea that one of the numbers added is more than 1/2, but the answer shown is less than 1/2.)

d. Read item C. (Signal.) *10/10 – 4/4 = 6/6.*
- Write why the answer is impossible. ✔
- (Call on a student.) Read your answer. (Accept answers that express the idea that all the fractions equal 1, but 1 minus 1 is zero, not 1.)

EXERCISE 7: EXPRESSIONS

INTERPRETATION

a. Last time, you worked with expressions that multiplied values.
(Display:)   [118:7A]

$$5 \, (346 - M)$$
$$\tfrac{1}{2} \, (346 - M)$$

- (Point to **5.**) Read this expression. (Signal.) *5 times the quantity, 346 – M.*
- (Point to **1/2.**) Read this expression. (Signal.) *1/2 times the quantity, 346 – M.*
- Which expression is more—5 times the quantity, 346 minus M or 1/2 times the quantity, 346 minus M? (Signal.) *5 times the quantity, 346 – M.*

b. If you multiply by more than 1, you end up with more than you start with.

- (Point to **5.**) Is 5 more than 1? (Signal.) *Yes.*
- So if you multiply 5 times the quantity, 346 minus M, will you end up with more than 346 minus M? (Signal.) *Yes.*
- If you multiply 346 minus M times 1/2, will you end up with more than 346 minus M? (Signal.) *No.*

=== TEXTBOOK PRACTICE ===

a. Find part 3 in your textbook. ✔
(Teacher reference:)

- Ⓣ $2(1149 + 357)$
- Ⓥ $3(1149 + 357)$
- Ⓧ $\tfrac{4}{3}(1149 + 357)$
- Ⓨ $\tfrac{1}{2}(1149 + 357)$
- Ⓩ $6(1149 + 357)$

Part 3		
a.	b.	c.

a. Write the letter of the expression that has the smallest value.

b. Write the letter of the expression that has the greatest value.

c. Write the letters of the expressions that are more than $\tfrac{4}{3}(1149 + 357)$.

d. Write the letter of the expression that is 3 times greater than $2(1149 + 357)$.

e. Write the letter of the expression that is less than $(1149 + 357)$.

f. Write the letters of all expressions that are more than $(1149 + 357)$.

> The quantity of $(1149 + 357)$ is 1506.

g. Figure out what $\tfrac{4}{3}(1149 + 357)$ equals.

h. Figure out what $\tfrac{1}{2}(1149 + 357)$ equals.

The value inside the parentheses is the same for all these expressions. I'll read each item. You'll write the answer.

b. Item A. Write the letter of the expression that has the smallest value. ✔
- Everybody, which expression has the smallest value? (Signal.) *Y.*

c. Item B: Write the letter of the expression that has the greatest value. ✔
- Everybody, which expression has the greatest value? (Signal.) *Z.*

d. Item C: Write the letter of the expressions that are more than 4/3 times the quantity, 1149 plus 357. ✔
- Everybody, which expressions are more than 4/3 times the quantity, 1149 plus 357? (Signal.) *T, V, and Z.*

e. Item D: Write the letter of the expression that is 3 times greater than 2 times the quantity, 1149 plus 357. ✔
- Everybody, which expression is 3 times greater than 2 times the quantity, 1149 plus 357? (Signal.) *Z.*

f. Item E: Write the letter of the expression that is less than the quantity, 1149 plus 357. ✔
- Everybody, which expression is less than the quantity, 1149 plus 357? (Signal.) *Y.*

g. Item F: Write the letters of all expressions that are more than the quantity, 1149 plus 357. ✔
- Everybody, which expressions are more than the quantity, 1149 plus 357? (Signal.) *T, V, X, and Z.*

h. The quantity 1149 plus 357 is 1506.
- Item G: Figure out what 4/3 times the quantity, 1149 plus 357 equals. ✔
- Everybody, what does 4/3 times the quantity, 1149 plus 357 equal? (Signal.) *2008.*

i. Item H: Figure out what 1/2 times the quantity, 1149 plus 357 equals. ✔
- Everybody, what does 1/2 times the quantity, 1149 plus 357 equal? (Signal.) *753.*

> Assign Independent Work, Textbook parts 4–11.

Optional extra math-fact practice worksheets are available on ConnectED.

Lesson 119

EXERCISE 1: MULTIPLICATION
REASONING

a. Open your workbook to Lesson 119 and find part 1. ✔
- Pencils down. ✔
(Teacher reference:)

$\frac{A}{B}$ equals a whole number.

$\frac{A}{B}\left(\frac{1}{2}\right) = $ ■

 $\left(\frac{1}{2}\right) = $

Answers may vary.

a. Will the answer be more than $\frac{A}{B}$, less than $\frac{A}{B}$, or equal to $\frac{A}{B}$? _____

b. If $\frac{A}{B}$ equals 5, will the answer be a whole number? _____

c. If $\frac{A}{B}$ equals 8, will the answer be a whole number? _____

_____

$\frac{A}{B}\left(\text{■}\right) = $ more than $\frac{A}{B}$

d. Do we multiply $\frac{A}{B}$ by more than 1 or less than 1? _____

_____

$\frac{A}{B}(3) = $ ■

e. Do we multiply $\frac{A}{B}$ by more than 1 or less than 1? _____

f. How many times greater than $\frac{A}{B}$ is the answer? _____

g. If $\frac{A}{B}$ equals 7, what will the answer be? _____

You're going to answer questions and show why your answers are correct.
The fact says that A over B equals a whole number.
The problem shows that you're multiplying A over B times 1/2.
Get ready to answer the questions.

b. Question A: Will the answer be more than A over B, less than A over B, or equal to A over B?
- What's the answer? (Signal.) *Less than A/B.*
- Question B: If A over B equals 5, will the answer be a whole number?
- What's the fraction answer? (Signal.) *5/2.*
- Is 5/2 a whole number? (Signal.) *No.*
- Question C: If A over B equals 8, will the answer be a whole number?
- What's the fraction answer? (Signal.) *8/2.*
- What whole number is 8 halves? (Signal.) *4.*
(Repeat until firm.)

c. Write answers to questions A through C. ✔
d. (Display:)   [119:1A]

$$\frac{A}{B}\left(\frac{1}{2}\right) = $$

$$10\left(\frac{1}{2}\right) = $$

Here's an example with A over B equal to 10. Let's see if the answer is less than 10.
- Can we simplify 10/2? (Signal.) *Yes.*
- What's the simplified value? (Signal.) *5.*
(Add to show:)   [119:1B]

$$\frac{A}{B}\left(\frac{1}{2}\right) = $$

$$\overset{5}{\cancel{10}}\left(\frac{1}{\cancel{2}}\right) = $$

- What's the answer? (Signal.) *5.*
(Add to show:)   [119:1C]

$$\frac{A}{B}\left(\frac{1}{2}\right) = $$

$$\overset{5}{\cancel{10}}\left(\frac{1}{\cancel{2}}\right) = \boxed{5}$$

e. Your turn: Below the equation, write a different whole number value for A over B so that the answer is a whole number. Multiply by 1/2 and show the answer.
(Observe students and give feedback.)
- (Call on different students. Ask:) What whole number did you start with? What's ___ times 1/2?
f. (Display:)   [119:1D]

$$\frac{A}{B}\left(\text{——}\right) = \text{more than } \frac{A}{B}$$

Here's the next problem. We want to multiply A over B by something and end up with more than A over B.

g. Question D says: Do we multiply A over B by more than 1 or less than 1?
- Write the answer. ✔
- Everybody, what's the answer? (Signal.) *More than 1.*

h. (Display:) [119:1E]

$$\frac{A}{B}(3) =$$

Here's the next problem: A over B times 3.

i. Question E: Do we multiply A over B by more than 1 or less than 1?
• Write the answer. ✔
• Everybody, what's the answer? (Signal.) *More than 1.*

j. F: How many times greater than A over B is the answer?
• Write the answer. ✔
• Everybody, what's the answer? (Signal.) *3 times greater.*

k. G: If A over B equals 7, what will the answer be?
• Write the answer. ✔
• Everybody, what's the answer? (Signal.) *21.*

EXERCISE 2: COORDINATE SYSTEM
RELATED RULES

a. Find part 2 in your workbook. ✔
(Teacher reference:)

You're going to draw lines for two rules. One rule gives the X value; the other rule gives the Y value.

b. Here are the rules for Line A: Add 3. Add 9.
• Start with zero and write the three values for X and the three values for Y. Show the comma and parentheses for each point. ✔
(Display:) [119:2A]

Line A
Add 3 Add 9
(0, 0)
(3, 9)
(6,18)
(x, y)

Here's what you should have.

c. Plot the points for line A, draw the line, and write the letter A above it.
(Observe students and give feedback.)
(Add to show:) [119:2B]

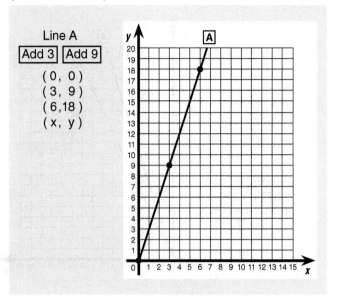

Here's what you should have.
• Make sure your line goes all the way to the top of the coordinate system. ✔

d. You're going to plot another line. It starts at zero, zero.
The X rule is: Add 4.
The Y rule is: Add 8.
• Write the three values for X and the three values for Y. Show the comma and parentheses for each point. ✔
(Display:) [119:2C]

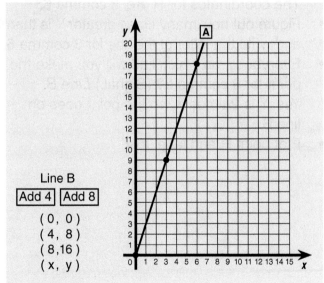

Here's what you should have.

e. Plot the points for line B, draw the line, and label it B.
(Observe students and give feedback.)
(Add to show:) [119:2D]

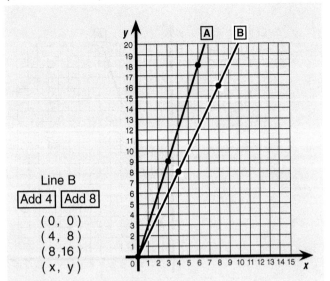

Line B
Add 4   Add 8
(0, 0)
(4, 8)
(8,16)
(x, y)

Here's what you should have.
- Make sure your line goes all the way to the top of the coordinate system. ✔
f. For line A, how many times greater is Y than X? (Signal.) *3 times.*
- For line B, how many times greater is Y than X? (Signal.) *2 times.*
(Repeat until firm.)
g. The items beside the coordinate system show coordinates for points. You're going to identify whether each point is on line A or line B. The coordinates for R are: 3 comma 6.
- Figure out how many times greater Y is than X, and write the letter of the line for 3 comma 6. ✔
- Everybody, on which line will you make the point for 3 comma 6? (Signal.) *Line B.* Yes, Y is 2 times X, so the point goes on line B.
- Plot point R and label it. ✔

(Change to show:) [119:2E]

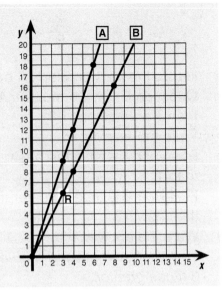

Here's what you should have.
h. The coordinates for P are: 5 comma 15.
- Figure out how many times greater Y is than X and the letter of the line for 5 comma 15. ✔
- Everybody, on which line will you make the point 5 comma 15? (Signal.) *Line A.* Yes, Y is 3 times X, so the point goes on line A.
- Plot point P. ✔
(Add to show:) [119:2F]

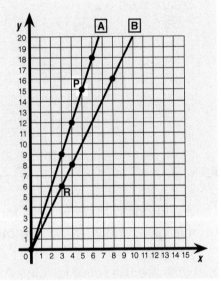

Here's what you should have.

i. Write the letter A or B for the rest of the points and plot the points.
(Observe students and give feedback.)
j. Check your work.
• Q is 5 comma 10. How many times 5 is 10? (Signal.) *2 times.*
• So on which line did you plot Q? (Signal.) *B.*
• T is 10, 20. How many times 10 is 20? (Signal.) *2 times.*
• So on which line did you plot T? (Signal.) *B.*
• V is 4,12. How many times 4 is 12? (Signal.) *3 times.*
• So on which line did you plot V? (Signal.) *A.*
(Add to show:) [119:2G]

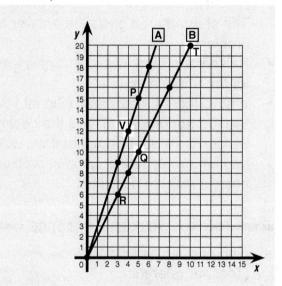

Here's what you should have for each point.

EXERCISE 3: FRACTION REPRESENTATION
PARTITIONS OF A GROUP

a. Find part 3 in your workbook. ✔
(Teacher reference:)

a. $\frac{2}{3} \times 15 =$ b. $\frac{3}{4} \times 12 =$

Squares   Eggs

You're going to make equal groups and shade some of the groups to show the answer to the problem.
b. Problem A: 2/3 × 15.
• How many equal groups will you make? (Signal.) *3.*
• How many groups will you shade? (Signal.) *2.*

c. Make the diagram with squares.
(Observe students and give feedback.)
(Display:) [119:3A]

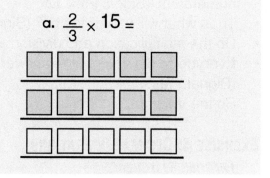

Here's what you should have.
• How many squares did you shade? (Signal.) *10.*
d. We can check the answer by working the problem 2/3 times 15. We'll multiply 2 times 15, then divide by 3. Don't simplify. Just work the problem. ✔
e. Check your work.
• What's 2/3 times 15? (Signal.) *10.*
(Add to show:) [119:3B]

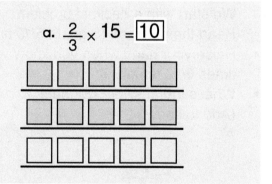

That's the answer we got by showing the groups.
f. Problem B: 3/4 × 12.
• Make the diagram with eggs. Shade 3/4 of the eggs.
(Observe students and give feedback.)
g. Check your work.
(Display:) [119:3C]

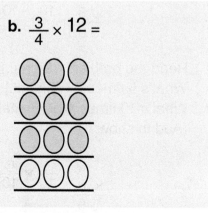

Here's what you should have.
There are 3 eggs in each group. You shaded 3 groups.
• How many eggs did you shade? (Signal.) *9.*

h. You'll do the multiplication and division to check your answer.
- Say the multiplication problem for the numerators. (Signal.) *3 × 12.*
- Then what will you divide by? (Signal.) *4.*
- Do the multiplication and division. ✔
- Everybody, did you get the answer of 9? (Signal.) *Yes.*
  So the diagram is correct.

EXERCISE 4: DECIMAL OPERATIONS
FRACTION EQUIVALENCE

a. You're going to work problems that multiply tenths. You'll show that you get the same answers if you write the values as fractions or as decimals.
   (Display:) [119:4A]

$$\begin{array}{r} .5 \\ \times\ .6 \\ \hline \end{array}$$

We start with a decimal problem.
- Read the problem. (Signal.) *5/10 times 6/10.*
- Raise your hand when you know what 5/10 times 6/10 equals. ✔
- What's the answer? (Signal.) *30 hundredths.*
  (Add to show:) [119:4B]

$$\begin{array}{r} .5 \\ \times\ .6 \\ \hline .3\,0 \end{array}$$

b. We get the same answer if we multiply fractions and work the problem without simplifying.
   (Add to show:) [119:4C]

$$\begin{array}{r} .5 \\ \times\ .6 \\ \hline .3\,0 \end{array} \qquad \frac{5}{10} \times \frac{6}{10} =$$

- Read the problem. (Signal.) *5/10 × 6/10.*
- What's 5 times 6? (Signal.) *30.*
- What's 10 times 10? (Signal.) *100.*
  (Add to show:) [119:4D]

$$\begin{array}{r} .5 \\ \times\ .6 \\ \hline .3\,0 \end{array} \qquad \frac{5}{10} \times \frac{6}{10} = \frac{30}{100}$$

- What's 5/10 times 6/10? (Signal.) *30/100.*
  That's the same answer we got when we worked the decimal problem.

c. (Add to show:) [119:4E]

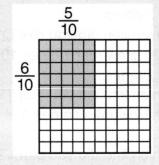

Here's a picture of 5/10 times 6/10. The shaded area gives the answer to the problem.
- Raise your hand when you know the fraction for the shaded area. ✔
- What's the shaded area? (Signal.) *30/100.*
  Yes, 30 hundredths. That's the decimal answer we got. So it doesn't matter if we work a fraction problem or a decimal problem. The answer is 30 hundredths.

━━━━━━━━━━ **WORKBOOK PRACTICE** ━━━━━━━━━━

a. Find part 4 in your workbook. ✔
   (Teacher reference:)

You can see the decimal problem and the fraction problem.
b. Read the decimal problem. (Signal.) *1 and 5 tenths times 4 tenths.*
- Work the problem. ✔
- Everybody, what's the answer? (Signal.) *60 hundredths.*
  Yes, 60 hundredths.
  (Add to show:) [119:4F]

$$\begin{array}{r} 1.5 \\ \times\ \ .4 \\ \hline .6\,0 \end{array} \qquad 1\frac{5}{10} \times \frac{4}{10}$$

c. We get the same answer if we work the fraction problem.
We change 1 and 5/10 into a fraction.
- Raise your hand when you know the fraction. ✔
- What fraction? (Signal.) *15/10.*
- Write the problem and the answer. Don't simplify. ✔
You worked the problem 15/10 times 4/10.
- What's 15 times 4? (Signal.) *60.*
- What's 10 times 10? (Signal.) *100.*
(Add to show:) [119:4G]

$$
\begin{array}{c}
1.5 \\
\times\ .4 \\
\hline
.6\,0
\end{array}
\qquad
1\frac{5}{10} \times \frac{4}{10}
$$
$$
\frac{15}{10} \times \frac{4}{10} = \frac{60}{100}
$$

Here's what you should have. We get the same answer—60/100.

d. Remember, you can work a decimal problem as a fraction problem. The answer is the same value.

EXERCISE 5: FRACTION COMPARISON
UNLIKE DENOMINATORS

a. Open your textbook to Lesson 119 and find part 1. ✔
(Teacher reference:)

a. $\frac{7}{5} \blacksquare \frac{13}{10}$ b. $\frac{7}{4} \blacksquare \frac{8}{5}$ c. $\frac{3}{6} \blacksquare \frac{5}{9}$ d. $\frac{3}{4} \blacksquare \frac{9}{12}$

For some of these problems, you have to rewrite both fractions. For others, you only rewrite one fraction.

b. Problem A. The fractions are 7/5 and 13/10.
- Copy the fractions. Do the rewriting and complete the original statement.
(Observe students and give feedback.)
Check your work.
- How many fractions did you rewrite? (Signal) *1.*
- Which fraction did you rewrite? (Signal.) *7/5.*
- What does 7/5 equal? (Signal.) *14/10.*
- Which is more, 7/5 or 13/10? (Signal.) *7/5.*
(Display:) [119:5A]

$$
\text{a.} \qquad \frac{7}{5} > \frac{13}{10}
$$
$$
\frac{7}{5}\left(\frac{2}{2}\right) = \frac{14}{10}
$$

Here's what you should have.

c. Work the rest of the problems in part 1.
(Observe students and give feedback.)
d. Check your work.
- Problem B: 7/4, 8/5.
- How many fractions did you rewrite? (Signal.) *2.*
- What does 7/4 equal? (Signal.) *35/20.*
- What does 8/5 equal? (Signal.) *32/20.*
- Which is more, 7/4 or 8/5? (Signal.) *7/4.*
(Display:) [119:5B]

$$
\text{b.} \qquad \frac{7}{4} > \frac{8}{5}
$$
$$
\frac{7}{4}\left(\frac{5}{5}\right) = \frac{35}{20}
$$
$$
\frac{8}{5}\left(\frac{4}{4}\right) = \frac{32}{20}
$$

Here's what you should have.

e. Problem C: 3/6, 5/9.
- How many fractions did you rewrite? (Signal.) *2.*
- What does 3/6 equal? (Signal.) *9/18.*
- What does 5/9 equal? (Signal.) *10/18.*
- Which is more, 3/6 or 5/9? (Signal.) *5/9.*
(Display:) [119:5C]

$$
\text{c.} \qquad \frac{3}{6} < \frac{5}{9}
$$
$$
\frac{3}{6}\left(\frac{3}{3}\right) = \frac{9}{18}
$$
$$
\frac{5}{9}\left(\frac{2}{2}\right) = \frac{10}{18}
$$

Here's what you should have.

f. Problem D: 3/4, 9/12.
- How many fractions did you rewrite? (Signal.) *1.*
- Which fraction did you change? (Signal.) *3/4.*
- What does 3/4 equal? (Signal.) *9/12.*
- What sign did you write? (Signal.) *Equals.*
(Display:) [119:5D]

$$
\text{d.} \qquad \frac{3}{4} = \frac{9}{12}
$$
$$
\frac{3}{4}\left(\frac{3}{3}\right) = \frac{9}{12}
$$

Here's what you should have.

EXERCISE 6: FRACTION OPERATIONS
REASONABLENESS OF ANSWERS

a. Find part 2 in your textbook. ✔
(Teacher reference:)

a. $\frac{3}{5} + \frac{2}{20} = \frac{7}{10}$ b. $\frac{3}{5} - \frac{2}{20} = \frac{5}{10}$

c. $\frac{3}{5} + \frac{6}{10} = \frac{18}{20}$ d. $\frac{3}{5} - \frac{2}{20} = \frac{11}{10}$

Some of these problems have answers that are possible. Some have answers that are impossible.
You're going to write **possible** or **impossible** for each item. If the answer is impossible, you'll write a sentence or two that tells why the answer is impossible.

b. Read item A. (Signal.) *3/5 + 2/20 = 7/10.*
- Is the answer more than 1/2 or less than 1/2? (Signal.) *More than 1/2.*
- Is the first number more or less than 1/2? (Signal.) *More than 1/2.*
- Is the other number more or less than 1/2? (Signal.) *Less than 1/2.*
The problem starts with a number that is more than 1/2 and adds. It has an answer that is more than 1/2.
- Is that answer possible? (Signal.) *Yes.*
So the answer is possible.
- Write the word possible. ✔

c. Read item B. (Signal.) *3/5 – 2/20 = 5/10.*
- Write the word possible or impossible to tell about the answer. ✔
- Everybody, what did you write? (Signal.) *Possible.* Yes, the answer is possible. You're starting with a number that is more than 1/2. You're subtracting, so the answer could be 1/2, which is 5/10.

d. Read problem C. (Signal.) *3/5 + 6/10 = 18/20.*
- Write possible or impossible. If the answer is impossible, tell why.
(Observe students and give feedback.)
- Everybody, did you write possible or impossible? (Signal.) *Impossible.*
- What did you write to explain why it is impossible? (Call on a student. Idea: *Both numbers added are more than 1/2, so the answer can't be less than 1.*)

e. Read problem D. (Signal.) *3/5 – 2/20 = 11/10.*
- Write possible or impossible. If the answer is impossible, tell why.
(Observe students and give feedback.)
- Everybody, is the answer possible or impossible? (Signal.) *Impossible.*
- Why is it impossible? (Call on a student. Idea: *You're starting with less than 1 and subtracting, so the answer can't be more than 1.*)

EXERCISE 7: SENTENCES
SINGLE-LETTER EQUATIONS

a. (Display:)   [119:7A]

> $\frac{5}{8}$ of the material weighed 40 pounds.

Listen: 5/8 of the material weighed 40 pounds. We're going to write an equation for this sentence. The equation will have only one letter.
- What do we write for 5/8 of the material? (Signal.) *5/8 M.*

b. The rest of the sentence tells that it weighed 40 pounds. For weighed 40 pounds, we write equals 40.
- What do we write for weighed 40 pounds? (Signal.) *Equals 40.*
(Add to show:)   [119:7B]

> $\frac{5}{8}$ of the material weighed 40 pounds.
>
> $$\frac{5}{8} m = 40$$

- Read the equation. (Signal.) *5/8 M = 40.*

c. (Display:)   [119:7C]

> $\frac{1}{7}$ of the bricks cost 75 dollars.

- New sentence. Read the sentence. (Signal.) *1/7 of the bricks cost 75 dollars.*
- Write the equation for the sentence in part A. ✔
(Add to show:)   [119:7D]

> $\frac{1}{7}$ of the bricks cost 75 dollars.
>
> $$\frac{1}{7} b = 75$$

Here's what you should have.
- Everybody, read the equation. (Signal.) *1/7 B = 75.*

TEXTBOOK PRACTICE

a. Find part 3 in your textbook. ✔
   (Teacher reference:)

   a. $\frac{4}{7}$ of the pens cost 88 dollars.

   b. $\frac{2}{3}$ of the gravel weighed 502 pounds.

   c. $\frac{6}{7}$ of the water was $\frac{5}{8}$ gallon.

   d. $\frac{3}{10}$ of the classroom was 500 square feet.

   For each sentence, you'll write an equation.

b. Item A: 4/7 of the pens cost 88 dollars.
• Write the equation. ✔
• Everybody, read equation A. (Signal.)
   4/7 P = 88.

c. Item B: 2/3 of the gravel weighed 502 pounds.
• Write the equation. ✔
• Everybody, read equation B. (Signal.)
   2/3 G = 502.

d. Item C: 6/7 of the water was 5/8 gallon.
• Write the equation. ✔
• Everybody, read equation C. (Signal.)
   6/7 W = 5/8.

e. Item D: 3/10 of the classroom was 500 square feet.
• Write the equation. ✔
• Everybody, read equation D. (Signal.)
   3/10 C = 500.

EXERCISE 8: INDEPENDENT WORK
WORD PROBLEMS

a. Find part 4 in your textbook. ✔
   (Teacher reference:)

   a. $\frac{3}{5}$ of the jeans cost 240 dollars. How much do all the jeans cost?

   b. Henry is $\frac{5}{4}$ the age of his barn. His barn is 60 years old. How old is Henry?

   c. The new car is $\frac{2}{3}$ the weight of the used car. The new car weighs 2800 pounds. What is the weight of the used car? How much do the two cars weigh together?

   d. The turtle was $\frac{7}{5}$ the age of the parrot. The parrot was 40 years old. How old was the turtle? How much older was the turtle than the parrot?

   e. $\frac{3}{4}$ of the eggs weigh 24 ounces. What do all the eggs weigh?

   Some of the equations you'll write for this part have two letters.

b. I'll read the part of each problem that tells about the equation you write.
• Problem A: 3/5 of the jeans cost $240.
   Say the letter equation. (Signal.) 3/5 J = 240.
• Problem B: Henry is 5/4 the age of his barn.
   Say the letter equation. (Signal.) H = 5/4 B.
• Problem C: The new car is 2/3 the weight of the used car.
   Say the letter equation. (Signal.) N = 2/3 U.
• Problem D: The turtle was 7/5 the age of the parrot.
   Say the letter equation. (Signal.) T = 7/5 P.
• Problem E: 3/4 of the eggs weigh 24 ounces.
   Say the letter equation. (Signal.) 3/4 E = 24.
   (Repeat until firm.)

c. You'll work the problems as part of your independent work.

> Assign Independent Work, Textbook parts 4–8 and Workbook part 5.

> Optional extra math-fact practice worksheets are available on ConnectED.

Lesson 120

EXERCISE 1: COORDINATE SYSTEM
RELATED RULES

a. Open your workbook to Lesson 120 and find part 1. ✔
   (Teacher reference:)

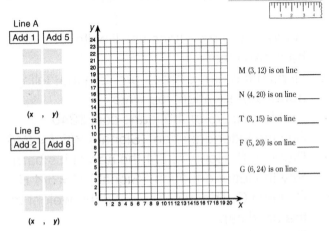

M (3, 12) is on line _____

N (4, 20) is on line _____

T (3, 15) is on line _____

F (5, 20) is on line _____

G (6, 24) is on line _____

   You're going to draw lines for two rules. One rule gives the X value; the other rule gives the Y value.

b. Here are the rules for line A: Add 1. Add 5.
   • Start with zero and write the three values for X and the three values for Y. Show the comma and parentheses for each point. ✔
   (Display:)   [120:1A]

Line A
Add 1   Add 5
(0, 0)
(1, 5)
(2,10)
(x, y)

   Here's what you should have.
   • Plot the points for line A, draw the line, and write the letter A above it.
   (Observe students and give feedback.)

(Add to show:)   [120:1B]

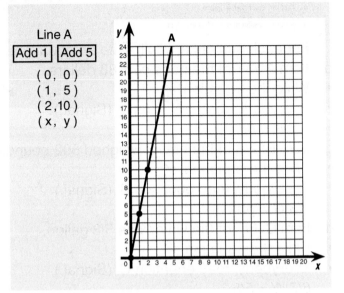

Line A
Add 1   Add 5
(0, 0)
(1, 5)
(2,10)
(x, y)

   Here's what you should have.

c. You're going to plot another line. It starts at zero, zero.
   The X rule is: Add 2. The Y rule is: Add 8.
   • Write the three values for X and the three values for Y. Show the comma and parentheses for each point. ✔
   (Display:)   [120:1C]

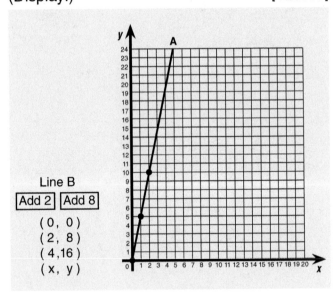

Line B
Add 2   Add 8
(0, 0)
(2, 8)
(4,16)
(x, y)

   Here's what you should have.

d. Plot the points for line B, draw the line, and label it B.
(Observe students and give feedback.)
(Add to show:) [120:1D]

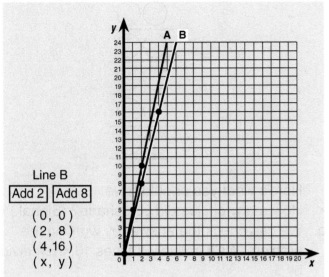

Line B
Add 2   Add 8
(0, 0)
(2, 8)
(4,16)
(x, y)

Here are the lines you should have.
- Make sure your lines go all the way to the top of the coordinate system. ✔
- Everybody, for line A, how many times greater is Y than X? (Signal.) *5 times.*
- For line B, how many times greater is Y than X? (Signal.) *4 times.*
(Repeat until firm.)
e. The items beside the coordinate system show coordinates for points. You're going to identify whether each point is on line A or line B.
- The coordinates for M are: 3 comma 12. Figure out how many times greater Y is than X, and write the letter of the line for 3 comma 12. ✔
- Everybody, on which line will you make the point 3 comma 12? (Signal.) *Line B.*
Yes, Y is 4 times X, so the point goes on line B.

f. Plot point M and label it. ✔
(Change to show:) [120:1E]

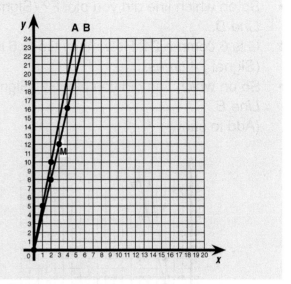

Here's what you should have.
g. The coordinates for N are: 4 comma 20. Figure out how many times greater Y is than X, and write the letter of the line for 4 comma 20. ✔
- Everybody, on which line do you plot 4 comma 20? (Signal.) *Line A.*
Yes, Y is 5 times X, so the point goes on line A.
- Plot point N. ✔
(Add to show:) [120:1F]

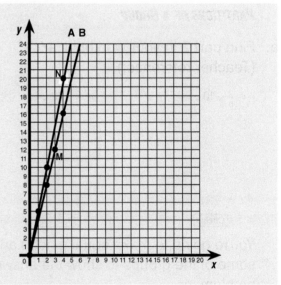

Here's what you should have.
h. Write the letter A or B for the rest of the points and plot the points.
(Observe students and give feedback.)
i. Check your work.
- T is 3 comma 15. How many times 3 is 15? (Signal.) *5 times.*
- So on which line did you plot T? (Signal.) *Line A.*

j. F is 5 comma 20. How many times 5 is 20? (Signal.) *4 times.*
•   So on which line did you plot F? (Signal.) *Line B.*
k. G is 6 comma 24. How many times 6 is 24? (Signal.) *4 times.*
•   So on which line did you plot G? (Signal.) *Line B.*
    (Add to show:)   [120:1G]

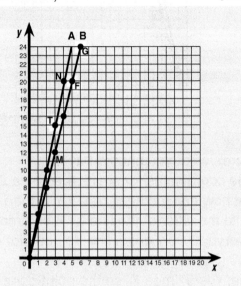

Here's what you should have for each point.

EXERCISE 2: FRACTION REPRESENTATION
PARTITIONS OF A GROUP

a. Find part 2 in your workbook. ✔
    (Teacher reference:)

 a. $\frac{3}{5} \times 10 =$ ☐   Circles

 b. $\frac{2}{3} \times 12 =$ ☐   Triangles

You're going to make equal groups and shade some of the groups to show the answer to the problem.

b. Problem A: 3/5 × 10.
•   How many equal parts will you make? (Signal.) *5.*
•   How many groups will you shade? (Signal.) *3.*
•   Make the diagram with circles and write the answer.
    (Observe students and give feedback.)

(Display:)   [120:2A]

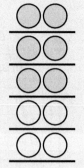

a. $\frac{3}{5} \times 10 = 6$

Here's what you should have.
•   How many circles did you shade? (Signal.) *6.*
c. We can check the answer by working the problem. We'll multiply 3 times 10, then divide by 5.
•   What's 3 × 10? (Signal.) *30.*
•   What's 30 ÷ 5? (Signal.) *6.*
    That's the answer we got by showing the groups.
d. Problem B. Make the diagram with triangles. Shade 2/3 of the triangles and write the answer.
    (Observe students and give feedback.)
e. Check your work.
    (Display:)   [120:2B]

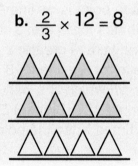

b. $\frac{2}{3} \times 12 = 8$

Here's what you should have.
There are 4 triangles in each group. You shaded 2 groups.
•   How many triangles did you shade? (Signal.) *8.*
f. You'll do the multiplication and division to check your answer.
•   Say the multiplication for the numerators. (Signal.) *2 × 12.*
•   Then what will you divide by? (Signal.) *3.*
•   Do the multiplication and division. ✔
g. Check your work.
•   What's 2 × 12? (Signal.) *24.*
•   What's 24 ÷ 3? (Signal.) *8.*
    So the diagram is correct.

EXERCISE 3: MULTIPLICATION
REASONING

a. Find part 3 in your workbook. ✔
* Pencils down. ✔
(Teacher reference:)

$\frac{A}{B}$ equals a whole number.

$$\frac{A}{B}\left(\frac{3}{2}\right) = \blacksquare$$

$$\left(\frac{\quad}{\quad}\right) = $$

a. Will the answer be more than $\frac{A}{B}$, less than $\frac{A}{B}$, or equal to $\frac{A}{B}$? _____

b. If $\frac{A}{B}$ equals 6, will the answer be a whole number? _____

c. If $\frac{A}{B}$ equals 7, will the answer be a whole number? _____

$$\frac{A}{B}\left(\frac{\blacksquare}{\quad}\right) = \text{less than } \frac{A}{B}$$

d. Do we multiply $\frac{A}{B}$ by more than 1 or less than 1? _____

$$\frac{A}{B}\left(\frac{8}{2}\right) = \blacksquare$$

e. Do we multiply $\frac{A}{B}$ by more than 1 or less than 1? _____

f. How many times greater than $\frac{A}{B}$ is the answer? _____

g. If $\frac{A}{B}$ equals 3, what will the answer be? _____

You're going to answer questions and show why your answers are correct.
The fact says that A over B equals a whole number. The problem shows that you're multiplying A over B times 3/2.
Get ready to answer the questions.

b. Question A: Will the answer be more than A over B, less than A over B, or equal to A over B?
* What's the answer? (Signal.) *More than A/B.*
c. Question B: If A over B equals 6, will the answer be a whole number?
* What's the fraction answer? (Signal.) *18/2.*
* Is 18 halves a whole number? (Signal.) *Yes.*
* What whole number is 18/2? (Signal.) *9.*
d. Question C: If A over B equals 7, will the answer be a whole number?
* What's the fraction answer? (Signal.) *21/2.*
* Is 21/2 a whole number? (Signal.) *No.*
(Repeat until firm.)
* Write answers to questions A through C. ✔
e. (Display:) [120:3A]

$$10\left(\frac{3}{2}\right) = $$

Here's an example with A over B equal to 10. Let's see if the answer is more than 10.
* Can we simplify 10/2? (Signal.) *Yes.*
* What's the simplified value? (Signal.) *5.*

(Add to show:) [120:3B]

$$\overset{5}{\cancel{10}}\left(\frac{3}{\cancel{2}}\right) = $$

* What's the answer to the problem? (Signal.) *15.*
(Add to show:) [120:3C]

$$\overset{5}{\cancel{10}}\left(\frac{3}{\cancel{2}}\right) = \boxed{15}$$

f. Your turn: Below the equation, write a different whole number that is a multiple of 2. Multiply by 3/2 and show the whole-number answer. (Observe students and give feedback.)
* (Call on different students. Ask:) What whole number did you start with? What's ___ times 3/2?
Explain why the answer is more than ___.
(Idea: *The problem multiplies by more than one.*)
Explain why the answer is a whole number.
(Idea: *___ over 2 simplifies to a whole number, so the answer is a whole number.*)
g. (Display:) [120:3D]

$$\frac{A}{B}\left(\frac{\quad}{\quad}\right) = \text{less than } \frac{A}{B}$$

Here's the next problem. We want to multiply A over B by something and end up with less than A over B.
* Question D says: Do we multiply A over B by more than 1 or less than 1? Write the answer. ✔
* Everybody, what's the answer? (Signal.) *Less than 1.*
h. (Display:) [120:3E]

$$\frac{A}{B}\left(\frac{8}{2}\right) = \blacksquare$$

Here's the next problem. A over B times 8/2.
i. Question E says: Do we multiply A over B by more than 1 or less than 1? Write the answer. ✔
* Everybody, what's the answer? (Signal.) *More than 1.*
* Is 8/2 a whole number? (Signal.) *Yes.*
j. Question F: How many times greater than A over B is the answer? Write the answer. ✔
* Everybody, what's the answer? (Signal.) *4 times greater.*

k. Question G: If A over B equals 3, what will the answer be? Write the answer. ✔
- Everybody, what's the answer? (Signal.) *12.*

EXERCISE 4: DECIMAL MULTIPLICATION
REPRESENTATION

a. Find part 4 in your workbook. ✔
(Teacher reference:)

$$\begin{array}{r} .9 \\ \times\ .4 \\ \hline \end{array}$$

You're going to make a diagram that shows multiplication of tenths.
- Read the problem. (Signal.) *9 tenths times 4 tenths.*
- Raise your hand when you know the answer. ✔
- What's the answer? (Signal.) *36 hundredths.*
- Write the answer. ✔
b. You're going to show the diagram for 9 tenths times 4 tenths.
- Start at the dot and make a square that has 10 parts on each side. Then stop. ✔
(Display:)   [120:4A]

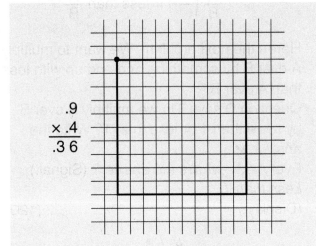

Here's what you should have.

c. Now shade a rectangle inside the square that is 9/10 times 4/10. ✔
(Add to show:)   [120:4B]

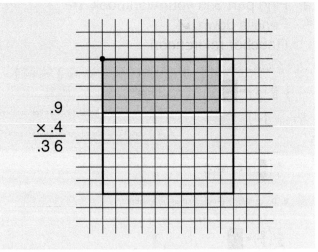

Here's what you should have.
d. Raise your hand when you know how many parts are shaded and how many parts are in the whole square. ✔
- How many parts are shaded? (Signal.) *36.*
- How many parts are in the whole square? (Signal.) *100.*
- So what's the fraction for the shaded area? (Signal.) *36 hundredths.*
The diagram shows the answer to the decimal problem.

EXERCISE 5: FRACTION OPERATIONS
REASONABLENESS OF ANSWERS

a. Open your textbook to Lesson 120 and find part 1. ✔
(Teacher reference:)

a. $\frac{1}{2} - \frac{3}{8} = \frac{4}{8}$ b. $\frac{7}{7} - \frac{9}{14} = \frac{10}{28}$

c. $\frac{9}{10} - \frac{1}{5} = \frac{22}{20}$ d. $\frac{2}{5} + \frac{9}{15} = \frac{30}{30}$

Some of these problems have answers that are possible. Some have answers that are impossible.
You're going to write **possible** or **impossible** for each item. If the answer is impossible, you'll write a sentence or two that tells why the answer is impossible.

b. Read item A. (Signal.) *1/2 − 3/8 = 4/8.*
• Write possible or impossible. If the answer is impossible, tell why.
  (Observe students and give feedback.)
c. Check your work.
• Did you write possible or impossible? (Signal.) *Impossible.*
• What did you write to explain why it is impossible? (Call on a student. Idea: *You're starting with 1/2 and subtracting, so the answer can't be 1/2.*)
d. Read problem B. (Signal.) *7/7 − 9/14 = 10/28.*
• Write possible or impossible. If the answer is impossible, tell why.
  (Observe students and give feedback.)
e. Check your work.
• Did you write possible or impossible? (Signal.) *Possible.*
• Who can tell me why the answer is possible? ✔
• (Call on a student. Idea: *You're starting with 1 and subtracting more than 1/2. The answer is less than 1/2. So it's possible.*)
f. Read item C. (Signal.) *9/10 − 1/5 = 22/20.*
• Write possible or impossible. If the answer is impossible, tell why.
  (Observe students and give feedback.)

g. Check your work.
• Did you write possible or impossible? (Signal.) *Impossible.*
• Why is it impossible? (Call on a student. Idea: *You're starting with less than 1 and subtracting, so the answer can't be more than 1.*)
h. Read item D. (Signal.) *2/5 + 9/15 = 30/30.*
• Write possible or impossible. If the answer is impossible, tell why.
  (Observe students and give feedback.)
i. Check your work.
• Did you write possible or impossible? (Signal.) *Possible.*
• Who can tell me why the answer is possible? (Call on a student. Idea: *You're adding less than 1/2 and more than 1/2, so you could end up with 1.*)

Assign Independent Work, Textbook parts 2–9 and Workbook part 5.

Optional extra math-fact practice worksheets are available on ConnectED.

Mastery Test 12

Teacher Presentation

a. Find Mastery Test 12 in your test booklet. ✔
b. Work parts 1 through 7 of the test by yourself. Read the directions carefully and do your best work. Put your pencil down when you've finished the test.
(Observe but do not give feedback.)

Scoring Notes

a. Collect test booklets. Use the *Answer Key* and Passing Criteria Table to score the tests.

Passing Criteria Table—Mastery Test 12			
Part	Score	Possible Score	Passing Score
1	1 for each equation	3	3
2	1 for each item	10	9
3	3 for each item (original problem, fraction problem, answer with unit name)	6	5
4	1 for each item	5	4
5	1 for each equivalent fraction 1 for each sign	10	8
6	3 for each item (letter equation, substitution, answer with unit name)	9	8
7	1 for each item	6	5
	Total	49	

b. Complete the Mastery Test 12 Remedy Summary Sheet to determine whether group remedies are needed. Reproducible Remedy Summary Sheets are at the back of the *Answer Key* and the back of the *Teacher's Guide*.

• If ¼ or more of the students did not pass a test part, present the remedy for that part. The Remedy Table follows and also appears at the end of the Mastery Test 12 *Answer Key*. Remedies worksheets follow Mastery Test 12 in the *Student Assessment Book*.

Remedy Table—Mastery Test 12					
Part	Test Items	Remedy Lesson	Ex.	Remedies Worksheet	Textbook
1	Sentences (Fraction Comparison)	111	5	—	Part 4
2	Quadrilaterals (Hierarchy)	109	2	Part A	—
		111	2	—	Part 2
3	Mixed-Number Multiplication (Word Problems)	111	4	—	
		112	4	—	Part 2
4	Line Plot	114	4	—	Part 2
		115	IW	—	Part 4
5	Fraction Comparison (Unlike Denominators)	114	1	Part B	—
		116	1	Part C	—
6	Word Problems (Fraction Comparison)	113	3	—	Part 1
		114	3	—	Part 1
7	Mental Math (Probability)	115	2	Part D	—
		116	6	—	Part 1

Retest

Retest individual students on any part failed.

Cumulative Test 2

Note: Cumulative Tests are group administered. Each student will need a pencil and the *Student Assessment Book.* Try to arrange students so they cannot look at other students' responses.

Note: Students will need a ruler for parts 5 and 10.

Cumulative Test 2 assesses the skills taught in Lessons 61–120 of *CMC Level F.* Presenting Cumulative Test 2 to students without having taught these lessons is not advised.

Teacher Presentation

a. Find Cumulative Test 2 in your test booklet. ✔ This is a test of some of the things you've learned in math this year. We'll work part of the test today and finish the test on the next lesson.

• Each part has directions that tell you what to do. Read them carefully. You may begin. (Observe but do not give feedback.)

Scoring

a. Collect the test booklets. Use the *Answer Key* to score the tests. A passing score is 85 percent.

b. If ¼ or more of the students did poorly on a test part, consider presenting the remedy for the corresponding Mastery Test part, referenced in the Cumulative Test 2 Scoring Chart.

Part	Score	Possible Score	Corresponding Mastery Test	Part
1	1 for each item	3	7	1
2	1 for each item	3	7	4
3	2 for each item (repeated multiplication, answer)	6	7	12
4	3 for each item (letter equation, multiplication problem, answer with unit name)	6	8	9
5	1 for each point 1 for the line	6	7	6
6	1 for each item	4	7	7
7	1 for each item	3	7	3
8	1 for each item	4	8	10
9	3 for the item (letter equation, number equation, answer with a unit name)	3	8	12
10	1 for each answer 1 for each axis label 1 for the line	7	8	1
11	2 for each item (original expression, answer)	4	9	8
12	1 for addition answer 1 for division problem 1 for answer with unit name	3	9	5
13	1 for each item	3	8	5
14	1 for each item	4	11	8
15	3 for each item (unit conversion, +/– problem, answer with unit name)	6	11	2
16	1 for each item	3	11	1
17	1 for each item	5	12	4
18	1 for each item	10	12	2
19	1 for each item	3	11	5
20	3 for each item (each digit in the whole-number part of the answer, fraction remainder)	6	11	11
21	2 for each item (substitution, answer)	4	9	3
22	1 for each item	3	7	8
23	4 for each item (letter equation, substitution, each answer)	8	10	9
24	3 for each item (letter equation, substitution, answer with unit name)	6	10	3
25	3 for each item (each fraction in the new problem, answer)	6	10	2

Part	Score	Possible Score	Corresponding Mastery Test	Part
26	3 for each item (division problem, mixed-number answer, unit name)	6	10	12
27	3 for the item (bottom stair, top stair, total with unit name)	3	10	7
28	2 for each item (fraction multiplication, answer with unit name)	4	12	3
29	1 for each item	3	9	7
30	3 for each item (letter equation, substitution, answer with unit name)	6	12	6
31	1 for each rewritten fraction 1 for each sign	8	12	5
32	3 for each item (original problem, answer, unit name)	6	10	5
	Total Points Possible:	155		
	Passing Score:	132(85%)		

Level F Correlation to Grade 5 Common Core State Standards for Mathematics

Operations and Algebraic Thinking (5.OA)

Write and interpret numerical expressions.

1. Use parentheses, brackets, or braces in numerical expressions, and evaluate expressions with these symbols.

Lesson	61	62	63	64	65	66	67	68	69	70
Exercise	61.6	62.2, 62.6	63.8	64.7	65.7	66.8	67.6	68.7	69.6, 69.7	70.7, 70.9

Lesson	71	72	73	74	75	76	77	78	79	83
Exercise	71.8	72.7	73.8	74.8	75.8	76.8	77.7	78.6	79.9	83.5

Lesson	84	85	86	87	88	89	90	91	92	93
Exercise	84.5, 84.9	85.4, 85.8	86.4, 86.9	87.7, 87.9	88.4, 88.7	89.4, 89.7	90.2, 90.8	91.8	92.7	93.8

Lesson	94	95	96	98	99	100	101	102	103	104
Exercise	94.6, 94.8	95.7	96.8	98.6, 98.7	99.7	100.5, 100.8	101.3, 101.5, 101.7	102.4, 102.7	103.3, 103.4, 103.7	104.3, 104.4, 104.7

Lesson	105	106	107	108	109	110	111	112	113	114
Exercise	105.4, 105.7	106.3, 106.6	107.5, 107.6	108.4, 108.6	109.4, 109.6	110.4, 110.6, 110.8	111.3, 111.6, 111.8	112.6	113.3, 113.6	114.3, 114.7

Lesson	115	116	117	118	119	120
Exercise	115.1, 115.5	116.1, 116.7	117.1, 117.7	118.5, 118.7	119.8	120.5

Operations and Algebraic Thinking (5.OA)

Write and interpret numerical expressions.

2. Write simple expressions that record calculations with numbers, and interpret numerical expressions without evaluating them. *For example, express the calculation "add 8 and 7, then multiply by 2" as 2 × (8 + 7). Recognize that 3 × (18932 + 921) is three times as large as 18932 + 921, without having to calculate the indicated sum or product.*

Lesson	87	88	89	90	91	92	93	117	118	119
Exercise	87.7	88.4	89.4	90.8	91.8	92.7	93.7	117.7	118.1, 118.7	119.2

Lesson	120
Exercise	120.3

Operations and Algebraic Thinking (5.OA)

Write and interpret numerical expressions.

*2.1 Express a whole number in the range 2–50 as a product of its prime factors. For example, find the prime factors of 24 and express 24 as 2 x 2 x 2 x 3.

Lesson	61	74	77
Exercise	61.6	74.8	77.7

*Denotes California-only content.

Operations and Algebraic Thinking (5.OA)

Analyze patterns and relationships.

3. Generate two numerical patterns using two given rules. Identify apparent relationships between corresponding terms. Form ordered pairs consisting of corresponding terms from the two patterns, and graph the ordered pairs on a coordinate plane. *For example, given the rule "Add 3" and the starting number 0, and given the rule "Add 6" and the starting number 0, generate terms in the resulting sequences, and observe that the terms in one sequence are twice the corresponding terms in the other sequence. Explain informally why this is so.*

Lesson	118	119	120
Exercise	118.2	119.2	120.1

Number and Operations in Base Ten (5.NBT)

Understand the place value system.

1. Recognize that in a multi-digit number, a digit in one place represents 10 times as much as it represents in the place to its right and 1/10 of what it represents in the place to its left.

Lesson	63	64	65	66	67	68	69	70	71	72
Exercise	63.6, 63.7	64.3	65.3	66.4	67.6	68.7	69.4, 69.7	70.3, 70.9	71.5	72.5, 72.7

Lesson	73	75	76	81	82	83	84	85	86	87
Exercise	73.2	75.8	76.1	81.2	82.2	83.3	84.3	85.3	86.2	87.9

Lesson	88	89	93	95	96	97	98	101	102	103
Exercise	88.7	89.7	93.7	95.7	96.7	97.8	98.7	101.2, 101.7	102.2	103.7

Lesson	104	105	106	108	109	110	111	112	113	115
Exercise	104.7	105.6	106.6	108.7	109.6	110.6, 110.7	111.7, 111.8	112.5, 112.6	113.6	115.7

Lesson	117
Exercise	117.7

Number and Operations in Base Ten (5.NBT)

Understand the place value system.

2. Explain patterns in the number of zeros of the product when multiplying a number by powers of 10, and explain patterns in the placement of the decimal point when a decimal is multiplied or divided by a power of 10. Use whole-number exponents to denote powers of 10.

Lesson	63	64	65	66	67	68	69	70	76	81
Exercise	63.6,63.7	64.3	65.3	66.4	67.6	68.7	69.4, 69.7	70.3, 70.9	76.1	81.2

Lesson	82	83	84	85	86	87	88	89	93	95
Exercise	82.2	83.3	84.3	85.3	86.2	87.9	88.7	89.7	93.7	95.7

Lesson	96	97	98	101	102	103	104	105	106	108
Exercise	96.7	97.8	98.7	101.2, 101.7	102.2	103.7	104.7	105.6	106.6	108.7

Lesson	109	110	111	112	113	115	117
Exercise	109.6	110.6	111.7, 111.8	112.5, 112.6	113.6	115.7	117.7

Number and Operations in Base Ten (5.NBT)

Understand the place value system.

3. Read, write, and compare decimals to thousandths.
   a. Read and write decimals to thousandths using base-ten numerals, number names, and expanded form, e.g., 347.392 = 3 × 100 + 4 × 10 + 7 × 1 + 3 × (1/10) + 9 × (1/100) + 2 × (1/1000).
   b. Compare two decimals to thousandths based on meanings of the digits in each place, using >, =, and < symbols to record the results of comparisons.

Lesson	61	62	63	64	65	67	68	69	70	72
Exercise	61.6	62.6	63.8	64.7	65.7	67.5, 67.6	68.5	69.7	70.7, 70.9	72.7

Lesson	74	75	76	77	78	79	80	81	82	83
Exercise	74.8	75.3, 75.8	76.2	77.2, 77.7	78.2	79.4	80.4	81.7	82.7	83.8

Lesson	84	85	88	92	93	95	97	98	100	102
Exercise	84.9	85.8	88.7	92.7	93.7	95.7	97.8	98.8	100.9	102.6

Lesson	103	104	105	107
Exercise	103.5	104.5	105.7	107.6

Number and Operations in Base Ten (5.NBT)

Understand the place value system.

4. Use place value understanding to round decimals to any place.

Lesson	67	68	69	70	71	72	73	74	75	76
Exercise	67.5	68.5	69.7	70.7, 70.9	71.7, 71.8	72.6	73.7	74.7, 74.8	75.7	76.6, 76.8

Number and Operations in Base Ten (5.NBT)

Perform operations with multi-digit whole numbers and with decimals to hundredths.

5. Fluently multiply with multi-digit whole numbers using the standard algorithm.

Number and Operations in Base Ten (5.NBT)

Perform operations with multi-digit whole numbers and with decimals to hundredths.

6. Find whole-number quotients of whole numbers with up to four-digit dividends and two-digit divisors, using strategies based on place value, the properties of operations, and/or the relationship between multiplication and division. Illustrate and explain the calculation by using equations, rectangular arrays, and/or area models.

Lesson	114	115	116	117	118	119	120
Exercise	114.7	115.7	116.7	117.6, 117.7	118.7	119.8	120.5

Number and Operations in Base Ten (5.NBT)

Perform operations with multi-digit whole numbers and with decimals to hundredths.

7. Add, subtract, multiply, and divide decimals to hundredths, using concrete models or drawings and strategies based on place value, properties of operations, and/or the relationship between addition and subtraction; relate the strategy to a written method and explain the reasoning used.

Lesson	61	62	63	64	65	66	67	68	69	70
Exercise	61.2, 61.6	62.5	63.3, 63.8	64.7	65.7	66.3, 66.8	67.2, 67.6	68.2, 68.7	69.2	70.9

Lesson	71	72	73	74	75	76	77	78	79	80
Exercise	71.8	72.7	73.6, 73.8	74.6, 74.8	75.6, 75.8	76.8	77.5, 77.7	78.6	79.7, 79.9	80.7, 80.9

Lesson	81	82	84	85	86	87	89	91	93	94
Exercise	81.7	82.7	84.3, 84.9	85.3	86.9	87.9	89.7	91.8	93.7	94.8

Lesson	95	96	97	98	99	101	102	103	104	105
Exercise	95.7	96.7	97.7, 97.8	98.7	99.8	101.7	102.2	103.7	104.7	105.6

Lesson	106	108	109	111	113	114	115	117	118	119
Exercise	106.6	108.7	109.6	111.8	113.2	114.2	115.6	117.3	118.4	119.4

Lesson	120
Exercise	120.4

Number and Operations—Fractions (5.NF)

Use equivalent fractions as a strategy to add and subtract fractions.

1. Add and subtract fractions with unlike denominators (including mixed numbers) by replacing given fractions with equivalent fractions in such a way as to produce an equivalent sum or difference of fractions with like denominators. *For example, 2/3 + 5/4 = 8/12 + 15/12 = 23/12. (In general, a/b + c/d = (ad + bc)/bd.)*

Lesson	78	80	84	91	92	93	94	95	96	97
Exercise	78.6	80.9	84.9	91.4	92.3, 92.7	93.7	94.8	95.7	96.5, 96.7	97.6, 97.8

Lesson	98	99	101	102	103	104	106	107	108	109
Exercise	98.8	99.8	101.6	102.5	103.7	104.7	106.6	107.6	108.7	109.6

Lesson	110	111	113	114	115	116	117	118	119
Exercise	110.8	111.8	113.6	114.7	115.7	116.7	117.3, 117.7	118.6, 118.7	119.8

Number and Operations—Fractions (5.NF)

Use equivalent fractions as a strategy to add and subtract fractions.

2. Solve word problems involving addition and subtraction of fractions referring to the same whole, including cases of unlike denominators, e.g., by using visual fraction models or equations to represent the problem. Use benchmark fractions and number sense of fractions to estimate mentally and assess the reasonableness of answers. *For example, recognize an incorrect result 2/5 + 1/2 = 3/7, by observing that 3/7 < 1/2.*

Lesson	78	80	81	82	83	84	85	86	87	90
Exercise	78.6	80.9	81.4	82.4	83.8	84.9	85.8	86.9	87.9	90.8

Lesson	92	96	97	98	100	101	102	103	104	105
Exercise	92.7	96.7	97.8	98.8	100.9	101.6	102.5	103.7	104.7	105.7

Lesson	106	107	108	109	110	111	113	114	115	116
Exercise	106.6	107.6	108.7	109.6	110.8	111.8	113.6	114.7	115.7	116.7

Lesson	117	118	119	120
Exercise	117.7	118.6, 118.7	119.6, 119.8	120.5

Number and Operations—Fractions (5.NF)

Apply and extend previous understandings of multiplication and division to multiply and divide fractions.

3. Interpret a fraction as division of the numerator by the denominator ($a/b = a \div b$). Solve word problems involving division of whole numbers leading to answers in the form of fractions or mixed numbers, e.g., by using visual fraction models or equations to represent the problem. *For example, interpret 3/4 as the result of dividing 3 by 4, noting that 3/4 multiplied by 4 equals 3, and that when 3 wholes are shared equally among 4 people each person has a share of size 3/4. If 9 people want to share a 50-pound sack of rice equally by weight, how many pounds of rice should each person get? Between what two whole numbers does your answer lie?*

Lesson	70	75	76	88	89	94	113	114	115	116
Exercise	70.5	75.1	76.1	88.1	89.1, 89.4	94.8	113.1	114.5	115.7	116.5, 116.7

Lesson	118
Exercise	118.3

Number and Operations—Fractions (5.NF)

Apply and extend previous understandings of multiplication and division to multiply and divide fractions.

4. Apply and extend previous understandings of multiplication to multiply a fraction or whole number by a fraction.
   a. Interpret the product (a/b) × q as a parts of a partition of q into b equal parts; equivalently, as the result of a sequence of operations $a \times q \div b$. *For example, use a visual fraction model to show (2/3) × 4 = 8/3, and create a story context for this equation. Do the same with (2/3) × (4/5) = 8/15. (In general, (a/b) × (c/d) = ac/bd.)*
   b. Find the area of a rectangle with fractional side lengths by tiling it with unit squares of the appropriate unit fraction side lengths, and show that the area is the same as would be found by multiplying the side lengths. Multiply fractional side lengths to find areas of rectangles, and represent fraction products as rectangular areas.

Lesson	61	62	63	64	65	66	67	68	69	70
Exercise	61.4, 61.6	62.2, 62.6	63.8	64.7	65.7	66.8	67.6	68.7	69.7	70.9

Lesson	71	72	73	74	75	76	77	78	79	80
Exercise	71.8	72.7	73.8	74.8	75.8	76.8	77.3, 77.4	78.4	79.6, 79.9	80.6

Lesson	81	82	83	84	85	86	87	88	89	90
Exercise	81.7	82.7	83.8	84.5, 84.9	85.8	86.4, 86.9	87.9	88.2, 88.7	89.2, 89.7	90.2, 90.7, 90.8

Lesson	91	92	93	94	95	96	97	98	99	100
Exercise	91.2, 91.7, 91.8	92.6	93.4, 93.7	94.4, 94.6, 94.8	95.7	96.7	97.8	98.2, 98.6, 98.8	99.2, 99.6	100.3, 100.5, 100.8

Lesson	101	102	103	104	105	106	107	108	109	110
Exercise	101.3, 101.5, 101.7	102.4, 102.7	103.4, 103.7	104.4, 104.7	105.7	106.6	107.3, 107.6	108.3, 108.7	109.3, 109.6	110.3, 110.8

Lesson	111	112	113	114	115	116	117	118	119	120
Exercise	111.4, 111.8	112.4, 112.6	113.3, 113.6	114.3, 114.7	115.4, 115.5. 115.7	116.3, 116.4, 116.7	117.2, 117.6, 117.7	118.1, 118.3, 118.5, 118.7	119.1, 119.3, 119.4, 119.5, 119.8	120.2, 120.5

Number and Operations—Fractions (5.NF)

Apply and extend previous understandings of multiplication and division to multiply and divide fractions.

5. Interpret multiplication as scaling (resizing), by:
   a. Comparing the size of a product to the size of one factor on the basis of the size of the other factor, without performing the indicated multiplication.
   b. Explaining why multiplying a given number by a fraction greater than 1 results in a product greater than the given number (recognizing multiplication by whole numbers greater than 1 as a familiar case); explaining why multiplying a given number by a fraction less than 1 results in a product smaller than the given number; and relating the principle of fraction equivalence $a/b = (n \times a)/(n \times b)$ to the effect of multiplying a/b by 1.

Lesson	61	62	63	64	65	66	67	68	69	70
Exercise	61.3, 61.6	62.4, 62.6	63.2, 63.8	64.3, 64.7	65.7	66.8	67.6	68.1, 68.7	69.1, 69.7	70.1, 70.9

Lesson	71	72	73	74	75	76	77	78	79	80
Exercise	71.1, 71.8	72.1, 72.7	73.1, 73.8	74.1, 74.8	75.1, 75.2	76.1, 76.8	77.7	78.6	79.1, 79.8, 79.9	80.2, 80.9

Lesson	81	82	83	84	85	87	88	89	90	91
Exercise	81.7	82.7	83.8	84.9	85.8	87.2, 87.9	88.2	89.2, 89.7	90.2	91.2, 91.8

Lesson	92	93	94	95	96	97	98	100	103	105
Exercise	92.6	93.4	94.4, 94.8	95.7	96.7	97.8	98.2, 98.7	100.9	103.7	105.7

Lesson	107	113	114	115	116	117	118	119	120
Exercise	107.6	113.5	114.1	115.1	116.1	117.1, 117.3	118.1, 118.5	119.5	120.5

Number and Operations—Fractions (5.NF)

Apply and extend previous understandings of multiplication and division to multiply and divide fractions.

6. Solve real world problems involving multiplication of fractions and mixed numbers, e.g., by using visual fraction models or equations to represent the problem.

Lesson	61	62	64	66	68	70	71	72	73	74
Exercise	61.6	62.6	64.7	66.8	68.7	70.9	71.8	72.7	73.8	74.8

Lesson	75	76	77	78	79	80	81	82	83	84
Exercise	75.8	76.8	77.3	78.4, 78.6	79.6, 79.9	80.6	81.7	82.7	83.8	84.9

Lesson	85	86	87	90	91	93	94	95	96	97
Exercise	85.8	86.9	87.9	90.7	91.7, 91.8	93.7	94.6	95.7	96.7	97.8

Lesson	98	99	100	103	104	106	107	108	109	110
Exercise	98.6, 98.8	99.6	100.5, 100.9	103.4, 103.7	104.4, 104.7	106.6	107.6	108.7	109.6	110.3, 110.8

Lesson	111	112	113	114	115	116	117	118	119	120
Exercise	111.4, 111.8	112.4, 112.6	113.3, 113.6	114.3, 114.7	115.5, 115.7	116.4, 116.7	117.6, 117.7	118.7	119.8	120.5

Number and Operations—Fractions (5.NF)

Apply and extend previous understandings of multiplication and division to multiply and divide fractions.

7. Apply and extend previous understandings of division to divide unit fractions by whole numbers and whole numbers by unit fractions.
   a. Interpret division of a unit fraction by a non-zero whole number, and compute such quotients. *For example, create a story context for (1/3) ÷ 4, and use a visual fraction model to show the quotient. Use the relationship between multiplication and division to explain that (1/3) ÷ 4 = 1/12 because (1/12) × 4 = 1/3.*
   b. Interpret division of a whole number by a unit fraction, and compute such quotients. *For example, create a story context for 4 ÷ (1/5), and use a visual fraction model to show the quotient. Use the relationship between multiplication and division to explain that 4 ÷ (1/5) = 20 because 20 × (1/5) = 4.*
   c. Solve real world problems involving division of unit fractions by non-zero whole numbers and division of whole numbers by unit fractions, e.g., by using visual fraction models and equations to represent the problem. *For example, how much chocolate will each person get if 3 people share 1/2 lb of chocolate equally? How many 1/3-cup servings are in 2 cups of raisins?*

Lesson	73	75	76	77	78	79	80	81	82	83
Exercise	73.3	75.4	76.3	77.4	78.5	79.5, 79.9	80.5, 80.9	81.3, 81.7	82.3	83.6, 83.8

Lesson	84	85	86	87	88	89	90	91	92	93
Exercise	84.6	85.5, 85.8	86.3	87.6	88.5	89.5, 89.7	90.6	91.6, 91.8	92.4	93.7

Lesson	94	95	96	97	98	113	114	115	116	117
Exercise	94.8	95.7	96.7	97.8	98.8	113.4	114.4	115.3, 115.7	116.2, 116.7	117.5

Measurement and Data (5.MD)

Convert like measurement units within a given measurement system.

1. Convert among different-sized standard measurement units within a given measurement system (e.g., convert 5 cm to 0.05 m), and use these conversions in solving multi-step, real world problems.

Lesson	90	91	92	93	94	95	96	97	98	99
Exercise	90.5	91.5	92.5	93.6	94.5	95.5	96.4	97.5	98.5	99.5

Lesson	100	101	102	103	104	105	106	107	108	109
Exercise	100.4	101.7	102.7	103.7	104.7	105.7	106.6	107.2, 107.6	108.5, 108.7	109.5, 109.6

Lesson	110	111	112	113	114	115	116
Exercise	110.5, 110.8	111.8	112.6	113.6	114.7	115.7	116.7

Measurement and Data (5.MD)

Represent and interpret data.

2. Make a line plot to display a data set of measurements in fractions of a unit (1/2, 1/4, 1/8). Use operations on fractions for this grade to solve problems involving information presented in line plots. *For example, given different measurements of liquid in identical beakers, find the amount of liquid each beaker would contain if the total amount in all the beakers were redistributed equally.*

Lesson	112	113	114	115	116	117	119	120
Exercise	112.1	113.4	114.4	115.3	116.2	117.5	119.8	120.5

Measurement and Data (5.MD)

Geometric measurement: understand concepts of volume and relate volume to multiplication and to addition.

3. Recognize volume as an attribute of solid figures and understand concepts of volume measurement.
   a. A cube with side length 1 unit, called a "unit cube," is said to have "one cubic unit" of volume, and can be used to measure volume.
   b. A solid figure which can be packed without gaps or overlaps using n unit cubes is said to have a volume of n cubic units.

Lesson	68	69
Exercise	68.3	69.6

Student Practice Software: Block 4 Activity 6, Block 5 Activities 3 and 6

Measurement and Data (5.MD)

Geometric measurement: understand concepts of volume and relate volume to multiplication and to addition.

4. Measure volumes by counting unit cubes, using cubic cm, cubic in, cubic ft, and improvised units.

Lesson	68	69
Exercise	68.3	69.6

Student Practice Software: Block 4 Activity 6, Block 5 Activities 3 and 6

Measurement and Data (5.MD)

Geometric measurement: understand concepts of volume and relate volume to multiplication and to addition.

5. Relate volume to the operations of multiplication and addition and solve real world and mathematical problems involving volume.
   a. Find the volume of a right rectangular prism with whole-number side lengths by packing it with unit cubes, and show that the volume is the same as would be found by multiplying the edge lengths, equivalently by multiplying the height by the area of the base. Represent threefold whole-number products as volumes, e.g., to represent the associative property of multiplication.
   b. Apply the formulas $V = l \times w \times h$ and $V = b \times h$ for rectangular prisms to find volumes of right rectangular prisms with whole-number edge lengths in the context of solving real world and mathematical problems.
   c. Recognize volume as additive. Find volumes of solid figures composed of two non-overlapping right rectangular prisms by adding the volumes of the non-overlapping parts, applying this technique to solve real world problems.

Lesson	68	69	70	71	72	73	74	76	77	80
Exercise	68.3	69.6	70.6	71.6	72.7	73.8	74.8	76.4	77.7	80.7

Lesson	82	84	86	89	93	94	95	96	97	98
Exercise	82.7	84.9	86.9	89.7	93.5, 93.7	94.7, 94.8	95.6	96.6	97.7	98.7

Lesson	99	101	102
Exercise	99.7	101.7	102.7

Geometry (5.G)

Graph points on the coordinate plane to solve real-world and mathematical problems.

1. Use a pair of perpendicular number lines, called axes, to define a coordinate system, with the intersection of the lines (the origin) arranged to coincide with the 0 on each line and a given point in the plane located by using an ordered pair of numbers, called its coordinates. Understand that the first number indicates how far to travel from the origin in the direction of one axis, and the second number indicates how far to travel in the direction of the second axis, with the convention that the names of the two axes and the coordinates correspond (e.g., x-axis and x-coordinate, y-axis and y-coordinate).

Lesson	62	102	104	106
Exercise	62.6	102.8	104.7	106.6

Student Practice Software: Block 2 Activity 5

Geometry (5.G)

Graph points on the coordinate plane to solve real-world and mathematical problems.

2. Represent real world and mathematical problems by graphing points in the first quadrant of the coordinate plane, and interpret coordinate values of points in the context of the situation.

Lesson	62	64	65	66	67	69	70	71	72	73
Exercise	62.6	64.2	65.2	66.2	67.3	69.3	70.4	71.3	72.2	73.4

Lesson	74	75	76	77	79	82	83	84	86	87
Exercise	74.4, 74.8	75.8	76.8	77.7	79.9	82.7	83.8	84.9	86.9	87.9

Lesson	88	91	94	96	99	101	105	109	111	118
Exercise	88.7	91.8	94.8	96.7	99.8	101.7	105.7	109.6	111.8	118.2

Lesson	119	120
Exercise	119.2	120.1

Geometry (5.G)

Classify two-dimensional figures into categories based on their properties.

3. Understand that attributes belonging to a category of two-dimensional figures also belong to all subcategories of that category. *For example, all rectangles have four right angles and squares are rectangles, so all squares have four right angles.*

Lesson	108	109	110	111	112	115	120
Exercise	108.2	109.2	110.2	111.2	112.2	115.7	120.5

Geometry (5.G)

Classify two-dimensional figures into categories based on their properties.

4. Classify two-dimensional figures in a hierarchy based on properties.

Lesson	108	109	110	111	112	115	120
Exercise	108.2	109.2	110.2	111.2	112.2	115.7	120.5

Standards for Mathematical Practice and Connecting Math Concepts

Connecting Math Concepts: Comprehensive Edition is a six-level series that is fully aligned with the Common Core State Standards for kindergarten through fifth grade. As its name implies, *Connecting Math Concepts* is designed to bring students to an understanding of mathematical concepts by making connections between central and generative concepts in the school mathematics curriculum, as defined by the Common Core State Standards (CCSS) for Mathematical Content. This document illustrates some of those connections as they relate to the eight CCSS for Mathematical Practices. Examples will be provided from each of the six levels of *Connecting Math Concepts (CMC)*, to illustrate how students engage in activities representative of the eight CCSS for Mathematical Practices.

MP1: MAKE SENSE OF PROBLEMS AND PERSEVERE IN SOLVING THEM.

CMC Level C (Grade 2): Comparison Story Problems

Students in the primary grades are just beginning to make sense of situations that can be expressed mathematically. Care must be taken to give them a way of thinking about situations that is not overly simplistic, resulting in a shallow understanding of the circumstance. For example, when considering two values in a statement such as *Sophia is 2 years older than Isabel,* students are likely to interpret the situation as implying addition because of the word *more.* Conversely, they are likely to interpret a statement such as *Isabel is 2 years younger than Sophia* as subtraction. The comparison statements do not imply either operation; they simply identify the larger and smaller of the two values being compared. Given one value, we can find the other by addition or subtraction. If the larger value is given, we subtract 2 (the difference) to find the smaller value. If the smaller value is given, we add the difference to find the larger value.

Students in *CMC* are taught to use this reasoning to make sense of word problems that compare. For example:

> Isabel is 2 years younger than Sophia.
> Isabel is 8 years old. How old is Sophia?

Rather than jumping into a solution attempt, students are taught to analyze and diagram the relationship between the two values, identify and substitute the value given in the problem, and base their solution strategy on the value that is not given. For this example, students first represent the relationship between the two values. Initial letters represent the values. The larger value is written at the end of a number-family arrow, with the smaller value and the difference shown on the arrow:

$$\underrightarrow{\quad 2 \qquad I \quad} S$$

The value given in the problem replaces the letter in the diagram:

$$\underrightarrow{\quad 2 \qquad \overset{8}{I} \quad} S$$

The smaller value is given, so we add to find Sophia's age:

$$\underrightarrow{\quad 2 \qquad \overset{8}{I} \quad} S \qquad \begin{array}{r} 2 \\ + 8 \\ \hline 1\,0 \end{array}$$

Here is the first part of the exercise from Level C Lesson 48, illustrating how the students apply the strategy.

(Teacher reference:)

a. Heidi has 17 more marbles than Bill has.
   Heidi has 48 marbles.
   How many marbles does Bill have?

b. Sarah made 10 more cupcakes than Maria made.
   Maria made 24 cupcakes.
   How many cupcakes did Sarah make?

c. Hank's car is 8 years older than Tim's car.
   Hank's car is 11 years old.
   How old is Tim's car?

d. Bob has 7 dollars less than Val has.
   Bob has 31 dollars.
   How many dollars does Val have?

You're going to work problems that tell about people and things. Remember, the first letter of each name is underlined.

b. Touch problem A. ✔
- Heidi has 17 more marbles than Bill has. Say the sentence. (Signal.) *Heidi has 17 more marbles than Bill has.*
- What's the letter for Heidi? (Signal.) *H.*
- What's the letter for Bill? (Signal.) *B.* Heidi has 17 more marbles than Bill has.
- Make the family with two letters and the number. ✔
  (Display:)   [48:8A]

  a. ___17___ ᴮ→H

Here's what you should have.

c. The next sentence in the problem says: Heidi has 48 marbles.
- Put a number for Heidi in the family. Then figure out how many marbles Bill has. Write that number in the family.
  (Observe students and give feedback.)

d. Check your work.
  (Add to show:)   [48:8B]

  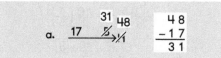

Here's what you should have.
- The problem asks: How many marbles does Bill have? What's the answer? (Signal.) *31.*

from Lesson 48, Exercise 8

Note that three different types of problems are worked in this problem set:

The smaller-value unknown with "more" language (items a and c), the bigger-value unknown with "more" language (item b), and the bigger-value unknown with "less" language (item d). These types are described in Table 1 in the *Common Core State Standards for Mathematics* document (excerpt below).

Bigger Unknown	Smaller Unknown
(Version with "more"): Julie has three more apples than Lucy. Lucy has two apples. How many apples does Julie Have?	(Version with "more"): Julie has three more apples than Lucy. Julie has five apples. How many apples does Lucy Have?
(Version with "fewer"): Lucy has 3 fewer apples than Julie. Lucy has two apples. How many apples does Julie have? $2 + 3 = ?, 3 + 2 = ?$	(Version with "fewer"): Lucy has 3 fewer apples than Julie. Julie has five apples. How many apples does Lucy have? $5 - 3 = ?, ? + 3 = 5$

Table 1 from the *Common Core State Standards for Mathematics* describes 15 different addition-subtraction situations, all of which students can solve with variations of a number-family strategy by the end of CMC Level C. Foundation work for number families is illustrated in Standard MP7.

MP2: REASON ABSTRACTLY AND QUANTITATIVELY.

CMC Level F (Grade 5): Algebraic Translation

As students progress thorough the levels of *CMC* they make sense of quantities and their relationships through a variety of representations. The earliest representations are pictorial and countable, but by Level C, students frequently use more abstract letter representations for quantities, relationships, and units named in problem situations. By Level E, these problem situations are extended to include the four basic operations, ratio and proportion, and unit conversion.

Students in *CMC Level F* make sense of quantities and their relationships through extensive work with algebraic translation, which they apply to a variety of word-problem contexts, including fractions of a group and probability.

For example:

- 2/3 of the cats are awake. 6 cats are sleeping. How many cats are there?
- There are 8 marbles in a bag; 5 are green. If you take 24 trials at pulling a marble from the bag without looking, how many times would you expect to draw a marble that is not green?

Students are well prepared to *decontextualize* and manipulate symbols independently through initial work with solving letter equations. They also can *contextualize* by attending carefully to the details of the problem to be solved in order to discriminate which specific letter equation a given problem requires.

Early work equips students with the algebraic skills to solve letter equations of the form 2/5 R = M with a substitution for either letter.

Here are the solution steps when R = 10:

$$\frac{2}{5} R = M$$

$$\frac{2}{5} (10) = M$$

$$\frac{20}{5} = M = 4$$

Here are the solution steps when M = 90:

$$\frac{2}{5} R = M$$

$$\frac{2}{5} R = 90$$

$$\left(\frac{5}{2}\right) \frac{2}{5} R = 90 \left(\frac{5}{2}\right)$$

$$R = \frac{450}{2} = 225$$

Students are first taught the definition and application of a reciprocal and the equality principle that states the following: If we change one side of an equation, we must change the other side in the same way.

Here is part of an exercise from Lesson 38.

h. (Display:) [38:3H]

$$\frac{5}{4} P = 10$$

$$1 P =$$

- Read the problem. (Signal.) *5/4 P = 10.*
  We have to figure out what 1 P equals.
- What do we change 5/4 into? (Signal.) *1.*
- So what do we multiply 5/4 by? (Signal.) *4/5.*
- What do we multiply the other side by? (Signal.) *4/5.*
  (Repeat until firm.)
  (Add to show:) [38:3I]

$$\left(\frac{4}{5}\right) \frac{5}{4} P = \frac{10}{1} \left(\frac{4}{5}\right)$$

$$1 P =$$

i. (Point right.) Say the problem for this side. (Signal.) *10 × 4/5.*
- Raise your hand when you know the fraction answer. ✔
- What's the fraction? (Signal.) *40/5.*
  (Add to show:) [38:3J]

$$\left(\frac{4}{5}\right) \frac{5}{4} P = \frac{10}{1} \left(\frac{4}{5}\right)$$

$$1 P = \frac{40}{5}$$

- Raise your hand when you know the number 1 P equals. ✔
- What does 1 P equal? (Signal.) *8.*
  (Add to show:) [38:3K]

$$\left(\frac{4}{5}\right) \frac{5}{4} P = \frac{10}{1} \left(\frac{4}{5}\right)$$

$$1 P = \frac{40}{5} = \boxed{8}$$

j. Remember, multiply both sides by the reciprocal. Then figure out what the letter equals.

from Lesson 38, Exercise 3

Connecting Math Concepts

Students have extensive practice with substitution and solving letter equations prior to the introduction of word problem applications. This level of proficiency enables them to manipulate symbols confidently once they have analyzed the problem and represented it with a letter equation.

Students first work with sentences that describe a fraction of a group, and write the letter equation. Here are two sentences and the corresponding equations from Lesson 46:

$\frac{3}{4}$ of the dogs were hungry.

$$\frac{3}{4} d = h$$

$\frac{1}{9}$ of the students wore coats.

$$\frac{1}{9} S = C$$

In subsequent lessons, students work the simplest type of word problem, where a number is given for one of the letters, and students solve for the other letter to answer the question the problem asks.

For example:

- 2/5 of the rabbits were white. There were 25 rabbits. How many white rabbits were there?

$$\frac{2}{5} r = w$$

$$\frac{2}{5}(25) = w = \frac{50}{5} = 10$$

$\boxed{\text{10 white rabbits}}$

Here's part of an exercise from Level F Lesson 78, where students work a complete problem that asks two questions. They assess their initial solution and refer back to the problem to ascertain which question relates to the letter solution, and which question remains to be answered.

b. Problem A: 2/7 of the dogs were sleeping. 8 dogs were sleeping. How many dogs were awake? How many dogs were there?
- Write the letter equation. Replace one of the letters with a number. Stop when you've done that much. ✔
- Everybody, read the letter equation. (Signal.) *2/7 D = S.*
- Read the equation with a number. (Signal.) *2/7 D = 8.*
 (Display:) [78:4C]

 a. $\frac{2}{7} d = s$

 $\frac{2}{7} d = 8$

 Here's what you should have.
- Work the problem and write the unit name in the answer. Remember to simplify before you multiply. (Observe students and give feedback.)
c. Check your work.
- What does D equal? (Signal.) *28.*
- Which question does that answer? (Signal.) *How many dogs were there?*
 (Add to show:) [78:4D]

 a. $\frac{2}{7} d = s$

 $\left(\frac{7}{2}\right)\frac{2}{7} d = \overset{4}{\cancel{8}}\left(\frac{7}{\cancel{2}}\right)$

 $d = 28$

 $\boxed{\text{28 dogs}}$

 Here's what you should have.
d. Now figure the answer to the other question. ✔ The other question is: How many dogs were awake?
- Everybody, say the subtraction problem you worked. (Signal.) *28 – 8.*
- What's the answer? (Signal.) *20.*
 (Display:) [78:4E]

 a. $\frac{2}{7} d = s$

 $\left(\frac{7}{2}\right)\frac{2}{7} d = \overset{4}{\cancel{8}}\left(\frac{7}{\cancel{2}}\right)$

 $d = 28$ $\begin{array}{r} 2\,8 \\ -\ \ 8 \\ \hline \end{array}$

 $\boxed{\text{28 dogs}}$ $\boxed{2\,0 \text{ awake dogs}}$

 Here's what you should have.

from Lesson 78, Exercise 4

On later lessons, students work more advanced problems where the problem gives a fraction for one part of the group but gives a number for the other part of the group. For example:

- 2/7 of the dogs were running. 20 dogs were *not* running. How many dogs were running? How many dogs were there in all?

Students must analyze the problem carefully to discriminate it from the earlier problem types that involve only two names (e.g., dogs and running dogs). The number given is for dogs *not running,* so rather than writing the basic equation 2/7 d = r, students write the complementary letter equation 5/7 d = n. (5/7 of the dogs were *not running.*)

Students first practice discriminating between sentence pairs that give two names and those that give three names.

For example:

- 3/8 of the children are boys. There are 15 boys. (2 names: 3/8 c = b)
- 3/8 of the children are boys. There are 15 girls. (3 names: 5/8 c = g)

Here is part of the exercise from Lesson 90, where students solve a complete problem involving three names.

g. (Display:)   [90:7C]

> $\frac{2}{7}$ of the dogs were running. 20 dogs were not running. How many dogs were running? How many dogs were there in all?

New problem: 2/7 of the dogs were running. 20 dogs were not running. How many dogs were running? How many dogs were there in all?
This problem asks two questions. You work it the same way you work the other problems with three names. You write the equation for the name that has a number and solve the equation. That answers one of the questions.

h. Write the equation and solve it. Write the answer with a unit name. Stop when you've done that much.
   (Observe students and give feedback.)

i. I'll read the questions the problem asks. You'll tell me which question you can now answer: How many dogs were running? How many dogs were there in all?
- Which question can you answer? (Signal.) *How many dogs were there in all?*
- What's the answer? (Signal.) *28 dogs.*

j. (Add to show:)   [90:7D]

> $\frac{2}{7}$ of the dogs were running. 20 dogs were not running. How many dogs were running? How many dogs were there in all?
>
> $$\frac{5}{7} d = n$$
> $$\left(\frac{7}{5}\right) \frac{5}{7} d = \overset{4}{\cancel{20}} \left(\frac{7}{\cancel{5}}\right)$$
> $$d = 28$$
>
> $\boxed{28 \text{ dogs}}$

Here's what you should have.

k. The other question is: How many dogs were running?
   Now you can figure out how many dogs were running.
- Raise your hand when you can say the problem. ✔
- Say the problem. (Signal.) *28 – 20.*

l. Figure out the answer and write the unit name—running dogs. ✔
- Everybody, how many dogs were running? (Signal.) *8 running dogs.*

from Lesson 90, Exercise 7

Once students have set up the correct equation, they follow familiar solution steps. They figure out how many dogs there were in all (which is the answer to the second question). They then subtract to figure out the number of dogs that were running.

The strategy builds the habit of a coherent representation of the problem at hand (a familiar equation form), requires students to consider the units for the quantities represented by the letters as they solve the equation, consider both questions the problem asks, and then determine which has been answered and which remains to be answered.

Coherent representation is further strengthened in later applications to new domains such as probability. Students first learn to create a probability fraction based on the composition of the set. This fraction gives the likelihood of a particular object being drawn from the set on any given trial. The following examples from lesson 96 require students to construct probability fractions for given sets of objects and to construct a set of objects given the probability fraction.

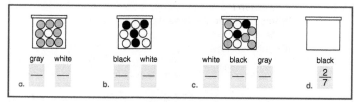

from Workbook Lesson 96, Part 1

For item a, students write the fraction 8/9 for gray and 1/9 for white. For item d, students draw 7 marbles, and shade 2 of them black.

With this background it is a simple transition to work problems that involve trials using the familiar equation form.

Here are two examples and the equations students work from Lesson 98.

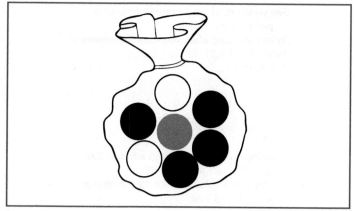

from Textbook Lesson 98, Part 2

- If you took trials until you pulled out four black marbles, about how many trials would you take?

$$\frac{4}{7} t = b$$

$$\frac{4}{7} t = 4$$

- If you take 98 trials, about how many black marbles would you expect to pull out?

$$\frac{4}{7} t = b$$

$$\frac{4}{7} (98) = b$$

Students already have the mathematical tools to solve the equations, and learning a wide range of applications for equation form builds flexibility in the application of those tools.

MP3: CONSTRUCT VIABLE ARGUMENTS AND CRITIQUE THE REASONING OF OTHERS.

CMC Level C (Grade 2): Column Addition/Subtraction

As students develop mathematically, they build a basis for constructing arguments to justify the mathematical procedures they use, or the steps they take to perform an algorithm. They are able to decide whether variations and deviations from the familiar progression of steps make sense. In a column addition or subtraction problem that involves regrouping, students should display an understanding of the underlying place value that conserves the numbers they are manipulating.

Students in *CMC Level C* work extensively with place value to build the conceptual basis for the steps they take in addition/subtraction regrouping problems, for example:

$$
\begin{array}{r}
\overset{1}{2}\,8 \\
+\,4\,5 \\
\hline
7\,3
\end{array}
\quad \text{or} \quad
\begin{array}{r}
\overset{8}{\cancel{9}}\overset{}{1}2 \\
-\,3\,6 \\
\hline
5\,6
\end{array}
$$

Students are familiar with identifying the tens digit and ones digit of two-digit numbers and with the place value of each digit. When applied to column addition, the students simply write a 2-digit answer to the problem for the ones column as the "tens digit" and the "ones digit" in the appropriate order and in the appropriate column. Here's part of an early exercise from Lesson 21.

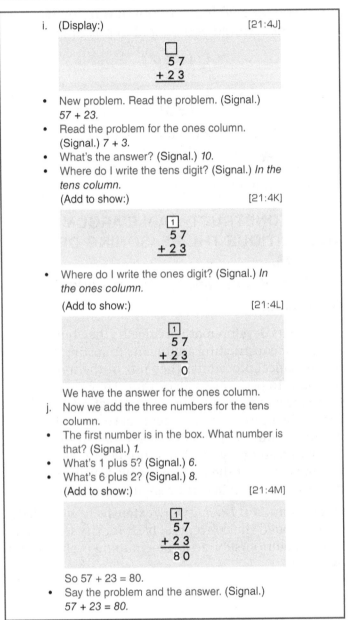

So 57 + 23 = 80.
- Say the problem and the answer. (Signal.) *57 + 23 = 80.*

from Lesson 21, Exercise 4

Having practiced this procedure to automaticity for more than 90 lessons, students are ready to discuss the steps they take and critique deviations from the procedure. Here is an exercise from Lesson 114.

a. (Display:) [114:4A]

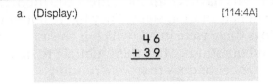

I'm going to ask you a lot of questions about working this problem.
- Read the problem. (Signal.) *46 + 39.*
- Read the problem for the ones. (Signal.) *6 + 9.*
- What's the answer? (Signal.) *15.*
b. What if I just put 15 in the ones column? (Add to show:) [114:4B]

$$\begin{array}{r} 4\,6 \\ +\,3\,9 \\ \hline 1\,5 \end{array}$$

- Can I do this? (Signal.) *No.*
- What's wrong with this? (Call on a student. Idea: *15 is too many to have in the ones column.*)
- What's the largest number we can have in the ones column? (Call on a student. *9.*)
c. The answer for the ones is 15. Everybody, say the place-value addition for 15. (Signal.) *10 + 5.*
- I can't write **both** digits in the ones column. Where do I write the **ones** digit of 15? (Signal.) *In the ones column.*
- Where do I write the **tens** digit of 15? (Signal.) *In the tens column.* (Change to show:) [114:4C]

$$\begin{array}{r} 4\,6 \\ +\,3\,9 \\ \hline 1\,5 \end{array}$$

- (Point to **1.**) Can I write the tens digit here? (Signal.) *No.*
- Why not? (Call on a student. Idea: *You have to add it to the tens.*)
d. Where do I write it? (Call on a student. Idea: *At the top of the tens column.*) (Change to show:) [114:4D]

$$\begin{array}{r} \overset{1}{4}\,6 \\ +\,3\,9 \\ \hline 5 \end{array}$$

- Everybody, read the problem for the tens. (Signal.) *1 + 4 + 3.*
- Raise your hand when you know the answer. ✔
- What's 1 + 4 + 3? (Signal.) *8.* (Add to show:) [114:4E]

$$\begin{array}{r} \overset{1}{4}\,6 \\ +\,3\,9 \\ \hline 8\,5 \end{array}$$

- Read the problem we started with and the answer. (Signal.) *46 + 39 = 85.*
- Say the place-value addition for the answer. (Signal.) *80 + 5 = 85.*

from Lesson 114, Exercise 4

A similar development is taught for subtraction. Before students work with column problems that require renaming the top number, student work on the conservation of two-digit values based on place value.

Here's an introductory exercise from Level C Lesson 26.

a. You're going to say the place-value fact for different numbers.
- Say the place-value fact for 53. (Signal.) *50 + 3 = 53.*
- Say the place-value fact for 29. (Signal.) *20 + 9 = 29.*
- Say the place-value fact for 70. (Signal.) *70 + 0 = 70.*
- Say the place-value fact for 71. (Signal.) *70 + 1 = 71.*
- Say the place-value fact for 17. (Signal.) *10 + 7 = 17.*
  (Repeat until firm.)

b. (Display:) [26:5A]

$$30 + 6 = 36$$
$$40 + 8 = 48$$
$$70 + 1 = 71$$
$$50 + 3 = 53$$

c. (Point to **30**.) You're going to subtract 10 from this number.
- What's 30 – 10? (Signal.) *20.*
d. (Point to **40**.) You're going to subtract 10 from this number.
- Say the problem. (Signal.) *40 – 10.*
- What's the answer? (Signal.) *30.*

e. (Point to **70**.) You're going to subtract 10 from this number.
- Say the problem. (Signal.) *70 – 10.*
- What's the answer? (Signal.) *60.*
f. (Point to **50**.) You're going to subtract 10 from this number.
- Say the problem. (Signal.) *50 – 10.*
- What's the answer? (Signal.) *40.*
  (Repeat until firm.)
g. You've subtracted 10 from the tens number. You're going to add 10 to the ones number.
- (Point to **6**.) What's 10 + 6? (Signal.) *16.*
- (Point to **8**.) What's 10 + 8? (Signal.) *18.*
- (Point to **1**.) What's 10 + 1? (Signal.) *11.*
- (Point to **3**.) What's 10 + 3? (Signal.) *13.*
h. This time you're going to subtract 10 from the tens number and add that 10 to the ones number.
i. (Point to **30**.) What's 30 – 10? (Signal.) *20.*
  (Change to show:) [26:5B]

$$30 + 6 = 36$$
$$20$$

- What's 10 + 6? (Signal.) *16.*
  (Add to show:) [26:5C]

$$30 + 6 = 36$$
$$20 + 16 = 36$$

- Read the new place-value fact for 36. (Signal.) *20 + 16 = 36.*
j. (Display:) [26:5D]

$$40 + 8 = 48$$

- (Point to **40**.) What's 40 – 10? (Signal.) *30.*
  (Add to show:) [26:5E]

$$40 + 8 = 48$$
$$30$$

- What's 10 + 8? (Signal.) *18.*
  (Add to show:) [26:5F]

$$40 + 8 = 48$$
$$30 + 18 = 48$$

- Read the new place-value fact for 48. (Signal.) *30 + 18 = 48.*

from Lesson 26, Exercise 5

Students then practice rewriting 2-digit numbers in isolation to show the "new fact." Here's part of the introduction from Lesson 33.

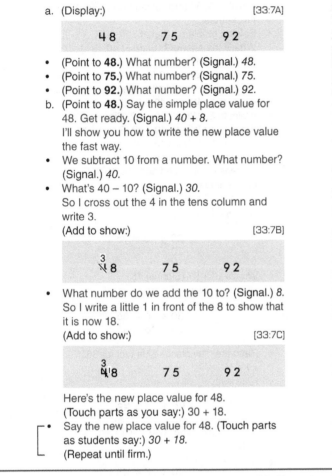

a. (Display:) [33:7A]

   48   75   92

- (Point to **48**.) What number? (Signal.) *48.*
- (Point to **75**.) What number? (Signal.) *75.*
- (Point to **92**.) What number? (Signal.) *92.*

b. (Point to **48**.) Say the simple place value for 48. Get ready. (Signal.) *40 + 8.*
I'll show you how to write the new place value the fast way.

- We subtract 10 from a number. What number? (Signal.) *40.*
- What's 40 – 10? (Signal.) *30.*
So I cross out the 4 in the tens column and write 3.
(Add to show:) [33:7B]

- What number do we add the 10 to? (Signal.) *8.*
So I write a little 1 in front of the 8 to show that it is now 18.
(Add to show:) [33:7C]

Here's the new place value for 48.
(Touch parts as you say:) 30 + 18.
- Say the new place value for 48. (Touch parts as students say:) *30 + 18.*
(Repeat until firm.)

from Lesson 33, Exercise 7

Given this background and practice that continues for 15 lessons, students have a conceptual framework to make sense of and justify the steps in the subtraction algorithm, which begins on Lesson 43.

After the algorithm is established and practiced for 70 lessons, students also discuss and critique the steps in the algorithm and deviations from the procedure. Here is an exercise from Lesson 115.

a. (Display:) [115:8A]

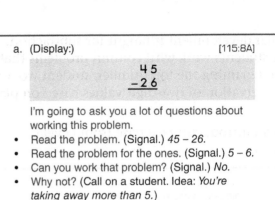

  45
 –26

I'm going to ask you a lot of questions about working this problem.
- Read the problem. (Signal.) *45 – 26.*
- Read the problem for the ones. (Signal.) *5 – 6.*
- Can you work that problem? (Signal.) *No.*
- Why not? (Call on a student. Idea: *You're taking away more than 5.*)

b. So I'll just change 5 to 15.
(Add to show:) [115:8B]

  4'5
  26

- Can I change the problem that way? (Signal.) *No.*
- Why not? (Call on a student. Idea: *The top number does not equal 45.*)
- Say the place-value addition for the top number. (Signal.) *40 + 15.*
- Does that equal 45? (Signal.) *No.*

c. What do I have to do to fix the top number? (Call on a student. Idea. *Take 10 from 40.*)
(Add to show:) [115:8C]

    3
  4'5
 –26

- Say the new place value for the top number. (Signal.) *30 + 15.*
- Does it equal 45? (Signal.) *Yes.*

d. Say the new problem for the ones. (Signal.) *15 – 6.*
- What's the answer? (Signal.) *9.*
(Add to show:) [115:8D]

    3
  4'5
 –26
   9

e. Say the new problem for the tens. (Signal.) *3 – 2.*
- What's the answer? (Signal.) *1.*
(Add to show:) [115:8E]

    3
  4'5
 –26
  19

- What's the top number we started with? (Signal.) *45.*
- Say the problem we started with and the answer. (Signal.) *45 – 26 = 19.*

from Lesson 115, Exercise 8

Students in Level C also check answers to addition and subtraction problems. For single-digit addition problems, students add the numbers in a different order to confirm that an answer is correct. Given the problem and (incorrect) answer:

$$\begin{array}{r} 2 \\ 4 \\ + 9 \\ \hline 1\,6 \end{array}$$

Students work the problem from the "bottom up" to check the answer. If this addition results in a different answer (in this case 15) students also work the problem from the "top down" to verify that the answer of 16 is incorrect.

Students also check answers to multi-digit problems using inverse operations. Given the problem and (incorrect) answer:

$$\begin{array}{r} 5\,8 \\ + 2\,1\,8 \\ \hline 2\,6\,6 \end{array}$$

Students use two of the numbers to work a subtraction problem (either 266 − 218, or 266 − 58) and observe whether the subtraction answer is the third number in the original problem. If it is not, they conclude that the answer to the original addition problem is wrong, and they rework the problem to figure out the correct answer.

These strategies enable students to check and justify their answers, or evaluate and critique the work of other students by approaching the problem a "different way." Once they have identified that there is an error, reworking the original problem provides an opportunity to figure out and explain precisely where an error was made.

MP4: MODEL WITH MATHEMATICS.
CMC Level D (Grade 3): Multiplication/ Division Story Problems

Students in *CMC Level D* learn to apply the mathematics they know to a range of word problems that represent situations arising in everyday life. This section will illustrate how students model problems solved by multiplication or division.

Students identify the important quantities in a situation and map their relationship with a number-family diagram showing one name that represents the product, and one name that represents a factor in a multiplicative relationship. This modeling parallels the work with addition/subtraction number families described in MP1 above.

Here's a basic problem:

Each box has 6 pencils. There are 24 pencils. How many boxes are there?

Students represent the relationship between the items named in the problem: boxes and pencils. Initial letters represent the values. The larger quantity is written at the end of the number-family arrow. There are 6 times more pencils than boxes, so *p* is at the end of the number family arrow:

The smaller quantity *(b)* is written on the arrow, and the relationship number (6) is the first number in the diagram.

The completed diagram shows that there are six times more pencils than boxes. With the relationship modeled in this way, when students substitute for the value that is given (24 pencils), they see that they must divide to find the smaller quantity. (Note that the diagram also resembles a division problem).

Conversely, when students substitute for the smaller quantity, they multiply to find the larger quantity. For example:

Each box has 6 pencils. There are 50 boxes. How many pencils are there?

The relationship that students map is the same as above:

Students substitute for b:

This diagram translates into a multiplication problem to figure out the larger quantity.

By carefully mapping the situation and relating the quantities before deciding which operation to perform, students can reliably solve problems that typically cause confusion for students. In the latter problem, students may read the word *each* and decide that the problem involves multiplication or division, but they are tempted to divide because the numbers given in the problem are from a familiar division fact (24 ÷ 6). By mapping the relationship before they work the problem, students in *CMC* are much less likely to make this mistake.

Students write number families from sentences for several lessons to establish the mapping strategy. They are then ready to work complete problems. Here's part of the exercise from Level D Lesson 72, where students work the following problems:

- Each dime is worth 10 cents. Tom had 8 dimes. How many cents did he have?
- Every chair has 4 legs. There are 12 legs. How many chairs are there?

b. Touch problem A in your textbook. ✔
I'll read it: Each dime is worth 10 cents. Tom had 8 dimes. How many cents did he have?
- Raise your hand when you know which sentence tells about each or every. ✔
- Read that sentence. (Signal.) *Each dime is worth 10 cents.*
- Is the big number dimes or cents? (Signal.) *Cents.*
  Yes, there are more cents than dimes.
- Write two letters and a number in the number family.
  (Observe students and give feedback.)
  (Display:) [72:6J]

Here's what you should have.
c. The problem gives another number for dimes.
- Look at the problem. Raise your hand when you know the number. ✔
- What's the number? (Signal.) *8.*
- Write the number for dimes in the family. ✔
  (Add to show:) [72:6K]

Here's what you should have.
d. Do you multiply or divide to find the missing number? (Signal.) *Multiply.*
- Start with the 2-digit number and say the problem you'll work. (Signal.) *10 × 8.*
- Write the problem next to the number family. Work the problem. Write the unit name in the answer.
  (Observe students and give feedback.)
- You worked the problem 10 times 8. What's the whole answer? (Signal.) *80 cents.*
  Yes, Tom had 80 cents.
e. Check your work.
  (Add to show:) [72:6L]

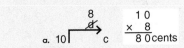

Here's what you should have.

f. Count down 4 lines on your lined paper. Write the letter B and a multiplication number family arrow. ✔
- Touch problem B. ✔
  I'll read the problem: Every chair has 4 legs. There are 12 legs. How many chairs are there?
- Raise your hand when you know which sentence tells about each or every. ✔
- Read that sentence. (Signal.) *Every chair has 4 legs.*
- Write two letters and a number in the family. (Observe children and give feedback.)
  (Display:) [72:6M]

Here's what you should have.
g. The problem gives a number for one of the letters.
  Look at the problem. Raise your hand when you know the number. ✔
- What's the number? (Signal.) *12.*
- Is that number for chairs or legs? (Signal.) *Legs.*
- Write that number in the family. ✔
  (Add to show:) [72:6N]

Here's what you should have.
h. Say the problem you'll work to find the number of chairs. (Signal.) *12 divided by 4.*
- Write the problem next to the number family. Work the problem. Write the unit name in the answer.
  (Observe students and give feedback.)
- Read the problem and the whole answer. (Signal.) *12 divided by 4 = 3 chairs.*
  Yes, there were 3 chairs.
i. Check your work.
  (Add to show:) [72:6O]

Here's what you should have.

from Lesson 72, Exercise 6

Note that the same strategy accommodates money problems that convert a coin value into cents and vice versa. (For example: Tom had 80 cents in dimes. How many dimes did he have?)

Later in Level D, students learn to work multiplication and division problems that use the word "times." Here are two items and the student work from Lesson 91.

a. The yellow snake was 9 times as long as the red snake.
The red snake was 10 inches long.
How long was the yellow snake?

$$\begin{array}{r} 1\,0 \\ \times\ \ 9 \\ \hline 9\,0\ feet \end{array}$$

b. There were 6 times as many spoons as forks.
There were 66 spoons.
How many forks were there?

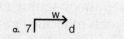

Item b results in a division problem. Again, representing the related values in the problem before choosing the operation reduces the likelihood that students will multiply instead of divide because the problem involves the word "times."

Another real-life application for the number-family mapping taught in Level D is unit conversion. Students make a number family for measurement facts. For example:

1 foot equals 12 inches.

Students identify that there are more inches than feet, so the number family relationship is:

Here are the facts that student map in Lesson 110:

- 1 week equals 7 days.
- 1 gallon equals 4 quarts.
- 1 quarter equals 25 cents.
- 1 pound equals 16 ounces.

Here's the first part of the exercise.

> **b.** You're going to make number families for measurement facts.
> - Touch and read fact A. (Signal.) *1 week = 7 days.*
> - Are there more weeks or days? (Signal.) *Days.*
> - So which is the big number? (Signal.) *Days.*
> - Make the family for the fact about weeks and days.
>   (Observe students and give feedback.)
> **c.** Check your work.
>   (Display:)   [110:4A]
>
> a. 7⌐──w──→ d
>
> Here's what you should have.
> 7 and weeks are small numbers. Days is the big number.
> **d.** Read fact B. (Signal.) *1 gallon = 4 quarts.*
> - Are there more gallons or quarts?
>   (Signal.) *Quarts.*
> - So which is the name for the big number in the family? (Signal.) *Quarts.*
> - Make the family.
>   (Observe students and give feedback.)
> **e.** Check your work.
>   (Display:)   [110:4B]
>
> b. 4⌐──G──→ q
>
> Here's what you should have.
> 4 and gallons are small numbers. Quarts is the big number.

from Lesson 110, Exercise 4

On later lessons, students work complete problems. For example:

- 1 nickel equals 5 cents. How many nickels is 150 cents?
- 1 day equals 24 hours. How many hours is 9 days?

Students divide to find the number of nickels in the first item, and multiply to find the number of hours in the second item. Traditionally these problem types cause confusion for elementary students because they are counterintuitive. To solve the problems correctly, students *multiply* when the problem asks about the *smaller* unit, and *divide* when the problem asks about the *larger* unit. The number-family analysis provides students in *CMC* with a consistent and reliable way to tackle these and the full range of other situations that call for multiplication or division.

MP5: USE APPROPRIATE TOOLS STRATEGICALLY.

CMC Level E (Grade 4): Geometry

CMC teaches students self-reliance through the extensive use of conceptual strategies that can be applied using paper and pencil. Students generate solutions to a wide range of problems presented in the Textbook and Workbook. Tasks often move systematically from a workbook (more highly structured) to a textbook setting as students develop the conceptual tools to tackle problems more independently. For example, initial work with area and perimeter appears in the Workbook and only requires students to write the multiplication or addition problem. Here is the Workbook part and student work from Lesson 8.

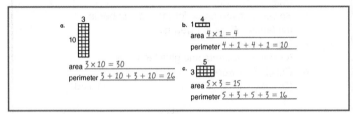

from Workbook Lesson 8 AK, Part 2

In later lessons, students respond to diagrams shown in the Textbook and work much more advanced problems. Below is a set of examples from Lesson 93 Textbook. For some problems, students find the area of a rectangle. For others they find the length of a side of the rectangle. They show their answers with appropriate linear or square units.

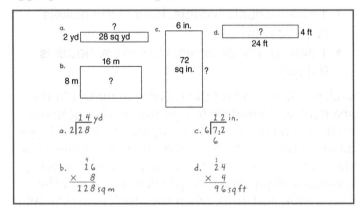

from Textbook Lesson 93 AK, Part 2

The work with area and perimeter culminates with story problems for which students sketch a rectangle to represent the problem. For example:

> a. A farmer wants to put a fence around a rectangular garden. The garden is 132 yards wide and 58 yards long. How many yards of fencing does the farmer need to put a fence around that garden?
>
> b. Jim has enough paint to cover 100 square feet of a wall. The wall he wants to paint is 8 feet tall. If he has just enough paint to cover the wall, how long is the wall?

from Textbook Lesson 126, Part 4

Here is part of the exercise.

> b. Read problem A. (Call on a student.) *A farmer wants to put a fence around a rectangular garden. The garden is 132 yards wide and 58 yards long. How many yards of fencing does the farmer need to put a fence around that garden?*
> * Tell me the length of the longest side of the rectangle. Get ready. (Signal.) *132 yards.*
> * What's the length of the other side? (Signal.) *58 yards.*
> * Make a sketch of the rectangle on your lined paper with the sides labeled. Make sure you label the longest side of your sketch with the biggest length.
> (Observe students and give feedback.)
> (Display:)   [126:6A]

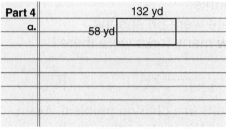

> Here's a sketch of the garden.
> * The question asks: How many yards of fencing does the farmer need to put a fence around that garden? Does that question ask about the area or perimeter? (Signal.) *Perimeter.*
> * Start with 132 and say the problem for finding the perimeter. Get ready. (Signal.) *132 + 58 + 132 + 58.*
>
> c. Read problem B. (Call on a student.) *Jim has enough paint to cover 100 square feet of a wall. The wall he wants to paint is 8 feet tall. If he has just enough paint to cover the wall, how long is the wall?*
> * The length of one of the sides of the wall is given. What's that length? (Signal.) *8 feet.*
> * Does the problem give another length or the area of the rectangle? (Signal.) *The area.*
> * What's the area of the rectangle? (Signal.) *100 square feet.*
> * Make a sketch of the rectangle on your lined paper with a side labeled and the area labeled. (Observe students and give feedback.)
> (Display:)   [126:6B]

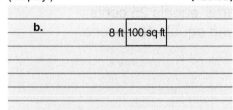

> Here's a sketch of the wall.
> * The question asks: How long is the wall? Say the problem for finding the length of the wall. Get ready. (Signal.) *100 ÷ 8.*
>
> d. Write the problem for each word problem below each rectangle and figure out the answer.
> (Observe students and give feedback.)

from Lesson 126, Exercise 6

Students work the addition or division problem and write the answer with a unit name.

Students are also familiar with tools appropriate to their grade, such as a protractor and ruler, which they use to explore angles, rays, line segments, and lines. They use a protractor first to measure given angles and then to construct angles. They use a ruler to complete the angle constructions and to graph straight lines on a coordinate system.

The protractor is introduced in Level E on Lesson 121. Before students physically use a protractor, they work with diagrams that show protractors properly placed. Students are often confused by the two rows of numbers on the protractor, so care is taken to teach students how to discriminate when one row or the other is used to measure an angle. Here is part of the introduction from Lesson 121.

---

a. In the last lesson, you learned about line segments and when they intersect.
- Think of a triangle. Are the sides of a triangle lines or line segments? (Signal.) *Line segments.*
- Do the line segments **intersect** at the corners or in the middle of a triangle? (Signal.) *At the corners.*
  (Repeat until firm.)
b. You're going to learn about a tool that's used to measure angles. The tool is called a protractor. Say **protractor.** (Signal.) *Protractor.*
- What's the name of a tool used to measure angles? (Signal.) *(A) protractor.*
(Display:) [121:5A]

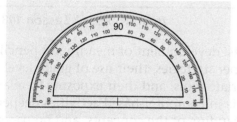

Here's a picture of a protractor. There are two sets of number scales on this protractor. The tens numbers on the inside scale start at zero here (touch 180 over zero) and go to 180 here (touch zero over 180). The tens numbers on the outside scale start at zero here (touch zero over 180) and go to 180 here (touch 180 over zero).
- Both number scales are the same for one number. What number? (Signal.) *90.* (Touch 90.)
- Everybody, what do you use a protractor for? (Signal.) *To measure angles.*
- What's the name of this tool? (Signal.) *(A) protractor.*

---

c. (Add to show:) [121:5B]

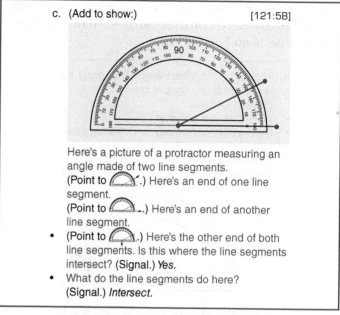

Here's a picture of a protractor measuring an angle made of two line segments.
(Point to ⌒ .) Here's an end of one line segment.
(Point to ⌒ .) Here's an end of another line segment.
- (Point to ⌒ .) Here's the other end of both line segments. Is this where the line segments intersect? (Signal.) *Yes.*
- What do the line segments do here? (Signal.) *Intersect.*

from Lesson 121, Exercise 5

Similarly, students learn the critical features of properly placing a protractor through positive and negative examples. The lines must intersect at the center marker, and one of the lines must go through zero. This instruction ensures that students are able to use the tool accurately. Here are the examples from Lesson 123.

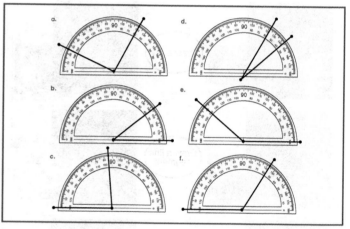

from Textbook Lesson 123, Part 2

Having measured angles for several lessons, students are well prepared to construct angles. Here is part of the exercise from Lesson 128.

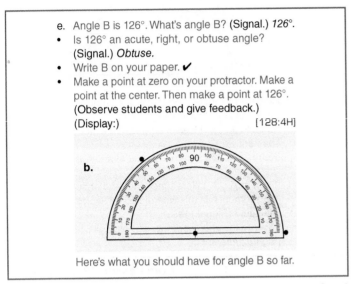

e. Angle B is 126°. What's angle B? (Signal.) *126°.*
- Is 126° an acute, right, or obtuse angle? (Signal.) *Obtuse.*
- Write B on your paper. ✔
- Make a point at zero on your protractor. Make a point at the center. Then make a point at 126°. (Observe students and give feedback.) (Display:)   [128:4H]

b.

Here's what you should have for angle B so far.

from Lesson 128, Exercise 4

Students also work with rays, lines, and line segments in the Practice Software. Here are three examples of a task students complete (using a line-drawing tool) to discriminate between a ray, a line, and a segment:

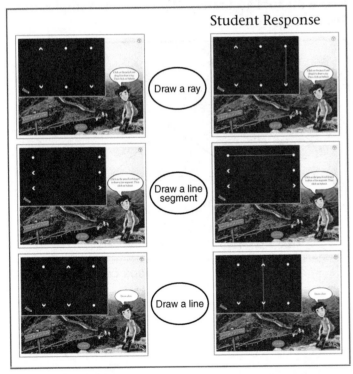

Student Response

Draw a ray

Draw a line segment

Draw a line

Following each example, students then make a line, line segment, or ray on the computer screen that is parallel or perpendicular to the first object they made.

Students also use a ruler to graph lines on the coordinate system. Here's the function table and coordinate system from Lesson 103.

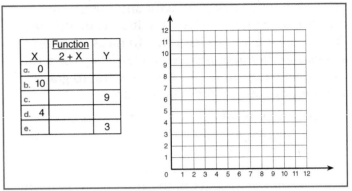

X	Function 2 + X	Y
a. 0		
b. 10		
c.		9
d. 4		
e.		3

from Workbook Lesson 103, Part 1

Students plot two points, draw the line, and inspect the line to complete the missing X or Y values in the table. Here is the table and coordinate system showing the student work.

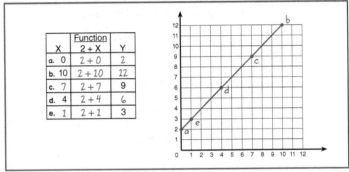

X	Function 2 + X	Y
a. 0	2 + 0	2
b. 10	2 + 10	12
c. 7	2 + 7	9
d. 4	2 + 4	6
e. 1	2 + 1	3

from Workbook Lesson 103 AK, Part 1

Students' development of meaningful pencil-and-paper strategies, their use of grade-level appropriate tools, and their exposure to related concepts in computer-based activities deepens their understanding of the concepts, and prepares them for more advanced applications in later grades.

MP6: ATTEND TO PRECISION.

CMC Level A (Kindergarten): Equality and Equations.

Precise communication in mathematics is necessary at every grade level, beginning in kindergarten. In *CMC Level A*, students are taught to understand the meaning of the mathematical symbols they use, including the equal sign. Students in *CMC Level A* learn a concise and clear definition of equality: "You must have the same number on one side of the equals that you have on the other side." This rule is applied first in the simplest context: $4 = \blacksquare$ or $\blacksquare = 7$. Here's part of the introduction:

a. (Write on the board:) [19:6A]

=

- (Point to =.) Everybody, what's this? (Touch.) *Equals.*
- Is equals a number? (Touch.) *No.*
  Right. It isn't.
- Here's the rule about equals. (Point to the space on the left.) You must have the same number on this side of the equals (point to the right side of the equals) and on this side of the equals.

b. Listen. (Point left.) If we have 4 on this side of the equals, (point right) we must have 4 on this side of the equals.
- (Point left.) Listen: If we have 4 on this side of the equals, (point right) how many must we have on this side of the equals? (Touch.) *4.*
  Yes, 4.

c. (Point right.) If we have 10 on this side of the equals, (point left) how many must we have on this side of the equals? (Touch.) *10.*
- (Point left.) If we have 15 on this side of the equals, (point right) how many must we have on this side of the equals? (Touch.) *15.*
  (Repeat steps b and c until firm.)

d. (Write to show:) [19:6B]

4 = ☐

- This says **4 equals box.** What does it say? (Touch.) *4 equals box.*
- There's a number on one side of the equals. What number? (Signal.) *4.*
  Yes, 4.

e. (Point to **4**.) If we have 4 on that side of the equals, (point to box) how many must there be on this side of the equals? (Touch.) *4.*
  Yes, that's the number that goes in the box.
- (Write to show:) [19:6C]

4 = ☐4☐

  (Point to 4 = ☐4☐.) Now this says (touch each symbol as you read) 4 equals 4.
- What does it say? (Touch symbols.) *4 equals 4.*

f. (Write on the board:) [19:6D]

☐ = 2

This says **box equals 2.**
- There's a number on one side of the equals. What number? (Signal.) *2.*
- (Point to 2.) If there are 2 on this side of the equals, (point to box) how many must there be on this side of the equals? (Touch.) *2.*
  Yes, that's the number that goes in the box.
- (Write to show:) [19:6E]

☐2☐ = 2

- (Point to ☐2☐ = 2.) What does this say now? (Touch symbols.) *2 equals 2.*

from Lesson 19, Exercise 6

By Lesson 25 students have learned to discriminate whether sides are equal by counting lines on one side of the equal sign and comparing that number with a number on the other side of the equal sign. If the sides are not equal, they cross out the equal sign.

Here are 3 examples from the student worksheet and part of the exercise:

$$5 = |||||| \qquad |||||||| = 9 \qquad |||| = 4$$

There are equals in the row next to the duck. A number is on one side of each equals, and lines are on the other side. You'll cross out an equals if the sides are not equal.
- Touch the first equals. ✔
  You'll touch and count the lines.
- Finger over the first line. ✔
- Get ready. (Tap 6.) *1, 2, 3, 4, 5, 6.*
- How many lines? (Signal.) *6.*
- Touch the number on the other side. ✔
- What number? (Signal.) *5.*
  It says 5 equals 6.
- So are the sides equal? (Signal.) *No.*
  So you cross out that equals.
c. Put your pencil on the big ball and make one cross-out line. (Observe children and give feedback.)
  (Teacher reference:)

  ✐5 ≠||||||

d. Touch the next equals. ✔
  You'll touch and count the lines.
- Finger over the first line. ✔
- Get ready. (Tap 7.) *1, 2, 3, 4, 5, 6, 7.*
- How many lines? (Signal.) *7.*
- Touch the number on the other side. ✔
- What number? (Signal.) *9.*
  It says 7 equals 9.
- So are the sides equal? (Signal.) *No.*
- Make one cross-out line through the equals. (Observe children and give feedback.)
e. Touch the last equals. ✔
  You'll touch and count the lines.
- Finger over the first line. ✔
- Get ready. (Tap 4.) *1, 2, 3, 4.*
- How many lines? (Signal.) *4.*
- Touch the number on the other side. ✔
- What number? (Signal.) *4.*
- What does it say? (Signal.) *4 equals 4.*
- So are the sides equal? (Signal.) *Yes.*
  The sides are equal, so you don't cross out the equals.

from Lesson 25, Exercise 8

Through these early exercises, the students learn the precise meaning of the symbol. If the conditions for equality are not met, they cross out the sign.

Next, students learn to complete equality statements, either by drawing lines for a numeral or by writing the numeral for the lines.

$$8 = \qquad ||| =$$

$$= |||||| \qquad = 2$$

This work lays the foundation for the operations that students perform in addition and subtraction. Students learn to modify sides of an equation to make sides equal:

For example:

$$3 = |||||$$

Students cross out two lines to make the sides equal:

$$3 = |||卄$$

Other items show lines crossed out:

$$||卄 = \qquad |||||||卄 =$$

Students record the number of lines not crossed out on the other side of the equals.

This work prepares students to incorporate the concept of equality into the subtraction strategy they will learn. For the problem $5 - 1 = \blacksquare$, students "start with 5 and take away 1."

They draw 5 lines and cross out 1 line:

$$5 - 1 = \blacksquare$$

$$||||卄$$

They count the lines not crossed out and apply the equality principle: They must have the same number on one side of the equals that they have on the other side. They write 4 in the box to make the sides equal. As each new problem type is introduced for addition or subtraction, the meaning of the equal sign is consolidated as students use the symbol consistently and appropriately

This careful foundation guards against a lack of understanding often displayed by older students who struggle with algebra. These students fail to develop a clear understanding of the equality principle. They may consider the equal sign to be more of a "punctuation mark" in a math statement than a symbol representing a powerful mathematical concept. This causes difficulty when they begin algebra, as the equality principle is the basis for modifying each side of an equation in the same way to conserve equality. The work in early levels of CMC prepares students to understand and apply the equality principle in a variety of contexts, in CMC and beyond.

MP7: LOOK FOR AND MAKE USE OF STRUCTURE.

CMC Levels B–D (Grades 1–3): Number family Addition/Subtraction Facts.

Students are taught to recognize the structure and organization of the number system in all levels of *Connecting Math Concepts*. They use patterns and relationships to master the basic facts in all four operations. For example, rather than learning the 200 basic addition/subtraction facts as isolated items of information, students learn related sets of facts through the concept of the number family.

The number family $\underset{\longrightarrow}{5 \qquad 2}7$ generates 4 facts:

$$5 + 2 = 7 \qquad 7 - 2 = 5$$
$$2 + 5 = 7 \qquad 7 - 5 = 2$$

Students refer to the largest of the three related numbers as the "big number," and the other two numbers as "small numbers." This standard vocabulary helps students see continuity in the number system as facts are systematically added to their repertoire. Seeing the same structure represented by the number family in many cases (small number + small number = big number; big number – small number = small number) helps them look for and make use of the same structure in new applications. Through verbal and written exercises, students can generate new facts from known facts because they have internalized the structure of the number family.

Here is the first part of an exercise from Level B Lesson 51 that illustrates how students make use of what they know from one family and apply it to a new, related family.

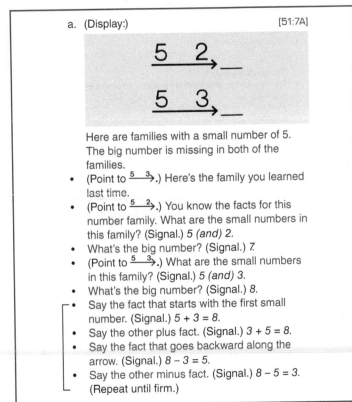

a. (Display:) [51:7A]

Here are families with a small number of 5. The big number is missing in both of the families.

- (Point to $\overset{5 \quad 3}{\longrightarrow}$.) Here's the family you learned last time.
- (Point to $\overset{5 \quad 2}{\longrightarrow}$.) You know the facts for this number family. What are the small numbers in this family? (Signal.) *5 (and) 2.*
- What's the big number? (Signal.) *7.*
- (Point to $\overset{5 \quad 3}{\longrightarrow}$.) What are the small numbers in this family? (Signal.) *5 (and) 3.*
- What's the big number? (Signal.) *8.*
- Say the fact that starts with the first small number. (Signal.) *5 + 3 = 8.*
- Say the other plus fact. (Signal.) *3 + 5 = 8.*
- Say the fact that goes backward along the arrow. (Signal.) *8 − 3 = 5.*
- Say the other minus fact. (Signal.) *8 − 5 = 3.*
- (Repeat until firm.)

from Lesson 51, Exercise 7

The number family table below incorporates the 200 basic addition and subtraction facts in 55 families. Students in *CMC Levels B–D* are systematically introduced to related sets of facts from columns in the table (as in the example above), rows in the table (e.g., all facts in the top row involve a "small number" of 1) or the diagonal (which shows "doubles").

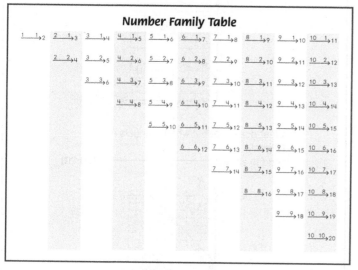

Number Family Table

Here are the number-family items and student work from Level B Lesson 44 Independent Work.

a.
$$6 \quad 2 \rightarrow 8$$
$$6 + 2 = 8$$
$$2 + 6 = 8$$
$$8 - 2 = 6$$
$$8 - 6 = 2$$

b.
$$3 \quad 2 \rightarrow 5$$
$$3 + 2 = 5$$
$$2 + 3 = 5$$
$$5 - 2 = 3$$
$$5 - 3 = 2$$

from Workbook Lesson 44, Part 6

Students are familiar with the number-family structure and generate the four related facts independently. By the end of first grade, students are well practiced with the commutative property, having applied it over the course of many lessons.

The work with number-family fact relationships continues in Levels C–E. Here are number-family items and student work from Lesson 30 of Level C.

Part 1

a.
$$3 \quad 3 \rightarrow \blacksquare$$
$$3 + 3 = 6$$
$$6 - 3 = 3$$

b.
$$4 \quad 3 \rightarrow \blacksquare$$
$$4 + 3 = 7$$
$$7 - 3 = 4$$

c.
$$5 \quad 3 \rightarrow \blacksquare$$
$$5 + 3 = 8$$
$$8 - 3 = 5$$

d.
$$6 \quad 3 \rightarrow \blacksquare$$
$$6 + 3 = 9$$
$$9 - 3 = 6$$

from Workbook Lesson 30, Part 1

The families are in sequence from the third row of the number-family table. For each item, students recognize that a missing big number requires addition (3 + 3, 4 + 3, 5 + 3, 6 + 3).Once the missing number in the family is identified, the students generate a related subtraction fact. This type of activity not only reduces the memory load for fact memorization, but it also prepares students for later work with inverse operations.

MP8: LOOK FOR AND EXPRESS REGULARITY IN REPEATED REASONING.
CMC Levels C–F: Renaming with Zeros
CMC Level E: Function Tables and the Coordinate System

In *CMC*, students work with repeated reasoning in a variety of situations. They learn to extend and apply what they have learned in one mathematical context to other, related contexts.

Subtraction: Renaming with Zeros

Students in several levels of the program apply the renaming strategy for the tens column to problems with zero in the minuend in an efficient and consistent way.

When renaming from the tens column, students learn a "new place value" for the top number. In the following example, 42 is 4 tens +2. It also equals 3 tens plus 12.

$$\begin{array}{r} \overset{3}{\cancel{4}}\,{}^{1}2 \\ -\ 7\,8 \\ \hline \end{array}$$

When a problem requires renaming from two columns, students apply the same reasoning and procedure:

402 is 40 tens plus 2. It also equals 39 tens plus 12.

$$\begin{array}{r} \overset{3\ 9}{\cancel{4}\cancel{0}}\,{}^{1}2 \\ -\ \ 7\,8 \\ \hline \end{array}$$

This same reasoning and procedure can be applied to any number of columns:

$$\begin{array}{r} \overset{3\ 9\ 9}{\cancel{4}\cancel{0}\cancel{0}}\,{}^{1}2 \\ -\ \ \ \ 7\,8 \\ \hline \end{array}$$
$$\begin{array}{r} \overset{3\ 9\ 9\ 9}{\cancel{4}\cancel{0}\cancel{0}\cancel{0}}\,{}^{1}2 \\ -\ \ \ \ \ 7\,8 \\ \hline \end{array}$$

The repeated reasoning provides students with an efficient algorithm, and a "shortcut" to solve any subtraction problem that requires renaming with zeros. In the last example above, students in *CMC* take one renaming step. In the example below, students not taught the method in *CMC* take four separate renaming steps to work the same problem.

$$\begin{array}{r} \overset{3\ 9\ 9\ 9}{\cancel{4}\cancel{0}\cancel{0}\cancel{0}}\,{}^{1}2 \\ -\ \ \ \ \ 7\,8 \\ \hline \end{array}$$

This less efficient method increases the likelihood that students will make an error or become confused.

Function Tables and the Coordinate System

Students in *CMC Level E* apply repeated reasoning to rows of a function table. The function table shows three columns: the X value, the function, and the Y value. For example:

X	Function 3 + X	Y
a. 7		10
b. 0		3
c. 2		
d.		4
e. 5		

from Workbook Lesson 99, Part 3

This table format enables students to see that the same transformation from the X value to the Y value is repeated for each point they will plot on the coordinate system. They also learn that this type of repeated transformation results in points that lie on a straight line. Given two X and Y values, students make two points and plot the line. They then refer to the line to identify where other points lie on the line, based on either the x or y coordinate. Finally, they perform the calculation based on the function rule to verify the accuracy of the points they have identified. Here's part of the exercise from Lesson 99.

(Add to show:) [99:3D]

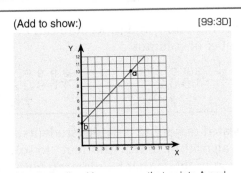

Here's the line. You can see that points A and B are on the line. You're going to figure out what Y equals for point C. Then we'll see if it's on the line.
- Complete row C in the function table. Put your pencil down when you can say the X and Y equation for point C.
  (Observe students and give feedback.)
- Look at row C. Say the equations for X and Y. Get ready. (Signal.) *X = 2 (and) Y = 5.*
  (Add to show:) [99:3E]

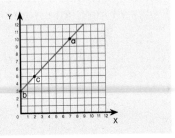

Here's the point for C.
- Is C on the line? (Signal.) *Yes.*

---

d. We're going to figure out the missing X or Y value for points D and E by using the line.
- Look at row D in the function table and say the equation for Y. Get ready. (Signal.) *Y = 4.*
  To find the point, I go to Y equals 4 on the arrow and go across to the line.
  (Add to show:) [99:3F]

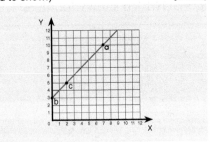

- What does X equal when Y equals 4? (Signal.) *1.*
  (Add to show:) [99:3G]

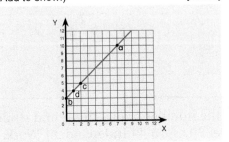

- Write what X equals for row D in the function table. ✔
f. Now we're going to check D to make sure it's right. The function for the table is 3 plus X.
- Raise your hand when you can say the problem for row D. ✔
- Say the problem for row D. (Signal.) *3 + 1.*
- What's the answer? (Signal.) *4.*
- Does Y equal 4 for row D? (Signal.) *Yes.*
  So the values for X and Y are right. Remember, these types of functions tell about all the points on a straight line.

from Lesson 99, Exercise 3

This exercise gives students the opportunity to apply repeated reasoning and also to evaluate the reasonableness of their results. If the point on the line does not correspond to the values in the table, they find the discrepancy and correct it.

On later lessons, students work with multiplication function tables. Here's an example from Lesson 102:

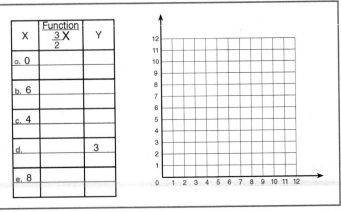

from Workbook Lesson 102, Part 3

Students deepen their understanding of functions and lines by repeating the same reasoning to these examples as they did to the add-subtract functions. They apply the function rule to figure out the y value for points a. and b., plot those points and draw the line.

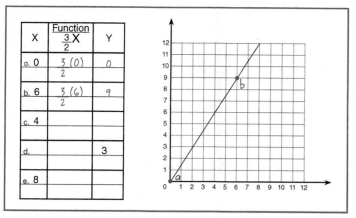

from Workbook Lesson 102, Part 3

They inspect the line to identify the missing x or y values for the remaining points.

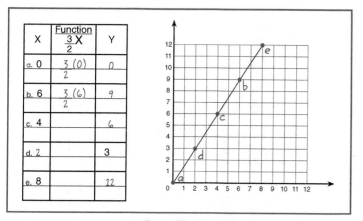

from Workbook Lesson 102, Part 3

Finally, students evaluate their answers for points c through e by multiplying each X value in the table by 3/2.

SUMMARY

For each mathematical concept or strategy introduced in *CMC*, students progress through a careful sequence of examples, move from highly structured activities to independent practice over several lessons, and revisit earlier concepts as they learn new applications or integrate elements of previously taught concepts. The *Comprehensive Edition* of *CMC* includes strategically designed instructional activities that fully meet the Common Core State Standards for Mathematical Content and the Standards for Mathematical Practice in kindergarten through fifth grade.